# De Gruyter Studies in Mathematical Physics



## Volume 50

Sergey Lychev and Konstantin Koifman

# Geometry of Incompatible Deformations

Differential Geometry in Continuum Mechanics

DE GRUYTER

**Physics and Astronomy Classification Scheme 2010**
Primary: 46.05.+b; Secondary: 02.40.Yy

**Authors**
Prof. Sergey Lychev
Ishlinsky Institute for Problems in Mechanics RAS (IPMech RAS)
lychevsa@mail.ru

Konstantin Koifman
Department of Applied Mathematics
Bauman Moscow State Technical University
koifman.bmstu@yandex.ru

ISBN 978-3-11-056201-9
e-ISBN (PDF) 978-3-11-056321-4
e-ISBN (EPUB) 978-3-11-056227-9
ISSN 2194-3532

**Bibliographic information published by the Deutsche Nationalbibliothek**
The Deutsche Nationalbibliothek lists this publication in the Deutsche Nationalbibliografie; detailed bibliographic data are available on the Internet at http://dnb.dnb.de.


Typesetting: le-tex publishing services GmbH, Leipzig
Printing and binding: CPI books GmbH, Leck

www.degruyter.com

To our families
and friends

# Preface

This book is intended to provide a systematic treatment of those parts of modern differential geometry that are essential for the modeling of incompatible finite deformations in solids. Included are discussions of generalized deformation and stress measures on smooth manifolds, geometrical formalization for structurally inhomogeneous bodies, various definitions for material connection, and evolution equations for them.

The incompatibility of deformations may be caused by a variety of physical phenomena; among them are distributed dislocations and disclinations, point defects, non-uniform thermal fields, shrinkage, growth, *etc.* Incompatible deformations result in residual stresses and distortion of geometrical shape. These factors are associated with critical parameters in modern high-precision technologies, particularly, in additive manufacturing, and are considered to be *ipso facto* essential constituents in the corresponding mathematical models. In this context, the development of methods for their quantitative description is the actual problem of modern solid mechanics. The methods in question are based on the representation of a body and physical space in terms of differentiable manifolds, namely material manifold and physical manifold. These manifolds are equipped with specific metrics and connections, non-Euclidian in general.

The book is divided into 14 chapters. The first one is an introduction. The second one briefly reviews various representations of the geometry of physical space and time, including Euclidean, Minkowski, and curved space time models. This review, on the one hand, leads to exhaustive definition of physical space in terms of smooth manifolds. On the other hand, it draws attention to the specific aspects of geometrical formalization for space and space time, which have much in common with their counterparts in the modeling of elastic bodies, such as smooth (material) manifolds.

The third chapter focuses on the essentials of conventional non-linear elasticity. In this chapter, those fundamentals of continuum mechanics that are rooted in the concept of absolute, Euclidean, space and absolute, Newtonian, time are discussed. Special attention is paid to the assumption that global stress-free shapes of any body exists. Rejection of this assumption reflects the main idea of the present book. Common notations for strain and stress measures, variational symmetries, and constitutive equations, which would be generalized in the rest of the book, are highlighted.

In the fourth chapter, we consider the issue of the physical interpretation of the non-Euclidean structure of the material manifold. It is shown that a two-dimensional rigid surface, which formalizes the curved substrate used in the deposition process, may serve as an example of a non-Euclidean physical manifold. Affine connection on the material manifold represents the intrinsic properties (inner geometry) of the body and is determined by the field of local uniform configurations performing its "assembly" of identical and uniform infinitesimal "bricks". Uniformity means that the response functional gives the same response on all admissible smooth deforma-

https://doi.org/10.1515/9783110563214-201

tions for them. As a result of assembling, one obtains a body that cannot be immersed in an undistorted state into the physical manifold. It is an essential feature of residual stressed bodies produced by additive processes. For this reason, it should benefit from immersion into a non-Euclidean space (material manifold with non-Euclidean material connection). To this end, it is convenient to formalize the body and physical space in terms of the theory of smooth manifolds. The deformation is formalized as embedding (or, in a special case, as immersion) of the former manifold into the latter one.

The fifth chapter is dedicated to generalization of relations for Cauchy–Green strain measures. They are generated by embedding of a Riemannian manifold, representing the body, into a Riemannian manifold, representing the space. We consider such issues as the transpose of the deformation gradient and generalization of the Cauchy decomposition theorem on smooth manifolds.

The sixth chapter covers the definition of motion, velocity, and acceleration fields in the framework of the theory of smooth manifolds. The motion is represented as a time-dependent flow.

The seventh chapter addresses the issues of the formalization of stress and power measures by fields defined on smooth manifolds that represent the body and physical space. It contains the systematic construction of a theory for the general non-Euclidean case. Forces are interpreted as covectors, *i.e.*, as a linear functionals, whose action on the velocity vectors of material points results in mechanical power. The abstract theory of integration based on exterior form formalism is adapted to the elements of this structure, which allows one to formulate the power balance equation of the material manifold (similarly to the reference description in the classical mechanics of compatible deformation) and of the physical one (similarly to the spatial description).

In the eighth chapter the response of hyperelastic solids on smooth manifolds is generalized. Only simple materials are considered: their response depends on the local configuration and material points. For such materials, the notion of material isomorphism is introduced. The main assumption that leads to the notion of a structurally inhomogeneous solid is the following: for each body of simple material, there exists a family of configurations, which index set is identical to the set of material points constituting the whole body. Each configuration maps its index point to the uniform state. This assumption is used for synthesizing the material metric. A method is proposed for describing a deformable body of variable composition as a family of Riemannian manifolds over which partition and joining operations are defined. These operations characterize the structural features of the inhomogeneities given by the additive process scenario.

In the ninth chapter, the various ways to specify the general form of affine connections on the material and physical manifolds are considered. Affine connection endows manifolds with geometric properties, in particular, with parallel translation rules for vector fields. For simple materials the parallel translation is an elegant mathematical formalization of the concept of a materially uniform (in particular, stress-free) non-Euclidean reference shape. In fact, one can obtain a connection on physical space by determining the parallel transport rule as a transformation of the tangent vector, which corresponds to the structure of the physical space containing shapes of the body. In turn, one can obtain affine connection on material manifold by defining a parallel transport rule as the transformation of the tangent vector, in which its inverse image with respect to locally uniform embeddings does not change. Utilizing the conception of material connections and the corresponding methods of non-Euclidean geometry may significantly simplify formulation of the initial boundary value problems of the theory of incompatible deformations, so the choice of connections is important. Connection on the physical manifold is compatible with the metric, and the Levi–Civita relation holds for it. Connection of the material manifold is considered in two alternative variants. The first leads to the Weitzenböck space (the space of absolute parallelism or teleparallelism, *i.e.*, space with zero curvature and non-metricity but with non-zero torsion) and gives a clear interpretation of the material connection in terms of the local linear transformations that return an elementary volume of simple material to the uniform state. The second one leads to the Riemannian space (space with zero non-metricity and torsion but with non-zero curvature). Examples for such generalizations are considered in the rest of the chapter.

The tenth chapter refers to the balance equations, which are obtained in terms of Cartan's exterior covariant derivative from the general principle of covariance.

The eleventh chapter contains the examples of inhomogeneous solids, which inhomogeneity was induced by some additive technological process. Various types of evolutionary problems, owning to different technological regimes, are considered. The calculations are illustrated by numerical computations and graphs.

In the last chapters, the required mathematical preliminaries adapted to the present book are considered.

May 2018, Moscow

Sergey Lychev
Konstantin Koifman

# Contents

# General Scheme of Notations

## Index of frequently used symbols

| Symbol | Name | Place of definition or first occurrence |
|---|---|---|
| $a$, $\boldsymbol{a}$ | Spatial acceleration | Section 6.3 |
| $[a_{ij}]$ | Matrix with elements $a_{ij}$ | Section 2.2.3 |
| $A$, $\boldsymbol{A}$ | Material acceleration | Section 6.3 |
| $\mathfrak{A}$ | Action | Section 2.6.6 |
| $\mathscr{B}$, $\boldsymbol{B}$, $B^i{}_j$ | Left Cauchy–Green tensor | Section 3.3 |
| $\mathfrak{B}$ | Body | Section 4.3 |
| $\mathscr{C}$, $\boldsymbol{C}$, $C^i{}_j$ | Right Cauchy–Green tensor | Section 3.3 |
| $\mathfrak{C}(\mathfrak{B};\mathscr{P})$ | Set of all configurations | Section 4.4 |
| $C^\infty(\mathfrak{M})$ | The algebra of all smooth functions on $\mathfrak{M}$ | Section 13.1.3 |
| $\mathrm{CVec}(\mathfrak{M})$ | The space of all smooth covector fields on $\mathfrak{M}$ | Section 13.3.5 |
| $\mathscr{D}$ | Cartesian coordinate map | (2.2) |
| $\mathscr{E}$ | Euclidean affine space | Section 2.1.1 |
| $\boldsymbol{F}$ | Deformation gradient | Section 3.3 |
| $\mathscr{F}$ | Configuration gradient | Section 5.5 |
| $\mathscr{g}$ | Spatial metric | Section 2.7 |
| $\mathscr{G}$ | Material metric | Section 4.3 |
| $\mathrm{GL}(s;\mathbb{R})$ | General linear group of degree $s$ | Section 12.6 |
| $\iota_X$ | Inclusion map of $X$ | (13.1) |
| $(\boldsymbol{i}_s)_{s=1}^m$ | Orthonormal basis | Section 2.1.1 |
| $\mathrm{Id}_X$ | Identity map of $X$ | (12.1) |
| $\mathrm{Inv}(\mathscr{V};\mathscr{W})$ | The set of all invertible linear mappings from $\mathscr{V}$ to $\mathscr{W}$ | Section 12.5.2 |
| $\boldsymbol{I}$ | Identity operator | Section 3.4 |
| $J$ | Jacobian of deformation | (3.8) |
| $\mathscr{K}_{\mathfrak{X}}$, $\boldsymbol{K}_{\mathfrak{X}}$ | Local configuration at point $\mathfrak{X}$ | Section 8.2 |
| $\mathrm{Lin}(\mathscr{V};\mathscr{W})$ | Vector space of linear mappings from $\mathscr{V}$ to $\mathscr{W}$ | Section 12.5.2 |
| $\mathrm{O}(s)$ | Orthogonal group in dimension $s$ | Section 12.6 |
| $\boldsymbol{p}$ | Position vector field | (2.1) |
| $P$ | First Piola–Kirchhoff covector-valued form | (7.3) |
| $\mathrm{Part}(\mathfrak{B})$ | Set of all parts of a body $\mathfrak{B}$ | Section 4.3 |
| $\boldsymbol{P}$ | First Piola–Kirchhoff stress tensor | Section 3.7 |
| $\mathscr{P}$ | Abstract physical space | Section 2.7 |
| $\mathfrak{P}$ | Part of a body | Section 4.3 |
| $\mathfrak{Q}$ | Nonmetricity | (9.10) |
| $\mathrm{rank}(\mathfrak{E})$ | Rank of vector bundle $\mathfrak{E}$ | Section 13.3.1 |
| $\mathfrak{R}$ | Riemann curvature tensor | (9.9) |
| $\mathrm{Sec}(\mathfrak{E})$ | The space of all smooth sections of vector bundle $\mathfrak{E}$ | Section 13.3.5 |
| $\mathrm{Sym}(\mathscr{V})$ | The set of all symmetric linear mappings from $\mathscr{V}$ to $\mathscr{V}$ | Section 12.5.4 |
| $T$ | Cauchy covector-valued form | (7.4) |
| $\boldsymbol{T}$ | Cauchy stress tensor | Section 3.7 |

https://doi.org/10.1515/9783110563214-202

| Symbol | Name | Place of definition or first occurrence |
|---|---|---|
| $\mathfrak{T}$ | Torsion | (9.8) |
| $T_p\mathfrak{M}$ | Tangent space to $\mathfrak{M}$ at $p \in \mathfrak{M}$ | Section 13.2.1 |
| $T_p^*\mathfrak{M}$ | Cotangent space to $\mathfrak{M}$ at $p \in \mathfrak{M}$ | Section 13.2.3 |
| $\boldsymbol{u}$ | Displacement vector field | Section 3.4 |
| $v, \boldsymbol{v}$ | Spatial velocity | Section 6.2 |
| $V, \boldsymbol{V}$ | Material velocity | Section 6.2 |
| $\mathscr{V}$ | Translation vector space | Section 2.1.1 |
| $\mathrm{Vec}(\mathfrak{M})$ | The space of all smooth vector fields on $\mathfrak{M}$ | Section 13.3.5 |
| $\gamma^i{}_{jk}, \Gamma^i{}_{jk}$ | Affine connection coefficients | Section 9.2 |
| $\Lambda^k(T_p^*\mathfrak{M})$ | The set of all exterior $k$-forms | Section 13.2.6 |
| $\Omega^k(\mathfrak{M})$ | The set of all differential $k$-forms on $\mathfrak{M}$ | Section 13.3.5 |

# Index of frequently used operations

| Operation | Name | Place of definition or first occurrence |
|---|---|---|
| $\cdot$ | Inner product | Section 12.5.4 |
| $\lrcorner, \llcorner$ | Contractions | Section 13.3.5 |
| $[\cdot, \cdot]$ | Lie bracket | (13.20) |
| $\otimes$ | Tensor product | Section 12.5.3 |
| $\wedge$ | Wedge (Exterior) product | Section 13.2.6 |
| $\dot{\wedge}$ | Inner-exterior product | Section 10.1.2 |
| $d$ | Exterior derivative | Section 13.3.9 |
| $\mathfrak{D}, \mathfrak{d}$ | Cartan's exterior derivative | Section 10.1.2 |

# 1 Introduction

Geometrical modeling methods of continuum mechanics are rooted in such parts of modern differential geometry as exterior calculus, theory of connections, theory of $G$-structures, *etc.* [1–3]. The overarching purpose of this book is to illustrate applications of geometrical methods in the modeling of structurally inhomogeneous solids. This kind of inhomogeneity appears in uniform solids due to incompatibility of local finite deformations. One can find the application of the proposed methods to the design of modern additive technologies, which is a rich field of research [4].

The elegance and effectiveness of geometrical language in continuum physics have been demonstrated in some excellent and comprehensive works [5–8]. The use of this language has a long history. The pioneering works of E. Beltrami, H. Poincaré, T. Levi-Civita, *etc.* [9] at the end of the nineteenth century laid the foundation for the application of non-Euclidean geometry in physics and mechanics. At the beginning of the twentieth century, A. Einstein, H. Weyl, H. Cartan, and others [10] used the geometric language in the framework of the general theory of relativity [11]. This gave impetus to the development of relativistic non-linear continuum mechanics. Relativistic models of deformable continua were obtained in terms of Riemannian and non-Riemannian geometry [12–15]. Herewith, the physical space containing a deformable body was equipped with a non-Euclidian structure.

The idea that the space, which contains the reference, unstressed shape of the body, should have a non-Euclidean structure was first proposed by Eckart[1] [16] and Kondo [17–20], although the first studies of incompatible deformations in solids go back to Weingarten [21] and Volterra [22]. Notice that the physical space containing the observed (actual) shape has a Euclidean structure. Speaking figuratively, a body from an imaginary non-Euclidean space is mapped into a physical, Euclidean space. This causes deformations of a special kind, which can be referred to as intrinsic (or local) deformations, since they do not disappear in the absence of external force fields. Kondo [20] demonstrated that such deformations fully characterize the fields of defects as sources of residual stresses. These sources, in turn, can be determined by specific, material, connection, and corresponding parallel transport in the space containing the reference shape. The material connection can be fully characterized by the relevant curvature and torsion.

In a series of classical works, J. Nye, K. Kondo, E. Kröner, and R. De Wit [18–20, 23, 24] demonstrated the effectiveness of the geometrical approach to the modeling of crystals with distributed defects (dislocations, disclinations, point defects, *etc.*) The

1 Carl Eckart, 1948: «*The traditional theory of the solid state rests on two false assumptions. One is the principle of a constant relaxed (or standard) state. The other is the principle of relaxability-in-the-large, first formulated mathematically by de Saint-Venant. His equations are essentially identical with Riemann's equations expressing the condition that a geometry be Euclidean-in-the-large*».

https://doi.org/10.1515/9783110563214-001

main feature of this approach lies in the specific geometry of space that represents the body itself. In general, this geometry is non-Euclidean. The torsion, curvature, and non-metricity of the connection, which determine its non-Euclidean properties, are closely related to the densities of distributed defects. Geometrical methods have been applied successfully to solve non-linear problems in the theories of surface growth and volumetric growth [25].

A few words should be said about the relevance of models for bodies with intrinsic deformations. At present, there is a variety of manufacturing processes that allow us to create a solid, especially thin films, by using layer-by-layer (LbL) technologies [26]. Stereolithography, electrochemical deposition, welding, *etc.* [4, 27] are among these. Such methods allow us to produce a body with a complicated shape that can be manufactured in non-conventional ways only. The essential mechanical feature of solids, produced by additive technologies, is that they always have intrinsic deformations. For example, such a phenomenon arises as a consequence of polymer shrinkage due to solidification. It leads to residual stresses and distortion of shape. In addition, it should be noted that non-linear effects manifest themselves significantly in additive technological processes. This is related to the high concentration of energy in the sintering zone (in laser welding processes) or to the complicated rheology of photochemical solidification processes. Therefore, general models should be formulated within non-linear approaches of continuum mechanics.

The typical feature of bodies generated during some additive process is the absence of a stress-free shape in the conventional sense. From the physical point of view, this means that the body stores elastic energy that cannot be released under any smooth deformation. This energy can be released only by cutting the body into an infinite number of disjoined parts. A similar situation arises in the theory of distributed defects (dislocations, disclinations, pores) [28]. When a stress-free shape does not exist, the specific mechanical and physical features of the body can be quantitatively represented in a variety of ways, for example, by measuring the deformation incompatibility [24], or the density of the distributed defects, or the intensity of residual stresses [29]. Such notions are closely related and can be described within the framework of the general idea of deformation as a general geometric transformation in non-Euclidian space.

Lastly, it should be noted that the topological properties of the body can be essentially non-trivial; in particular, the body may have the structure of a non-orientable surface, for instance, a Klein surface. The "physical" implementation of such a body can be represented by a self-intersecting grid shell [30]. In light of the rapid development of additive manufacturing technologies, such "exotic" bodies can be easily implemented in the material.

# 2 Geometry of Physical Space

A theme running through this book is the proposition that the incompatible local deformations, arising in elastic solids for a variety of reasons, can be mathematically modeled as an embedding of a smooth manifold, representing a body, into a smooth manifold, representing a physical space. In order to take an advantage of such formalism, the general guidelines for the geometrical description of actual objects should be defined. Our point of departure will be the review of various alternatives for space and time formalization. The mathematical structure of space-time is the subject of geometrodynamics, a science that studies the fundamentals of space, time, and gravitation. The reader may find a comprehensive view of these issues in [31–34].

This book is not meant to be consistent presentation of geometrodynamics and only uses some of its results to show the affinity of geometric approaches in theoretical physics and non-linear elasticity. The present chapter is aimed to set the stage for the definition of physical space in terms of a smooth manifold. In this regard, we discuss in detail some particular ways for the mathematical description of flat and curved two- and three-dimensional physical space as well as four-dimensional space-time.A flat three-dimensional model for physical space determines a relatively simple (but not at all elementary) formal structure that contains elements that can be considered as prototypes of more sophisticated ones, defined in non-Euclidean spaces. Apart from this, it is intended to serve as an arena for the embodiment of three-dimensional bodies in the form of their shapes within the framework of classical non-linear elasticity. Curved two-dimensional (Riemannian) space provides an opportunity for the modeling of material surfaces and curvilinear membranes. Curved four-dimensional space-time gives the example of a more general formal structure that has much in common with the non-Euclidian structure of the material manifold discussed in Chapter 4.

In this book, we systematically use diagrams that illustrate relations between mappings, their domains, and codomains[1]. For example, if $A, B, C, D$ are sets and $f\colon A \to B, g\colon B \to C, h\colon A \to D, j\colon D \to C$ are mappings, then the diagram

$$\begin{array}{ccc} A & \xrightarrow{\;f\;} & B \\ {\scriptstyle h}\downarrow & & \downarrow{\scriptstyle g} \\ D & \xrightarrow[\;j\;]{} & C \end{array}$$

illustrates the relations between the sets and the maps. Such a diagram is called *commutative* if $g \circ f = j \circ h$.

1 A more detailed outline of the theory of mappings is given in Chapter 12, Section 12.1.

https://doi.org/10.1515/9783110563214-002

We use ordered $n$-tuples, which consist of sets and mappings, for defining geometric structures such as affine space, Riemannian space, *etc.* If it does not lead to any misunderstanding, after the first reference to such ordered tuples we omit its components and simply write the underlying set.

## 2.1 Euclidean Affine Space

The naive definition of Euclidean space comes down to the definition of the space of "directed line segments". Technically, it can be given in terms of a collection of mutually perpendicular unit length directed segments, namely unit vectors. Herein, the notion of inner product has already been inherent in pure geometric properties of unit vectors. However, such an approach usually neglects the distinction between points and translation vectors (radius-vectors) that bring the origin to those points. In the following, we apply a more rigorous approach based on Weyl axioms. In the framework of this approach, the points and translation vectors are treated as elements of different sets, whose properties and relations are axiomatically defined. This would enable us to determine the structure of classical flat physical space in more detail and to undertake a more comprehensive comparison of flat space with curved space.

We would like to highlight that space, according to [35], is "a form of existence of the real world". The experience gained in the terrestrial (local) laboratory suggests the axiomatics for this form, which was outlined in general terms by Greek geometers [36]. It is synthetic geometry, the key tool of Greek mathematics, that leads to the modern formulation of affine-Euclidean space. It is based on the principle of congruence, which establishes the equivalence of geometric objects, *i.e.*, figures, and bodies, which can be transformed one into the other by a combination of rigid motions, namely a translation, a rotation, and a reflection. Together, these motions form a Euclidean group of isometries.

We begin with a general definition.

### 2.1.1 General Definitions

**Affine space.** Affine space is a triplet $(\mathscr{E}, \mathscr{V}, \psi)$, where[2]

(i) $\mathscr{E}$ is a set whose elements are referred to as *spatial points*;

(ii) $\mathscr{V}$ is a real vector space whose elements are referred to as *translation vectors*;

(iii) $\psi : \mathscr{E} \times \mathscr{E} \to \mathscr{V}$ is a mapping, which assigns to each ordered pair $(a, b) \in \mathscr{E} \times \mathscr{E}$ some vector from $\mathscr{V}$ (denoted by $\overrightarrow{ab}$ or $b - a$), such that

2 Here we briefly introduce the notions that are connected with affine space. The definition of affine space and a listing of its properties are presented in Chapter 12, Section 12.7.

$W_1$) for every points $a, b, c \in \mathscr{E}$ the following relation holds:

$$\overrightarrow{ab} + \overrightarrow{bc} + \overrightarrow{ca} = \mathbf{0} \in \mathscr{V} ;$$

$W_2$) for any point $a \in \mathscr{E}$ and for any vector $\boldsymbol{v} \in \mathscr{V}$ there exists a unique point $b \in \mathscr{E}$, such that $\overrightarrow{ab} = \boldsymbol{v}$.

The set of two axioms $W_1$) and $W_2$) is usually associated with the name of H. Weyl [35]. The relation in axiom $W_1$) is associated with the name Chasles or with the "parallelogram rule".

According to axiom $W_2$), for any fixed point $a \in \mathscr{E}$ the mapping

$$\psi_a : \mathscr{E} \to \mathscr{V} , \quad \psi_a(x) := \overrightarrow{ax} ,$$

is a bijection. This allows one to define the external operation

$$+ : \mathscr{E} \times \mathscr{V} \to \mathscr{E} , \quad + : (a, \boldsymbol{v}) \mapsto a + \boldsymbol{v} := \psi_a^{-1}(\boldsymbol{v}) ,$$

which assigns to each ordered pair $(a, \boldsymbol{v})$ a unique point $b \in \mathscr{E}$, such that $\overrightarrow{ab} = \boldsymbol{v}$.

**Dimension of affine space.** The dimension of $(\mathscr{E}, \mathscr{V}, \psi)$ is defined as the dimension of $\mathscr{V}$, *i.e.*, $\dim \mathscr{E} := \dim \mathscr{V} = m$. In the following, only finite dimensions will be considered.

**Affine coordinate system.** An affine coordinate system can be determined by an origin $o \in \mathscr{E}$ and a basis $(\boldsymbol{e}_i)_{i=1}^m$ for $\mathscr{V}$. It is represented by a doublet $(o, (\boldsymbol{e}_1, \dots, \boldsymbol{e}_m))$. According to $W_2$), the mapping

$$\boldsymbol{p} : \mathscr{E} \to \mathscr{V} , \quad \boldsymbol{p}(x) := x - o , \tag{2.1}$$

is a bijection. In mechanics, the vector $\boldsymbol{p}(x)$ is known primarily as the *position vector* (radius-vector or place vector) of a point $x$ [37–39].

**Euclidean affine space.** Euclidean affine space is an 4-tuple $(\mathscr{E}, \mathscr{V}, (\cdot), \psi)$, where $(\mathscr{E}, \mathscr{V}, \psi)$ is an affine space and $(\cdot)$ is an inner product on $\mathscr{V}$ (see Chapter 12, Section 12.5.4).

**Rectilinear coordinates.** In order to obtain numerical values for the components of the position vector one has to define the dual basis $(\boldsymbol{e}^i)_{i=1}^m$ for $\mathscr{V}$ from the system of linear equations $\boldsymbol{e}^i \cdot \boldsymbol{e}_j = \delta^i_j$, $i, j = 1, \dots, m$. Then, any vector $\boldsymbol{u} \in \mathscr{V}$ can be represented as the decomposition $\boldsymbol{u} = u^k \boldsymbol{e}_k$, where $u^k = \boldsymbol{e}^k \cdot \boldsymbol{u}$. Thus,

$$\forall x \in \mathscr{E} : \quad x = o + x^k \boldsymbol{e}_k , \quad \text{where} \quad x^k = \boldsymbol{e}^k \cdot \boldsymbol{p}(x), k = 1, \dots, m .$$

We refer to the $m$-tuple $(x^i)_{i=1}^m$ as *rectilinear coordinates.*

Rectilinear coordinates can be represented as images of the bijective mapping

$$\mathscr{D}_{(o,(\boldsymbol{e}_i)_{i=1}^m)} : \mathscr{E} \to \mathbb{R}^m , \quad \mathscr{D}_{(o,(\boldsymbol{e}_i)_{i=1}^m)}(x) := \left(\boldsymbol{e}^1 \cdot \boldsymbol{p}(x), \dots, \boldsymbol{e}^m \cdot \boldsymbol{p}(x)\right) , \tag{2.2}$$

which is completely determined by the $(m+1)$-tuple $(o, (\boldsymbol{e}_1, \dots, \boldsymbol{e}_m))$. The $m$-tuple $(x^k)_{k=1}^m = \mathscr{D}_{(o,(\boldsymbol{e}_i)_{i=1}^m)}(x)$ represents coordinates of a point $x$ relative to $(o, (\boldsymbol{e}_1, \dots, \boldsymbol{e}_m))$. We refer to the map $\mathscr{D}_{(o,(\boldsymbol{e}_i)_{i=1}^m)}$ as a *Cartesian coordinate map*. In the following, we omit the lower index of $\mathscr{D}_{(o,(\boldsymbol{e}_i)_{i=1}^m)}$, since we do not consider different affine coordinates simultaneously.

The relations between $\mathscr{D}$ and $\boldsymbol{p}$ are illustrated in the following commutative diagram:

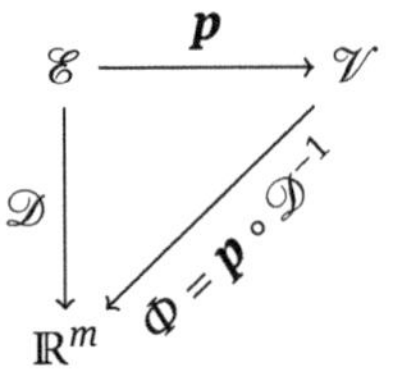


In the diagram, $\Phi\colon \mathscr{V} \to \mathbb{R}^m$ is a vector space isomorphism.

**Cartesian coordinates.** A particular case of the affine coordinate system is the *Cartesian coordinate system* $(o, (\boldsymbol{i}_1, \dots, \boldsymbol{i}_m))$, where $(\boldsymbol{i}_k)_{k=1}^m$ is an orthonormal basis of $\mathscr{V}$. This basis is defined by the relations $\boldsymbol{i}_k \cdot \boldsymbol{i}_j = \delta_{kj}$, $k, j = 1, \dots, m$. For any vector $\boldsymbol{u} \in \mathscr{V}$, one has $\boldsymbol{u} = u^k \boldsymbol{i}_k$, where $u^k = \boldsymbol{i}_k \cdot \boldsymbol{u}$.

**Multivectors.** The inner product allows one to calculate lengths and angles. At the same time, for volume measurements one needs some extra tool. Such a tool is provided by the wedge (exterior) product $\wedge$ on vectors and is intimately related with the notion of the determinant. This product has the following properties[3]:

(i) $\forall \boldsymbol{u}, \boldsymbol{v} \in \mathscr{V}: \boldsymbol{u} \wedge \boldsymbol{v} = -\boldsymbol{v} \wedge \boldsymbol{u}$;

(ii) $\forall \boldsymbol{u}, \boldsymbol{v}, \boldsymbol{w} \in \mathscr{V} \quad \forall a, b \in \mathbb{R}$:

$$\boldsymbol{u} \wedge (a\boldsymbol{v} + b\boldsymbol{w}) = a\boldsymbol{u} \wedge \boldsymbol{v} + b\boldsymbol{u} \wedge \boldsymbol{w}\,;$$

(iii) $\forall \boldsymbol{u}, \boldsymbol{v}, \boldsymbol{w} \in \mathscr{V}: (\boldsymbol{u} \wedge \boldsymbol{v}) \wedge \boldsymbol{w} = \boldsymbol{u} \wedge (\boldsymbol{v} \wedge \boldsymbol{w})$.

For example, let $m = 2$ and $(o, (\boldsymbol{i}_1, \boldsymbol{i}_2))$ be a Cartesian system. In classical synthetic geometry, the area is defined as the specific property of the rectangular $\Pi_0$ with vertices $o$, $o + \boldsymbol{i}_1$, $o + \boldsymbol{i}_2$, $o + \boldsymbol{i}_1 + \boldsymbol{i}_2$. Consider the "elementary" bivector $\boldsymbol{i}_1 \wedge \boldsymbol{i}_2$ as a basic (reference) object with the geometric sense of the unit area, *i.e.*, the unit area of $\Pi_0$. Now let $\boldsymbol{u}, \boldsymbol{v} \in \mathscr{V}$ be vectors. Consider the parallelogram $\Pi$ with vertices $o$, $o + \boldsymbol{u}$, $o + \boldsymbol{v}$, $o + \boldsymbol{u} + \boldsymbol{v}$. If one has $\boldsymbol{u} = u^k \boldsymbol{i}_k$, $\boldsymbol{v} = v^k \boldsymbol{i}_k$, then the value of the determinant

$$\Delta(\boldsymbol{u}, \boldsymbol{v}) = \begin{vmatrix} u^1 & v^1 \\ u^2 & v^2 \end{vmatrix}$$

3 Here we present the sketch of multivector and wedge product notions. A comprehensive treatment can be found in [40, 41].

gives the dimensionless oriented area of $\Pi$. The wedge product (bivector) $\boldsymbol{u} \wedge \boldsymbol{v}$ is defined as

$$\boldsymbol{u} \wedge \boldsymbol{v} := \Delta(\boldsymbol{u}, \boldsymbol{v})\boldsymbol{i}_1 \wedge \boldsymbol{i}_2 ,$$

and gives the oriented area of $\Pi$.

Developing the example, consider the case $m = 3$. Let $(o, (\boldsymbol{i}_1, \boldsymbol{i}_2, \boldsymbol{i}_3))$ be a Cartesian system. Then, one has three "elementary" bivectors $\boldsymbol{i}_2 \wedge \boldsymbol{i}_3$, $\boldsymbol{i}_1 \wedge \boldsymbol{i}_3$, and $\boldsymbol{i}_1 \wedge \boldsymbol{i}_2$ that give mutually perpendicularly oriented unit areas of the corresponding parallelograms. Again, let $\boldsymbol{u}, \boldsymbol{v} \in \mathscr{V}$ be vectors and $\Pi$ the parallelogram, like in the preceding example. Then, for $\boldsymbol{u} = u^k \boldsymbol{i}_k$, $\boldsymbol{v} = v^k \boldsymbol{i}_k$, the square root

$$\sqrt{\begin{vmatrix} u^2 & v^2 \\ u^3 & v^3 \end{vmatrix}^2 + \begin{vmatrix} u^1 & v^1 \\ u^3 & v^3 \end{vmatrix}^2 + \begin{vmatrix} u^1 & v^1 \\ u^2 & v^2 \end{vmatrix}^2}$$

is the dimensionless area of $\Pi$. Define bivector $\boldsymbol{u} \wedge \boldsymbol{v}$ as

$$\boldsymbol{u} \wedge \boldsymbol{v} := \begin{vmatrix} u^2 & v^2 \\ u^3 & v^3 \end{vmatrix} \boldsymbol{i}_2 \wedge \boldsymbol{i}_3 + \begin{vmatrix} u^1 & v^1 \\ u^3 & v^3 \end{vmatrix} \boldsymbol{i}_1 \wedge \boldsymbol{i}_3 + \begin{vmatrix} u^1 & v^1 \\ u^2 & v^2 \end{vmatrix} \boldsymbol{i}_1 \wedge \boldsymbol{i}_2 .$$

The geometric sense of $\boldsymbol{u} \wedge \boldsymbol{v}$ is, again, the oriented area of $\Pi$. To the bivector $\boldsymbol{u} \wedge \boldsymbol{v}$ there corresponds the vector $\boldsymbol{u} \times \boldsymbol{v}$, namely, the *cross-product of* $\boldsymbol{u}$ *and* $\boldsymbol{v}$. This vector is defined as having the same components as the bivector, *i.e.*,

$$\boldsymbol{u} \times \boldsymbol{v} := e_{ijk} u^i v^j \boldsymbol{i}^k , \tag{2.3}$$

where $e_{ijk}$ is the *alternator*:

$$e_{ijk} = e^{ijk} := \begin{cases} 1, & \text{if } (i, j, k) \text{ is an even permutation of } (1, 2, 3) , \\ -1, & \text{if } (i, j, k) \text{ is an odd permutation of } (1, 2, 3) , \\ 0, & \text{if } (i, j, k) \text{ is not a permutation of } (1, 2, 3) . \end{cases} \tag{2.4}$$

Similarly to the notion of area, in classical synthetic geometry, the volume is defined as the specific property of the parallelepiped $P_0$ with vertices $o$, $o + \boldsymbol{i}_1$, $o + \boldsymbol{i}_2$, $o + \boldsymbol{i}_1 + \boldsymbol{i}_2$, $o + \boldsymbol{i}_3$, $o + \boldsymbol{i}_1 + \boldsymbol{i}_3$, $o + \boldsymbol{i}_2 + \boldsymbol{i}_3$ and $o + \boldsymbol{i}_1 + \boldsymbol{i}_2 + \boldsymbol{i}_3$. The "elementary" three-vector $\boldsymbol{i}_1 \wedge \boldsymbol{i}_2 \wedge \boldsymbol{i}_3$ is defined as a basic (reference) object with the geometric sense of the unit volume, *i.e.*, the unit volume of $P_0$. Now, let $\boldsymbol{u}, \boldsymbol{v}, \boldsymbol{w} \in \mathscr{V}$. Consider the parallelepiped $P$ with vertices $o$, $o + \boldsymbol{u}$, $o + \boldsymbol{v}$, $o + \boldsymbol{u} + \boldsymbol{v}$, $o + \boldsymbol{w}$, $o + \boldsymbol{u} + \boldsymbol{w}$, $o + \boldsymbol{v} + \boldsymbol{w}$ and $o + \boldsymbol{u} + \boldsymbol{v} + \boldsymbol{w}$. Its dimensionless oriented volume is given by

$$V(\boldsymbol{u}, \boldsymbol{v}, \boldsymbol{w}) = \begin{vmatrix} u^1 & v^1 & w^1 \\ u^2 & v^2 & w^2 \\ u^3 & v^3 & w^3 \end{vmatrix} ,$$

where $\boldsymbol{u} = u^k \boldsymbol{i}_k$, $\boldsymbol{v} = v^k \boldsymbol{i}_k$, $\boldsymbol{w} = w^k \boldsymbol{i}_k$. Define the three-vector $\boldsymbol{u} \wedge \boldsymbol{v} \wedge \boldsymbol{w}$ as

$$\boldsymbol{u} \wedge \boldsymbol{v} \wedge \boldsymbol{w} := V(\boldsymbol{u}, \boldsymbol{v}, \boldsymbol{w})\boldsymbol{i}_1 \wedge \boldsymbol{i}_2 \wedge \boldsymbol{i}_3 .$$

Thus, $\boldsymbol{u} \wedge \boldsymbol{v} \wedge \boldsymbol{w}$ is the oriented volume of $P$.

Generalization of the notions of bivector and three-vector lead to the notion for multivector [40]. The examples considered illustrate the close relations between multivectors, determinants, and volume measurements. That is, the inner product $(\cdot)$ and wedge product $\wedge$ represent the complete set of tools for metric measurements on $\mathscr{E}$.

**Remark 2.1.** Above we defined the Euclidean affine space axiomatically and we will systematically use the axiomatic approach to define more general spaces further. However, it is important to note that, since for the last two thousand years the structure of Euclidean space has been studied in detail within the framework of synthetic (Greek) geometry, it becomes possible to construct a structure, equivalent to Euclidean space equipped with a Cartesian coordinate system, relying only on the concept of congruence. Referring again to the argument of H. Weyl [35], we show a way of a synthetic (coordinate-free) reasoning that leads to the idea of a Cartesian space. The primary concepts of synthetic geometry are a point, a curve, and a surface. The congruence principle allows one to choose the straight lines of all the curves by the following condition: straight lines coincide with their images under arbitrary translations along themselves and, thus, can be defined by any two distinct points on them. Such translations correspond to rotations and screw motions of the ambient space, which gives a kinematic interpretation of the straight line as the rotation axes.

Any point on a line divides it into three parts, the point and two open half-lines. These half-lines are congruent, because they can be superimposed on each other by movement (rotation around the point). Two different half-lines define an angle; among all the angles is a special angle that is congruent with its complementary one. This is a right angle, and the continuation of its sides are perpendicular straight lines (note that the notion of perpendicularity is introduced here without using the metric, in spirit of the arguments of synthetic geometry).

The collection of all lines passing through a given point $a$ and perpendicular to a given line $l$, containing this point, defines a new geometrical object, namely the plane $\Pi_l^a$, which has rotational symmetry with respect to $l$. The translation of the plane $\Pi_l^a$ along the line $l$ preserves the property of its perpendicularity to the line $l$. Therefore, one can obtain the family $\{\Pi_l^\alpha\}_{\alpha \in l}$ of planes parallel to $\Pi_l$. Choosing a straight line $j$ lying in some $\Pi_l^\alpha$, and repeating the procedure described above, we obtain a family of planes $\Pi_j^\beta$ that are perpendicular to the planes $\Pi_l^\alpha$. Finally, choosing a line $k$ lying in $\Pi_l^{\alpha'}$, $\alpha' \in l$ and in some $\Pi_j^\beta$, $\beta \in j$, we obtain the third family $\Pi_k^\gamma$ which elements are perpendicular to the elements of family $\Pi_l^\alpha$ and family $\Pi_j^\beta$. The pairwise intersections of such planes are straight lines. These lines can be "graded" by means of a sequence of translations of unit intervals $i$, $j$, and $k$, which can be aligned with each other by rotations. As a result, we obtain spatial forests of graded lines, which constructively define the Cartesian net and the corresponding coordinate system.

Bearing in mind the fact that later we will move on from affine Euclidean space to spaces with a more general geometry, let us explain how the advanced analytical features that usually are pointed out in the definition of general smooth manifolds can be defined in the Euclidean approach.

### 2.1.2 Topology on $\mathscr{E}$

The Weyl axioms, establishing the relation between $\mathscr{E}$ and $\mathscr{V}$, as well as the algebraic structure of the Euclidean space $\mathscr{V}$, define a particular kind of topology on $\mathscr{E}$, namely, the metric topology. Let us explain this in more detail.

**$\mathscr{E}$ as a metric space.** The inner product $(\cdot)$ induces the norm $\|\cdot\|$ on $\mathscr{V}$:

$$\forall \boldsymbol{u} \in \mathscr{V}: \|\boldsymbol{u}\| := \sqrt{\boldsymbol{u} \cdot \boldsymbol{u}}\,. \tag{2.5}$$

The norm (2.5), in turn, induces a metric on $\mathscr{E}$ as follows:

$$d: \mathscr{E} \times \mathscr{E} \to \mathbb{R}\,, \quad d(a, b) := \|\overrightarrow{ab}\|\,. \tag{2.6}$$

Thus, $\mathscr{E}$ is endowed with the *metric topology*[4]. The mapping $\mathscr{D}$ (2.2) is a homeomorphism relatively to this topology and the topology of $\mathbb{R}^m$.

The metric topology turns $\mathscr{E}$ into Hausdorff topology space $(\mathscr{E}, \mathfrak{S})$ with basis $\mathscr{B} \subset \mathfrak{S}$ that is defined by the collection of open balls with rational radii and centers:

$$(A \in \mathscr{B}) \Leftrightarrow \left(\exists o \in \widetilde{\mathscr{E}}\ \exists \alpha \in \mathbb{Q}\ \forall a \in A: \|a - o\| < \alpha\right)\,,$$

where $\widetilde{\mathscr{E}} = \mathscr{D}^{-1}(\mathbb{Q}^m)$ is a dense countable subset of $\mathscr{E}$. Since $\mathscr{B}$ is countable, the topological space $\mathscr{E}$ is second countable.

**Atlas on $(\mathscr{E}, \mathfrak{S})$.** The ordered pair $(\mathscr{E}, \mathscr{D})$ represents a chart on $\mathscr{E}$. Such a chart covers the whole $\mathscr{E}$, and the minimal atlas $A_{\text{affine}} = \{(\mathscr{E}, \mathscr{D})\}$ is trivial. Thus, $\mathscr{E}$ is a smooth manifold with smooth structure generated by $A_{\text{affine}}$.

### 2.1.3 Smooth Mappings on $\mathscr{E}$

The affine-Euclidean space, which we presented as a topological space $(\mathscr{E}, \mathfrak{S})$ equipped with the trivial atlas $A_{\text{affine}}$, can be used to define a smooth manifold, if we define a smooth structure on it. To this end, we introduce the notions of smooth[5] curves, functions, and fields as follows.

**Curves.** Define a curve on $\mathscr{E}$ as a mapping $\chi: \mathbb{I} \to \mathscr{E}$, where $\mathbb{I} \subset \mathbb{R}$ is an interval. A curve $\chi$ has the coordinate representation $\tilde{\chi} = \mathscr{D} \circ \chi: \mathbb{I} \to \mathbb{R}^m$ relatively to the atlas $A_{\text{affine}}$. The curve $\chi$ is said to be smooth (belongs to class $C^\infty$) if the mapping $\tilde{\chi}$ is smooth, *i.e.*, $\tilde{\chi} \in C^\infty(\mathbb{I}; \mathbb{R})$.

4 Definitions of a metric, a topology, and related notions are given in Chapter 13, Section 13.1.1.

5 Smooth mappings between manifolds are briefly considered in Chapter 13, Section 13.1.3.

**Functions.** Define a function on $\mathscr{E}$ as a mapping $f\colon O \to \mathbb{R}$, where $O \subset \mathscr{E}$ is an open set. Its coordinate representation has the form: $\tilde{f} = f \circ \mathscr{D}^{-1}\colon \mathscr{D}(O) \to \mathbb{R}$. That is, $\tilde{f}$ is a real valued function of $m$ real arguments. The function $f$ is said to be smooth (belongs to class $C^\infty$) if the mapping $\tilde{f}$ is smooth, *i.e.*, $\tilde{f} \in C^\infty(\mathscr{D}(O); \mathbb{R})$.

**Vector fields.** A vector field on $\mathscr{E}$ is a mapping $\boldsymbol{u}\colon O \to \mathscr{V}$, where $O \subset \mathscr{E}$ is an open set. It can be represented relatively to $A_{\text{affine}}$ as a vector valued function of $m$ real arguments:

$$\tilde{\boldsymbol{u}} = \boldsymbol{u} \circ \mathscr{D}^{-1}\colon \mathscr{D}(O) \to \mathscr{V}\,,$$

or as a mapping

$$\tilde{\tilde{\boldsymbol{u}}} = \Phi \circ \boldsymbol{u} \circ \mathscr{D}^{-1}\colon \mathscr{D}(O) \to \mathbb{R}^m\,,$$

where $\Phi = \mathscr{D} \circ \boldsymbol{p}^{-1}\colon \mathscr{V} \to \mathbb{R}^m$ is a vector space isomorphism. The singleton $A_{\text{vect}} = \{(\mathscr{V}, \Phi)\}$ is a trivial atlas on $\mathscr{V}$ and, thus, $\mathscr{V}$ is a smooth manifold with smooth structure generated by $A_{\text{vect}}$. The smoothness of a vector field $\boldsymbol{u}$ is defined by the smoothness of its coordinate representation $\tilde{\tilde{\boldsymbol{u}}}$ relatively to the atlases $A_{\text{affine}}$ and $A_{\text{vect}}$.

The following commutative diagrams illustrate relations between functions, vector fields, and their coordinate representations:

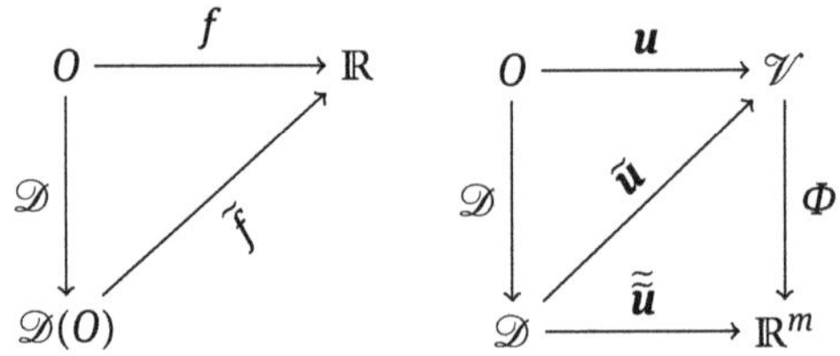


**Covectors.** The affine-Euclidean space possesses all the facilities for calculating the components of translation vectors with respect to the chosen basis. Indeed, they can be found as a result of a generally non-orthogonal projection that is analytically determined by a scalar product with elements of a dual basis. Moreover, the scalar product defines the concept of the angle and, together with the definition of vector product, represents the complete framework of elementary geometry. Thus, the requirements of the geometer are completely satisfied. However, this cannot be said about the requirements of a physicist! In order to determine pair vector-like objects, for example, energetically conjugate velocity-force pairs, one has to consider the space of linear functionals (covectors) over $\mathscr{V}$, namely, dual space $\mathscr{V}^*$. The dual space satisfies all the axioms of the vector space (see Chapter 12, Section 12.5.1), and, moreover, if the dimension of $\mathscr{V}$ is finite, then the dimension of $\mathscr{V}^*$ coincides with it (for proof of this assertion, see [42]).

It should be noted that in certain cases, covectors without loss of generality can be represented by vectors dual to them. However, such an identification of pair objects can lead to a loss of physical meaning. For this reason, we will further strictly distinguish between vectors and covectors.

More detailed definitions are given as follows. The elements of $\mathscr{V}^* := \mathrm{Lin}(\mathscr{V}; \mathbb{R})$, the dual vector space of $\mathscr{V}$, are linear functionals over $\mathscr{V}$ (covectors). The dimension of dual space is equal to the dimension of $\mathscr{V}$. The value of a covector $\nu \in \mathscr{V}^*$ on a vector $\boldsymbol{u} \in \mathscr{V}$ is often written as $\langle \nu, \boldsymbol{u} \rangle := \nu(\boldsymbol{u})$. It is convenient to represent covectors in the basis $(e^i)_{i=1}^m$ for $\mathscr{V}^*$ that satisfy the property

$$\left\langle e^i, \boldsymbol{e}_j \right\rangle = \delta^i_j \,, \quad i, j = 1, \ldots, m \,.$$

The $m$-tuple $(e^i)_{i=1}^m$ is a dual basis in the dual space. In the following we will refer to it in a shorter manner as a dual basis. In the dual basis, covectors can be represented by their components:

$$\forall \nu \in \mathscr{V}^* \; \exists \nu_1, \ldots, \nu_m \in \mathbb{R}: \quad \nu = \nu_k e^k \,.$$

The norm $\| \cdot \|$ on $\mathscr{V}$ induces the norm on $\mathscr{V}^*$ as follows:

$$\forall \nu \in \mathscr{V}^*: \quad \|\nu\| := \sup_{\|\boldsymbol{u}\|=1} |\nu(\boldsymbol{u})| \,.$$

With such a norm the dual space $\mathscr{V}^*$ is a smooth manifold; its smooth structure is generated by the trivial atlas $A_{\text{covect}} = \{(\mathscr{V}^*, \Psi)\}$. Here,

$$\Psi: \mathscr{V}^* \to \mathbb{R}^m \,, \quad \Psi: \nu = \nu_k e^k \mapsto (\nu_1, \ldots, \nu_m) \,,$$

is the vector space isomorphism.

**Remark 2.2.** Despite the fact that in classic works on continuum mechanics, the distinction between vectors and covectors is rarely taken into account, distinguishing them as objects from different spaces seems to be useful from the physical standpoint. Indeed, if all quantities that possess a vector nature are defined in the same vector space, which, as a rule, is equipped with a scalar product, then a completely justified desire arises to treat, for example, the velocity vector and the force vector[6] as elements of one vector space. Their scalar product results in a scalar (a power). On the other hand, since they belong to the same space, one can formally consider their sum, which, of course, has no physical meaning. This incorrect situation can be redressed by the following consideration: the force had to be formalized as a covector, while the velocity naturally has to be represented by a vector. The covector (a force), as a linear functional, acts on the velocity vector, which results in a scalar (a power). Of course, the sum of a vector and any covector is not defined.

It should be noted that the difference in vector-like quantities, such as force and displacement, was already discussed in the pioneering works on vector and tensor calculus [35].

6 The concept of force is one of the fundamentals in modern physics. However, this concept does not have a strict and universal definition. This fact was described figuratively by Max Jammer in his monograph [43]: *To introduce the term "force" as an explanatory element in the theory of physical science means to develop a misleading vocabulary.*

**Musical isomorphisms.** Recall that $\mathscr{V}$ is equipped with scalar product $(\cdot)$. According to the Riesz theorem[7] for any covector $v \in \mathscr{V}^*$ there exists the unique vector $\boldsymbol{w} \in \mathscr{V}$, such that

$$\forall \boldsymbol{u} \in \mathscr{V} : v(\boldsymbol{u}) = \boldsymbol{w} \cdot \boldsymbol{u} .$$

The mapping $(\cdot)^\sharp : \mathscr{V}^* \to \mathscr{V}$, which assigns to each covector $v \in \mathscr{V}^*$ the corresponding (due to the Riesz theorem) vector $(v)^\sharp = \boldsymbol{w} \in \mathscr{V}$, is a vector space isomorphism. Its inverse is denoted by $(\cdot)^\flat : \mathscr{V} \to \mathscr{V}^*$. Thus, for a vector $\boldsymbol{v} \in \mathscr{V}$, one has

$$\forall \boldsymbol{u} \in \mathscr{V} : \boldsymbol{v}^\flat(\boldsymbol{u}) = \boldsymbol{v} \cdot \boldsymbol{u} .$$

The mutually inverse isomorphisms $(\cdot)^\flat$ and $(\cdot)^\sharp$ are referred to as *musical isomorphisms*[8].

**Covector fields.** A covector field on $\mathscr{E}$ is a mapping $v : O \to \mathscr{V}^*$, where $O \subset \mathscr{E}$ is an open set. Relatively to the atlas $A_{\text{affine}}$, a covector field $v$ is represented by

$$\tilde{v} = v \circ \mathscr{D}^{-1} : \mathscr{D}(O) \to \mathscr{V}^* .$$

In atlases $A_{\text{affine}}$ and $A_{\text{covect}}$ one has the following representation:

$$\tilde{\tilde{v}} = \Psi \circ v \circ \mathscr{D}^{-1} : \mathscr{D}(O) \to \mathbb{R}^m .$$

The smoothness of a covector field $v$ is defined by the smoothness of its coordinate representation $\tilde{\tilde{v}}$.

**Tensor fields.** Tensors are elements of tensor products of vector spaces[9] $\mathscr{V}, \mathscr{V}^*$. Thus, mixed tensors of type $(k, l)$ are elements of

$$T^{(k,l)}(\mathscr{V}) := \underbrace{\mathscr{V} \otimes \cdots \otimes \mathscr{V}}_{k \text{ copies}} \otimes \underbrace{\mathscr{V}^* \otimes \cdots \otimes \mathscr{V}^*}_{l \text{ copies}} .$$

The field $\mathbb{R}$, the vector spaces $\mathscr{V}$, and $\mathscr{V}^*$ are particular cases of mixed tensor spaces:

$$T^{(0,0)}(\mathscr{V}) = \mathbb{R} , \quad T^{(1,0)}(\mathscr{V}) = \mathscr{V} , \quad T^{(0,1)}(\mathscr{V}) = \mathscr{V}^* .$$

Equivalently, a $(k, l)$-type tensor can by viewed as a multilinear map, the element of

$$\mathrm{Lin}_{k+l}(\underbrace{\mathscr{V}^*, \ldots, \mathscr{V}^*}_{k \text{ copies}}, \underbrace{\mathscr{V}, \ldots, \mathscr{V}}_{l \text{ copies}}; \mathbb{R}) .$$

Thus, one can introduce a norm on $T^{(k,l)}(\mathscr{V})$ as follows:

$$\|\boldsymbol{T}\| := \sup_{\substack{\|v_1\|=1,\ldots,\|v_k\|=1, \\ \|\boldsymbol{u}_1\|=1,\ldots,\|\boldsymbol{u}_l\|=1}} |\boldsymbol{T}(v_1, \ldots, v_k, \boldsymbol{u}_1, \ldots, \boldsymbol{u}_l)| ,$$

7 The formulation and proof can be found in Chapter 12, Section 12.5.4.

8 These isomorphisms are introduced in Chapter 12, Section 12.5.4. Their generalization for smooth manifolds is given in Chapter 13, Section 13.3.10.

9 The formal definition of an abstract tensor product is given in Chapter 12, Section 12.5.3.

for each $\boldsymbol{T} \in T^{(k,l)}(\mathscr{V})$. A basis $(\boldsymbol{e}_i)_{i=1}^m$ and its dual, $(\boldsymbol{e}^i)_{i=1}^m$, establish vector space isomorphism $\Theta\colon T^{(k,l)}(\mathscr{V}) \to \mathbb{R}^{m^{k+l}}$,

$$\Theta\colon \boldsymbol{T} = T^{i_1\ldots i_k}{}_{j_1\ldots j_l}\boldsymbol{e}_{i_1} \otimes \cdots \otimes \boldsymbol{e}_{i_k} \otimes \boldsymbol{e}^{j_1} \otimes \cdots \otimes \boldsymbol{e}^{j_l} \mapsto \left(T^{i_1\ldots i_k}{}_{j_1\ldots j_l}\right)_{1\le i_1,\ldots,i_k\le m\,,1\le j_1,\ldots,j_l\le m} .$$

The vector space $T^{(k,l)}(\mathscr{V})$ is endowed with the smooth manifold structure. The smooth structure is generated by the trivial atlas $A_{\text{tensor}} = \{(T^{(k,l)}(\mathscr{V}), \Theta)\}$.

A $(k, l)$-type tensor field on $\mathscr{E}$ is a map $\boldsymbol{T}\colon O \to T^{(k,l)}(\mathscr{V})$, where $O \subset \mathscr{E}$ is an open set. Relatively to the atlas $A_{\text{affine}}$, a tensor field $\boldsymbol{T}$ is represented by

$$\widetilde{\boldsymbol{T}} = \boldsymbol{T} \circ \mathscr{D}^{-1}\colon \mathscr{D}(O) \to T^{(k,l)}(\mathscr{V}) .$$

In atlases $A_{\text{affine}}$ and $A_{\text{tensor}}$, one has the following representation:

$$\widetilde{\widetilde{\boldsymbol{T}}} = \Theta \circ \boldsymbol{T} \circ \mathscr{D}^{-1}\colon \mathscr{D}(O) \to \mathbb{R}^{m^{k+l}} .$$

As in the previous cases, the smoothness of a tensor field $\boldsymbol{T}$ is defined by smoothness of its coordinate representation $\widetilde{\widetilde{\boldsymbol{T}}}$.

### 2.1.4 Parallel Transport

The concept of parallel transport plays a crucial role in geometrodynamics, as well as in the geometrical theory of incompatible deformations. In this regard, special attention will be given to this concept within the present book. The most simple definition for it is given in the Euclidean geometry.

The inner product $(\cdot)$ on $\mathscr{V}$ allows one to introduce the cosine mapping

$$\cos\colon \mathscr{V} \setminus \{0\} \times \mathscr{V} \setminus \{0\} \to [-1, 1] , \quad \cos(\boldsymbol{u}_1, \boldsymbol{u}_2) := \frac{\boldsymbol{u}_1 \cdot \boldsymbol{u}_2}{\|\boldsymbol{u}_1\|\|\boldsymbol{u}_2\|} . \tag{2.7}$$

The value of $\cos(\boldsymbol{u}_1, \boldsymbol{u}_2)$ on a pair $(\boldsymbol{u}_1, \boldsymbol{u}_2)$ of vectors has the sense of a cosine of angle between them. The mapping cos allows one to introduce the notion of parallelism as follows.

Vectors $\boldsymbol{u}, \boldsymbol{v} \in \mathscr{V}$ are said to be *parallel*, if $|\cos(\boldsymbol{u}_1, \boldsymbol{u}_2)| = 1$. A vector field $\boldsymbol{u}\colon O \to \mathscr{V}$ is said to be weakly parallel transported along a curve $\chi\colon [a, b] \to O$ if

$$\forall x \in [a, b]\colon \quad |\cos(\boldsymbol{u}(\chi(a)), \boldsymbol{u}(\chi(x)))| = 1 ,$$

and is said to be strongly parallel transported if

$$\forall x \in [a, b]\colon \quad \boldsymbol{u}(\chi(a)) = \boldsymbol{u}(\chi(x)) .$$

The condition of weakly parallel transport is equal to the representation $\boldsymbol{u}(x) = f(x)\boldsymbol{u}(\chi(a))$, for all $x \in \chi([a, b])$. Here, $f\colon \chi([a, b]) \to \mathbb{R}$ is a function.

### 2.1.5 Total Derivatives of Vector and Tensor Fields

In order to apply calculus one has to introduce the notion of derivatives for vector and tensor fields. We begin with the following definition.

**Total derivative of a function.** Suppose that $O \subset \mathscr{E}$ is an open set and $f\colon O \to \mathbb{R}$ is a function. It is called differentiable at a point $a \in O$ if there exists a covector $v_a \in \mathscr{V}^*$, such that for any $\boldsymbol{h} \in \mathscr{V}$, $a + \boldsymbol{h} \in O$, the following identity holds [44]:

$$f(a + \boldsymbol{h}) = f(a) + \langle v_a, \boldsymbol{h}\rangle + \varphi(\boldsymbol{h})\|\boldsymbol{h}\| \,, \tag{2.8}$$

where $\lim_{\boldsymbol{h}\to\boldsymbol{0}, a+\boldsymbol{h}\in O} \varphi(\boldsymbol{h}) = 0$. In terms of the E. Landau symbol $o$, one can write

$$f(a + \boldsymbol{h}) = f(a) + \langle v_a, \boldsymbol{h}\rangle + o(\|\boldsymbol{h}\|) \quad \text{as} \quad \boldsymbol{h} \to \boldsymbol{0} \,.$$

The mapping $v_a$ is called the *total derivative of $f$ at point $a$*, and since it is the unique mapping that satisfies (2.8), one of the following notations can be used: $Df(a)$, $f'(a)$. The uniqueness follows from the fact that the value of $\operatorname{grad} f(a)$ is the directional derivative:

$$\forall \boldsymbol{h} \in \mathscr{V}\colon \langle f'(a), \boldsymbol{h}\rangle = \lim_{O_{a;\boldsymbol{h}} \ni t \to 0} \frac{f(a + t\boldsymbol{h}) - f(a)}{t} \,,$$

where $O_{a;\boldsymbol{h}} = \{t \in \mathbb{R} \mid t \neq 0,\, a + t\boldsymbol{h} \in O\}$.

If $f$ is differentiable at each point $x \in O$ of the set $O$, then the mapping

$$f'\colon O \to \mathscr{V}^* \,, \qquad x \mapsto f'(x)$$

is defined correctly and is called the total derivative of $f$.

**Gradient.** Due to Riesz theorem there exists the unique vector

$$\operatorname{grad} f(a) := (f'(a))^\sharp \,, \tag{2.9}$$

such that $\langle f'(a), \boldsymbol{h}\rangle = \boldsymbol{h} \cdot \operatorname{grad} f(a)$. The vector $\operatorname{grad} f(a)$ is called the *gradient $f$ at point $a$*.

In the basis $(\boldsymbol{e}_i)_{i=1}^m$ of $\mathscr{V}$ and the dual basis $(e^i)_{i=1}^m$ of $\mathscr{V}^*$, the covector $f'(a)$ has the representation:

$$f'(a) = \partial_i f(a) e^i \,, \quad \text{where} \quad \partial_i f(a) = \big\langle f'(a), \boldsymbol{e}_i \big\rangle \,, \quad i = 1, \ldots, m \,.$$

**Total derivative of a vector field.** Suppose that $O \subset \mathscr{E}$ is an open set and $\boldsymbol{u}\colon O \to \mathscr{V}$ is a vector field. It is called differentiable at a point $a \in O$, if there exists a linear map (a $(1,1)$-type tensor) $\boldsymbol{L}_a \in \operatorname{Lin}(\mathscr{V};\mathscr{V})$, such that for any $\boldsymbol{h} \in \mathscr{V}$, $a + \boldsymbol{h} \in O$, the following identity holds [44]:

$$\boldsymbol{u}(a + \boldsymbol{h}) = \boldsymbol{u}(a) + \boldsymbol{L}_a[\boldsymbol{h}] + \boldsymbol{o}(\|\boldsymbol{h}\|) \quad \text{as} \quad \boldsymbol{h} \to \boldsymbol{0} \,. \tag{2.10}$$

The mapping $\boldsymbol{L}_a$ is called the *total derivative of $\boldsymbol{u}$ at the point $a$*, and since it is the unique mapping that satisfies (2.10), the following notations are used: $D\boldsymbol{u}(a)$, $\boldsymbol{u}'(a)$.

As in the case of a function, the uniqueness follows from the fact that the value of $\boldsymbol{Du}(a)$ is the directional derivative:

$$\forall \boldsymbol{h} \in \mathscr{V}: \quad \boldsymbol{Du}(a)[\boldsymbol{h}] = \lim_{O_{a;\boldsymbol{h}} \ni t \to 0} \frac{\boldsymbol{u}(a + t\boldsymbol{h}) - \boldsymbol{u}(a)}{t},$$

where $O_{a;\boldsymbol{h}} = \{t \in \mathbb{R} \mid t \neq 0, a + t\boldsymbol{h} \in O\}$.

If $\boldsymbol{u}$ is differentiable at each point $x \in O$, then the mapping

$$\boldsymbol{Du}: \quad O \to \operatorname{Lin}(\mathscr{V}; \mathscr{V}), \quad x \mapsto \boldsymbol{Du}(x)$$

is defined correctly and is called the total derivative of $\boldsymbol{u}$.

In the basis $(\boldsymbol{e}_i)_{i=1}^m$ of $\mathscr{V}$, the vector field $\boldsymbol{u}$ has the representation:

$$\forall x \in O: \quad \boldsymbol{u}(x) = u^i(x)\boldsymbol{e}_i .$$

Here the mappings $u^i: O \to \mathbb{R}$, $i = 1, \ldots, m$, are component functions. The differentiability of $u$ at $a \in O$ implies the differentiability of each $u^i$ at $a$. In the basis $(\boldsymbol{e}_i)_{i=1}^m$ and the dual basis $(e^i)_{i=1}^m$ of $\mathscr{V}^*$, the linear map $\boldsymbol{Du}(a)$ has the representation:

$$\boldsymbol{Du}(a) = \partial_j u^i(a)\boldsymbol{e}_i \otimes e^j, \quad \text{where} \quad \partial_j u^i(a) = \left\langle Du^i(a), \boldsymbol{e}_j \right\rangle, \quad i, j = 1, \ldots, m .$$

The total derivative of a covector field, and, generally, of a tensor field is defined in an analogous way. In all such cases, the smoothness in terms of the total derivative coincides with the smoothness in terms of smooth manifolds. The smoothness of a vector, covector, and, generally, tensor fields is equivalent to the smoothness of all its component functions.

**Remark 2.3.** We considered the cases of a function and vector field separately. That allowed us to pick out individual features of their derivatives. In analysis, the notion of the total derivative (Fréchet derivative) at a point is introduced in a uniform way as follows [44]. Suppose that $(\mathscr{A}_1, \mathscr{V}_1, \psi_1)$ and $(\mathscr{A}_2, \mathscr{V}_2, \psi_2)$ are finite dimensional affine spaces with translation spaces $\mathscr{V}_1$ and $\mathscr{V}_2$, respectively[10]. That is, $\mathscr{A}_1$, $\mathscr{A}_2$ are sets, $\mathscr{V}_1$, $\mathscr{V}_2$ are finite dimensional real vector spaces and $\psi_1: \mathscr{A}_1 \times \mathscr{A}_1 \to \mathscr{V}_1$, $\psi_2: \mathscr{A}_2 \times \mathscr{A}_2 \to \mathscr{V}_2$ are mappings that satisfy the Weyl axioms $W_1$) and $W_2$). In addition, suppose that translation spaces $\mathscr{V}_1$ and $\mathscr{V}_2$ are endowed with norms $\|\cdot\|_1$ and $\|\cdot\|_2$. These norms turn $\mathscr{A}_1$ and $\mathscr{A}_2$ into metric spaces by defining metrics on them in the same manner as in Section 2.1.2 of the present chapter.

Let $X \subset \mathscr{A}_1$ be a non-empty set and $a \in X$ be an interior point[11] of $X$. A mapping $f: X \to \mathscr{A}_2$ is called differentiable at the point $a$, if there exists a linear map $\mathscr{L}_a \in \operatorname{Lin}(\mathscr{V}_1; \mathscr{V}_2)$, such that for any $h \in \mathscr{V}_1$, $a + h \in X$, the following identity holds[12]:

$$f(a + h) = f(a) + \mathscr{L}_a[h] + \varphi(h)\|h\|_1 , \tag{2.11}$$

10 Recall that the definition of affine space and its properties are given in Chapter 12, Section 12.7.

11 That is, $a$ is a center of some open ball in $X$.

12 Here, one needs to use the external operations $+_i: \mathscr{A}_i \times \mathscr{V}_i \to \mathscr{A}_i$, $i = 1, 2$, which are constructed by virtue of the axiom $W_2$). For readability we omit the indices 1 and 2.

where $\lim_{h\to 0, a+h\in X}\varphi(h) = 0$. The mapping $\mathscr{L}_a$ is called the total derivative of $f$ at point $a$, and since it is the unique mapping that satisfies (2.11), the following notation can be used: $f'(a) := \mathscr{L}_a$.

The value of $f'(a)$ at an arbitrary vector is defined as follows:

$$\forall h \in \mathscr{V}_1 : f'(a)[h] = \lim_{X_{a;h}\ni t\to 0} \frac{f(a+th)-f(a)}{t} ,$$

where $X_{a;h} = \{t \in \mathbb{R} \mid t \neq 0,\, a+th \in X\}$. Such a formula establishes the uniqueness of $f'(a)$.

Since $\mathbb{R}$, $\mathscr{V}$, $\mathscr{V}^*$ and, generally, $T^{(k,l)}(\mathscr{V})$ are affine spaces, the general notion of the total derivative is applied directly to them.

### 2.1.6 Integration

Let $\mathscr{D}$ be generated by the Cartesian coordinate system $(o, (\boldsymbol{i}_k)_{k=1}^m)$. An elementary volume in the Euclidean space $\mathscr{E}$ is represented by $m$-brick $\mathscr{I}_m := \mathscr{D}^{-1}(I_m)$, where $I_m = [a_1, b_1]\times\cdots\times[a_m, b_m]$ and $[a_i, b_i] \subset \mathbb{R}$, $i = 1,\ldots,m$. If $f : \mathscr{E} \to \mathbb{R}$ is a function, then we define the integral over $\mathscr{I}_m$ as

$$\int_{\mathscr{I}_m} f := \int_{I_m} \tilde{f} = \int_{a_1}^{b_1} dx^1 \cdots \int_{a_m}^{b_m} dx^m\, \tilde{f} ,$$

where repeated integrals are treated as Riemannian or Lebesque integrals.

Let $\boldsymbol{T} : \mathscr{E} \to T^{(k,l)}(\mathscr{V})$ be a $(k,l)$-type tensor field. Define the integral over $\mathscr{I}_m$ similarly to the scalar case:

$$\int_{\mathscr{I}_m} \boldsymbol{T} := \int_{I_m} \tilde{\boldsymbol{T}} = \int_{a_1}^{b_1} dx^1 \cdots \int_{a_m}^{b_m} dx^m\, \tilde{\boldsymbol{T}} .$$

In a basis $(\boldsymbol{e}_i)_{i=1}^m$ and its dual, $(\boldsymbol{e}^i)_{i=1}^m$, one has

$$\boldsymbol{T} = T^{i_1\ldots i_k}{}_{j_1\ldots j_l}\, \boldsymbol{e}_{i_1}\otimes\cdots\otimes\boldsymbol{e}_{i_k}\otimes\boldsymbol{e}^{j_1}\otimes\cdots\otimes\boldsymbol{e}^{j_l} ,$$

where $T^{i_1\ldots i_k}{}_{j_1\ldots j_l} : \mathscr{E} \to \mathbb{R}$ are component functions. Thus, the integral can be represented in the form

$$\int_{\mathscr{I}_m} \boldsymbol{T} = \boldsymbol{e}_{i_1}\otimes\cdots\otimes\boldsymbol{e}_{i_k}\otimes\boldsymbol{e}^{j_1}\otimes\cdots\otimes\boldsymbol{e}^{j_l} \int_{a_1}^{b_1} dx^1 \cdots \int_{a_m}^{b_m} dx^m\, \tilde{T}^{i_1\ldots i_k}{}_{j_1\ldots j_l} .$$

The integral of a tensor field was reduced to the integral of scalar mappings (it is assumed that all transformations are well defined).

### 2.1.7 Symmetrization and Antisymmetrization

Suppose that $\mathscr{A}$ is a finite-dimensional affine space with a distance function generated by the norm on the translation space $\mathscr{V}$ via (2.6). Consider a smooth vector field, $u$, and a covector field, $v$, defined on some open set $O \subset \mathscr{A}$. If $(e_i)$ is some basis for $\mathscr{V}$ and $(e^i)$ is its dual basis for $\mathscr{V}^*$, then

$$Du = \partial_k u^s e_s \otimes e^k \,, \quad Dv = \partial_k v_s e^s \otimes e^k \,.$$

In the linear approach adopted in convectional continuum mechanics, the symmetric and antisymmetric parts of the gradient of a vector field play the role of infinitesimal strains and vorticity tensors. Here, we do not presuppose that any Euclidean structure is defined on $\mathscr{A}$. Thus, we have to introduce an operation similar to transposition without reference to any inner product. To this end, we define the operation $(\cdot)^\dagger$, that is, the transposition of the elements in dyads: $(a \otimes b)^\dagger := b \otimes a$. Such definition coincides with the definition for transposition of the second-rank tensors in the Cartesian description of Euclidean space. Performing this operation $(\cdot)^\dagger$ for $v$, we observe that expressions

$$\frac{1}{2}(Dv + Dv^\dagger) \,, \quad \frac{1}{2}(Dv - Dv^\dagger)$$

are well defined, but at the same time the expressions

$$\frac{1}{2}(Du + Du^\dagger) \,, \quad \frac{1}{2}(Du - Du^\dagger) \,,$$

for $u$ are ill defined, since $e_s \otimes e^k$ and $e^k \otimes e_s$ belong to different tensor spaces. The situation can be improved by introducing an inner product $(\cdot)$ on $\mathscr{A}$ that turns it into Euclidean affine space $\mathscr{E}$. Then, the transposed tensor $\boldsymbol{L}^{\mathrm{T}} \in \mathrm{Lin}(\mathscr{V};\mathscr{V})$ to a tensor $\boldsymbol{L} \in \mathrm{Lin}(\mathscr{V};\mathscr{V})$ can be defined as[13]

$$\forall \boldsymbol{u}, \boldsymbol{v} \in \mathscr{V} : \boldsymbol{v} \cdot \boldsymbol{L}[\boldsymbol{u}] = \boldsymbol{u} \cdot \boldsymbol{L}^{\mathrm{T}}[\boldsymbol{v}] \,.$$

The symmetrization and antisymmetrization of $\boldsymbol{u}$ are well defined and can be written using the notion of the transpose:

$$\frac{1}{2}(\boldsymbol{Du} + \boldsymbol{Du}^{\mathrm{T}}) \,, \quad \frac{1}{2}(\boldsymbol{Du} - \boldsymbol{Du}^{\mathrm{T}}) \,.$$

### 2.1.8 Divergence

In order to achieve a complete analytical toolkit for field analysis, one has to supplement the above definitions with divergence and curl operators.

Let $O \subset \mathscr{E}$ be an open set.

13 The general definition of transposed tensor is given in Chapter 12, Section 12.5.4 by formula (12.3).

**Divergence of the vector field.** If a vector field $\boldsymbol{u}\colon O \to \mathscr{V}$ is differentiable at $a \in O$, then the divergence of $\boldsymbol{u}$ at $a$ is the scalar

$$\operatorname{div} \boldsymbol{u}(a) := \operatorname{tr}[\boldsymbol{u}'(a)]\,. \tag{2.12}$$

With respect to the orthonormal basis $(\boldsymbol{i}_k)_{k=1}^{s}$, one has $\operatorname{div} \boldsymbol{u}(a) = \frac{\partial u^i(a)}{\partial x^i}$. If $\boldsymbol{u}$ is differentiable at $O$, then one can define the divergence field

$$\operatorname{div} \boldsymbol{u}\colon O \to \mathbb{R}\,, \quad x \mapsto \operatorname{div} \boldsymbol{u}(x)\,.$$

Gauss' theorem gives the following geometrical interpretation for $\operatorname{div} \boldsymbol{u}(a)$ in case $m = 3$:

$$\operatorname{div} \boldsymbol{u}(a) = \lim_{\delta \to 0} \frac{1}{\operatorname{Vol}(V_\delta)} \int_{\partial V_\delta} \boldsymbol{u} \cdot \boldsymbol{n}\, dS\,.$$

Here, $V_\delta$ is an open ball with a center at $a$ and a radius $\delta$; $\operatorname{Vol}(V_\delta)$ is its volume and $\boldsymbol{n}$ is an outward unit normal.

**Divergence of the tensor field.** Suppose that a $(1,1)$-tensor field $\boldsymbol{T}\colon O \to T^{(1,1)}(\mathscr{V})$ is differentiable at $a \in O$. Here, $T^{(1,1)}(\mathscr{V}) = \operatorname{Lin}(\mathscr{V};\mathscr{V})$. Then, the field $\boldsymbol{T}^{\mathrm{T}}\colon O \to T^{(1,1)}(\mathscr{V})$ is also differentiable at $a \in O$. The divergence of $\boldsymbol{T}$ at $a$ is a unique vector $\operatorname{div} \boldsymbol{T}(a)$ that is defined by the following relation:

$$\boldsymbol{u} \cdot \operatorname{div} \boldsymbol{T}(a) = \operatorname{div}\{\boldsymbol{T}^{\mathrm{T}}(a)[\boldsymbol{u}]\}\,, \tag{2.13}$$

for every vector $\boldsymbol{u} \in \mathscr{V}$. If $\boldsymbol{T} = T^k{}_l \boldsymbol{i}_k \otimes \boldsymbol{i}^l$, then $\operatorname{div} \boldsymbol{T}(a) = \frac{\partial T^k{}_l}{\partial x^l} \boldsymbol{i}_k$. If $\boldsymbol{T}$ is differentiable at $O$, then one has the divergence field

$$\operatorname{div} \boldsymbol{T}\colon O \to \mathscr{V}\,, \quad x \mapsto \operatorname{div} \boldsymbol{T}(x)\,.$$

Suppose that $f\colon O \to \mathbb{R}$, $\boldsymbol{u}\colon O \to \mathscr{V}$, $\boldsymbol{L}\colon O \to \operatorname{Lin}(\mathscr{V};\mathscr{V})$ are differentiable scalar, vector, and tensor fields, respectively. The following identities hold:

$$\operatorname{div}(f\boldsymbol{u}) = \boldsymbol{u} \cdot \operatorname{grad} f + f \operatorname{div} \boldsymbol{u}\,, \tag{2.14}$$

$$\operatorname{div}(\boldsymbol{L}[\boldsymbol{u}]) = \boldsymbol{u} \cdot \operatorname{div} \boldsymbol{L}^{\mathrm{T}} + \operatorname{tr}(\boldsymbol{L}\boldsymbol{u}')\,. \tag{2.15}$$

### 2.1.9 Curl

Again, let $O \subset \mathscr{E}$ be an open set.

**Curl of the vector field.** A curl of a vector field at some point is a vector that is associated with the antisymmetric part of its gradient. More precisely, suppose that a vector field $\boldsymbol{u}\colon O \to \mathscr{V}$ is differentiable at $a \in O$. The curl of $\boldsymbol{u}$ at $a$ is the vector $\operatorname{curl} \boldsymbol{u}(a)$, which is defined by the following relation:

$$(D\boldsymbol{u}(a) - D\boldsymbol{u}^{\mathrm{T}}(a))\boldsymbol{v} = (\operatorname{curl} \boldsymbol{u}(a)) \times \boldsymbol{v}\,, \tag{2.16}$$

for every constant vector $\boldsymbol{v}$. If $\boldsymbol{u} = u^k \boldsymbol{i}_k$, then direct calculations with (2.3) show that $\operatorname{curl} \boldsymbol{u}(a) = e^{skq}\delta_{ql}\partial_k u^l(a)\boldsymbol{i}_s$. Stokes' theorem gives the following geometric interpretation for $\operatorname{curl} \boldsymbol{u}(a)$ in the case $m = 3$:

$$\boldsymbol{n} \cdot \operatorname{curl} \boldsymbol{u}(a) = \lim_{\delta \to 0} \frac{1}{\operatorname{Area}(D_\delta)} \int_{\partial D_\delta} \boldsymbol{u} \cdot \boldsymbol{\tau}\, dA\,,$$

where $D_\delta$ is a two-dimensional disc centered at $a$ with unit normal $\boldsymbol{n}$; $\operatorname{Area}(D_\delta)$ is its area and $\boldsymbol{\tau}$ is the unit tangent vector to $\partial D_\delta$.

If $\boldsymbol{u}$ is differentiable everywhere at $O$, then we define the field

$$\operatorname{curl} \boldsymbol{u} \colon O \to \mathscr{V}\,, \quad x \mapsto \operatorname{curl} \boldsymbol{u}(x)\,.$$

**Curl of the tensor field.** Let a $(1,1)$-tensor field $\boldsymbol{T} \colon O \to T^{(1,1)}(\mathscr{V})$ be differentiable at $a \in O$. Then, the curl of $\boldsymbol{T}$ at $a$ is a unique $(1,1)$-tensor $\operatorname{curl} \boldsymbol{T}(a)$ with the property:

$$(\operatorname{curl} \boldsymbol{T}(a))\boldsymbol{u} = \operatorname{curl}\{\boldsymbol{T}^{\mathrm{T}}(a)[\boldsymbol{u}]\}\,, \tag{2.17}$$

for every vector $\boldsymbol{u} \in \mathscr{V}$. If $\boldsymbol{T} = T^s{}_k \boldsymbol{i}_s \otimes i^k$, then $\operatorname{curl} \boldsymbol{T}(a) = e^{srq}\delta_{kl}\partial_r T^l{}_q(a)\boldsymbol{i}_s \otimes i^k$.

If $\boldsymbol{T}$ is differentiable everywhere at $O$, then we define the field

$$\operatorname{curl} \boldsymbol{T} \colon O \to \operatorname{Lin}(\mathscr{V};\mathscr{V})\,, \quad x \mapsto \operatorname{curl} \boldsymbol{T}(x)\,.$$

## 2.2 Curvilinear Coordinates

Curvilinear coordinates are defined as a local reparametrization of the Cartesian chart on the open subset of affine-Euclidean space. In fact, a new coordinate chart is added to the atlas $A_{\text{affine}}$. This chart is related with the Cartesian one by smooth transition mappings. As a result, the points of $\mathscr{E}$ are identified directly with a tuple of coordinates. Thus, formally, the translation vectors are no longer required, and the Euclidean properties of $\mathscr{E}$, declared by H. Weyl axioms, result from specific properties of the metric. Certainly, it is worth remembering that these specific properties of the metric are a consequence of the fact that it is calculated by reparametrization of the Cartesian (unit) metric, which, in turn, is obtained by H. Weyl axiomatic consideration. It should be noted that such reparametrization results in a structure that formally defines the Riemannian smooth manifold, although, of course, all non-Euclidean features are fictional, because they are the consequence of the change of variables only.

### 2.2.1 Reparametrization

Let $(\mathscr{E}, \mathscr{V}, (\cdot), \psi)$ be an $m$-dimensional Euclidean affine space. Suppose that $U \subset \mathscr{E}$ is an open set. Its points can be uniquely described by the affine coordinates $(x^i)_{i=1}^m$ in the coordinate system $(o, (\boldsymbol{e}_i)_{i=1}^m)$. Let us assume that there is an open set $G \subset \mathbb{R}^m$ and a $C^1$-diffeomorphism[14] $\xi\colon G \to U$. Then, each point $x \in U$ can be uniquely described by $m$-tuple

$$(q^1, \ldots, q^m) = \xi^{-1}(x) .$$

Numbers $q^i$ are called *curvilinear coordinates* of the point $x$. The ordered pair $(U, \xi^{-1})$ represents a *curvilinear* chart on $\mathscr{E}$. The whole set $\mathscr{E}$ may be covered by a number of charts $(U_k, \xi_k^{-1})$, $\xi_k\colon G_k \to U_k$, which together constitute an atlas $A_{\text{curvilin}} = \{(U_k, \xi_k^{-1})\}_{k=1}^N$.

The coordinate representation of each map $\xi_k$, $k = 1, \ldots, N$, is the following:

$$\tilde{\xi}_k = \mathscr{D} \circ \xi_k\colon G \to \mathbb{R}^m , \quad \tilde{\xi}_k\colon (q^1, \ldots, q^m) \mapsto (x^1, \ldots, x^m) .$$

For each point $x \in G$, the Jacobian $J = \det(\frac{\partial x^i}{\partial q^j})$ is not equal to zero.

A point $x \in \mathscr{E}$ is represented in curvilinear coordinates as

$$x = o + \tilde{x}_k^i(q^1, \ldots, q^m)\boldsymbol{e}_i , \quad \text{where} \quad (\tilde{x}_k^1, \ldots, \tilde{x}_k^m) = \tilde{\xi}_k ,$$

for some $k = 1, \ldots, N$. Each field (scalar or, generally, tensor) $u\colon O \to T^{(r,l)}(\mathscr{V})$, where $O \subset \mathscr{E}$ is open, can be parameterized locally as $\tilde{u}_{q;k} = u \circ \xi_k$. In particular, for the position vector field (2.1), one has

$$\tilde{\boldsymbol{p}}_k = \boldsymbol{p}|_{U_k} \circ \xi_k\colon G_k \to \mathscr{V} .$$

In terms of the functions $\tilde{x}_k^s$, the value of such a field can be represented as follows:

$$\tilde{\boldsymbol{p}}_k(q^1, \ldots, q^m) = \tilde{x}_k^s(q^1, \ldots, q^m)\boldsymbol{e}_s . \tag{2.18}$$

The relation between $\xi_k$, $\boldsymbol{p}$, and $\tilde{\boldsymbol{p}}_k$ is illustrated on the diagram

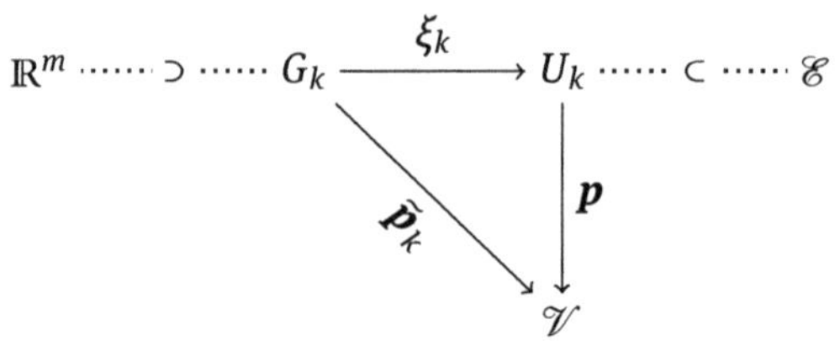


Table 2.1 contains some familiar examples of curvilinear coordinates in three-dimensional space.

14 That is, $\xi$ is a bijection of class $C^1$ with its inverse.

**Tab. 2.1:** Curvilinear coordinates, $m = 3$.

| | Coordinates | $\tilde{x}^s$ |
|---|---|---|
| Cylindrical | $(r, \varphi, z)$ | $\tilde{x}^1(r, \varphi, z) = r\cos\varphi$,<br>$\tilde{x}^2(r, \varphi, z) = r\sin\varphi$,<br>$\tilde{x}^3(r, \varphi, z) = z$ |
| Spherical | $(r, \theta, \varphi)$ | $\tilde{x}^1(r, \theta, \varphi) = r\sin\theta\cos\varphi$,<br>$\tilde{x}^2(r, \theta, \varphi) = r\sin\theta\sin\varphi$,<br>$\tilde{x}^3(r, \theta, \varphi) = r\cos\theta$ |

### 2.2.2 Coordinate Curves

We will work with a single chart $(U, \xi^{-1}) \in A_{\text{curvilin}}$. Suppose that $x \in U$ has curvilinear coordinates $\xi^{-1}(x) = (q_0^1, \ldots, q_0^m)$. The family $\{\chi_{x;i}\}_{i=1}^m$ of $m$ curves that pass through $x$ can be defined as follows:

$$\chi_{x;i} \colon \mathbb{I}_i \to \mathscr{E}\,, \quad \chi_{x;i} = \xi\left(q_0^1, \ldots, q_0^{i-1}, \cdot, q_0^{i+1}, \ldots, q_0^m\right)\,.$$

Here, $\mathbb{I}_i = \{y \in \mathbb{R} \mid (q_0^1, \ldots, q_0^{i-1}, y, q_0^{i+1}, \ldots, q_0^m) \in B_x\}$ is an interval; $B_x \subset G$ is an open ball centered at $(q_0^1, \ldots, q_0^m)$. A curve with index $i$ is obtained by fixing all arguments of $\xi$, except for the argument with number $i$. Carrying out such a procedure at each point $x \in U$ results in a net of curves $\{\chi_{x;i}\}_{i=1,\ldots,m,x\in U}$. We refer to each such curve as a *coordinate curve*.

### 2.2.3 Local Basis

**Definition.** The mapping (2.18) or, similarly, coordinate curves, generate $m$ vectors at $x \in U$:

$$\boldsymbol{\varepsilon}_s|_x := \frac{\partial \tilde{\boldsymbol{p}}\left(q_0^1, \ldots, q_0^m\right)}{\partial q^s} = \frac{\partial \chi_{x;s}\left(q_0^1, \ldots, q_0^m\right)}{\partial q^s}\,, \qquad s = 1, \ldots, m\,.$$

Since $\det\left(\frac{\partial \tilde{x}^i}{\partial q^j}\right) \neq 0$, these vectors are linearly independent and form a basis of $\mathscr{V}$. Thus, one has the fields $\boldsymbol{\varepsilon}_s \colon U \to \mathscr{V}$, $s = 1, \ldots, m$, the values of which form a basis of $\mathscr{V}$. We refer to the collection $(\boldsymbol{\varepsilon}_s)_{s=1}^m$ as a *local basis.*

The transformation from the constant basis $(\boldsymbol{e}_s)_{s=1}^m$ into the local basis $(\boldsymbol{\varepsilon}_s)_{s=1}^m$ is provided by a field $[\Omega^i{}_j] \colon U \to \mathrm{GL}(m;\mathbb{R})$ of non-degenerate $m \times m$ matrices. At a point $x \in U$, the transformation $(\boldsymbol{e}_s)_{s=1}^m \mapsto (\boldsymbol{\varepsilon}_s|_x)_{s=1}^m$ is defined by

$$\boldsymbol{\varepsilon}_s|_x = \Omega^i{}_s|_x \boldsymbol{e}_i\,, \quad \text{where} \quad \Omega^i{}_s|_x = \left.\frac{\partial \tilde{x}^i}{\partial q^s}\right|_{\xi^{-1}(x)}\,.$$

The inverse transformation, $(\boldsymbol{\varepsilon}_s|_x)_{s=1}^m \mapsto (\boldsymbol{e}_s)_{s=1}^m$, is given by

$$\boldsymbol{e}_s = \mho^i{}_s|_x \boldsymbol{\varepsilon}_i|_x\,.$$

Here, $[\mho^i{}_s] = [\Omega^i{}_j]^{-1}$, or, pointwise, at $x \in U$, $\mho^i{}_s|_x \Omega^s{}_j|_x = \delta^i_j$, $i, j = 1, \ldots, m$. Thus, at a point $x \in U$,

$$\mho^i{}_s|_x = \left.\frac{\partial \tilde{q}^i}{\partial x^s}\right|_x ,$$

where $\tilde{q}^i : (x^1, \ldots, x^m) \mapsto (q^1, \ldots, q^m)$ is the coordinate representation of $\xi^{-1}$.

Consider the relation between $\boldsymbol{\varepsilon}_i|_x$ and $\boldsymbol{\varepsilon}_i|_y$, where $x, y \in U$. Using inverse transformation one can write $\mho^i{}_s|_x \boldsymbol{\varepsilon}_i|_x = \mho^i{}_s|_y \boldsymbol{\varepsilon}_i|_y$. Multiplying both sides by $\Omega^s{}_j|_y$ and summing over $s$ gives

$$\boldsymbol{\varepsilon}_i|_y = \Omega^s{}_i|_y \mho^j{}_s|_x \boldsymbol{\varepsilon}_j|_x , \quad i = 1, \ldots, m . \tag{2.19}$$

The dual basis $(\varepsilon^s|_x)_{s=1}^m$ consists of covector fields $\varepsilon^s : U \to \mathscr{V}^*$, which are related with the local basis as follows:

$$x \in U: \ \left\langle \varepsilon^i|_x, \boldsymbol{\varepsilon}_j|_x \right\rangle = \delta^i_j , \quad i, j = 1, \ldots, m .$$

If $(e^s)_{s=1}^m$ is the dual basis that corresponds to $(\boldsymbol{e}_s)_{s=1}^m$, then one has

$$\varepsilon^i|_x = \mho^i{}_s|_x e^s .$$

**Representations in a local basis.** Consider representations of geometric objects in terms of the local basis. For a point $x \in U$, one obtains

$$x = o + \tilde{x}^i_k(q^1, \ldots, q^m) \mho^j{}_i|_x \boldsymbol{\varepsilon}_j|_x .$$

If $\boldsymbol{v} = v^i \boldsymbol{e}_i$ is a vector from $\mathscr{V}$, then for each $x \in U$ one arrives at

$$\boldsymbol{v} = v^i \mho^j{}_i|_x \boldsymbol{\varepsilon}_j|_x = w^j|_x \boldsymbol{\varepsilon}_j|_x ,$$

where $w^j|_x = v^i \mho^j{}_i|_x$. Note that the numbers $v^i$ transform into numbers $w^j|_x$, which change from point to point, *i.e.*, numerical components of a vector are represented by functions on $U$!

Let $\boldsymbol{v}: U \to \mathscr{V}$ be a vector field. It is represented by the decomposition $\boldsymbol{v} = v^i \boldsymbol{e}_i$, where $v^i : U \to \mathbb{R}$ are functions. Representing it in the local basis:

$$\forall x \in U: \quad \boldsymbol{v}|_x = w^j|_x \boldsymbol{\varepsilon}_j|_x , \quad \text{where} \quad w^j|_x = v^i|_x \mho^j{}_i|_x .$$

### 2.2.4 From Pointwise to the Whole

In the previous arguments we used pointwise notation, where one writes relations between values of the tensor fields. Such a notation is quite cumbersome, since one needs to write arguments in every slot. Further, we use pointwise notation only if there is a danger of confusion. If $f$ is some function and $\boldsymbol{u}$ is a vector field, then $f\boldsymbol{u}$ means the vector field $x \mapsto f|_x \boldsymbol{u}|_x$. The relations between bases $(\boldsymbol{e}_s)_{s=1}^m$ and $(\boldsymbol{\varepsilon}_s)_{s=1}^m$ are written as

$$\boldsymbol{\varepsilon}_s = \Omega^i{}_s \boldsymbol{e}_i , \quad \boldsymbol{e}_s = \mho^i{}_s \boldsymbol{\varepsilon}_i .$$

The decomposition $\boldsymbol{u}_x = u^i|_x \boldsymbol{\varepsilon}_i|_x$ of a vector field $\boldsymbol{u}$ is represented by $\boldsymbol{u} = u^i \boldsymbol{\varepsilon}_i$.

### 2.2.5 Metric Tensors

**The metric tensor on $\mathscr{V}$.** The scalar product $(\cdot)$ on $\mathscr{V}$ generates a constant tensor field $\boldsymbol{g}\colon \mathscr{E} \to \mathscr{V}^* \otimes \mathscr{V}^*$, referred to as *metric tensor*:

$$\forall \boldsymbol{u}, \boldsymbol{v} \in \mathscr{V}: \quad \boldsymbol{g}(\boldsymbol{u}, \boldsymbol{v}) := \boldsymbol{u} \cdot \boldsymbol{v} .$$

In a constant basis $(\boldsymbol{e}_i)_{i=1}^m$, the metric tensor can be decomposed as $\boldsymbol{g} = g_{ij} e^i \otimes e^j$, where $g_{ij} := \boldsymbol{e}_i \cdot \boldsymbol{e}_j$. If, in addition, $(\boldsymbol{e}_i)_{i=1}^m$ is orthonormal, then $g_{ij} = \delta_{ij}$. Hence, the components of $\boldsymbol{g}$, namely $g_{ij}$, are constant functions in rectilinear coordinates. Meanwhile, if one uses curvilinear coordinates, the components of $\boldsymbol{g}$ in local basis $(\boldsymbol{\varepsilon}_i)_{i=1}^m$ become non-constants: $g_{ij} = \boldsymbol{\varepsilon}_i \cdot \boldsymbol{\varepsilon}_j \neq \text{const}$.

**Metric coefficients.** Components $\delta_{ij}$ of metric tensors in Cartesian coordinates and components $g_{ij}$ of metric tensors in curvilinear coordinates are related as follows. Since,

$$\boldsymbol{\varepsilon}_s|_x = \left.\frac{\partial \tilde{x}^k}{\partial q^s}\right|_{\xi^{-1}(x)} \boldsymbol{i}_k ,$$

we obtain the explicit expression for calculating the metric coefficients $g_{kl} = \boldsymbol{\varepsilon}_k \cdot \boldsymbol{\varepsilon}_l$:

$$g_{kl}|_x = \boldsymbol{\varepsilon}_k|_x \cdot \boldsymbol{\varepsilon}_l|_x = \delta_{ij} \left.\frac{\partial \tilde{x}^i}{\partial q^k}\right|_{\xi^{-1}(x)} \left.\frac{\partial \tilde{x}^j}{\partial q^l}\right|_{\xi^{-1}(x)} .$$

**Geometrical measurements using metric tensors.** Let $\boldsymbol{u}\colon U \to \mathscr{V}$ be a vector field. At each point $x \in U$ one can calculate the norm $\|\boldsymbol{u}_x\|$ of its value at that point. Thus, the function $x \mapsto \|\boldsymbol{u}_x\|$ is defined. This function can be expressed as $\|\boldsymbol{u}\| = \sqrt{\boldsymbol{g}(\boldsymbol{u}, \boldsymbol{u})}$, or, in curvilinear coordinates, $\|\boldsymbol{u}\| = \sqrt{g_{ij} u^i u^j}$, where $\boldsymbol{u} = u^i \boldsymbol{\varepsilon}_i$.

For vector fields $\boldsymbol{u}\colon U \to \mathscr{V}$, $\boldsymbol{v}\colon U \to \mathscr{V}$, one can obtain their scalar product, *i.e.*, the mapping $x \mapsto \boldsymbol{u}_x \cdot \boldsymbol{v}_x$, using the metric tensor: $\boldsymbol{u} \cdot \boldsymbol{v} = g_{ij} u^i v^j$, where $\boldsymbol{u} = u^i \boldsymbol{\varepsilon}_i$, $\boldsymbol{v} = v^i \boldsymbol{\varepsilon}_i$. Finally, angles between $\boldsymbol{u}, \boldsymbol{v}$ are measured pointwise. In terms of the cosine map,

$$\cos(\boldsymbol{u}, \boldsymbol{v}) = \frac{g_{ij} u^i v^j}{\sqrt{g_{ij} u^i u^j} \sqrt{g_{ij} v^i v^j}} .$$

**Metric tensors on $\mathscr{V}^*$.** The metric tensor $\boldsymbol{g}$ and the musical isomorphisms induce the *associated metric tensor* $\overset{*}{\boldsymbol{g}}$ on $\mathscr{V}^*$ as follows:

$$\overset{*}{\boldsymbol{g}} := \boldsymbol{g}^\sharp = g^{ij} \boldsymbol{\varepsilon}_i \otimes \boldsymbol{\varepsilon}_j .$$

The matrices of components of these tensors are mutually inverse: $[g^{ij}] = [g_{ij}]^{-1}$.

### 2.2.6 Parallel Transport

Consider the cosine map cos (2.7). Let $\boldsymbol{u}$ be a vector field on $U$, $\boldsymbol{u}_x = u^i|_x \boldsymbol{\varepsilon}_i|_x$, where $x \in U$. Then, for $x, y \in U$,

$$\cos(\boldsymbol{u}_x, \boldsymbol{u}_y) = \frac{u^i|_x u^j|_y \Xi_{i,j}(x, y)}{\sqrt{g_{kl}|_x u^k|_x u^l|_x}\sqrt{g_{kl}|_y u^k|_y u^l|_y}} ,$$

where $\Xi_{i,j}(x, y) = \boldsymbol{\varepsilon}_i|_x \cdot \boldsymbol{\varepsilon}_j|_y = \boldsymbol{e}_k \cdot \boldsymbol{e}_l \, \Omega^k{}_i|_x \Omega^l{}_j|_y$. Thus, the vector field $\boldsymbol{u}$ is weakly parallel transported along a curve $\chi\colon [a, b] \to U$ if

$$\forall x \in [a, b]\colon \left| \frac{u^i|_{\chi(a)} u^j|_{\chi(x)} \Xi_{i,j}(\chi(a), \chi(x))}{\sqrt{g_{kl}|_{\chi(a)} u^k|_{\chi(a)} u^l|_{\chi(a)}}\sqrt{g_{kl}|_{\chi(x)} u^k|_{\chi(x)} u^l|_{\chi(x)}}} \right| = 1 .$$

The equivalent representation, $\boldsymbol{u}(x) = f(x)\boldsymbol{u}(\chi(a))$, is represented in coordinates as

$$u^i|_x = f(x) u^j|_a \Omega^s{}_j|_y \mho^i{}_s|_x ,$$

where the relation (2.19) was used.

### 2.2.7 Total Derivatives

**The case of the vector field.** Suppose that $\boldsymbol{u}\colon U \to \mathscr{V}$ is a differentiable vector field. One has $D\boldsymbol{u} = \partial_j u^i \boldsymbol{e}_i \otimes \boldsymbol{e}^j$. Consider the similar field in the local basis $(\boldsymbol{\varepsilon}_s)_{s=1}^m$: $\boldsymbol{u} = u^i \boldsymbol{\varepsilon}_i$. Applying the product rule gives[15]

$$D\boldsymbol{u} = D\left(u^i \boldsymbol{\varepsilon}_i\right) = \boldsymbol{\varepsilon}_i \otimes D(u^i) + u^i D(\boldsymbol{\varepsilon}_i) .$$

**Remark 2.4.** Let us obtain this formula using the definition of the total derivative. Suppose that $a \in U$. Choose some $\boldsymbol{h} \in \mathscr{V}$ such that $a + \boldsymbol{h} \in U$. Then,

$$\begin{aligned} \boldsymbol{u}(a + \boldsymbol{h}) - \boldsymbol{u}(a) &= u^i(a + \boldsymbol{h})\boldsymbol{\varepsilon}_i(a + \boldsymbol{h}) - u^i(a)\boldsymbol{\varepsilon}_i(a) \\ &= \{u^i(a) + Du^i(a)[\boldsymbol{h}] + o(\|\boldsymbol{h}\|)\}\{\boldsymbol{\varepsilon}_i(a) + D\boldsymbol{\varepsilon}_i(a)[\boldsymbol{h}] + \boldsymbol{o}(\|\boldsymbol{h}\|)\} - u^i(a)\boldsymbol{\varepsilon}_i(a) \\ &= u^i(a) D\boldsymbol{\varepsilon}_i(a)[\boldsymbol{h}] + \boldsymbol{\varepsilon}_i(a) Du^i(a)[\boldsymbol{h}] + \boldsymbol{o}(\|\boldsymbol{h}\|) . \end{aligned}$$

Since total derivative is unique, one obtains

$$D\boldsymbol{u}(a)[\boldsymbol{h}] = u^i(a) D\boldsymbol{\varepsilon}_i(a)[\boldsymbol{h}] + \boldsymbol{\varepsilon}_i(a) Du^i(a)[\boldsymbol{h}] ,$$

and thus,

$$D\boldsymbol{u}(a) = u^i(a) D\boldsymbol{\varepsilon}_i(a) + \boldsymbol{\varepsilon}_i(a) \otimes Du^i(a) .$$

15 The mapping $(\lambda, \boldsymbol{x}) \mapsto \lambda \boldsymbol{x}$ is bilinear and one applies the product rule to it.

We will consider the total derivative $\boldsymbol{Du}$, in which $u^i$ and $\boldsymbol{\varepsilon}_i$ are thought of as functions of $(q^i)$. Applying the chain rule gives

$$D(u^i) = \frac{\partial u^i}{\partial x^j} e^j = \frac{\partial u^i}{\partial q^k} \varepsilon^k ,$$
$$D(\boldsymbol{\varepsilon}_i) = \frac{\partial \boldsymbol{\varepsilon}_i}{\partial x^j} \otimes e^j = \frac{\partial \boldsymbol{\varepsilon}_i}{\partial q^k} \otimes \varepsilon^k .$$

The vector field $\frac{\partial \boldsymbol{\varepsilon}_i}{\partial q^k}$ can be represented as

$$\frac{\partial \boldsymbol{\varepsilon}_i}{\partial q^k} = \Gamma^l{}_{ik} \boldsymbol{\varepsilon}_l . \tag{2.20}$$

The functions $\Gamma^l{}_{ik} : U \to \mathbb{R}$ are referred to as the *Christoffel symbols (of the second kind*[16]*)*. Finally, $\boldsymbol{Du}$ has the form

$$\boldsymbol{Du} = \left( \frac{\partial u^i}{\partial q^k} + u^j \Gamma^i{}_{jk} \right) \boldsymbol{\varepsilon}_i \otimes \varepsilon^k .$$

**Christoffel symbols.** Consider some properties of functions $\Gamma^l{}_{ik}$. From now on we suppose that $\xi$ is a $C^2$-diffeomorphism. Since

$$\frac{\partial \boldsymbol{\varepsilon}_i}{\partial q^k} = \frac{\partial^2 \tilde{\boldsymbol{p}}}{\partial q^k \partial q^i} = \frac{\partial^2 \tilde{\boldsymbol{p}}}{\partial q^i \partial q^k} = \frac{\partial \boldsymbol{\varepsilon}_k}{\partial q^i} ,$$

one obtains that Christoffel symbols are symmetric in the lower two indices:

$$\Gamma^l{}_{ik} = \Gamma^l{}_{ki} , \quad i, k, l = 1, \ldots, m .$$

Next, one can obtain the relation between Christoffel symbols and functions $g_{ij} = \boldsymbol{\varepsilon}_i \cdot \boldsymbol{\varepsilon}_j$.

Differentiating the relation $g_{ij} = \boldsymbol{\varepsilon}_i \cdot \boldsymbol{\varepsilon}_j$ with respect to $q^k$ one arrives at

$$\partial_k g_{ij} = \boldsymbol{\varepsilon}_j \cdot \partial_k \boldsymbol{\varepsilon}_i + \boldsymbol{\varepsilon}_i \cdot \partial_k \boldsymbol{\varepsilon}_j .$$

Using the formula (2.20) gives

$$\partial_k g_{ij} = g_{lj} \Gamma^l{}_{ik} + g_{il} \Gamma^l{}_{jk} .$$

16 There are two types of Christoffel symbols. The first one is represented by functions $\Gamma_{lik} = g_{ls} \Gamma^s{}_{ik}$, which are called Christoffel symbols of the first kind. The second type is represented by functions $\Gamma^l{}_{ik}$, which are called Christoffel symbols of the second kind. We deal only with the latter kind of Christoffel symbols. In this regard, we further omit the words “second kind”.

By a cyclic permutation we obtain the expressions

$$\partial_k g_{ij} = g_{lj}\Gamma^l{}_{ik} + g_{il}\Gamma^l{}_{jk} ,$$
$$\partial_i g_{jk} = g_{lk}\Gamma^l{}_{ji} + g_{jl}\Gamma^l{}_{ki} ,$$
$$\partial_j g_{ki} = g_{li}\Gamma^l{}_{kj} + g_{kl}\Gamma^l{}_{ij} .$$

Summing the latter two equations and subtracting the former gives

$$2\Gamma^s{}_{ij} g_{sk} = \partial_i g_{kj} + \partial_j g_{ik} - \partial_k g_{ij} ,$$

where the symmetry of $g_{ij}$ and $\Gamma^l{}_{ik}$ was taken into account. Finally, one arrives at

$$\Gamma^l{}_{ij} = \frac{g^{lk}}{2} \left(\partial_i g_{kj} + \partial_j g_{ik} - \partial_k g_{ij}\right) . \tag{2.21}$$

**The Voss–Weyl formula.** The Voss–Weyl formula

$$\partial_k \sqrt{g} = \sqrt{g}\,\Gamma^l{}_{lk} , \quad g = \det[g_{ij}] , \tag{2.22}$$

is used in the following for the coordinate representation of divergence. In order to derive it, put $i = l$ in (2.21):

$$\Gamma^l{}_{lk} = \frac{g^{lr}}{2} \partial_k g_{lr} = \frac{1}{2} \partial_k \ln g .$$

To prove that $g^{lr}\partial_k g_{lr}$ is equal to $\partial_k \ln g$, consider the function $f\colon \mathrm{GL}(m;\mathbb{R}) \to \mathbb{R}$, which acts as follows:

$$f(A) = \ln|\det A| .$$

Let $\delta A$ be an increment of $A$. Then, for sufficiently small $\delta A$,

$$f(A + \delta A) - f(A) = \ln(\det(I + A^{-1}\delta A)) .$$

Using the expansion $\det(I + \alpha) = 1 + \operatorname{tr}\alpha + O(\alpha^2)$, we obtain

$$f(A + \delta A) - f(A) \approx \ln(1 + \operatorname{tr}(A^{-1}\delta A)) .$$

This means that $\delta f = \operatorname{tr}(A^{-1}\delta A)$. Applying this result to $A = [g_{ij}]$, we arrive at

$$\delta \ln g = g^{ij}\delta g_{ij} ,$$

and this implies the equality $g^{lr}\partial_k g_{lr} = \partial_k \ln g$. Note that

$$\partial_k \ln g = \frac{\partial_k g}{g} = \frac{2}{\sqrt{g}} \partial_k \sqrt{g} .$$

This observation leads to the desired Voss–Weyl formula (2.22).

Table 2.2 contains non-zero Christoffel symbols and the matrices of metric for cylindrical and spherical coordinates.

**Tab. 2.2:** Christoffel symbols and metric.

| Coordinates | Metric | Christoffel symbols |
|---|---|---|
| $(r, \varphi, z)$ | $\operatorname{diag}\{1, r^2, 1\}$ | $\Gamma^1_{\ 22} = -r, \Gamma^2_{\ 21} = \Gamma^2_{\ 12} = \frac{1}{r}$ |
| $(r, \theta, \varphi)$ | $\operatorname{diag}\{1, r^2, r^2 \sin^2\theta\}$ | $\Gamma^1_{\ 22} = -r, \Gamma^1_{\ 33} = -r\sin^2\theta,$ $\Gamma^2_{\ 21} = \Gamma^2_{\ 12} = \Gamma^3_{\ 13} = \Gamma^3_{\ 31} = \frac{1}{r},$ $\Gamma^2_{\ 33} = -\cos\theta\sin\theta, \Gamma^3_{\ 23} = \Gamma^3_{\ 32} = \cot\theta$ |

**Dual basis.** Consider the dual basis $(\varepsilon^i)$. Since each element of the dual basis is a mapping with values in $\mathscr{V}^*$, one has that $\partial_k \varepsilon^i$ similarly takes values in $\mathscr{V}^*$. Let $C^i_{\ jk}\colon U \to \mathbb{R}$ be smooth mappings such that $\partial_k \varepsilon^i = C^i_{\ jk}\varepsilon^j$. In fact, $C^i_{\ jk} = -\Gamma^i_{\ jk}$. To prove this, it is sufficient to differentiate the equality $\langle \varepsilon^i, \boldsymbol{\varepsilon}_i\rangle = \delta^i_j$ with respect to $q^k$. Indeed,

$$0 = \left\langle \partial_k \varepsilon^i, \boldsymbol{\varepsilon}_j \right\rangle + \left\langle \varepsilon^i, \partial_k \boldsymbol{\varepsilon}_j \right\rangle = C^i_{\ jk} + \Gamma^i_{\ jk} \,.$$

Thus,

$$\frac{\partial \varepsilon^i}{\partial q^k} = -\Gamma^i_{\ jk}\varepsilon^j \,. \tag{2.23}$$

**(1, 1)-tensor field.** Suppose that $\boldsymbol{T}\colon U \to \operatorname{Lin}(\mathscr{V};\mathscr{V})$ is a differentiable tensor field. One has the decomposition $\boldsymbol{T} = T^i_{\ j}\boldsymbol{\varepsilon}_i \otimes \varepsilon^j$. Let $a \in U$ be a point and let $\boldsymbol{h} \in \mathscr{V}$ be a vector such that $a + \boldsymbol{h} \in U$. One has

$$D\boldsymbol{T}(a)[\boldsymbol{h}] = \boldsymbol{\varepsilon}_i(a) \otimes \varepsilon^j(a) DT^i_{\ j}(a)[\boldsymbol{h}] + T^i_{\ j}(a) D\boldsymbol{\varepsilon}_i(a)[\boldsymbol{h}] \otimes \varepsilon^j(a) + T^i_{\ j}(a)\boldsymbol{\varepsilon}_i(a) \otimes D\varepsilon^j(a)[\boldsymbol{h}] \,.$$

**Remark 2.5.** This formula can be obtained in a similar manner as for the vector field. That is, one needs to consider the increment $\boldsymbol{T}(a + \boldsymbol{h}) - \boldsymbol{T}(a)$:

$$\begin{aligned}\boldsymbol{T}(a + \boldsymbol{h}) - \boldsymbol{T}(a) &= T^i_{\ j}(a + \boldsymbol{h})\boldsymbol{\varepsilon}_i(a + \boldsymbol{h}) \otimes \varepsilon^j(a + \boldsymbol{h}) - T^i_{\ j}(a)\boldsymbol{\varepsilon}_i(a) \otimes \varepsilon^j(a) = \\ &= \boldsymbol{\varepsilon}_i(a) \otimes \varepsilon^j(a) DT^i_{\ j}(a)[\boldsymbol{h}] + T^i_{\ j}(a) D\boldsymbol{\varepsilon}_i(a)[\boldsymbol{h}] \otimes \varepsilon^j(a) + \\ &\quad + T^i_{\ j}(a)\boldsymbol{\varepsilon}_i(a) \otimes D\varepsilon^j(a)[\boldsymbol{h}] + \boldsymbol{o}(\|\boldsymbol{h}\|) \,.\end{aligned}$$

The latter equality and the uniqueness of total derivative give the desired formula.

Let $h^k = \langle \varepsilon^k, \boldsymbol{h}\rangle$, $k = 1, \ldots, m$. Since

$$DT^i_{\ j} = \frac{\partial T^i_{\ j}}{\partial q^k}\varepsilon^k \,, \quad D(\boldsymbol{\varepsilon}_i) = \frac{\partial \boldsymbol{\varepsilon}_i}{\partial q^k} \otimes \varepsilon^k \,, \quad D\varepsilon^j = \frac{\partial \boldsymbol{\varepsilon}^j}{\partial q^k} \otimes \varepsilon^k \,,$$

one obtains (here we omit the argument $a$)

$$D\boldsymbol{T}[\boldsymbol{h}] = h^k \left\{ \frac{\partial T^i_{\ j}}{\partial q^k}\boldsymbol{\varepsilon}_i \otimes \varepsilon^j + T^i_{\ j}\frac{\partial \boldsymbol{\varepsilon}_i}{\partial q^k} \otimes \varepsilon^j + T^i_{\ j}\varepsilon_i \otimes \frac{\partial \boldsymbol{\varepsilon}^j}{\partial q^k} \right\} \,.$$

Taking into account the relations (2.20) and (2.23), one obtains

$$DT[\boldsymbol{h}] = h^k \left\{ \frac{\partial T^i{}_j}{\partial q^k} + T^l{}_j \Gamma^i{}_{lk} - T^i{}_l \Gamma^l{}_{jk} \right\} \boldsymbol{\varepsilon}_i \otimes \varepsilon^j \,.$$

Thus, finally,

$$DT = \left\{ \frac{\partial T^i{}_j}{\partial q^k} + T^l{}_j \Gamma^i{}_{lk} - T^i{}_l \Gamma^l{}_{jk} \right\} \boldsymbol{\varepsilon}_i \otimes \varepsilon^j \otimes \varepsilon^k \,.$$

### 2.2.8 Integration

Suppose that $\mathscr{D}$ is generated by the Cartesian coordinate system $(o, (\boldsymbol{i}_k)_{k=1}^m)$ and $\mathscr{I}_m$ is an $m$-brick (see Section 2.1.6). Suppose that $\mathscr{I}_m \subset U$ and $f\colon U \to \mathbb{R}$ is a function, for which the integral $\int_{\mathscr{I}_m} f$ exists. Using the change of variables theorem one arrives at

$$\int_{\mathscr{I}_m} f = \int_{\xi^{-1}(I_m)} Jf \circ \xi \,,$$

where $J = \det(\frac{\partial x^i}{\partial q^j})$.

### 2.2.9 The Physical Basis

Suppose that the local basis $(\boldsymbol{\varepsilon}_s)_{s=1}^m$ is orthogonal:

$$g_{ij} = \boldsymbol{\varepsilon}_i \cdot \boldsymbol{\varepsilon}_j = 0 \,,$$

if $i \neq j$. Then, one can introduce the collection of fields $(\boldsymbol{\varepsilon}_{\langle s \rangle})_{s=1}^m$,

$$\boldsymbol{\varepsilon}_{\langle s \rangle} := \frac{1}{\|\boldsymbol{\varepsilon}_s\|} \boldsymbol{\varepsilon}_s \,, \quad \text{where} \quad \|\boldsymbol{\varepsilon}_s\| = \sqrt{g_{ss}} \,.$$

The values of the obtained fields $(\boldsymbol{\varepsilon}_{\langle s \rangle})_{s=1}^m$ form a basis of $\mathscr{V}$ at every point of $U$. In general, these fields are not tangent to coordinate curves. The basis $(\boldsymbol{\varepsilon}_{\langle s \rangle})_{s=1}^m$ is called the *physical basis*.

## 2.3 Nabla Operator Formalism in 3D-Space

The symbolic vector $\nabla$ is heavily used in continuum mechanics [45]. It is alternative way to obtain differential expressions for scalar, vector, and tensor fields. Let $\mathscr{E}$ be a three-dimensional Euclidean affine space.

### 2.3.1 Definitions of Nabla

**Nabla in Cartesian coordinates.** Introduce a symbolic vector $\nabla$, which is represented in Cartesian coordinates as follows [46]:

$$\nabla = \boldsymbol{i}^k \frac{\partial}{\partial x^k} ,$$

where $(\boldsymbol{i}^k)_{k=1}^3$ is the dual vector basis that coincides with the orthonormal basis $(\boldsymbol{i}_k)_{k=1}^3$, *i.e.*, $\boldsymbol{i}^k = \boldsymbol{i}_k$, for all $k = 1, 2, 3$. From this point of view, applied to the ordered pair $(\mathscr{V}, \mathscr{V}^*)$, vectors $\boldsymbol{i}^k$ may be regarded as covectors $i^k$, *i.e.*, elements of the dual space $\mathscr{V}^*$.

**Nabla in curvilinear coordinates.** Suppose that some curvilinear coordinates $(q^i)_{i=1}^3$ are chosen on some open set $U \subset \mathscr{E}$. If $(\boldsymbol{\varepsilon}_s)_{s=1}^3$ is the local basis, then let $(\boldsymbol{\varepsilon}^s)_{s=1}^3$ denote the dual vector basis,

$$\boldsymbol{\varepsilon}^i \cdot \boldsymbol{\varepsilon}_j = \delta^i_j , \quad i, j = 1, 2, 3 .$$

The chain rule implies that nabla operator would take the form

$$\nabla = \boldsymbol{\varepsilon}^k \frac{\partial}{\partial q^k} .$$

**Nabla in the physical basis.** If the basis $(\boldsymbol{\varepsilon}_s)_{s=1}^3$ is orthogonal, then in the physical basis $(\boldsymbol{\varepsilon}_{\langle s \rangle})_{s=1}^3$ one has

$$\nabla = \sum_{s=1}^{3} \frac{\boldsymbol{\varepsilon}_{\langle s \rangle}}{\sqrt{g_{ss}}} \frac{\partial}{\partial q^s} .$$

Table 2.3 contains expressions for $\nabla$ in the case of cylindrical and spherical coordinates.

**Tab. 2.3:** Nabla operator in physical basis.

| **Coordinates** | $\nabla$ |
|---|---|
| $(r, \varphi, z)$ | $\nabla = \boldsymbol{\varepsilon}_{\langle r \rangle} \frac{\partial}{\partial r} + \frac{\boldsymbol{\varepsilon}_{\langle \varphi \rangle}}{r} \frac{\partial}{\partial \varphi} + \boldsymbol{\varepsilon}_{\langle z \rangle} \frac{\partial}{\partial z}$ |
| $(r, \theta, \varphi)$ | $\nabla = \boldsymbol{\varepsilon}_{\langle r \rangle} \frac{\partial}{\partial r} + \frac{\boldsymbol{\varepsilon}_{\langle \theta \rangle}}{r} \frac{\partial}{\partial \theta} + \frac{\boldsymbol{\varepsilon}_{\langle \varphi \rangle}}{r \sin \theta} \frac{\partial}{\partial \varphi}$ |

### 2.3.2 Operations in Euclidean Space

The application of $\nabla$ presupposes the identification of vectors and covectors (due to musical isomorphisms). Let $(\boldsymbol{e}_i)_{i=1}^3$ be a basis for $\mathscr{V}$ and $(\boldsymbol{e}^i)_{i=1}^3$ be the corresponding dual vector basis. For a vector $\boldsymbol{u} \in \mathscr{V}$, one has the decompositions

$$\boldsymbol{u} = u^i \boldsymbol{e}_i = u_i \boldsymbol{e}^i .$$

Introduce the following particular definition of tensor product (this definition holds only in Euclidean spaces). If $\boldsymbol{u}, \boldsymbol{v} \in \mathscr{V}$, then [47]

$$\boldsymbol{u} \otimes \boldsymbol{v} \colon \mathscr{V} \to \mathscr{V} , \quad \boldsymbol{u} \otimes \boldsymbol{v}(\boldsymbol{w}) := (\boldsymbol{v} \cdot \boldsymbol{w})\boldsymbol{u} .$$

It is linear map, an element of $\mathrm{Lin}(\mathscr{V};\mathscr{V})$.

Left and right "dot" actions of $\boldsymbol{u} \otimes \boldsymbol{v}$ on a vector $\boldsymbol{w}$ are defined via

$$(\boldsymbol{u} \otimes \boldsymbol{v}) \cdot \boldsymbol{w} := (\boldsymbol{v} \cdot \boldsymbol{w})\boldsymbol{u} , \quad \boldsymbol{w} \cdot (\boldsymbol{u} \otimes \boldsymbol{v}) := (\boldsymbol{u} \cdot \boldsymbol{w})\boldsymbol{v} .$$

The "cross"-action of the dyad $\boldsymbol{u} \otimes \boldsymbol{v}$ on a vector $\boldsymbol{w}$ is defined as follows:

$$(\boldsymbol{u} \otimes \boldsymbol{v}) \times \boldsymbol{w} := \boldsymbol{u} \otimes (\boldsymbol{v} \times \boldsymbol{w}) , \quad \boldsymbol{w} \times (\boldsymbol{u} \otimes \boldsymbol{v}) := (\boldsymbol{w} \times \boldsymbol{u}) \otimes \boldsymbol{v} .$$

In Euclidean space, second-order tensors of all four types are identified. Let $(\boldsymbol{e}_i)_{i=1}^3$ be a basis for $\mathscr{V}$ and $(\boldsymbol{e}^i)_{i=1}^3$ be the corresponding dual vector basis. Then each of

$$(\boldsymbol{e}_i \otimes \boldsymbol{e}_j) , \quad (\boldsymbol{e}_i \otimes \boldsymbol{e}^j) , \quad (\boldsymbol{e}^i \otimes \boldsymbol{e}_j) , \quad (\boldsymbol{e}^i \otimes \boldsymbol{e}^j) ,$$

form a basis for $\mathrm{Lin}(\mathscr{V};\mathscr{V})$. Any element $\boldsymbol{T} \in \mathrm{Lin}(\mathscr{V};\mathscr{V})$ has the decompositions

$$\boldsymbol{T} = T^{ij} \boldsymbol{e}_i \otimes \boldsymbol{e}_j = T^i{}_j \boldsymbol{e}_i \otimes \boldsymbol{e}^j = T_i{}^j \boldsymbol{e}^i \otimes \boldsymbol{e}_j = T_{ij} \boldsymbol{e}^i \otimes \boldsymbol{e}^j .$$

If $(\boldsymbol{e}_i)_{i=1}^3$ is a basis for $\mathscr{V}$ and $(\boldsymbol{e}^i)_{i=1}^3$ is the corresponding dual vector basis, then introduce quantities $\epsilon_{ijk}$, $\epsilon^{ijk}$, such that

$$\boldsymbol{e}_i \times \boldsymbol{e}_j = \epsilon_{ijk} \boldsymbol{e}^k , \quad \boldsymbol{e}^i \times \boldsymbol{e}^j = \epsilon^{ijk} \boldsymbol{e}_k .$$

In classical books it is shown that [45]

$$\epsilon_{ijk} = \sqrt{g}\, e_{ijk} , \quad \epsilon^{ijk} = \frac{1}{\sqrt{g}} e^{ijk} . \tag{2.24}$$

Recall that $g = \det[g_{ij}]$, $g_{ij} = \boldsymbol{e}_i \cdot \boldsymbol{e}_j$.

### 2.3.3 Nabla and Mappings

**Nabla and scalar functions.** Consider a differentiable function $f \colon U \to \mathbb{R}$. One has

$$\nabla f = \boldsymbol{i}^k \frac{\partial f}{\partial x^k} ,$$

and in this regard, $\nabla f = \operatorname{grad} f$ (2.9).

**Nabla and vector fields.** One can form the following combinations with $\nabla$ and operations $\otimes$, $(\cdot)$, $\times$:

$$\nabla\otimes = \boldsymbol{i}^k \otimes \frac{\partial}{\partial x^k}\,, \quad \nabla\cdot = \boldsymbol{i}^k \cdot \frac{\partial}{\partial x^k}\,, \quad \nabla\times = \boldsymbol{i}^k \times \frac{\partial}{\partial x^k}\,. \tag{2.25}$$

Consider a differentiable vector field $\boldsymbol{u}\colon U \to \mathscr{V}$. Applying $\nabla\otimes$ on $\boldsymbol{u} = u^s \boldsymbol{i}_s$ gives

$$\nabla \otimes \boldsymbol{u} = \frac{\partial u^s}{\partial x^k} \boldsymbol{i}^k \otimes \boldsymbol{i}_s\,.$$

Let $a \in U$ be a point. Then, for a vector $\boldsymbol{h} \in \mathscr{V}$, such that $a + \boldsymbol{h} \in U$, one has two possible notations

$$\begin{aligned} \boldsymbol{u}(a+\boldsymbol{h}) &= \boldsymbol{u}(a) + D\boldsymbol{u}(a)[\boldsymbol{h}] + \boldsymbol{o}(\|\boldsymbol{h}\|) && \text{as} \quad \boldsymbol{h} \to \boldsymbol{0}\,, \\ \boldsymbol{u}(a+\boldsymbol{h}) &= \boldsymbol{u}(a) + \boldsymbol{h} \cdot \nabla \otimes \boldsymbol{u}|_{x=a} + \boldsymbol{o}(\|\boldsymbol{h}\|) && \text{as} \quad \boldsymbol{h} \to \boldsymbol{0}\,. \end{aligned}$$

Thus,

$$D\boldsymbol{u}(a) = \{\nabla \otimes \boldsymbol{u}|_{x=a}\}^{\mathrm{T}}\,.$$

Applying $\nabla\cdot$ on $\boldsymbol{u} = u^s \boldsymbol{i}_s$ gives the divergence of $\boldsymbol{u}$ (2.12):

$$\operatorname{div} \boldsymbol{u} = \nabla \cdot \boldsymbol{u}\,.$$

Finally, consider $\nabla\times$. This gives the curl of $\boldsymbol{u}$ (2.16):

$$\operatorname{curl} \boldsymbol{u} = \nabla \times \boldsymbol{u}\,.$$

**Nabla and $(1,1)$-tensor fields.** Again, one can consider combinations (2.25). Consider a differentiable tensor field $\boldsymbol{T}\colon U \to \operatorname{Lin}(\mathscr{V};\mathscr{V})$. At a point $a \in U$, one has

$$\begin{aligned} \boldsymbol{T}(a+\boldsymbol{h}) &= \boldsymbol{T}(a) + D\boldsymbol{T}(a)[\boldsymbol{h}] + \boldsymbol{o}(\|\boldsymbol{h}\|) && \text{as} \quad \boldsymbol{h} \to \boldsymbol{0}\,, \\ \boldsymbol{T}(a+\boldsymbol{h}) &= \boldsymbol{T}(a) + \boldsymbol{h} \cdot \nabla \boldsymbol{T}|_{x=a} + \boldsymbol{o}(\|\boldsymbol{h}\|) && \text{as} \quad \boldsymbol{h} \to \boldsymbol{0} \end{aligned}$$

Like for vector fields one has the equality

$$\operatorname{div} \boldsymbol{T} = \nabla \cdot \boldsymbol{T}\,,$$

for divergence (2.13) and the equality

$$\operatorname{curl} \boldsymbol{T} = \nabla \times \boldsymbol{T}\,,$$

for curl (2.17).

**Remark 2.6.** The curl operation is used especially in the linear theory of elasticity [47]. In this book one will meet this operation in arguments on incompatibility, where, among general considerations, the linear approach is analyzed. In the rest of the book, we use the generalization of the curl, which is based on the exterior differential.

### 2.3.4 Coordinate Representations for Divergence

In Section 2.1.8, we obtained representations for divergences of the smooth vector $\boldsymbol{u}$ and the tensor $\boldsymbol{T}$ fields in Cartesian coordinates $(x^i)$:

$$\operatorname{div}\boldsymbol{u} = \partial_{x^i} u^i\,, \quad \operatorname{div}\boldsymbol{T} = \partial_{x^l} T^k{}_l \boldsymbol{i}_k\,.$$

Operator $\nabla$ gives a convenient way to calculate divergences in curvilinear coordinates $(q^k)$.

**The vector field.** Let $\boldsymbol{u} = u^i \boldsymbol{\varepsilon}_i$ be a smooth vector field represented in curvilinear coordinates. Then, $\operatorname{div}\boldsymbol{u} = \nabla \cdot \boldsymbol{u}$, and

$$\operatorname{div}\boldsymbol{u} = \boldsymbol{\varepsilon}^k \cdot \frac{\partial}{\partial q^k}\left(u^i \boldsymbol{\varepsilon}_i\right) = \boldsymbol{\varepsilon}^k \cdot \left(\partial_k u^i + u^j \Gamma^i{}_{jk}\right)\boldsymbol{\varepsilon}_i\,,$$

from where it follows that

$$\operatorname{div}\boldsymbol{u} = \partial_i u^i + u^j \Gamma^i{}_{ji}\,,$$

since $\boldsymbol{\varepsilon}^k \cdot \boldsymbol{\varepsilon}_i = \delta^k_i$. Using the Voss–Weyl formula (2.22), $\partial_j \sqrt{g} = \sqrt{g}\,\Gamma^i{}_{ij}$, one obtains $\operatorname{div}\boldsymbol{u} = \partial_i u^i + (1/\sqrt{g}) u^i \partial_i \sqrt{g}$, and, finally,

$$\operatorname{div}\boldsymbol{u} = \frac{1}{\sqrt{g}} \frac{\partial}{\partial q^i}(u^i \sqrt{g})\,. \tag{2.26}$$

Suppose that $(\boldsymbol{\varepsilon}_i)$ is orthogonal. Consider $\operatorname{div}\boldsymbol{u}$ in the physical basis $(\boldsymbol{\varepsilon}_s)$. Let $\boldsymbol{u} = u^{\langle i \rangle} \boldsymbol{\varepsilon}_{\langle i \rangle}$. Here, $u^{\langle i \rangle} = u^i \sqrt{g_{ii}}$. One can obtain the expression for $\operatorname{div}\boldsymbol{u}$ by direct differentiation or by changing $u^i$ in (2.26). In all cases, one obtains

$$\operatorname{div}\boldsymbol{u} = \frac{1}{\sqrt{g}} \sum_{i=1}^{3} \frac{\partial}{\partial q^i}\left(u^{\langle i \rangle} \sqrt{\frac{g}{g_{ii}}}\right)\,. \tag{2.27}$$

**The tensor field.** Let $\boldsymbol{T} = T^{ij} \boldsymbol{\varepsilon}_i \otimes \boldsymbol{\varepsilon}_j$ be a smooth tensor field. One has

$$\operatorname{div}\boldsymbol{T} = \partial_k T^{kj} \boldsymbol{\varepsilon}_j + T^{kj}\left(\boldsymbol{\varepsilon}^i \cdot \partial_i \boldsymbol{\varepsilon}_k\right)\boldsymbol{\varepsilon}_j + T^{kj} \partial_k \boldsymbol{\varepsilon}_j\,. \tag{2.28}$$

Since $\boldsymbol{\varepsilon}^i \cdot \partial_i \boldsymbol{\varepsilon}_k = \Gamma^i{}_{ik}$, one can use the Voss–Weyl formula (2.22). Thus, from (2.28), one obtains

$$\operatorname{div}\boldsymbol{T} = \partial_k T^{kj} \boldsymbol{\varepsilon}_j + T^{kj} \frac{\partial_k \sqrt{g}}{\sqrt{g}} \boldsymbol{\varepsilon}_j + T^{kj} \partial_k \boldsymbol{\varepsilon}_i\,.$$

Finally, one obtains

$$\operatorname{div}\boldsymbol{T} = \frac{1}{\sqrt{g}} \frac{\partial}{\partial q^k}\left\{\sqrt{g}\, T^{kj} \boldsymbol{\varepsilon}_j\right\}\,.$$

There is an alternative formula. Starting from (2.28), one arrives at

$$\operatorname{div}\boldsymbol{T} = \left(\partial_k T^{kj} + T^{lj} \Gamma^k{}_{lk} + T^{kl} \Gamma^j{}_{lk}\right)\boldsymbol{\varepsilon}_j\,.$$

Using the Voss–Weyl formula (2.22) and the relation (2.21), one has

$$\operatorname{div} \boldsymbol{T} = \left( \partial_k T^{kj} + T^{lj} \frac{\partial_l \sqrt{g}}{\sqrt{g}} + \frac{T^{kl} g^{js}}{2} \left(\partial_l g_{sk} + \partial_k g_{ls} - \partial_s g_{lk}\right) \right) \boldsymbol{\varepsilon}_j \,.$$

Assume that $(\boldsymbol{\varepsilon}_i)$ is an orthogonal basis. Consider div $\boldsymbol{T}$ in the physical basis $(\boldsymbol{\varepsilon}_s)$. Since $T^{\langle kj \rangle} = T^{kj} \sqrt{g_{kk}} \sqrt{g_{jj}}$, one has

$$\operatorname{div} \boldsymbol{T} = \frac{1}{\sqrt{g}} \sum_{k=1}^{3} \frac{\partial}{\partial q^k} \left\{ T^{\langle kj \rangle} \sqrt{\frac{g}{g_{kk}}} \, \boldsymbol{\varepsilon}_{\langle j \rangle} \right\} \,.$$

### 2.3.5 Coordinate Representations for Curl

**The vector field.** Let $\boldsymbol{u} = u_i \boldsymbol{e}^i$ be a smooth vector field. Then,

$$\nabla \times \boldsymbol{u} = \boldsymbol{e}^k \times \boldsymbol{e}^i \left( \partial_k u_i - u_j \Gamma^j_{\ ik} \right) \,.$$

Using (2.24), one obtains

$$\nabla \times \boldsymbol{u} = \epsilon^{kil} \boldsymbol{e}_l \left( \partial_k u_i - u_j \Gamma^j_{\ ik} \right) \,.$$

Consider the term $\epsilon^{kil} \Gamma^j_{\ ik}$, where $l$ and $j$ are fixed. Since $\epsilon^{kil} = -\epsilon^{ikl}$ and $\Gamma^j_{\ ik} = \Gamma^j_{\ ki}$, one arrives at

$$\epsilon^{kil} \Gamma^j_{\ ki} = -\epsilon^{ikl} \Gamma^j_{\ ik} = -\epsilon^{kil} \Gamma^j_{\ ki} \,,$$

and this implies $\epsilon^{kil} \Gamma^j_{\ ki} = 0$. Finally,

$$\operatorname{curl} \boldsymbol{u} = \epsilon^{kil} \partial_k u_i \boldsymbol{e}_l \,.$$

**The tensor field.** Consider the tensor field $\boldsymbol{T} = T_{ij} \boldsymbol{e}^i \otimes \boldsymbol{e}^j$. One has

$$\begin{aligned} \nabla \times \boldsymbol{T} &= \boldsymbol{e}^k \times \left( \partial_k T_{ij} \boldsymbol{e}^i \otimes \boldsymbol{e}^j - T_{ij} \Gamma^i_{\ lk} \boldsymbol{e}^l \otimes \boldsymbol{e}^j + T_{ij} \boldsymbol{e}^i \otimes \partial_k \boldsymbol{e}^j \right) \\ &= \epsilon^{kil} \boldsymbol{e}_l \partial_k (T_{ij} \boldsymbol{e}^j) - T_{ij} \epsilon^{kls} \Gamma^i_{\ lk} \boldsymbol{e}_s \otimes \boldsymbol{e}^j \,. \end{aligned}$$

Like in the case of vector field, one obtains $\epsilon^{kls} \Gamma^i_{\ lk} = 0$, and, thus one arrives at the final expression

$$\operatorname{curl} \boldsymbol{T} = \epsilon^{kil} \boldsymbol{e}_l \partial_k \left( T_{ij} \boldsymbol{e}^j \right) \,.$$

## 2.4 Riemannian Space

### 2.4.1 General Definition

From general point of view, the Riemannian space of a dimension $m$ is a 5-tuple

$$(\mathscr{P}, \mathfrak{S}, D, \mathscr{g}, \mu) \,,$$

where $\mathscr{P}$ is a set whose elements are referred to as *spatial points*, $\mathfrak{S}$ is a topology on $\mathscr{P}$, $D$ is a smooth structure on $\mathscr{P}$, $g$ is a Riemannian metric on $\mathscr{P}$, and $\mu$ is a volume form on $\mathscr{P}$. The topological space $(\mathscr{P}, \mathfrak{S})$ is required to be Hausdorff and second countable[17]. The triple $(\mathscr{P}, \mathfrak{S}, D)$ is a smooth $m$-dimensional manifold.

The smooth structure $D$ is introduced by a chosen equivalence class of smooth atlases and defines the maximal atlas $A_{\max}$, *i.e.*, the union of all atlases from this equivalence class[18]. Any chart $(U, \varphi) \in A_{\max}$, where $U \subset \mathscr{P}$ is an open set, establishes local coordinates for all points from $U$. We use the following terminology: any chart from the maximal atlas is called a *smooth chart*. If two charts $(U, \varphi), (V, \psi) \in A_{\max}$ overlap, *i.e.*, $U \cap V \neq \emptyset$, then the transition map $\psi \circ \varphi^{-1}$ is a diffeomorphism between open subsets of $\mathbb{R}^m$. This map represents the transformation law between local coordinates.

The Riemannian metric[19] $g$ represents a field of inner products on tangent spaces $T_p\mathscr{P}$. It is a (0, 2)-type tensor field on $\mathscr{P}$, or, formally, a smooth section of the vector bundle $T^*\mathscr{P} \otimes T^*\mathscr{P}$. For any point $p \in \mathscr{P}$, one has the bilinear mapping $g_p$, which is symmetric and positive definite.

The volume form $\mu$ is a differential $m$-form, the element of $\Omega^m(\mathscr{P})$ that satisfies the requirement: $\mu_p \neq 0$, for each $p \in \mathscr{P}$. Since $\mathscr{P}$ admits a volume form, it is orientable[20]. We choose the orientation that is generated by $\mu$.

The structure of Riemannian space is more "flexible" than the structure of Euclidean space; one can define topology, local coordinates, and the metric directly. In Euclidean space, these features are induced by the structure of the translation vector space.

**Remark 2.7.** Suppose that we are given an $s$-dimensional topological manifold $\mathfrak{M}$. Let # denote the number of $C^\infty$ manifolds that one can make out from $\mathfrak{M}$ up to a diffeomorphism. Such a number depends on the dimension, $s$, of the manifold. Table 2.4 illustrates such dependence [48].

**Tab. 2.4:** Dimension and number of smooth structures.

| $s$ | # |
|---|---|
| 1, 2, 3 | 1 |
| 4 | in general, uncountably many |
| 5, 6, 7 | finite |

17 The definitions of the Hausdorff and second countability properties are given in Chapter 13, Section 13.1.1.

18 The notion of smooth structure is introduced in Chapter 13, Section 13.1.2.

19 The definition is given in Chapter 13, Section 13.3.10.

20 See Chapter 13, Section 13.4.

### 2.4.2 Smooth Mappings on $\mathscr{P}$

We repeat the definitions given in Section 2.1.3, but now they are adapted to the Riemannian geometry.

**Curves.** A curve on $\mathscr{P}$ is a mapping $\chi\colon \mathbb{I} \to \mathscr{P}$, where $\mathbb{I} \subset \mathbb{R}$ is an interval. In a smooth chart $(U, \varphi)$ a curve $\chi$ has the coordinate representation $\tilde{\chi} = \varphi \circ \chi\colon \mathbb{I} \to \mathbb{R}^m$. The smoothness of the curve $\chi$ is understood in the sense of the smoothness of its coordinate representation $\tilde{\chi}$.

**Functions.** A function on $\mathscr{P}$ is a mapping $f\colon \mathscr{P} \to \mathbb{R}$. Its coordinate representation relative to a smooth chart $(U, \varphi)$ on $\mathscr{P}$ has the form: $\tilde{f} = f \circ \varphi^{-1}\colon \varphi(U) \to \mathbb{R}$. The smoothness of $f$ is defined by the smoothness of its coordinate representation $\tilde{f}$; we say that $f$ is of class $C^\infty$ at a point $p \in \mathscr{P}$, if for some smooth chart that contains $p$, the mapping $\tilde{f}$ is of $C^\infty$. We deal with functions that are smooth everywhere in $\mathscr{P}$. Such functions form the real algebra $C^\infty(\mathscr{P})$ under pointwise addition and multiplication[21]. If $O \subset \mathscr{P}$ is an open subset, then it is considered as an open submanifold of $\mathscr{P}$, and one can introduce the algebra $C^\infty(O)$ of smooth functions on it.

**Vector fields.** Unlike the Euclidean case, one has no translation vector space for the space $\mathscr{P}$. A tangent vector[22] at a point $p \in \mathscr{P}$ can be viewed as an equivalence class of curves that pass through $p$, or, equivalently, as a derivation on the space of germs at $p$, or, as a derivation of the algebra $C^\infty(\mathscr{P})$ at $p$ (see Chapter 13, Section 13.2.1). For a point $p \in \mathscr{P}$, the tangent space to $\mathscr{P}$ at $p$ is denoted by $T_p\mathscr{P}$. Suppose that $f \in C^\infty(\mathscr{P})$ and $u_p \in T_p\mathscr{P}$ is a vector. One can calculate the directional derivative at $p$:

$$\mathscr{L}_{u_p}(f) = \frac{d}{dt}(f \circ \chi)(t)\bigg|_{t=0},$$

where $\chi\colon \mathbb{I} \to \mathscr{P}$ is a smooth curve from $u_p$ (which is an equivalence class); $\chi(0) = p$. Here, $f \circ \chi\colon \mathbb{I} \to \mathbb{R}$. The first-order Taylor expansion holds for $h \in \mathbb{I}$:

$$(f \circ \chi)(h) = (f \circ \chi)(0) + h\mathscr{L}_{u_p}(f) + o(h) \quad \text{as} \quad h \to 0\,.$$

In some smooth chart $(U, \varphi)$ on $\mathscr{P}$, the vector $u_p$ has the representation $u_p = u^i \partial_i|_p$, where

$$(u^1, \ldots, u^m) = \frac{d}{dt}(\varphi \circ \chi)(t)\bigg|_{t=0}.$$

The coordinate representation of $f$ is the mapping

$$\tilde{f}\colon (x^1, \ldots, x^m) \mapsto \tilde{f}(x^1, \ldots, x^m)\,.$$

---

21 The space $C^\infty(\mathscr{P})$ is defined in Chapter 13, Section 13.1.3. The notion of algebra is defined in Chapter 12, Section 12.5.2.

22 Three equivalent definitions of tangent vector are given in Chapter 13, Section 13.2.1.

Thus, $\mathscr{L}_{u_p}(f)$ is represented by

$$\mathscr{L}_{u_p}(f) = u^i \left.\frac{\partial \tilde{f}}{\partial x^i}\right|_{\varphi(p)} .$$

The collection of tangent spaces $T_p\mathscr{P}$ at all points $p \in \mathscr{P}$ forms the tangent bundle $T\mathscr{P} = \bigsqcup_{p\in\mathscr{P}} T_p\mathscr{P}$. The latter is a $2m$-dimensional smooth manifold[23]. A vector field is a global section $u\colon \mathscr{P} \to T\mathscr{P}$ of the tangent bundle or a local section $u\colon O \to T\mathscr{P}$ of the tangent bundle[24]. Here, $O$ is an open subset of $\mathscr{P}$. The term *section* means that for any $p$ from $\mathscr{P}$ (or from $O$), one has $u_p \in T_p\mathscr{P}$.

Since a vector field is a mapping between manifolds, its smoothness is established as the smoothness of a mapping. If $(U, \varphi)$ is a smooth chart on $\mathscr{P}$, then one can introduce the coordinate frame[25] $(\partial_i)_{i=1}^m$, $U \ni p \mapsto \partial_i|_p \in T_p\mathscr{P}$. For a smooth vector field $u$, one has $u = u^i\partial_i$. The set of all smooth vector fields on $\mathscr{P}$ is denoted by $\mathrm{Vec}(\mathscr{P})$.

Suppose that $u \in \mathrm{Vec}(\mathscr{P})$. Since $u_p$ is a derivation on $C^\infty(\mathscr{P})$, for any $f \in C^\infty(\mathscr{P})$ one can define the mapping $uf \in C^\infty(\mathscr{P})$ as $uf\colon p \mapsto u_p(f)$. Another vector field $v \in \mathrm{Vec}(\mathscr{P})$ may act on such a function: $v(uf) \in C^\infty(\mathscr{P})$. Similarly, $u(vf) \in C^\infty(\mathscr{P})$. Both vector fields define the *Lie bracket* operation (13.20) $[u, v]f = u(vf) - v(uf)$, where $[u, v] \in \mathrm{Vec}(\mathscr{P})$. The properties of the Lie bracket are given in Chapter 13, Section 13.3.5.

**Covector fields.** A tangent covector at $p \in \mathscr{P}$ is an element of the cotangent space $T_p^*\mathscr{P} := (T_p\mathscr{P})^*$. The set $T^*\mathscr{P} = \bigsqcup_{p\in\mathscr{P}} T_p^*\mathscr{P}$ represents the cotangent bundle. As the tangent bundle, it is a $2m$-dimensional smooth manifold[26]. A covector field is a global section $v\colon \mathscr{P} \to T^*\mathscr{P}$ of the cotangent bundle, or a local section $v\colon O \to T^*\mathscr{P}$ of the cotangent bundle. Here, $O$ is an open subset of $\mathscr{P}$.

Any covector field is a mapping between manifolds, and its smoothness is established as the smoothness of a mapping. If $(U, \varphi)$ is a smooth chart on $\mathscr{P}$, then one can introduce the coordinate coframe $(dx^i)_{i=1}^m$, $U \ni p \mapsto dx^i|_p \in T_p^*\mathscr{P}$. The coordinate frame and coframe are related via

$$\left\langle dx^i, \partial_j \right\rangle = \delta^i_j , \quad i, j = 1, \ldots, m .$$

For a smooth covector field $v$, one has $v = v_i dx^i$. The set of all smooth covector fields on $\mathscr{P}$ is denoted by $\mathrm{CVec}(\mathscr{P})$.

23 The smooth atlas is introduced for $T\mathscr{P}$ in Chapter 13, Section 13.3.2.

24 The notion of section is discussed in Chapter 13, Section 13.3.5.

25 Coordinate frames for different tensor fields are introduced in Chapter 13, Section 13.3.8.

26 Smooth atlas on $T^*\mathscr{P}$ is introduced in the analogous manner as for $T\mathscr{P}$. See Chapter 13, Section 13.3.2.

**Tensor fields.** A mixed $(k,l)$-tensor is an element of the vector bundle

$$T^{(k,l)}(T\mathscr{P}) = \bigsqcup_{p\in\mathscr{P}} T^{(k,l)}(T_p\mathscr{P}) .$$

As in the case of the vector and the covector field, a tensor field is a global or local section of this bundle. The coordinate frame is represented by the collection

$$\left(\partial_{i_1}\otimes\cdots\otimes\partial_{i_k}\otimes dx^{j_1}\otimes\cdots\otimes dx^{j_l}\right)_{1\le i_1,\dots,i_k\le s\,,1\le j_1,\dots,j_l\le s} .$$

For a smooth mixed $(k,l)$-tensor $\mathscr{T}$, one has

$$\mathscr{T} = T^{i_1\dots i_k}{}_{j_1\dots j_l}\partial_{i_1}\otimes\cdots\otimes\partial_{i_k}\otimes dx^{j_1}\otimes\cdots\otimes dx^{j_l} .$$

### 2.4.3 Geometric Measures and Geodesics

**Geometric measures.** The machinery that allows one to determine lengths and angles is given by the Riemannian metric $g$. For a tangent vector $v \in T_p\mathscr{P}$, where $p \in \mathscr{P}$ is some point, the square of its length is defined by the formula

$$\|v\|_g^2 = g(v,v) .$$

If $v$ is a velocity vector (13.18) of a curve $\chi$, then $\|v\|_g$ has the sense of the *speed* of this curve.

If $\chi\colon\, ]a,b[\, \to \mathscr{P}$ is a smooth curve on $\mathscr{P}$, then its length, $l_g(\chi)$, is, by definition,

$$l_g(\chi) = \int_a^b \sqrt{g_{\chi(t)}(\chi'(t),\chi'(t))}\, dt .$$

Here, $\chi'(t)$ is defined by (13.18).

Let $\chi_1\colon\, ]a,b[\to \mathscr{P}$, $\chi_2\colon\, ]c,d[\to \mathscr{P}$ be two smooth curves that intersect at $p = \chi_1(t_1) = \chi_2(t_2)$. Denote the velocity vector of the curve $\chi_i$, $i = 1,2$ by $u_i\colon t \mapsto u_i(t) = \chi_i'(t)$. The angle $\angle(\chi_1,\chi_2)_p$ between the curves $\chi_1$ and $\chi_2$ at the point $p$ is the angle between their velocity vectors. That is,

$$\cos\angle(\chi_1,\chi_2)_p = \frac{g_p(u_1(t_1),u_2(t_2))}{\|u_1(t_1)\|_g\|u_2(t_2)\|_g} .$$

**Geodesics.** For a smooth curve $\chi\colon\, ]a,b[\, \to \mathscr{P}$ in a smooth chart, one has $g = g_{ij}dx^i\otimes dx^j$, $\chi(t) = (\chi^1(t),\dots,\chi^m(t))$ and $\chi'(t) = \dot\chi^i(t)\partial_i|_{\chi(t)}$ (13.18). Then, one arrives at the coordinate expression for $l_g(\chi)$:

$$l_g(\chi) = \int_a^b L(\chi^i,\dot\chi^i)\, dt ,$$

where $L(\chi^i, \dot{\chi}^i) = \sqrt{g_{ij}(\chi^1, \ldots, \chi^m)\dot{\chi}^i\dot{\chi}^j}$ is the Lagrangian density. A *geodesic* is an extremal of the functional $l_g(\chi)$: $\delta l_g(\chi) = 0$. The system of the Euler–Lagrange equations has the form

$$\frac{\partial L}{\partial \chi^k} - \frac{d}{dt}\left(\frac{\partial L}{\partial \dot{\chi}^k}\right) = 0\,, \quad k = 1, \ldots, m\,.$$

Derivation gives the equation of geodesics:

$$\ddot{\chi}^q + \frac{g^{qk}}{2}(\partial_i g_{kj} + \partial_j g_{ik} - \partial_k g_{ij})\dot{\chi}^i\dot{\chi}^j = 0\,, \quad q = 1, \ldots, m\,. \tag{2.29}$$

Here, $[g^{ij}] = [g_{ij}]^{-1}$.

**Remark 2.8.** In order to derive the equation of geodesics one calculates the derivatives of Lagrangian $L(\chi^i, \dot{\chi}^i) = \sqrt{g_{ij}(\chi^1, \ldots, \chi^m)\dot{\chi}^i\dot{\chi}^j}$ on two variables $\chi^i$ and $\dot{\chi}^i$, *i.e.*,

$$\frac{\partial L}{\partial \chi^k} = \frac{\partial_k g_{ij}\dot{\chi}^i\dot{\chi}^j}{2L}\,,$$

$$\frac{\partial L}{\partial \dot{\chi}^k} = \frac{g_{ik}\dot{\chi}^i}{L}\,, \quad \frac{d}{dt}\left(\frac{\partial L}{\partial \dot{\chi}^k}\right) = \left(\frac{1}{L}\right)^{\cdot} g_{ik}\dot{\chi}^i + \frac{\partial_j g_{ik}\dot{\chi}^i\dot{\chi}^j + g_{ik}\ddot{\chi}^i}{L}\,.$$

The derivation with respect to an arbitrary parameter $t$ will lead to cumbersome expressions. In order to simplify the calculations, suppose without loss of generality that the geodesic curve can be parameterized by a natural parameter, such that $\|\chi'(t)\|_g = 1$. Then, for all $\chi^i$, the Lagrangian $L = 1$ (but not its derivative) and the Euler–Lagrange equations can be written as

$$\frac{1}{2}\partial_k g_{ij}\dot{\chi}^i\dot{\chi}^j - \partial_j g_{ik}\dot{\chi}^i\dot{\chi}^j - g_{ik}\ddot{\chi}^i = 0\,, \quad k = 1, \ldots, m\,.$$

Since $\partial_j g_{ik}\dot{\chi}^i\dot{\chi}^j = (\partial_j g_{ik} + \partial_i g_{kj})\dot{\chi}^i\dot{\chi}^j/2$, the above equation transforms into the following one:

$$\frac{1}{2}(\partial_k g_{ij} - \partial_j g_{ik} - \partial_i g_{kj})\dot{\chi}^i\dot{\chi}^j - g_{ik}\ddot{\chi}^i = 0\,, \quad k = 1, \ldots, m\,.$$

Finally, multiplying both sides of this equation by $g^{qk}$ and summing the left-hand side over $k$, we obtain (2.29).

### 2.4.4 Parallel Transport

**Connection.** An affine connection $\nabla$ on $\mathscr{P}$ is a mapping (see Chapter 9)

$$\nabla\colon \operatorname{Vec}(\mathscr{P}) \times \operatorname{Vec}(\mathscr{P}) \to \operatorname{Vec}(\mathscr{P})\,, \quad \nabla\colon (u, v) \mapsto \nabla_u v\,.$$

Its coefficients relative to a smooth chart on $\mathscr{P}$ are denoted by ${\gamma^i}_{jk}$. One has the following coordinate representation for $\nabla_u v$:

$$\nabla_u v = u^c \left\{\partial_c v^b + v^a {\gamma^b}_{ca}\right\} \partial_b \,.$$

A smooth curve $\chi\colon \mathbb{I} \to \mathscr{P}$ is said to be *autoparallel* if $\nabla_{\chi'}\chi' = 0$. In a smooth chart, this equation reduces to the system

$$\ddot{\chi}^q + {\gamma^q}_{ij}\dot{\chi}^i\dot{\chi}^j = 0\,, \quad q = 1, \ldots, m\,.$$

Require from the connection $\nabla$ that *autoparallel curves are exactly geodesics*. This implies that in a coordinate frame $(\partial_i)_{i=1}^m$ connection coefficients must have the following representations:

$$ {\gamma^i}_{jk} = \frac{g^{im}}{2}(\partial_j g_{mk} + \partial_k g_{mj} - \partial_m g_{jk})\,,$$

where $g_{ij} = g(\partial_i, \partial_j)$, $[g^{ij}] = [g_{ij}]^{-1}$. Note that such an expression literally coincides with the expression (2.21) for Christoffel symbols. The connection $\nabla$ satisfies the conditions:

(i) $\forall i, j, k = 1, \ldots, m\colon {\gamma^i}_{jk} = {\gamma^i}_{kj}$ (torsion free);
(ii) $\nabla g = 0$, or, in components, $\partial_i g_{jk} - {\Gamma^m}_{ij} g_{mk} - {\Gamma^m}_{ik} g_{mj} = 0$ (metric compatible).

The connection $\nabla$ is called the *Levi–Civita connection.*

**Parallel transport.** A vector field $u \in \mathrm{Vec}(\mathscr{P})$ is said to be parallel transported along a smooth curve $\chi\colon \mathbb{I} \to \mathscr{P}$, if $\nabla_{\chi'} u = 0$. In a smooth chart on $\mathscr{P}$,

$$\dot{u}^b + u^c \dot{\chi}^a {\gamma^b}_{ca} = 0\,, \quad b = 1, \ldots, m\,.$$

Let $p = \chi(t_0)$. Then, one has the following initial condition: $u_p = u_0$. Thus, locally, there exists a unique solution for the Cauchy problem

$$\begin{aligned} \dot{u}^b + u^c \dot{\chi}^a {\gamma^b}_{ca} &= 0\,, \quad b = 1, \ldots, m\,, \\ u_p^b &= u_0^b\,, \quad b = 1, \ldots, m\,. \end{aligned} \tag{2.30}$$

One can define the *parallel transport map* $P_{t_0}^t\colon T_{\gamma(t_0)}\mathscr{P} \to T_{\gamma(t)}\mathscr{P}$ as follows: with each element $u_0 \in T_{\gamma(t_0)}\mathscr{P}$ it assigns the value $u(t)$ of the solution $u$ for the Cauchy problem (2.30); $P_{t_0}^t\colon u_0 \mapsto u(t)$.

Suppose that $p \in \mathscr{P}$ and $\chi$ is a smooth curve on $\mathscr{P}$, such that $\chi(0) = p$. Using the parallel transport map one can express the value of covariant derivative in terms of directional derivative:

$$\nabla_u v|_p := \lim_{h\to 0} \frac{P_h^0 v_{\chi(h)} - v_{\chi(0)}}{h} = \frac{d}{dt} P_t^0 v_{\chi(t)}\Big|_{t=0}\,.$$

Here, $P_t^0 v_{\chi(t)}$ belongs to $T_{\chi(0)}\mathscr{P}$, and it means that the difference $P_t^0 v_{\chi(t)} - v_{\chi(0)}$ is well defined and belongs to $T_{\chi(0)}\mathscr{P}$.

**Curvature.** The Riemann curvature tensor

$$\mathfrak{R}\colon \mathrm{Vec}(\mathscr{P}) \times \mathrm{Vec}(\mathscr{P}) \times \mathrm{Vec}(\mathscr{P}) \to \mathrm{Vec}(\mathscr{P}),$$

is defined as follows:

$$\mathfrak{R}(u, v, w) = \nabla_u \nabla_v w - \nabla_v \nabla_u w - \nabla_{[u,v]} w.$$

Due to the definition, the Riemann curvature tensor can be represented by the dyadic decomposition $\mathfrak{R} = R^t{}_{ijk} \partial_t \otimes dx^i \otimes dx^j \otimes dx^k$, where

$$R^t{}_{ijk} = \partial_i \Gamma^t{}_{jk} - \partial_j \Gamma^t{}_{ik} + \Gamma^l{}_{jk} \Gamma^t{}_{il} - \Gamma^l{}_{ik} \Gamma^t{}_{jl}.$$

One can introduce the purely covariant version of $\mathfrak{R}$. It is represented by the dyadic decomposition $\mathfrak{R}^\flat = R_{tijk} dx^t \otimes dx^i \otimes dx^j \otimes dx^k$, where $R_{tijk} = g_{tl} R^l{}_{ijk}$. The Riemann curvature tensor defines the Ricci curvature tensor $\mathrm{Ric} = R_{ij} dx^i \otimes dx^j$ [15], where $R_{ij} = R^l{}_{ilj}$. In turn, the Ricci tensor defines the Ricci scalar (scalar curvature) $S = g^{ij} R_{ij}$.

The geometrical meaning of the Riemann curvature tensor is the following [13, 49]. The difference between the final and the original directions of a vector that was parallel transported around an infinitesimal parallelogram is of second order and depends on the value of $R^t{}_{ijk}$ at the starting point and on the curve.

### 2.4.5 Integration

There exists the Riemannian volume form $dV_g$ on $\mathscr{P}$, which belongs to the orientation defined by the volume form $\mu$. In a smooth chart,

$$dV_g = \sqrt{g}\, dx^1 \wedge \cdots \wedge dx^m.$$

If $\mathscr{S} \subset \mathscr{P}$ is some smooth compact oriented submanifold of $\mathscr{P}$, then its volume, $\mathrm{Vol}_g(\mathscr{S})$, is calculated via

$$\mathrm{Vol}_g(\mathscr{S}) = \int_{\mathscr{S}} dV_{\iota_{\mathscr{S}}^* g},$$

where $\iota_{\mathscr{S}}^* g$ is the pullback of $g$ to the space $\mathscr{S}$, and $dV_{\iota_{\mathscr{S}}^* g}$ is the corresponding volume form. Here, $\iota_{\mathscr{S}}\colon \mathscr{S} \to \mathscr{P}$ is the inclusion map, that is, the restriction of the identity map $\mathrm{Id}_{\mathscr{P}}$ to $\mathscr{S}$.

## 2.5 Newtonian Space-Time

Formalized ways to describe the notions of space and time have attracted the interest of innovative individuals since ancient times. A retrospective vision of these concepts is well expressed in [50]. One tends to assume that, from the standpoint of conventional continuum mechanics, the space is usually regarded as "absolute", *i.e.*, independent of the observer, 3D Euclidean space, while the time is represented by real numbers that are in accordance with a time scale depending on the chosen "absolute" clock. Nevertheless, already from the beginning of the twentieth century, *i.e.*, almost simultaneously with the formulation of the foundations on special and general relativity, works on the four-dimensional formulation of continuum mechanics equations began to appear [35, 51]. In this respect, it is worth recalling that one of the first recovered exact solutions of the Einstein equations describes the flow of perfect relativistic fluid that represents our Universe at large scale. Moreover, in the middle of the twentieth century, it was established that 4D formalism allows concise formulation of governing equations for elastic continuum with distributed defects (this issue will be discussed in more detail below). In this regard, 4D representations are no longer unaccustomed in the framework of continuum mechanics. In the following, we briefly touch upon a few of fundamentals of both Newtonian mechanics and general relativity. This, in our opinion, allows us to be more concise in the first place, and secondly, points to the close affinity of Newtonian and Einsteinian mechanics.

We begin with brief consideration of the space-time structure, accepted in conventional continuum mechanics. In common life, we recognize ourselves and surrounding objects as occupying places. Places are represented by points in a three-dimensional *space*. As was mentioned above, important notions of classical understanding of space are *translation* and *congruence*. These notions lead to Euclidean point space. The geometric structure of the latter allows one to define points and vectors (*translations*), to introduce position and displacement vector fields, to introduce a global Cartesian frame, and to identify vectors and covectors by virtue of the scalar product.

Changes that we observe in ourselves and in the surrounding world are considered as occurring at specific *instants*. The latter are represented as points in one-dimensional space, independent of the space of places. Introducing the affine-Euclidean structure and orientation on such space allows one to fix the coordinate frame and to appreciate future and past. Thus, the coordinate of an instant is *time*, and difference between coordinates is the *time lapse*.

### 2.5.1 Newtonian Space-Time Manifold

Place and instance are combined to define an *event*. The collection of all events form *space-time*. Now, we switch to precise language. An event is considered to be a primitive entity, like a point in Euclidean geometry, and we give the following definition.

A *Newtonian space-time* is a 5-tuple[27]

$$(\mathscr{M}, \mathfrak{S}, D, \nabla, t)\,,$$

where $\mathscr{M}$ is a set whose elements are referred to as *events*, $\mathfrak{S}$ is a topology on $\mathscr{M}$, $D$ is a smooth structure on $\mathscr{M}$, $\nabla$ is a connection on $\mathscr{M}$, and $t\colon \mathscr{M} \to \mathbb{R}$ is an *absolute time* smooth function. The following axioms are stated:

$N_1$) The triple $(\mathscr{M}, \mathfrak{S}, D)$ is a smooth 4-dimensional manifold.
$N_2$) $\forall p \in \mathscr{M}: (dt)_p \neq 0$ (time never stops).
$N_3$) $\forall p \in \mathscr{M}: (\nabla dt)_p = 0$ (time flows uniformly).
$N_4$) $\nabla$ is torsion free.

Hereinafter the symbol $d$ in $N_2$) and $N_3$) denote the exterior differentiation operation[28]. We assume that $t(\mathscr{M}) = \mathbb{R}$. For any $\tau \in \mathbb{R}$ denote

$$\mathscr{S}_\tau = \{p \in \mathscr{M} \mid t(p) = \tau\}\,.$$

The axiom $N_2$) states that $\mathscr{S}_{\tau_1}$ and $\mathscr{S}_{\tau_2}$ with distinct $\tau_1, \tau_2 \in \mathbb{R}$ do not intersect. Thus, $\mathscr{M}$ is represented as the disjoint union

$$\mathscr{M} = \bigsqcup_{\tau \in \mathbb{R}} \mathscr{S}_\tau\,.$$

Each $\mathscr{S}_\tau$ is called the *absolute space at time* $\tau$.

**Remark 2.9.** The absolute space may have a Euclidean affine structure (conventional 3D mechanics of a continuum) or may have a Riemannian structure (2D mechanics of material surfaces).

Suppose that $p \in \mathscr{M}$. By virtue of $N_2$) one can make the following classification of tangent vectors. A vector $u \in T_p\mathscr{M}$ is called

- future directed, if $dt(u) > 0$;
- spatial, if $dt(u) = 0$;
- past directed, if $dt(u) < 0$.

### 2.5.2 Newton's Laws

Smooth curves on $\mathscr{M}$ are called *world lines* of particles. Since one knows $\nabla$, the autoparallel world line can be determined. This curve plays the role of a "straight line", on which a particle moves uniformly when no forces act. The Newton I is formalized as

27 The present and subsequent paragraphs are based on Frederic P. Schuller's Lecture 9 at the International Winter School on Gravity and Light (2015) https://www.youtube.com/watch?v=IBlCu1zgD4Y.
28 See Chapter 13, Section 13.3.9.

follows: *the world line of a particle with the influence of no force is an autoparallel curve with future directed velocity at every point*. If $\chi$ is such world line, then, by definition:

i) $\nabla_{\chi'}\chi' = 0$;
ii) $dt(\chi') > 0$.

Suppose that $\chi$ is a world line. The Newton II law states that the deviation of $\nabla_{\chi'}\chi' = 0$ from zero is caused by forces that act on the particle. Formally,

$$\nabla_{\chi'}\chi' = \frac{1}{m}f\,,$$

where $f \in \mathrm{Vec}(\mathscr{M})$ is a force that acts on the particle. It is assumed to be a spatial vector field, *i.e.*, $dt(f) = 0$, since the force is supposed to accelerate in spatial directions. The coefficient $m$ is a positive constant, a *mass* of the particle. Thus, $\nabla_{\chi'}\chi'$ is the *acceleration* of the particle.

Assume that one can choose an atlas $\mathscr{A}$ from the smooth structure $D$ that has the following form. For any chart $(U, \varphi)$, a coordinate mapping $\varphi\colon x \mapsto (x^0, x^1, x^2, x^3)$ is such that $\varphi^0\colon x \mapsto x^0$ coincides with the absolute time function $t$: $\varphi^0 = t|_U$. The postulate $N_3$) takes the form

$$\nabla_{\partial_i}dx^0 = 0\,, \quad i = 0, 1, 2, 3\,,$$

which, since $(\nabla_{\partial_i}dx^0)_j = -\Gamma^0{}_{ji}$, means that $\Gamma^0{}_{ji} = 0$, for all $i, j = 0, 1, 2, 3$. Such an atlas is called stratified.

Consider the Newton II law in the atlas $\mathscr{A}$:

$$(\chi^0)'' + \Gamma^0{}_{ij}(\chi^i)'(\chi^j)' = 0\,,$$
$$(\chi^q)'' + \Gamma^q{}_{ij}(\chi^i)'(\chi^j)' = \frac{1}{m}f^q\,, \quad q = 1, 2, 3\,.$$

The right-hand side of the first equation is zero, since $F$ is spatial. Since $\Gamma^0{}_{ij} = 0$, one has $(\chi^0)'' = 0$. Thus, there are constants $a, b \in \mathbb{R}$, such that $\chi^0(\lambda) = a\lambda + b$. The next three equations can be represented as follows:

$$(\chi^q)'' + \Gamma^q{}_{\gamma\delta}(\chi^\gamma)'(\chi^\delta)' + \Gamma^q{}_{00}(\chi^0)'(\chi^0)' + 2\Gamma^q{}_{\gamma 0}(\chi^\gamma)'(\chi^0)' = \frac{1}{m}f^q\,,$$

where $q, \gamma, \delta = 1, 2, 3$. Here, the postulate $N_4$) was used. Now, parameterize the world line by the absolute time. One has $d/d\lambda = a\,d/d\tau$. Changing primes by dots one can obtain

$$\ddot{\chi}^q + \Gamma^q{}_{\gamma\delta}\dot{\chi}^\gamma\dot{\chi}^\delta + \Gamma^q{}_{00}\dot{\chi}^0\dot{\chi}^0 + 2\Gamma^q{}_{\gamma 0}\dot{\chi}^\gamma\dot{\chi}^0 = \frac{1}{ma^2}f^q\,, \quad q, \gamma, \delta = 1, 2, 3\,.$$

The left-hand side of this equation is the $q$-th component of the acceleration vector. None of the terms of this side transform as a vector, but they have the following meaning:

- $\Gamma^q{}_{\gamma\delta}\dot{\chi}^\gamma\dot{\chi}^\delta$ is a coordinate correction term;
- $\Gamma^q{}_{00}\dot{\chi}^0\dot{\chi}^0$ is gravity or centrifugal term;
- $2\Gamma^q{}_{\gamma 0}\dot{\chi}^\gamma\dot{\chi}^0$ is Coriolis term.

### 2.5.3 Rigid Frame

A *rigid frame* [52] is a bijective mapping $\varphi: \mathscr{M} \to \mathscr{E}\times\mathbb{T}$, where $\mathscr{E}$ is a three-dimensional Euclidean affine space and $\mathbb{T} \subset \mathbb{R}$ is an open interval, referred to as a *chronometric manifold*. We presume that $\varphi$ satisfies two conditions[29]:

(i) for any $t \in \mathbb{T}$ there exists $s \in \mathbb{T}$, such that $\varphi(\mathcal{S}_t) = \mathscr{E} \times \{s\}$;

(ii) for any $s \in \mathbb{T}$ there exists $t \in \mathbb{T}$, such that $\varphi^{-1}(\mathscr{E} \times \{s\}) = \mathcal{S}_t$.

**Remark 2.10.** We claimed that the chronometric manifold $\mathbb{T}$, whose points correspond to instants of time (time points), is assumed to be diffeomorphic to $\mathbb{R}$. To justify the correspondence of this hypothesis to our relationship to the physical world, we follow the reasonings of H. Weyl [35]. The real world appears to us as a sequence of intentional objects of the human mind. In this regard, time can be interpreted as the primary form (*Urform*) of the flow of observations. Because of the nature of human consciousness, this flow is continuous and directed from the "past" to the "future". Furthermore, in the observable world, there are processes that occur (to our perception) cyclically. These cyclic processes, which might be termed as a clock, allow one to calibrate the flow of observations, *i.e.*, perform its arithmetization. Similar considerations (reflecting the philosophical position of intuitionism) lead to a hypothesis $\mathbb{T} \cong \mathbb{R}$. More strictly, $\mathbb{T}$ can be identified with an one-dimensional vector space over the field $\mathbb{R}$, the only element of whose basis is a vector characterizing the unit of time (which is reflected, in particular, in its "physical dimension").

### 2.5.4 Change of Frame

Let

$$\varphi: \mathscr{M} \to \mathscr{E} \times \mathbb{T}\,, \quad \varphi: e \mapsto (x, t)\,,$$
$$\varphi^*: \mathscr{M} \to \mathscr{E} \times \mathbb{T}\,, \quad \varphi^*: e \mapsto (x^*, t^*)\,,$$

be two rigid frames. Then the bijection

$$\varphi_{12} = \varphi^* \circ \varphi^{-1}: \mathscr{E} \times \mathbb{T} \to \mathscr{E} \times \mathbb{T}\,, \quad (x, t) \mapsto (x^*, t^*)$$

is a *change of frame* [52]. In component functions, it is represented as

$$(x^*, t^*) = \varphi_{12}(x, t) = (\alpha(x, t), \beta(t))\,,$$

where $\alpha: \mathscr{E} \times \mathbb{T} \to \mathscr{E}$, and $\beta: \mathbb{T} \to \mathbb{T}$.

**Remark 2.11.** The value of the mapping $\varphi_{12}$ at any element $(x, t) \in \mathscr{E} \times \mathbb{T}$ has the form

$$\varphi_{12}(x, t) = (\alpha(x, t), \widetilde{\beta}(x, t))\,,$$

29 In each rigid frame simultaneous events stay simultaneous.

where $\alpha\colon \mathscr{E}\times\mathbb{T} \to \mathscr{E}$ and $\widetilde{\beta}\colon \mathscr{E}\times\mathbb{T} \to \mathbb{T}$ are component functions. Factually, $\widetilde{\beta}$ does not depend on $x$. Fix $t \in \mathbb{T}$ and take arbitrary $x, x' \in \mathscr{E}$. Since the points $(x, t)$, $(x', t)$ are simultaneous, according to (ii) there exists $\tau \in \mathbb{T}$ for which $\varphi^{-1}(x, t), \varphi^{-1}(x', t) \in \mathcal{S}_\tau$. According to (i), for $\varphi^*$ there exists $s_t \in \mathbb{T}$, such that $\varphi_{12}(x, t), \varphi_{12}(x', t) \in \mathscr{E} \times \{s_t\}$. This implies

$$\widetilde{\beta}(x, t) = \widetilde{\beta}(x', t) = \beta(t) ,$$

where $\beta\colon \mathbb{T} \to \mathbb{T}$, $\beta\colon t \mapsto s_t$.

Consider a special type of change of frame. This type is defined by Galilean transformations and based on the assumption that distances between simultaneous events and time lapses between events do not depend on a rigid frame.

**Remark 2.12.** The notions of the rigid frame and the Galilean transformation group, which generates a change of frame, are necessary to postulate the principle of material frame-indifference. The latter principle gives certain physical restrictions on possible constitutive relations.

Let $e, f \in \mathscr{M}$ be two simultaneous events. A *distance $d(e, f; \varphi)$ between $e$ and $f$ relative to a rigid frame* $\varphi\colon \mathscr{M} \to \mathscr{E} \times \mathbb{T}$ is a Euclidean distance $\sqrt{\overrightarrow{xy} \cdot \overrightarrow{xy}}$, where $(x, t) = \varphi(e)$, $(y, t) = \varphi(f)$. That is,

$$d(e, f; \varphi) = \sqrt{\overrightarrow{xy} \cdot \overrightarrow{xy}} .$$

Let $e, f \in \mathscr{M}$ be two events. A *time lapse $T(e, f; \varphi)$ between $e$ and $f$ relative to a rigid frame* $\varphi\colon \mathscr{M} \to \mathscr{E} \times \mathbb{T}$ is the difference $t_2 - t_1$, where $(x, t_1) = \varphi(e)$, $(y, t_2) = \varphi(f)$. That is,

$$T(e, f; \varphi) = t_2 - t_1 .$$

We state the following postulates:

($P_1$) Let $e, f \in \mathscr{M}$ be two simultaneous events. For any rigid frames $\varphi$, $\varphi^*$, the following equality holds:

$$d(e, f; \varphi) = d(e, f; \varphi^*) .$$

($P_2$) Let $e, f \in \mathscr{M}$ be two events. For any rigid frames $\varphi$, $\varphi^*$, the following equality holds:

$$T(e, f; \varphi) = T(e, f; \varphi^*) .$$

From ($P_1$) and ($P_2$) it follows that for any change of frame,

$$(x, t) \mapsto \varphi_{12}(x, t) = (x^*, t^*) ,$$

there exists a mapping $a\colon \mathbb{T} \to \mathscr{E}$, an origin $b \in \mathscr{E}$, a field $\boldsymbol{Q}\colon \mathbb{T} \to O(3)$ of orthogonal tensors[30], and a number $c \in \mathbb{T}$, such that [52]

$$x^* = a(t) + \boldsymbol{Q}(t)(x - b)\,, \tag{2.31}$$

$$t^* = t - c\,. \tag{2.32}$$

## 2.6 Relativistic Space-Time

We have come to the most complicated model of physical space considered in this book, but at the same time, to the most bewitching of them. This is relativistic space-time. In order to become familiar with it, one has to extend the definition of Riemannian space given above somewhat.[31]

### 2.6.1 Lorentzian Manifolds

Suppose that $\mathscr{M}$ is a smooth $m$-dimensional manifold. A *pseudo-Riemannian metric*[32] is a section $g \in \mathrm{Sec}(T^*\mathscr{M} \otimes T^*\mathscr{M})$, such that for every point $p \in \mathscr{M}$ the following axioms hold:

$L_1$) $\forall u, v \in T_p\mathscr{M}\colon\ \ g_p(u, v) = g_p(v, u)$;

$L_2$) $\forall u \in T_p\mathscr{M}\colon\ \ (\forall v \in T_p\mathscr{M}\colon g_p(u, v) = 0) \Leftrightarrow (u = 0)$.

In a smooth chart $(U, \varphi)$, the pseudo-Riemannian metric is represented as $g = g_{ij}dx^i \otimes dx^j$. At each point $p \in \mathscr{M}$, the matrix $[g_{ij}(p)]$ has $m$ real non-zero eigenvalues. The amounts $m_+$, $m_-$ of positive and negative eigenvalues are independent of the choice of $p$; their sum is equal to $m_+ + m_- = m$. The pair $(m_+, m_-)$ is thus point independent and is called the *signature of the metric* $g$. In the framework of such a definition, a Riemannian metric is a pseudo-Riemannian metric with signature $(m, 0)$.

Suppose that $\mathscr{M}$ is a smooth four-dimensional manifold. A pseudo-Riemannian metric $g$ with signature $(1, 3)$ is called a *Lorentzian metric*. The signature for the Lorentzian metric is often denoted by $(+, -, -, -)$. Let $p \in \mathscr{M}$ be a point. One can

---

30 Recall that orthogonal tensor is such a linear map that preserves inner product; see Chapter 12, Section 12.5.4.

31 The present and subsequent paragraphs are based on Frederic P. Schuller's Lectures 13–15 in the International Winter School on Gravity and Light (2015) https://www.youtube.com/watch?v=iFAxSEoj6Go.

32 If one compares the definition of pseudo-Riemannian metric with the definition of Riemannian metric (see Chapter 13, Section 13.3.10) then one observes that the properties $L_1$) and $g_1$) are identical, but the properties $g_2$) and $g_3$) have changed to $L_2$). That is, a pseudo-Riemannian metric is a field of bilinear symmetric and non-degenerate forms. These forms are not positive definite, in general.

choose a basis $(e_i)_{i=0}^3$ in $T_p\mathcal{M}$ with the property[33]:

$$g_p(e_i, e_j) = \eta_{ij}\,, \quad \text{where} \quad [\eta_{ij}] = \operatorname{diag}\{1, -1, -1, -1\}\,.$$

For any vector $u \in T_p\mathcal{M}$, $u = u^i e_i$, one has

$$g_p(u, u) = (u^0)^2 - (u^1)^2 - (u^2)^2 - (u^3)^2\,.$$

Now, consider the vector space $\mathbb{R}^4$ of all 4-tuples $(u^0, u^1, u^2, u^3)$. The set

$$C_{T_p\mathcal{M}} = \{u \in T_p\mathcal{M} \mid g_p(u, u) = 0\}$$

is represented in $\mathbb{R}^4$ by the set

$$C_{\mathbb{R}^4} = \{(u^0, u^1, u^2, u^3) \in \mathbb{R}^4 \mid (u^0)^2 - (u^1)^2 - (u^2)^2 - (u^3)^2 = 0\}\,,$$

which is a circular double cone with the apex $(0, 0, 0, 0)$.

### 2.6.2 Time Orientation

Suppose that $\mathcal{M}$ is a smooth four-dimensional manifold with the Lorentzian metric $g$. The field $g$ establishes the distribution $\{C_{T_p\mathcal{M}}\}_{p\in\mathcal{M}}$ of double cones in tangent spaces. Recall that in the case of Newtonian space-time, one has the absolute time function $t$. The condition $dt(u) = 0$ generates a hyperplane in tangent space, and future directed vectors are situated on one side of it. Thus, the absolute time function allows us to specify a distribution of half-spaces of tangent spaces. In relativistic physics, no absolute time is presupposed. According to one of relativistic postulates, velocities of mass particles must be situated inside the cones in the future directed positions. In other words, one needs a rule that allows one to specify a distribution $\{C^{1/2}_{T_p\mathcal{M}}\}_{p\in\mathcal{M}}$ of circular cones from the given distribution $\{C_{T_p\mathcal{M}}\}_{p\in\mathcal{M}}$ of double cones. This means that for every point $p \in \mathcal{M}$, one needs to choose exactly one of two cones that constitute $C_{T_p\mathcal{M}}$. This choice can be made using time orientation.

A *time orientation* on $\mathcal{M}$ is a vector field $T \in \mathrm{Vec}(\mathcal{M})$ with the properties [53]:

$T_1$) $T \neq 0$

$T_2$) $g(T, T) > 0$.

By definition, for each point $p \in \mathcal{M}$ the value $T(p)$ belongs to one of two cones that constitute $C_{T_p\mathcal{M}}$. Thus, it is exactly the desired rule of specifying a distribution $\{C^{1/2}_{T_p\mathcal{M}}\}_{p\in\mathcal{M}}$ of circular cones. The role of the absolute time function is now played by the pair $(g, T)$ of the Lorentzian metric and time orientation.

33 In the case of Lorentzian manifolds, we consider values of indices from the set $\{0, 1, 2, 3\}$.

### 2.6.3 Definition of Relativistic Space-Time

A *relativistic space-time* is a 7-tuple [31, 33]

$$(\mathscr{M}, \mathfrak{S}, D, g, \nabla, T, \mu) ,$$

where $\mathscr{M}$ is a set whose elements are referred to as *events*, $\mathfrak{S}$ is a topology on $\mathscr{M}$, $D$ is a smooth structure on $\mathscr{M}$, $g \in \operatorname{Sec}(T^*\mathscr{M} \otimes T^*\mathscr{M})$ is a section, $\nabla$ is a connection on $\mathscr{M}$, $T \in \operatorname{Vec}(\mathscr{M})$ is a vector field on $\mathscr{M}$, and $\mu$ is a volume form on $\mathscr{M}$. The following axioms are stated:

$R_1$) The triple $(\mathscr{M}, \mathfrak{S}, D)$ is a smooth four-dimensional manifold.
$R_2$) $g$ is a Lorentzian metric.
$R_3$) $\nabla$ is the Levi–Civita connection.
$R_4$) $T$ is a time orientation.

Smooth curves on $\mathscr{M}$ are called *world lines* of particles.

In contrast to Newtonian space-time, the relativistic space-time admits massive and massless particles. They are subjected the following two postulates:

$P_1$) The world line $\chi: \mathbb{I} \to \mathscr{M}$ of a *massive particle* satisfies:
- (i) $\forall t \in \mathbb{I}: g_{\chi(t)}(\chi'(t), \chi'(t)) > 0$ (*i.e.*, the velocity lies inside the double cone $C_{T_{\chi(t)}\mathscr{M}}$);
- (ii) $\forall t \in \mathbb{I}: g_{\chi(t)}(T_t, \chi'(t)) > 0$ (*i.e.*, the velocity lies exactly in that cone which is specified by $T$).

$P_2$) The world line $\chi: \mathbb{I} \to \mathscr{M}$ of a *massless particle* satisfies:
- (i) $\forall t \in \mathbb{I}: g_{\chi(t)}(\chi'(t), \chi'(t)) = 0$ (*i.e.*, the velocity lies in the boundary of the double cone $C_{T_{\chi(t)}\mathscr{M}}$);
- (ii) $\forall t \in \mathbb{I}: g_{\chi(t)}(T_t, \chi'(t)) > 0$ (*i.e.*, the velocity belongs to the boundary of such cone which is specified by $T$).

### 2.6.4 Observers

Both a time and the velocity of a particle are observer-dependent quantities. This notion is formalized as follows.

An *observer* is a world line $\gamma: \mathbb{J} \to \mathscr{M}$ of a massive particle together with a choice of the basis

$$(e_0(t), e_1(t), e_2(t), e_3(t))$$

on each $T_{\gamma(t)}\mathscr{M}$, $t \in \mathbb{I}$. This collection of bases needs to satisfy the following conditions:

(i) $\forall t \in \mathbb{I}: e_0(t) = \gamma'(t)$;
(ii) $\forall t \in \mathbb{I}: g_{\gamma(t)}(e_i(t), e_j(t)) = \eta_{ij}$, $i, j = 0, 1, 2, 3$.

Any observer is subjected to the following postulates:

$P_1$) A *clock* carried by a specific observer $(\gamma, e)$ will measure a time

$$\tau := \int_{t_1}^{t_2} \sqrt{g_{\gamma(t)}(\gamma'(t), \gamma'(t))}\, dt$$

between the two "events":

- $\gamma(t_1)$ – "start the clock";
- $\gamma(t_2)$ – "stop the clock".

The number $\tau$ is called the *proper time.*

$P_2$) Let $(\gamma, e)$ be an observer and $\chi\colon \mathbb{I} \to \mathcal{M}$ be a massive particle world line, which is parameterized such that $\forall t \in \mathbb{I}$: $g_{\chi(t)}(\chi'(t), \chi'(t)) = 1$. Suppose that the observer and the particle meet somewhere in space-time. That is, $\chi(t_1) = \gamma(t_2)$, for some $t_1 \in \mathbb{I}$, $t_2 \in \mathbb{J}$. The observer measures the 3-*velocity* (*spatial velocity*) of this particle as

$$v := e^{\alpha}_{t_2}\left(\chi'(t_1)\right) e_{\alpha}(t_2)\,, \quad \alpha = 1, 2, 3\,,$$

where $(e^{i}_{t_2})^{3}_{i=0}$ is the unique dual basis of $(e_i(t_2))^{3}_{i=0}$.

### 2.6.5 Lorentz Transformations

Let $(\gamma, e)$ and $(\widetilde{\gamma}, \widetilde{e})$ be observers with $\widetilde{\gamma}(\widetilde{t}) = p = \gamma(t)$. Reparametrizing the curves, if necessary, one may assume that $\widetilde{t} = t = 0$. Thus, $(e_i(0))^{3}_{i=0}$ and $(\widetilde{e}_i(0))^{3}_{i=0}$ are bases of the similar tangent space $T_p\mathcal{M}$. There exists a matrix $\Omega \in \mathrm{GL}(4; \mathbb{R})$, such that

$$\widetilde{e}_i(0) = \Omega^{j}{}_{i} e_j(0)\,.$$

Since $g_p(\widetilde{e}_i(0), \widetilde{e}_j(0)) = \eta_{ij}$, $g_p(e_k(0), e_l(0)) = \eta_{kl}$ and using the relation between "twiddled" and "non-twiddled" vectors, one obtains the Lorentz transformation

$$\Omega^{k}{}_{i}\Omega^{l}{}_{j}\eta_{kl} = \eta_{ij}\,.$$

Thus, the Lorentz transformation relates the frames of two observers at the same point. One has $\Omega \in O(1, 3)$.

### 2.6.6 Matter

There are two types of matter: point matter and field matter. Assume that $g$ and $T$ are given.

**Point matter.** Postulates $P_1$) and $P_2$) about massive and massless particles give constraints on the possible particle world lines. To obtain world lines one needs to use law of motion. Consider actions of particles in the case of no acting external fields.

The action of a massive particle with world line $\chi$ has the form

$$\mathfrak{A}_{\text{massive}}[\chi] = m \int \sqrt{g_{\chi(t)}(\chi'(t), \chi'(t))}\, dt\,.$$

Here, $m > 0$ is a constant, a mass of the particle. Dynamical laws of motion (Euler – Lagrange equations) are obtained from the condition $\delta\mathfrak{A}_{\text{massive}}[\chi] = 0$. The possible world lines $\chi$ need to satisfy the restriction $g_{\chi(t)}(T_t, \chi'(t)) > 0$, for all $t \in \mathbb{I}$.

The action of a massless particle with world line $\chi$ has the form

$$\mathfrak{A}_{\text{massless}}[\chi, \lambda] = \int \lambda(t)\sqrt{g_{\chi(t)}(\chi'(t), \chi'(t))}\, dt\,.$$

Here, $\lambda(t)$ is a Lagrange multiplier. The partial variation of $\mathfrak{A}_{\text{massless}}[\chi, \lambda]$ with respect to $\lambda$ gives the condition $g_{\chi(t)}(\chi'(t), \chi'(t)) = 0$, for all $t$. Dynamical laws of motion (Euler – Lagrange equations) are obtained from the partial variation of $\mathfrak{A}_{\text{massless}}[\chi, \lambda]$ with respect to $\chi$.

The reason for describing equations of motions by actions is that composite systems have an action that is the sum of the actions of the parts of this system possibly including interaction terms. Consider an example. Suppose that a massive particle interacts with the electromagnetic field that is generated by a covector field $A \in \mathrm{CVec}(\mathcal{M})$ (electromagnetic potential). Let $q$ be a charge of this field. Then, the action has the form

$$\mathfrak{A}[\chi; A] = \int \left(m\sqrt{g_{\chi(t)}(\chi'(t), \chi'(t))} + qA(\chi'(t)\right)\, dt\,.$$

Here, the semicolon means that $A$ is fixed. The Euler–Lagrange equation has the form

$$m(\nabla_{\chi'}\chi')^i = -qF^i_j\dot{\chi}^j\,,$$

where $F_{ij} = \frac{\partial A_i}{\partial x^j} - \frac{\partial A_j}{\partial x^i}$ are components of the Faraday tensor. The expression $-qF^i_j\dot{\chi}^j$ represents the Lorentz force on a charge particle in the electromagnetic field. Note that the action $\mathfrak{A}[\chi; A]$ has the form

$$\mathfrak{A}[\chi; A] = \mathfrak{A}_{\text{massive}}[\chi] + \mathfrak{A}_{\text{interact}}[\chi; A]\,,$$

where $\mathfrak{A}_{\text{interact}}[\chi; A]$ is the interaction term.

**Field matter.** A classical field matter is any tensor field on $\mathcal{M}$ whose equations of motions are derived from an action. Consider an example. The action for electromagnetic field, in assumption that $\mathcal{M}$ is a trivial manifold, has the form (in the atlas with one chart)

$$\mathfrak{A}_{\text{Maxwell}}[A] = \frac{1}{4}\int_{\mathcal{M}} \sqrt{-g}F_{ab}F_{cd}g^{ac}g^{bd}\, d^4x\,,$$

where $g = \det[g_{ij}] < 0$.

**Remark 2.13.** Note that the Lagrangian density of $\mathfrak{A}_{\text{Maxwell}}$ depends on the first partial derivatives of the electromagnetic potential $A$. A similar situation holds for *simple materials* in continuum mechanics. The response functional for the simple material depends on the deformation gradient which, in a basis, is represented by first-order partial derivatives of actual coordinates.

**The energy-momentum tensor of matter fields.** Suppose that the metric $g$ is unknown, and one intends to obtain it. Thus, one needs to write down the action $\mathfrak{A}_{\text{grav}}[g]$ for the metric tensor. This action will be added to any action $\mathfrak{A}_{\text{matter}}[A, \dots]$ in order to describe the total system. In particular,

$$\mathfrak{A}[A, g] = \mathfrak{A}_{\text{grav}}[g] + \mathfrak{A}_{\text{Maxwell}}[A, g] .$$

The partial variation of $\mathfrak{A}[A, g]$ with respect to $\delta A$ gives the Maxwell equations. Only $\mathfrak{A}_{\text{Maxwell}}$ gives contributions for it. In the case of the partial variation of $\mathfrak{A}[A, g]$ with respect to $\delta g$, both $\mathfrak{A}_{\text{grav}}$ and $\mathfrak{A}_{\text{Maxwell}}$ give contributions. Thus, the Euler–Lagrange equations have the form

$$\underbrace{\frac{1}{16\pi G_{\text{Newt}}} G^{ij}}_{\text{contribution from } \mathfrak{A}_{\text{grav}}} + \underbrace{\left(-\frac{1}{2} T^{ij}\right)}_{\text{contribution from } \mathfrak{A}_{\text{Maxwell}}} = 0 ,$$

where $G_{\text{Newt}}$ is Newton's gravitation constant. In another form, this equation is represented as

$$G^{ij} = 8\pi G_{\text{Newt}} T^{ij} .$$

This is the Einstein equation. The quantities $T^{ij}$ form the *energy-momentum tensor.* In general, if $\mathfrak{A}_{\text{matter}}[\Phi, g]$ is some matter action with the Lagrangian density $L_{\text{matter}}$, then the energy-momentum tensor is

$$T^{ij} := -\frac{2}{\sqrt{-g}} \left( \frac{\partial L_{\text{matter}}}{\partial g_{ij}} - \partial_s \frac{\partial L_{\text{matter}}}{\partial \partial_s g_{ij}} + \cdots \right) .$$

### 2.6.7 The Einstein Equation

In Newtonian space-time, Poisson's law $\Delta\varphi = 4\pi\rho G_{\text{Newt}}$ can be formulated in terms of the Ricci curvature as

$$R_{00} = 4\pi\rho G_{\text{Newt}} .$$

Based in this equation, Einstein suggested that the relativistic field equations for the metric $g$ of relativistic space-time should be as follows:

$$R_{ij} = 8\pi G_{\text{Newt}} T_{ij} .$$

However, this equation is wrong because $T_{ij}$ are components of the energy-momentum tensor that is derived from an action. The divergence of the right-hand side of the equation is zero. The divergence of the left-hand side of the equation is not zero, in general, which leads to a contradiction. These equations cannot be upheld. In order to find the appropriate left-hand side of the equation Hilbert suggested considering the action

$$\mathfrak{A}_{\text{Hilbert}}[g] = \int_{\mathcal{M}} \sqrt{-g} R_{ij} g^{ij} \, d^4x \,.$$

The stationary points of the functional $\mathfrak{A}_{\text{Hilbert}}$ can be found as the solutions of the corresponding Euler–Lagrange equations. In order to obtain them, derive the variation of $\mathfrak{A}_{\text{Hilbert}}$:

$$\delta\mathfrak{A}_{\text{Hilbert}}[g] = \int_{\mathcal{M}} \left\{ \delta\sqrt{-g} R_{ab} g^{ab} + \sqrt{-g} R_{ab} \delta g^{ab} + \sqrt{-g} g^{ab} \delta R_{ab} \right\} d^4x \,. \tag{2.33}$$

Using formula for derivative of the determinant (see the derivation of the Voss–Weyl formula), we obtain

$$\delta\sqrt{-g} = \frac{1}{2\sqrt{-g}}(-g) g^{ab} \delta g_{ab} = \frac{1}{2}\sqrt{-g} g^{ab} \delta g_{ab} \,.$$

Taking into account the identity $g^{ab} g_{bc} = \delta^a_c$, which leads to the following equation:

$$\delta(g^{ab}) g_{bc} + g^{ab} \delta g_{bc} = 0 \,,$$

we arrive at

$$\delta g^{ab} = -g^{am} g^{bn} \delta g_{mn} \,.$$

In order to simplify further derivations suppose that the coordinates, defined by the chosen chart, are normal with respect to point $p$. Then, all $\Gamma^i{}_{jk}$ (but not their derivatives) vanish at $p$. Thus, the variation of the Ricci tensor can be calculated as follows:

$$\delta R_{ab} = \delta\partial_b \Gamma^m{}_{am} - \delta\partial_m \Gamma^m{}_{ab} = \partial_b \delta\Gamma^m{}_{am} - \partial_m \delta\Gamma^m{}_{ab} = \nabla_b \delta\Gamma^m{}_{am} - \nabla_m \delta\Gamma^m{}_{ab} \,,$$

because $\delta$ and $\partial$ commute, all terms including $\Gamma$'s are reduced to zero, and partial derivatives (at point $p$) coincide with corresponding covariant derivatives. Keeping in mind that $\mathcal{M}$ is a pseudo-Riemannian manifold, and, consequently, $\nabla_c g^{ab} = 0$, we obtain

$$\begin{aligned} \sqrt{-g} g^{ab} \delta R_{ab} &= \sqrt{-g} g^{ab} \nabla_b \delta\Gamma^m{}_{am} - \sqrt{-g} g^{ab} \nabla_m \delta\Gamma^m{}_{ab} \\ &= \sqrt{-g} \nabla_b \left( g^{ab} \delta\Gamma^m{}_{am} \right) - \sqrt{-g} \nabla_m \left( g^{ab} \delta\Gamma^m{}_{ab} \right) . \end{aligned}$$

Finally, using the Voss–Weyl formula (2.22), we get

$$\sqrt{-g} g^{ab} \delta R_{ab} = \partial_b(\sqrt{-g} \delta A^b) - \partial_m(\sqrt{-g} \delta B^m) = \partial_b(\sqrt{-g}(\delta A^b - \delta B^b)) \,,$$

where $\delta A^b = g^{ab} \delta\Gamma^m{}_{am}$, $\delta B^b = g^{am} \delta\Gamma^b{}_{am}$.

Collecting all items in (2.33) one arrives at the following expression:

$$\delta\mathfrak{A}_{\text{Hilbert}}[\mathcal{g}] = \int\limits_{\mathcal{M}} \left\{ \frac{1}{2}\sqrt{-g}g^{mn}\delta g_{mn}R_{ab}g^{ab} - \sqrt{-g}g^{am}g^{bn}\delta g_{mn}R_{ab} \right.$$
$$\left. + \partial_b(\sqrt{-g}(\delta A^b - \delta B^b)) \right\} d^4x$$
$$= \int\limits_{\mathcal{M}} \left\{ \sqrt{-g}\delta g_{mn}\left(\frac{1}{2}g^{mn}S - R^{mn}\right)\right\} d^4x + \int\limits_{\mathcal{M}} \partial_b(\sqrt{-g}(\delta A^b - \delta B^b))\, d^4x\,.$$

Here, $S$ is the scalar curvature. The last item is an integral of divergence and, following standard argumentation (see [32]), we put it equal to zero. Thus, in view of the functional independence of $g_{mn}$,

$$(\delta\mathfrak{A}_{\text{Hilbert}}[\mathcal{g}] = 0) \Leftrightarrow \left(\frac{1}{2}g^{mn}S - R^{mn} = 0\right)\,. \tag{2.34}$$

We use the following notation for the Einstein tensor:

$$G^{ij} = R^{ij} - \frac{1}{2}g^{ij}S\,, \quad G_{ij} = g_{ia}g_{jb}G^{ab}\,.$$

Then, the equations (2.34) can be rewritten as

$$\nabla_i G^{ij} = 0\,. \tag{2.35}$$

Note that from a purely geometrical point of view, the Einstein tensor satisfies the Bianchi identity, which formally coincides with (2.35).

If the space is filled with interacting matter, then the action $\mathfrak{A}_{\text{Hilbert}}$ is more complex. For example, if we take into account electromagnetic interaction, then $\mathfrak{A}_{\text{Hilbert}}$ depends on the metric and the electromagnetic potential, which leads to the so-called Einstein–Maxwell theory. Thus, in the case of interacting matter, the equations are the following:

$$R_{ij} - \frac{1}{2}Sg_{ij} = 8\pi G_{\text{Newt}}T_{ij}\,. \tag{2.36}$$

These equations were obtained by Hilbert. Using physical arguments, Einstein came to similar equations [54]. The equations (2.36) are called the *Einstein equations*, and the action $\mathfrak{A}_{\text{Hilbert}}$ is called the *Einstein–Hilbert action*. Einstein equations are partial differential equations. Some particular solutions are given below.

**The Schwarzschild solution.** Let $\mathcal{M}$ be a flat (Minkowski) space equipped with the pseudo-Cartesian chart $(x^0, x^1, x^2, x^3)$ and let $\mathcal{B}$ be an open 3D ball of radius $R$ with center at the origin of the pseudo-Cartesian frame $(o, (\boldsymbol{i}_k)_{k=0}^3)$, *i.e.*,

$$\mathcal{M} \supset \mathcal{B} := \left\{ o + x^1\boldsymbol{i}_1 + x^2\boldsymbol{i}_2 + x^3\boldsymbol{i}_3 \mid \sqrt{(x^1)^2 + (x^2)^2 + (x^3)^2} < R \right\}\,.$$

Let $\mathcal{U} = \mathcal{M} \setminus \overline{\mathcal{B}}$ be an open subset of $\mathcal{M}$ that represents the exterior of $\mathcal{B}$. Following K. Schwarzschild, find the curvature of $\mathcal{M}$ caused by the filling of the region $\mathcal{B}$

with non-interacting matter (dust) of total mass $M$. To do this, we use a semi-inverse method. Keeping in mind spherical symmetry with respect to spatial coordinates choose the chart $(\mathcal{U}, h)$ on $\mathcal{M}$ that relates with the pseudo-Cartesian chart by the following transition functions:

$$\begin{aligned} x^0 &= t\,, \\ x^1 &= r\cos\varphi\sin\theta\,, \\ x^2 &= r\sin\varphi\sin\theta\,, \\ x^3 &= r\cos\theta\,. \end{aligned} \tag{2.37}$$

The components of the metric tensor induced by (2.37) and its inverse have the form

$$\begin{aligned} &\mathring{g}_{00} = 1\,, \quad \mathring{g}_{11} = -1\,, \quad \mathring{g}_{22} = -r^2\,, \quad \mathring{g}_{33} = -r^2\sin^2\theta\,, \\ &\mathring{g}^{00} = 1\,, \quad \mathring{g}^{11} = -1\,, \quad \mathring{g}^{22} = -\frac{1}{r^2}\,, \quad \mathring{g}^{33} = -\frac{1}{r^2\sin^2\theta}\,. \end{aligned}$$

The square of the interval, $d\mathring{s}^2$, can be written as (hereinafter we consider speed of light $c = 1$)

$$\mathring{g} = dt\otimes dt - dr\otimes dr - r^2 d\theta\otimes d\theta - r^2\sin^2\theta d\varphi\otimes d\varphi\,. \tag{2.38}$$

Now, assume that the region $\mathcal{B}$ is filled with massive non-interacting matter that brings flat $\mathcal{M}$ to some curved space-time. In order to determine the metric of curved $\mathcal{M}$, construct the generalization of (2.38),

$$g = Udt\otimes dt - Vdr\otimes dr - Wr^2 d\theta\otimes d\theta - Xr^2\sin^2\theta d\varphi\otimes d\varphi\,,$$

where $U$, $V$, $W$, $X$ are some functions of coordinates. Taking into account
(i) spherical symmetry with respect to spatial coordinates
(ii) the static character of problem

we conclude that $U$, $V$, $W$, $X$ do not depend on $\theta$, $\varphi$; moreover, $W = X = 1$, and these functions do not depend on $t$. Thus,

$$ds^2 = U(r)dt\otimes dt - V(r)dr\otimes dr - r^2 d\theta\otimes d\theta - r^2\sin^2\theta d\varphi\otimes d\varphi\,.$$

Consequently,

$$g_{00} = U(r)\,, \qquad g_{11} = -V(r)\,, \qquad g_{22} = -r^2\,, \qquad g_{33} = -r^2\sin^2\theta\,, \tag{2.39}$$

$$g^{00} = \frac{1}{U(r)}\,, \qquad g^{11} = -\frac{1}{V(r)}\,, \qquad g^{22} = -\frac{1}{r^2}\,, \qquad g^{33} = -\frac{1}{r^2\sin^2\theta}\,. \tag{2.40}$$

For the given metric components (2.39), we can calculate connection coefficients $\Gamma^i{}_{jk}$:

$$\begin{aligned} &\Gamma^0{}_{01} = \Gamma^0{}_{10} = \frac{U'}{2U}\,, && \Gamma^1{}_{00} = \frac{U'}{2V}\,, && \Gamma^1{}_{11} = \frac{V'}{2V}\,, \\ &\Gamma^1{}_{22} = -\frac{r}{V}\,, && \Gamma^1{}_{33} = -\frac{r}{V}\sin^2\theta\,, && \Gamma^2{}_{12} = \Gamma^2{}_{21} = \frac{1}{r}\,, \\ &\Gamma^2{}_{33} = -\cos\theta\sin\theta\,, && \Gamma^3{}_{13} = \Gamma^3{}_{31} = \frac{1}{r}\,, && \Gamma^3{}_{23} = \Gamma^3{}_{32} = \cot\theta\,. \end{aligned} \tag{2.41}$$

Now, we can derive the components of the Ricci tensor:

$$
\begin{aligned}
R_{00} &= -\frac{U''}{2V} + \frac{U'V'}{4V^2} + \frac{(U')^2}{4UV} - \frac{1}{r}\frac{U'}{V}\,, \\
R_{11} &= \frac{U''}{2U} - \frac{(U')^2}{4U^2} - \frac{U'V'}{4UV} - \frac{1}{r}\frac{V'}{V}\,, \\
R_{22} &= \frac{rU'}{2UV} + \frac{1}{V} - \frac{rV'}{2V^2} - 1\,, \\
R_{33} &= R_{22}\sin^2\theta\,,
\end{aligned}
$$

and the scalar curvature $R = g^{ij}R_{ij}$,

$$
S = -\frac{U''}{UV} + \frac{U'V'}{2UV^2} + \frac{(U')^2}{2U^2V} - \frac{2}{r}\frac{U'}{UV} + \frac{2}{r}\frac{V'}{V^2} + \frac{2}{r^2}\left(1 - \frac{1}{V}\right).
$$

The tensorial Einstein equation reduces to the following system of three non-linear equations:

$$
\begin{aligned}
\frac{1}{r}\frac{V'}{V} + \frac{1}{r^2}\left(1 - \frac{1}{V}\right) &= 0\,, \\
-\frac{U'}{rUV} + \frac{1}{r^2}\left(1 - \frac{1}{V}\right) &= 0\,, \\
-\frac{U'}{U} + \frac{V'}{V} - \frac{rU''}{U} + \frac{rU'}{2}\frac{V'}{UV} + \frac{r(U')^2}{2U^2} &= 0\,,
\end{aligned}
\tag{2.42}
$$

which admits the solution[34]

$$
V = \frac{1}{1 - A/r}\,, \qquad U = B\left(1 - \frac{A}{r}\right),
$$

where $A$ and $B$ are the constants of integration. These constants have to be chosen such that obtained metric in weak-field approximation corresponds to the Newtonian law of gravity [55]:

$$
A = 2MG_{\text{Newt}}\,, \qquad B = 1\,.
$$

Thus, the Schwarzschild metric $g$ has the form

$$
g = \left(1 - \frac{2MG_{\text{Newt}}}{r}\right)dt \otimes dt - \left(1 - \frac{2MG_{\text{Newt}}}{r}\right)^{-1} dr \otimes dr - r^2(d\theta \otimes d\theta + \sin^2 d\varphi \otimes d\varphi)\,.
$$

Note that Schwarzschild metric is

(i) spherically symmetric;
(ii) asymptotically flat (when $r \to \infty$ it tends to flat space-time, Minkowski, metrics).

The constant $A$ is denoted by $r_s$ and is called the Schwarzschild radius. That is, $r_s = 2MG_{\text{Newt}}$.

34 The function $V$ can be found separately from the first equation of (2.42), while the function $U$ is determined as the solution of the second equation of (2.42), with $V$ already found. It is easy to check that the so-obtained functions $U$ and $V$ satisfy the third equation (2.42).

**The Kerr solution.** Assume that there is a massive body with mass $M$ that rotates with angular momentum $J$. Introduce the coordinates $(t, r, \theta, \varphi)$, which are related with pseudo-Cartesian coordinates by the following expressions [56]:

$$\begin{aligned} x^0 &= t\,, \\ x^1 &= \sqrt{r^2+a^2}\cos\varphi\sin\theta\,, \\ x^2 &= \sqrt{r^2+a^2}\sin\varphi\sin\theta\,, \\ x^3 &= r\cos\theta\,, \end{aligned} \tag{2.43}$$

where $a = J/M$. In coordinates $(t, r, \theta, \varphi)$ (2.43), the Kerr metric is written as [57]

$$\begin{aligned} g = &\left(1-\frac{r_s r}{\rho^2}\right) dt\otimes dt - \frac{\rho^2}{\Delta} dr\otimes dr - \rho^2 d\theta\otimes d\theta \\ &- \left(r^2+a^2+\frac{r_s r a^2}{\rho^2}\sin^2\theta\right)\sin^2\theta d\varphi\otimes d\varphi + \frac{r_s r a}{\rho^2}\sin^2\theta(d\varphi\otimes dt + dt\otimes d\varphi)\,, \end{aligned}$$

where $r_s$ is the Schwarzschild radius, and

$$\rho^2 = r^2 + a^2\cos^2\theta\,, \quad \Delta = r^2 - r_s r + a^2\,.$$

Introduce the tetrad fields $h_i = h_i^{\ j}\partial_j$ and $h^i = h^i_{\ j}dx^j$ with the property $g(h_i, h_j) = \eta_{ij}$, $i, j = 0, 1, 2, 3$. That is,

$$\eta_{ij} = g_{kl} h_i^{\ k} h_j^{\ l}\,.$$

One has the following inverse relation:

$$g_{ij} = \eta_{kl} h^k_{\ i} h^l_{\ j}\,.$$

Using this relation one can find the components of the Kerr tetrad [57]:

$$[h^i_{\ j}] = \begin{pmatrix} \gamma_{00} & 0 & 0 & \eta \\ 0 & \gamma_{11}\sin\theta\cos\varphi & \gamma_{22}\cos\theta\cos\varphi & -\beta\sin\varphi \\ 0 & \gamma_{11}\sin\theta\sin\varphi & \gamma_{22}\cos\theta\sin\varphi & \beta\cos\varphi \\ 0 & \gamma_{11}\cos\theta & -\gamma_{22}\sin\theta & 0 \end{pmatrix},$$

where $\gamma_{00} = \sqrt{g_{00}}$, $\gamma_{ii} = \sqrt{-g_{ii}}$, $\beta^2 = \eta^2 - g_{33}$ and $\eta = g_{03}/\gamma_{00}$. Next,

$$[h_i^{\ j}] = \begin{pmatrix} \gamma_{00}^{-1} & 0 & 0 & 0 \\ -\beta g^{03}\sin\varphi & \gamma_{11}^{-1}\sin\theta\cos\varphi & \gamma_{22}^{-1}\cos\theta\cos\varphi & -\beta^{-1}\sin\varphi \\ \beta g^{03}\cos\varphi & \gamma_{11}^{-1}\sin\theta\sin\varphi & \gamma_{22}^{-1}\cos\theta\sin\varphi & \beta^{-1}\cos\varphi \\ 0 & \gamma_{11}^{-1}\cos\theta & -\gamma_{22}^{-1}\sin\theta & 0 \end{pmatrix}.$$

In the framework of teleparallel gravity a gravitation is represented by torsion, not by curvature. The Weitzenböck connection gives an example of the affine connection, which gives non-zero torsion but zero curvature and non-metricity[35]. Its connection coefficients are related with the tetrad $(h_i)$ as

$$\Gamma^i_{\ jk} = h_l^{\ i}\partial_k h^l_{\ j}\,. \tag{2.44}$$

Let us draw attention to the structure of the expression (2.44). Such a structure arises hereinafter in the definition of material connection (Chapter 9, Section 9.2.7). The role tetrads play is local deformations (implants). It also should be noted that both mentioned expressions are a particular case for the general one, which defines the calibration procedure for connection on the principal bundle (Chapter 14, Section 14.3.4).

**The Reissner–Nordström metric.** Suppose that charged spherical massive body is given. Then, in coordinates $(t, r, \theta, \varphi)$, one has

$$\mathscr{g} = \left(1 - \frac{r_s}{r} + \frac{r_q^2}{r^2}\right) dt \otimes dt - \left(1 - \frac{r_s}{r} + \frac{r_q^2}{r^2}\right)^{-1} dr \otimes dr - r^2 (d\theta \otimes d\theta + \sin^2 d\varphi \otimes d\varphi)\,,$$

where $r_s$ is the Schwarzschild radius and $r_q^2 = \frac{q^2 G_{\mathrm{Newt}}}{4\pi\varepsilon_0}$. Here, $q$ is the charge and $1/(4\pi\varepsilon_0)$ is Coulomb's constant.

The list of exact solutions of the Einstein equation could be continued, however, the above collection is sufficient for our purposes. Note the close affinity between them and exact universal solutions, obtained for non-linear equations of elasticity (see Chapter 3, Section 3.10).

## 2.7 Concluding Remarks

In the rest of the book, a *physical space* $\mathscr{P}$ is a set, whose elements $X, x, Y, y, \ldots,$ are referred to as *places*. We suppose that such a set is endowed with Riemannian space structure $(\mathscr{P}, \mathfrak{S}, D, \mathscr{g}, \mu)$. It is assumed that $\dim \mathscr{P} = m \le 3$. Coefficients of the Levi–Civita connection $\nabla$ are denoted by $\gamma^i_{\ jk}$. Intuitively, the physical space represents an "ideal laboratory", where one observes places and provides measurements. The particular case of the physical space is the Euclidean affine space $\mathscr{E}$ introduced in 2.1.

35 The Weitzenböck connection is introduced by applying the moving frame method to tetrad $(h_i)$. See Chapter 9, Section 9.2.6 for details.

# 3 Essentials of Non-Linear Elasticity Theory

## 3.1 Shapes and Deformation

In the framework of conventional continuum mechanics, a physical space is modeled by a three-dimensional Euclidean affine space $\mathscr{E}$ with the translation vector space $\mathscr{V}$ and the inner product $(\cdot)$. Solids are modeled as subsets of $\mathscr{E}$ with special properties. As a rule, one requires these subsets to be open in $\mathscr{E}$ and to have regular boundary[1] (in the sense of O. D. Kellogg [58]). In the present chapter, such subsets are called *shapes*. Let us consider their properties in more detail.

**Shapes.** Recall that, since $\mathscr{E}$ is Euclidean affine space, one can define the metric $d\colon \mathscr{E} \times \mathscr{E} \to \mathbb{R}$ according to (2.6). Thus, $(\mathscr{E}, d)$ is a metric space and it makes sense to consider open subsets.

The notion of the regular boundary is introduced sequentially, beginning from one-dimensional sets [58]:

1. A regular arc is a set $\mathfrak{a} \subset \mathscr{E}$, such that for some Cartesian coordinate system it admits a representation

$$\mathfrak{a} = \{x \in \mathscr{E} \mid x^2 = f(x^1), x^3 = g(x^1), x^1 \in [a, b]\} ,$$

   where $a < b$ and $f, g \in C^\infty([a, b]; \mathbb{R})$.
2. A regular curve is a set $\mathfrak{c} \subset \mathscr{E}$, which consists of a finite number of regular arcs arranged in order and such that the terminal point of each arc (other than the last one) is the initial point of the following arc. The arcs have no other points in common, except that the terminal point of the last arc may be the initial point of the first one (in this case, $\mathfrak{c}$ is a closed regular curve).
3. A regular surface element is a non-empty set $\mathfrak{s}$, such that for some Cartesian coordinate system, it admits the representation

$$\mathfrak{s} = \{x \in \mathscr{E} \mid x^3 = f(x^1, x^2), (x^1, x^2) \in K\} ,$$

   where $K \subset \mathbb{R}^2$ is a compact connected set, such that $\partial K$ is a closed regular curve and $f \in C^\infty(K; \mathbb{R})$.
4. The union of a finite number of regular surface elements is called a regular surface, if:
   (a) the intersection of any two of the elements is either empty, a single point which is a vertex for both, or a single regular arc which is an edge for both;
   (b) the intersection of any collection of three or more elements consists at most of vertices;

1 The requirement of regularity (in the sense of Kellogg) is due to the fact that one needs to use Stokes' theorem for deriving the differential balance equations.

https://doi.org/10.1515/9783110563214-003

(c) any two of the elements are the first and last of a chain such that each has an edge in common with the next;
(d) all elements having a vertex in common form a chain such that each has an edge, terminating in that vertex, in common with the next; the last may, or may not, have an edge in common with the first.

The term "edge" here refers to one of the (finite number of) regular arcs comprising the boundary of a regular surface element, while a vertex is a point at which two edges meet. If all the edges of a regular surface each belong to two of its surface elements, the surface is a closed regular surface.

5. A non-empty set $\mathcal{S} \subset \mathcal{E}$ has a regular boundary, if $\partial\mathcal{S}$ is a closed regular surface.

All is ready for the following definition. A *shape* is an open non-empty set $\mathcal{S} \subset \mathcal{E}$, whose boundary $\partial\mathcal{S}$ is regular.

One privileged shape $\mathcal{S}_R$ is called a *reference shape*. Its points are used as "labels" in consideration of deformation. Let $\mathcal{S}$ be another shape; we refer to it as the *actual shape*. Following classical notations, we denote points of $\mathcal{S}_R$ by the Latin majuscules $X, Y, \ldots$ and points of $\mathcal{S}$ by the Latin minuscules $x, y, \ldots$. The gradient and divergence operators, which are introduced by (2.9), (2.12), and (2.13), are denoted by $\text{grad}_R$ and $\text{div}_R$, if they are defined on $\mathcal{S}_R$.

**Deformation.** A deformation is a mapping

$$y \colon \mathcal{S}_R \to \mathcal{S}\,, \quad y \colon X \mapsto x\,,$$

that satisfies the following postulates [59]:

($y_1$) **Postulate of impenetrability.** Two particles cannot occupy the same place at the same time.

($y_2$) Any point from $\mathcal{S}$ is a place for some point from $\mathcal{S}_R$.

($y_3$) **Postulate of continuity.** Deformation is as many times continuously differentiable as required.

The mathematical formalization of these postulates is: $y$ is an injective and surjective $C^\infty$-mapping.

We considered $\mathcal{S}_R$ a non-deformed shape and $\mathcal{S}$ a deformed one. However, this point of view may be reversed. One can recognize $\mathcal{S}$ as a reference shape and $\mathcal{S}_R$ as an actual one. Since both points of view are equal, the mapping $y^{-1} \colon \mathcal{S} \to \mathcal{S}_R$ should be required to be smooth. Thus, one obtains that $y$ is a $C^\infty$-diffeomorphism. In the following, we use such an assumption.

Finally, by a *deformation* we mean a $C^\infty$-diffeomorphism $y \colon \mathcal{S}_R \to \mathcal{S}$. The inverse diffeomorphism $y^{-1} \colon \mathcal{S} \to \mathcal{S}_R$ is called the *inverse deformation*.

## 3.2 Shape Coordinates and Basis

Since shapes $\mathcal{S}_R$ and $\mathcal{S}$ are open subsets of $\mathcal{E}$, they are open three-dimensional submanifolds of $\mathcal{E}$. The scalar product $(\cdot)$ induces metric tensors $\boldsymbol{G}$ and $\boldsymbol{g}$ on $\mathcal{S}_R$ and $\mathcal{S}$.

**The Cartesian chart.** Denote by $\mathcal{D}_{\mathcal{S}_R} : \mathcal{S}_R \to \mathbb{R}^3$ and $\mathcal{D}_{\mathcal{S}} : \mathcal{S} \to \mathbb{R}^3$ the restrictions of $\mathcal{D}$ to the shapes $\mathcal{S}_R$ and $\mathcal{S}$. Thus, $\{(\mathcal{S}_R, \mathcal{D}_{\mathcal{S}_R})\}$ and $\{(\mathcal{S}, \mathcal{D}_{\mathcal{S}})\}$ are atlases. The coordinate representation of $y$ is the following:

$$y_d = \mathcal{D}_{\mathcal{S}} \circ y \circ \mathcal{D}_{\mathcal{S}_R}^{-1} : \mathcal{D}(\mathcal{S}_R) \to \mathbb{R}^3 , \tag{3.1}$$

Here, $\mathcal{D}(\mathcal{S}_R)$ is an open subset of $\mathbb{R}^3$. The relation between $y$ and $y_d$ is illustrated by the diagram

$$\begin{array}{ccc}
\mathcal{E} & & \mathbb{R}^3 \\
\cup & & \cup \\
\mathcal{S} & \xrightarrow{\;\mathcal{D}_{\mathcal{S}}\;} & \mathcal{D}(\mathcal{S}) \\
\uparrow y & & \uparrow y_d \\
\mathcal{S}_R & \xrightarrow{\;\mathcal{D}_{\mathcal{S}_R}\;} & \mathcal{D}(\mathcal{S}_R) \\
\cap & & \cap \\
\mathcal{E} & & \mathbb{R}^3
\end{array}$$

In the following, a triple $(X^i)_{i=1}^3$ denotes local coordinates induced on the shape $\mathcal{S}_R$ by $\mathcal{D}_{\mathcal{S}_R}$, and a triple $(x^i)_{i=1}^3$ denotes local coordinates induced on the shape $\mathcal{S}$ by $\mathcal{D}_{\mathcal{S}}$.

**The curvilinear chart.** Suppose that one introduced different curvilinear coordinate systems $A_1 = \{(U_\alpha, \sigma_\alpha)\}_{\alpha \in I}$ and $A_2 = \{(V_\beta, \psi_\beta)\}_{\beta \in J}$ on $\mathcal{E}$, which are associated with reference and actual shapes[2]. These systems, being induced to the shapes $\mathcal{S}_R$ and $\mathcal{S}$, result in atlases $A_{\mathcal{S}_R} = \{(\mathcal{S}_R \cap U_\alpha, \widetilde{\sigma}_\alpha)\}_{\alpha \in I_{\mathcal{S}_R}}$ and $A_{\mathcal{S}} = \{(\mathcal{S} \cap V_\beta, \widetilde{\psi}_\beta)\}_{\beta \in J_{\mathcal{S}}}$. Here, $\widetilde{\sigma}_\alpha$ and $\widetilde{\psi}_\beta$ are restrictions of the coordinate mappings $\sigma_\alpha$ and $\psi_\beta$. Consider the representation of the deformation $y$ in these curvilinear coordinates. Let $X \in \mathcal{S}_R$.

Choose $\alpha \in I_{\mathcal{S}_R}$, $\beta \in J_{\mathcal{S}}$, such that $X \in U_\alpha$ and $y(X) \in V_\beta$. Denoting $U_{\alpha\beta} = U_\alpha \cap y^{-1}(V_\beta)$ one arrives at the coordinate representation

$$y_c = \widetilde{\psi}_\beta \circ y \circ \widetilde{\sigma}_\alpha^{-1} : \mathbb{R}^3 \supset \sigma_\alpha(U_{\alpha\beta}) \to \mathbb{R}^3 . \tag{3.2}$$

---

2 For example, if $\mathcal{S}_R$ is an open ball and $\mathcal{S}$ is an open cube, then one can introduce spherical and Cartesian coordinates on $\mathcal{E}$.

The relation between $y$ and $y_c$ is shown in the diagram

$$
\begin{array}{ccc}
\mathscr{E} & & \mathbb{R}^3 \\
\vdots & & \vdots \\
\cup & & \cup \\
\vdots & & \vdots \\
y(U_{\alpha\beta}) & \xrightarrow{\quad \widetilde{\psi}_\beta \quad} & \psi_\beta \circ y(U_{\alpha\beta}) \\
\uparrow y & & \uparrow y_c \\
U_{\alpha\beta} & \xrightarrow{\quad \widetilde{\sigma}_\alpha \quad} & \sigma_\alpha(U_{\alpha\beta}) \\
\vdots & & \vdots \\
\cap & & \cap \\
\vdots & & \vdots \\
\mathscr{E} & & \mathbb{R}^3
\end{array}
$$

**Local bases and metrics.** Let $(Q^i)_{i=1}^3$ denote curvilinear coordinates induced on the shape $\mathscr{S}_R$ by $\widetilde{\sigma}_\alpha$ and $(q^i)_{i=1}^3$ denote curvilinear coordinates induced on the shape $\mathscr{S}$. Using the relations established in Chapter 2, Section 2.2 we write

$$\boldsymbol{E}_i = \frac{\partial X^k}{\partial Q^i}\boldsymbol{i}_k \,, \quad E^i = \frac{\partial Q^i}{\partial X^k} i^k \,, \tag{3.3}$$

for the local basis $(\boldsymbol{E}_i)_{i=1}^3$ and the dual basis $(E^i)_{i=1}^3$ on $\mathscr{S}_R$, and

$$\boldsymbol{e}_i = \frac{\partial x^k}{\partial q^i}\boldsymbol{i}_k \,, \quad e^i = \frac{\partial q^i}{\partial x^k} i^k \,, \tag{3.4}$$

for the local basis $(\boldsymbol{e}_i)_{i=1}^3$ and the dual basis $(e^i)_{i=1}^3$ on $\mathscr{S}$. The metric tensor $\boldsymbol{G}$ and its associated, $\overset{*}{\boldsymbol{G}}$, have the decompositions

$$\boldsymbol{G} = G_{ij} E^i \otimes E^j \,, \quad G_{ij} = \boldsymbol{E}_i \cdot \boldsymbol{E}_j \,,$$
$$\overset{*}{\boldsymbol{G}} = G^{ij} \boldsymbol{E}_i \otimes \boldsymbol{E}_j \,,$$

while the metric tensor $\boldsymbol{g}$ and its associated, $\overset{*}{\boldsymbol{g}}$,

$$\boldsymbol{g} = g_{ij} e^i \otimes e^j \,, \quad g_{ij} = \boldsymbol{e}_i \cdot \boldsymbol{e}_j \,,$$
$$\overset{*}{\boldsymbol{g}} = g^{ij} \boldsymbol{e}_i \otimes \boldsymbol{e}_j \,.$$

Here, $[G^{ij}] = [G_{ij}]^{-1}$ and $[g^{ij}] = [g_{ij}]^{-1}$.

**The relation between local bases.** Since the local bases $(\boldsymbol{E}_i)$ and $(\boldsymbol{e}_i)$ in (3.3) and (3.4) take values in the space $\mathscr{V}$, there exists a smooth field $[\Omega^i{}_j] : \mathscr{S}_R \to \mathrm{GL}(3; \mathbb{R})$ of square matrices, such that

$$\boldsymbol{E}_i = \Omega^j{}_i \boldsymbol{e}_j \,.$$

Using the definitions of $(\boldsymbol{E}_i)$ and $(\boldsymbol{e}_i)$, one can obtain the explicit formula for $[\Omega^i{}_j]$. Indeed, since

$$\frac{\partial X^k}{\partial Q^i} = \Omega^j{}_i \frac{\partial x^k}{\partial q^j}\,, \qquad k = 1, 2, 3\,,$$

one obtains

$$\Omega^j{}_i = \frac{\partial q^j}{\partial x^k}\frac{\partial X^k}{\partial Q^i}\,.$$

For the dual bases $(E^i)$ and $(e^i)$ in (3.3) and (3.4), one can similarly write

$$E^i = \mho^i{}_j e^j\,,$$

where $[\mho^i{}_j]\colon \mathcal{S}_R \to \mathrm{GL}(3;\mathbb{R})$ is a smooth field of matrices. From the relation $\langle E^i, \boldsymbol{E}_j\rangle = \delta^i_j$ one obtains $[\mho^i{}_j] = [\Omega^i{}_j]^{-1}$.

**The relation between representations of $y$.** Mappings (3.1) and (3.2) are related as follows:

$$y_d = \mathcal{D}_{\mathcal{S}} \circ \widetilde{\psi}_\beta^{-1} \circ y_c \circ \widetilde{\sigma}_\alpha \circ \mathcal{D}_{\mathcal{S}_R}^{-1}\,. \tag{3.5}$$

Here,

$$\mathcal{D}_{\mathcal{S}_R} \circ \widetilde{\sigma}_\alpha^{-1} : (Q^i) \mapsto (X^i)\,,$$
$$\mathcal{D}_{\mathcal{S}} \circ \widetilde{\psi}_\beta^{-1} : (q^i) \mapsto (x^i)$$

are transition functions.

**A particular case.** Let $A_{\mathcal{S}_R} = \{(U_\alpha, \sigma_\alpha)\}_{\alpha\in I_{\mathcal{S}_R}}$ be a smooth atlas on $\mathcal{S}_R$. The deformation $y$ transfers this atlas to $\mathcal{S}$ in such a way that points $x = y(X)$ have similar coordinates to points $X$. Formally, one can introduce the atlas $A_{\mathcal{S}} = \{(y(U_\alpha), \psi_\alpha)\}_{\alpha\in I_{\mathcal{S}_R}}$ on $\mathcal{S}$. Here, $\psi_\alpha = \sigma_\alpha \circ y^{-1}$. The coordinate representation $y_c$, relative to these atlases, is just the identity map:

$$y_c = \psi_\alpha \circ y \circ \sigma_\alpha^{-1} = \sigma_\alpha \circ \sigma_\alpha^{-1} = \mathrm{Id}\,.$$

**Two descriptions.** Lagrangian (material) and Eulerian (spatial) descriptions are traditional in conventional continuum mechanics. In the material one, the behavior of an individual point $X \in \mathcal{S}_R$ is described. In this case, one uses the mappings defined on $\mathcal{S}_R$. In particular, the actual position of a point $X \in \mathcal{S}_R$ is expressed as a value of a function of $X$, the deformation $y$. In a spatial description, points from $\mathcal{S}$ are treated as independent variables, and one considers fields on $\mathcal{S}$. In particular, one uses the inverse mapping $y^{-1}$ to represent points in the reference shape as values of a function defined on actual positions.

## 3.3 Deformation Gradient and Strain Measures

### 3.3.1 Deformation Gradient

Suppose that $y\colon \mathcal{S}_R \to \mathcal{S}$ is a deformation.

**The deformation gradient.** The *deformation gradient* is the mapping

$$y'\colon \mathcal{S}_R \to \mathrm{Lin}(\mathcal{V};\mathcal{V}), \quad y'\colon X \mapsto y'(X),$$

where linear map $y'(X)$ is the total derivative of $y$ at point $X \in \mathcal{S}_R$. That is, for any $\boldsymbol{h} \in \mathcal{V}$ such that $X + \boldsymbol{h} \in \mathcal{S}_R$, the following expression holds (see Remark 2.3):

$$y(X + \boldsymbol{h}) = y(X) + y'(X)[\boldsymbol{h}] + \boldsymbol{o}(\|\boldsymbol{h}\|) \quad \text{as} \quad \boldsymbol{h} \to \boldsymbol{0}. \tag{3.6}$$

**Remark 3.1.** In the classical works (for example, *cf.* [37, 45, 52]) the deformation gradient is denoted by $\nabla y$. We avoid such notation for the reason that the deformation gradient can be confused with the covariant derivative.

Following classical notation, we will frequently use the symbols $\boldsymbol{F}$ for $y'$ and $\boldsymbol{F}_X$ for $y'(X)$.

In a Cartesian coordinate system $(o, (\boldsymbol{i}_s)_{s=1}^3)$, the deformation $y$ has the representation $y(X) = o + y^s(X)\boldsymbol{i}_s$ for any point $X \in \mathcal{S}_R$, where

$$y^s\colon \mathcal{S}_R \to \mathbb{R},$$

are component functions. Thus, the deformation gradient has the following dyadic decomposition at point $X$:

$$\boldsymbol{F}_X = \left.\frac{\partial x^s}{\partial X^k}\right|_X \boldsymbol{i}_s \otimes \boldsymbol{i}^k. \tag{3.7}$$

In this decomposition,

$$\left.\frac{\partial x^s}{\partial X^k}\right|_X := \lim_{\mathcal{S}_{R;X;\boldsymbol{i}_k} \ni t \to 0} \frac{y^s(X + t\boldsymbol{i}_k) - y^s(X)}{t},$$

are partial derivatives of $y_d$. Here, $\mathcal{S}_{R;X;\boldsymbol{i}_k} = \{t \in \mathbb{R} \mid t \neq 0, X + t\boldsymbol{i}_k \in \mathcal{S}_R\}$ is a punctured neighborhood of the point $t = 0$ underlying in the set $\mathcal{S}_R$. We usually express (3.7) as

$$\boldsymbol{F} = F^s{}_k \boldsymbol{i}_s \otimes \boldsymbol{i}^k, \quad \text{where} \quad F^s{}_k = \frac{\partial x^s}{\partial X^k}.$$

Hereafter, occasionally, we will refer to the components of second-rank tensors as the elements of rectangular matrices. We will denote such matrices by placing the symbol of a tensor into square brackets. For example, $[F^i{}_j]$ denotes the matrix of $\boldsymbol{F}$'s components. In order to establish correspondence between the indexed components and the matrix elements, we adopt the convention: the first index determines a row, while second one determines a column of the matrix. For example,

$$[F^i_{\ j}] = \begin{pmatrix} F^1_{\ 1} & F^1_{\ 2} & F^1_{\ 3} \\ F^2_{\ 1} & F^2_{\ 2} & F^2_{\ 3} \\ F^3_{\ 1} & F^3_{\ 2} & F^3_{\ 3} \end{pmatrix} .$$

It should be noted that the correspondence between second-rank tensors and their matrices is not one-to-one, because the passage from tensors to matrices results in the loss of information about the type of tensor.

We adopt the notation $J$ for the determinant of $\boldsymbol{F}$, *i.e.*,

$$J = \det \boldsymbol{F} . \tag{3.8}$$

By virtue of the injectivity of $y$,

$$\forall X \in \mathcal{S}_R : (J_X \neq 0) .$$

In other words, the deformation gradient $\boldsymbol{F}$ can be considered as a field $\boldsymbol{F} : \mathcal{S}_R \to \text{Inv}(\mathscr{V}; \mathscr{V})$ of invertible linear mappings.

Let $X$ and $Y$ be any points of $\mathcal{S}_R$. The difference $Y - X$ is a vector, which, if $Y$ is "close" to $X$, may be considered as an infinitesimal fiber in the reference state. After deformation, this fiber translates to the fiber $y(Y) - y(X)$. The equality (3.6) states that the deformed fiber $y(Y) - y(X)$ with accuracy up to $\boldsymbol{o}(\|Y - X\|)$ is equal to $\boldsymbol{F}_X[Y - X]$.

In the following, we assume that $y$ satisfies the postulate:

($y_4$) **Orientation preserving.** For any non-coplanar fibers in $\mathcal{S}_R$, under action of $\boldsymbol{F}_X$ an ordered triple of tangent vectors to them transforms to another ordered triple with the same orientation.

In mathematical language, this requirement can be stated as

$$\forall X \in \mathcal{S}_R : (J_X > 0) .$$

**The deformation gradient as three 1-forms.** The deformation gradient can be viewed as a triple of 1-forms [8, 60]. Suppose that $y : \mathcal{S}_R \to \mathcal{S}$ is a deformation. Each of the sets $\mathcal{S}_R$, $\mathcal{S}$ is an open submanifold of $\mathscr{E}$. Let us identify the tangent spaces $T_X\mathcal{S}_R$, $T_x\mathcal{S}$ with the tangent spaces $T_X\mathscr{E}$, $T_x\mathscr{E}$ (in Chapter 4, Section 4.5.4 this identification is considered in more detail), for all $X \in \mathcal{S}_R$, $x \in \mathcal{S}$. Then, one can identify the deformation gradient $\boldsymbol{F}$ with the tangent map $Ty$. Thus, $\boldsymbol{F} = F^i_{\ j} \partial_{x^i} \otimes dX^j$, and one arrives at the relation

$$\boldsymbol{F} = \partial_{x^i} \otimes F^i , \quad \text{where} \quad F^i = F^i_{\ j} dX^j , \quad i = 1, 2, 3 . \tag{3.9}$$

**The inverse deformation gradient.** The inverse deformation gradient $(y^{-1})'(x)$ is equal to $\boldsymbol{F}_X^{-1}$. In the Cartesian coordinate system $(o, (\boldsymbol{i}_s)_{s=1}^3)$, the following dyadic decomposition holds:

$$\boldsymbol{F}_X^{-1} = \left.\frac{\partial X^m}{\partial x^n}\right|_{y(X)} \boldsymbol{i}_m \otimes \boldsymbol{i}^n .$$

**Deformations of higher order.** Since $y$ is smooth, one can use the Taylor formula [44]:

$$y(X+\boldsymbol{h}) = y(X) + y'(X)[\boldsymbol{h}] + \cdots + \frac{y^{(r)}(X)[\boldsymbol{h}^r]}{r!} + \boldsymbol{o}(\|\boldsymbol{h}\|^r) \quad \text{as} \quad \boldsymbol{h} \to \boldsymbol{0},$$

where the $k$-linear map[3] $y^{(k)}(X) \in \mathrm{Lin}_k(\mathscr{V}, \ldots, \mathscr{V}; \mathscr{V})$ is a $k$-th order total derivative ($k = 1, \ldots, r$) of $y$ at point $X$, and $y^{(k)}(X)[\boldsymbol{h}^k] := y^{(k)}(X)[\boldsymbol{h}, \ldots, \boldsymbol{h}]$. Thus, one has deformation gradients of high order:

$$y(X+\boldsymbol{h}) = y(X) + \boldsymbol{F}_X[\boldsymbol{h}] + \cdots + \frac{\boldsymbol{F}_X^{(r)}[\boldsymbol{h}^r]}{r!} + \boldsymbol{o}(\|\boldsymbol{h}\|^r) \quad \text{as} \quad \boldsymbol{h} \to \boldsymbol{0}, \tag{3.10}$$

where $\boldsymbol{F}_X^{(k)} := y^{(k)}(X)$ for all $k = 1, \ldots, r$.

Referring again to the material fibers, one can say that deformed fiber $y(Y) - y(X)$ with accuracy up to $\boldsymbol{o}(\|Y - X\|^r)$ is defined by the expression

$$\boldsymbol{F}_X[Y - X] + \cdots + \frac{\boldsymbol{F}_X^{(r)}[Y - X, \ldots, Y - X]}{r!}.$$

The choice of $r$ defines the desired accuracy and the corresponding theory.

### 3.3.2 Strain Measures

Since $\mathscr{V}$ is Euclidean, one can obtain the expression for length of the deformed fiber. Let $X \in \mathcal{S}_R$ be a point in the reference shape. For a point $Y \in \mathcal{S}_R$ that lies near $X$, one has the expression

$$y(Y) - y(X) = \boldsymbol{F}_X[Y - X] + \boldsymbol{o}(\|Y - X\|),$$

for the deformed fiber, $y(Y) - y(X)$. After multiplying this expression by itself, one arrives at

$$\|y(Y) - y(X)\|^2 = \boldsymbol{F}_X[Y - X] \cdot \boldsymbol{F}_X[Y - X] + \boldsymbol{o}(\|Y - X\|^2) \quad \text{as} \quad Y \to X.$$

Using the definition (12.3) of the transposed linear map, which in the case of $\boldsymbol{F}_X$ takes the form

$$\forall \boldsymbol{u}, \boldsymbol{v} \in \mathscr{V}: \left(\boldsymbol{F}_X[\boldsymbol{u}] \cdot \boldsymbol{v} = \boldsymbol{u} \cdot \boldsymbol{F}_X^{\mathrm{T}}[\boldsymbol{v}]\right), \tag{3.11}$$

where $\boldsymbol{F}_X^{\mathrm{T}} \in \mathrm{Lin}(\mathscr{V}; \mathscr{V})$, one obtains the formula

$$\|y(Y) - y(X)\|^2 = (Y - X) \cdot \boldsymbol{F}_X^{\mathrm{T}} \boldsymbol{F}_X[Y - X] + \boldsymbol{o}(\|Y - X\|^2) \quad \text{as} \quad Y \to X.$$

3 $k$-linear maps are defined in Chapter 12, Section 12.5.2.

The tensor $\boldsymbol{C}_X = \boldsymbol{F}_X^{\mathrm{T}}\boldsymbol{F}_X \in \mathrm{Sym}(\mathscr{V})$ is called the *right Cauchy–Green strain*. Thus, the last equality may be written in the form

$$\|y(Y) - y(X)\|^2 = (Y - X) \cdot \boldsymbol{C}_X[Y - X] + \boldsymbol{o}(\|Y - X\|^2) \quad \text{as} \quad Y \to X\,. \tag{3.12}$$

With an accuracy up to $\boldsymbol{o}(\|Y - X\|^2)$, the square of deformed fiber length is equal to $(Y - X) \cdot \boldsymbol{C}_X[Y - X]$.

Tensors $\boldsymbol{C}_X, X \in \mathcal{S}_R$ define the smooth mapping (tensor field)

$$\boldsymbol{C}\colon \mathcal{S}_R \to \mathrm{Sym}(\mathscr{V})\,, \quad X \mapsto \boldsymbol{C}_X\,,$$

called the *right Cauchy–Green tensor field* (right Cauchy–Green strain).

Consider the inverse deformation, $y^{-1}$. Let $x = y(X)$. One has

$$y^{-1}(y) - y^{-1}(x) = \boldsymbol{F}_X^{-1}[y - x] + \boldsymbol{o}(\|y - x\|) \quad \text{as} \quad y \to x\,.$$

Multiplying this relation by itself and denoting $\boldsymbol{B}_X = \boldsymbol{F}_X\boldsymbol{F}_X^{\mathrm{T}}$, one obtains

$$\|y^{-1}(y) - y^{-1}(x)\|^2 = (y - x) \cdot \boldsymbol{B}_X^{-1}[y - x] + \boldsymbol{o}(\|y - x\|^2) \quad \text{as} \quad y \to x\,. \tag{3.13}$$

With an accuracy up to $\boldsymbol{o}(\|y - x\|^2)$, the square of reference fiber length is equal to $(y - x) \cdot \boldsymbol{B}_X^{-1}[y - x]$. The tensor $\boldsymbol{B}_X = \boldsymbol{F}_X\boldsymbol{F}_X^{\mathrm{T}}$ is called the *left Cauchy–Green strain tensor*. One has the field

$$\boldsymbol{B}\colon \mathcal{S}_R \to \mathrm{Sym}(\mathscr{V})\,, \quad X \mapsto \boldsymbol{B}_X\,,$$

called the *left Cauchy–Green tensor field* (left Cauchy–Green strain).

Cauchy–Green measures define the following measures: $\boldsymbol{C}^{-1}$ (Finger strain) and $\boldsymbol{B}^{-1}$ (Almansi strain).

### 3.3.3 Deformation Gradient in Curvilinear Coordinates

In the following paragraph, we obtain curvilinear coordinate representations using the results established in Section 3.2. Let $y\colon \mathcal{S}_R \to \mathcal{S}$ be a deformation.

**The general case.** Let $(Q^i)_{i=1}^3$ be curvilinear coordinates on $\mathcal{S}_R$ and $(q^i)_{i=1}^3$ be curvilinear coordinates on $\mathcal{S}$. Using the relation (3.5) between the Cartesian representation $y_d$ (3.1) of the deformation $y$ and the curvilinear representation $y_c$ (3.2) of the same deformation $y$, the chain rule and relations (3.3) and (3.4), from (3.7) one obtains

$$\boldsymbol{F}_X = \left.\frac{\partial q^i}{\partial Q^j}\right|_{\tilde{\sigma}_\alpha(X)} \boldsymbol{e}_i|_{y(X)} \otimes \boldsymbol{E}^j|_X\,. \tag{3.14}$$

Here, $(\partial q^i)/(\partial Q^j)$ are partial derivatives of $y_c\colon (Q^1, Q^2, Q^3) \mapsto (q^1, q^2, q^3)$. One should note from the relation (3.14) that $\boldsymbol{F}_X$ is not a tensor in the classical sense. It is an example of a so-called two-point tensor, which is considered in detail in Chapter 5, Section 5.4.

To obtain the relation for the transpose map $\boldsymbol{F}^{\mathrm{T}}$ let $\boldsymbol{F} = F^i{}_j \boldsymbol{e}_i \otimes E^j$ and $\boldsymbol{F}^{\mathrm{T}} = (\boldsymbol{F}^{\mathrm{T}})^k{}_l \boldsymbol{E}_k \otimes e^l$. Applying $\boldsymbol{u} = \boldsymbol{E}_j$, $\boldsymbol{v} = \boldsymbol{e}_l$ in (3.11), we obtain

$$g_{il} F^i{}_j = (\boldsymbol{F}^{\mathrm{T}})^k{}_l G_{jk} \,.$$

Multiplying both sides of this relation by $G^{js}$ and summing over $j$, we finally obtain

$$(\boldsymbol{F}^{\mathrm{T}})^s{}_l = g_{il} G^{js} F^i{}_j \,. \tag{3.15}$$

The expression (3.14) can be represented in dyadic forms like $\boldsymbol{e}_i \otimes e^j$ and $\boldsymbol{E}_i \otimes E^j$ as follows. One needs to use relations $\boldsymbol{E}_i = \Omega^j{}_i \boldsymbol{e}_j$ and $E^i = \mho^i{}_j e^j$, established in Section 3.2. Thus,

$$\boldsymbol{F}_X = \mho^j{}_k|_X \left.\frac{\partial q^i}{\partial Q^j}\right|_{\tilde{\sigma}_\alpha(X)} \boldsymbol{e}_i|_{y(X)} \otimes e^k|_{y(X)} \,,$$

$$\boldsymbol{F}_X = \mho^k{}_i|_X \left.\frac{\partial q^i}{\partial Q^j}\right|_{\tilde{\sigma}_\alpha(X)} \boldsymbol{E}_k|_X \otimes E^j|_X \,.$$

**A particular case.** If $y_c = \mathrm{Id}$, then $(\partial q^i)/(\partial Q^j) = \delta^i_j$, and one has

$$F^i{}_j = \delta^i_j \,, \quad (\boldsymbol{F}^{\mathrm{T}})^s{}_l = g_{il} G^{is} \,.$$

Since $\boldsymbol{E}^i = G^{is} \boldsymbol{E}_s$ and $e_i = g_{il} e^l$, one obtains the following decompositions:

$$\boldsymbol{F}_X = \boldsymbol{e}_i|_{y(X)} \otimes E^i|_X \,, \quad \boldsymbol{F}^{\mathrm{T}}_X = \boldsymbol{E}^i|_X \otimes e_i|_{y(X)} \,.$$

### 3.3.4 Strain Measures in Curvilinear Coordinates

We obtain the representation for the right Cauchy–Green strain in curvilinear coordinates. Since $\boldsymbol{C} = \boldsymbol{F}^{\mathrm{T}} \boldsymbol{F}$, one has $\boldsymbol{C} = (\boldsymbol{F}^{\mathrm{T}})^k{}_i F^i{}_j \boldsymbol{E}_k \otimes E^j$. This relation and (3.15) imply

$$\boldsymbol{C}_X = g_{si}|_{y(X)} G^{kl}|_X \left.\frac{\partial q^s}{\partial Q^l}\right|_{\tilde{\sigma}_\alpha(X)} \left.\frac{\partial q^i}{\partial Q^j}\right|_{\tilde{\sigma}_\alpha(X)} \boldsymbol{E}_k|_X \otimes E^j|_X \,.$$

Analogously, one can obtain the representation for the left Cauchy–Green strain. Since $\boldsymbol{B} = \boldsymbol{F}\boldsymbol{F}^{\mathrm{T}}$, one obtains $\boldsymbol{B} = (\boldsymbol{F}^{\mathrm{T}})^k{}_l F^i{}_k \boldsymbol{e}_i \otimes e^l$. From the latter, it follows that

$$\boldsymbol{B}_X = g_{sl}|_{y(X)} G^{jk}|_X \left.\frac{\partial q^s}{\partial Q^j}\right|_{\tilde{\sigma}_\alpha(X)} \left.\frac{\partial q^i}{\partial Q^k}\right|_{\tilde{\sigma}_\alpha(X)} \boldsymbol{e}_i|_{y(X)} \otimes e^l|_{y(X)} \,.$$

Cauchy–Green strains are related with metrics by the following geometric relations. Applying the musical isomorphism $(\cdot)^{\flat_b}$, associated with the metric tensor $\boldsymbol{G}$, to

$\boldsymbol{C}$ we obtain the field of bilinear forms $\boldsymbol{C}^{\flat_b}$,

$$\boldsymbol{C}_X^{\flat_b} = g_{si}|_{y(X)} \left.\frac{\partial q^s}{\partial Q^m}\right|_{\tilde{\sigma}_\alpha(X)} \left.\frac{\partial q^i}{\partial Q^j}\right|_{\tilde{\sigma}_\alpha(X)} E^m|_X \otimes E^j|_X \,.$$

The latter expression implies that

$$\boldsymbol{C}^{\flat_b} = y^* \boldsymbol{g} \,.$$

Let $(\cdot)^{\sharp_p}$ denote the musical isomorphism associated with the metric tensor $\boldsymbol{g}$. For the tensor $\boldsymbol{B}$, we have

$$\boldsymbol{B}^{\sharp_p} = G^{jk}|_X \left.\frac{\partial q^m}{\partial Q^j}\right|_{\tilde{\sigma}_\alpha(X)} \left.\frac{\partial q^i}{\partial Q^k}\right|_{\tilde{\sigma}_\alpha(X)} \boldsymbol{e}_i|_{y(X)} \otimes \boldsymbol{e}_m|_{y(X)} \,.$$

Thus,

$$\boldsymbol{B}^{\sharp_p} = y_*(\overset{*}{\boldsymbol{G}}) \,.$$

## 3.4 Displacement Field

For any point $X \in \mathcal{S}_R$ and a deformation $y$, a *displacement vector* is a vector $\boldsymbol{u}(X) := y(X) - X$. Thus, one has the smooth mapping

$$\boldsymbol{u}\colon \mathcal{S}_R \to \mathcal{V} \,, \quad \boldsymbol{u}\colon X \mapsto y(X) - X \,,$$

called a *displacement vector field.*

The deformation gradient $\boldsymbol{F}$ is related with the displacement vector field $\boldsymbol{u}$ by the following relation:

$$\boldsymbol{F}_X = \boldsymbol{u}'(X) + \boldsymbol{I} \,, \tag{3.16}$$

where $\boldsymbol{I} \in \mathrm{Lin}(\mathcal{V};\mathcal{V})$ is *identity operator, i.e.,*

$$\boldsymbol{I} := \mathrm{Id}_{\mathcal{V}} \,.$$

The total derivative $\boldsymbol{u}'(X)$ is called the *displacement gradient*, which is often denoted by $\nabla\boldsymbol{u}(X)$ and may be used in the classical theory instead of[4] $\boldsymbol{F}_X$.

The right and left Cauchy–Green strain measures $\boldsymbol{C}_X = \boldsymbol{F}_X^{\mathrm{T}}\boldsymbol{F}_X$ and $\boldsymbol{B}_X = \boldsymbol{F}_X\boldsymbol{F}_X^{\mathrm{T}}$ can be represented in terms of displacements. Using the expression (3.16), one obtains

$$\begin{aligned} \boldsymbol{C}_X &= \boldsymbol{I} + \boldsymbol{u}'(X) + [\boldsymbol{u}'(X)]^{\mathrm{T}} + [\boldsymbol{u}'(X)]^{\mathrm{T}}\boldsymbol{u}'(X) \,, \\ \boldsymbol{B}_X &= \boldsymbol{I} + \boldsymbol{u}'(X) + [\boldsymbol{u}'(X)]^{\mathrm{T}} + \boldsymbol{u}'(X)[\boldsymbol{u}'(X)]^{\mathrm{T}} \,. \end{aligned} \tag{3.17}$$

4 This is not true in the case of non-affine physical space. In the latter one does not have machinery to define difference between points.

## 3.5 Motion

A *motion* is an one-parametric family of deformations: $M = \{y_t\}_{t\in\mathbb{T}}$, where $\mathbb{T} \subset \mathbb{R}$ is an open interval, and $y_t : \mathcal{S}_R \to \mathcal{S}_t$ is a deformation. Choosing a point $X \in \mathcal{S}_R$, one can define the curve

$$\chi_{M;X} : \mathbb{T} \to \mathcal{E}\,, \quad \chi_{M;X}(t) := y_t(X)\,,$$

which represents a spatial trajectory of the point $X$. It is convenient to consider the motion $M$ as the mapping

$$y : \mathcal{S}_R \times \mathbb{T} \to \mathcal{E}\,, \quad y(X, t) := y_t(X)\,.$$

of two variables. Assume that for any $X \in \mathcal{S}_R$ and $t \in \mathbb{T}$ the partial mappings

$$y(\cdot, t) : \mathcal{S}_R \to \mathcal{E}\,, \quad y(X, \cdot) : \mathbb{T} \to \mathcal{E}$$

are smooth. In this case, the motion $M$ is called a *smooth motion*. For the smooth motion, one can define the deformation gradient

$$\boldsymbol{F} : \mathcal{S}_R \times \mathbb{T} \to \mathrm{Lin}(\mathcal{V}; \mathcal{V})\,, \quad \boldsymbol{F}(X, t) := \frac{\partial y(X, t)}{\partial X}\,,$$

as well as the fields of velocity $\boldsymbol{V}$ and acceleration $\boldsymbol{A}$,

$$\boldsymbol{V} : \mathcal{S}_R \times \mathbb{T} \to \mathcal{V}\,, \quad \boldsymbol{V}(X, t) := \lim_{\mathbb{T}_t \ni s \to 0} \frac{y(X, t+s) - y(X, t)}{s}\,,$$
$$\boldsymbol{A} : \mathcal{S}_R \times \mathbb{T} \to \mathcal{V}\,, \quad \boldsymbol{A}(X, t) := \lim_{\mathbb{T}_t \ni s \to 0} \frac{\boldsymbol{V}(X, t+s) - \boldsymbol{V}(X, t)}{s}\,,$$

where $\mathbb{T}_t = \{s \in \mathbb{R} \mid (s \neq 0) \wedge (t + s \in \mathbb{T})\}$.

## 3.6 Compatibility Conditions

The relation (3.9) shows that $\boldsymbol{F}$ can be considered as a triple of vector-valued 1-forms on $\mathcal{S}_R$. This observation allows one to involve such terms as *exact form*, *closed form*, *etc.*, to study the compatibility question.

### 3.6.1 Review on de Rham Cohomology

Let $\mathfrak{M}$ be a smooth $s$-manifold. A smooth differential $k$-form $\omega \in \Omega^k(\mathfrak{M})$ is *closed*, if $d\omega = 0$, and is *exact*, if $\omega = d\eta$ for some smooth differential $(k-1)$-form $\eta \in \Omega^{k-1}(\mathfrak{M})$. Since $d \circ d = 0$ one has that each exact form is closed. The converse assertion is not true, in general. Its truth value depends on the topology of the underlying manifold and is studied in the framework of de Rham cohomology theory.

Let $1 \le k \le s$. Denote

$$Z^k(\mathfrak{M}) = \ker(d\colon \Omega^k(\mathfrak{M}) \to \Omega^{k+1}(\mathfrak{M})) ,$$
$$B^k(\mathfrak{M}) = \mathrm{Im}(d\colon \Omega^{k-1}(\mathfrak{M}) \to \Omega^k(\mathfrak{M})) .$$

The vector space $Z^k(\mathfrak{M})$ is the space of all closed $k$-forms, and the vector space $B^k(\mathfrak{M})$ is the vector space of all exact $k$-forms. Since $B^k(\mathfrak{M}) \subset Z^k(\mathfrak{M})$, one can introduce the following equivalence relation:

$$\forall \omega, \eta \in \Omega^k(\mathfrak{M}): \quad (\omega \sim \eta) \Leftrightarrow (\exists \tau \in \Omega^{k-1}(\mathfrak{M}): \quad \omega - \eta = d\tau) .$$

The corresponding quotient vector space is denoted by $H^k_{dR}(\mathfrak{M}) := Z^k(\mathfrak{M})/B^k(\mathfrak{M})$ and is referred to as the *de Rham cohomology group in degree p of* $\mathfrak{M}$ [3, 61].

By the construction, the statement "each closed $k$-form on $\mathfrak{M}$ is exact" is equivalent to the equality $H^k_{dR}(\mathfrak{M}) = 0$. In particular, one has (see [3], for instance):

- If $\mathfrak{M}$ is a contractible smooth manifold, *i.e.*, $\mathrm{Id}_{\mathfrak{M}}$ is smoothly homotopic[5] to a constant map, then $H^k_{dR}(\mathfrak{M}) = 0$;
- If $\mathfrak{M}$ is smooth manifold, then each point of $\mathfrak{M}$ has a neighborhood in which every closed form is exact.

### 3.6.2 Necessary and Sufficient Conditions for Compatibility

**Compatibility and exactness.** Let $y\colon \mathcal{S}_R \to \mathcal{S}$ be a deformation. Since each component mapping $y^i$, $i = 1, 2, 3$, is a smooth real-valued function, one can apply the exterior differential to it. This implies

$$dy^i = F^i \in \Omega^1(\mathcal{S}_R) , \quad i = 1, 2, 3 .$$

So, the obtained $F^i$ are exact forms on $\mathcal{S}_R$. Conversely, if $F^i \in \Omega^1(\mathcal{S}_R)$, $i = 1, 2, 3$, are some exact 1-forms, then one can find three smooth functions $y^i\colon \mathcal{S}_R \to \mathbb{R}$, such that $F^i = dy^i$. These functions can be considered as component functions of some smooth mapping $y$. For this mapping, $Ty = \partial_{x^i} \otimes F^i$.

Let $F^i \in \Omega^1(\mathcal{S}_R)$, $i = 1, 2, 3$, be 1-forms. If there exists a smooth mapping $y\colon \mathcal{S}_R \to \mathcal{E}$, for which $F^i = dy^i$, then these forms are called *compatible*. One obtains the following necessary and sufficient condition: 1-forms $F^i \in \Omega^1(\mathcal{S}_R)$, $i = 1, 2, 3$, are compatible *iff* they all are exact.

5 Two smooth mappings $f_0$ and $f_1$ between smooth manifolds $\mathfrak{M}$ and $\mathfrak{N}$ are smoothly homotopic if there exists a smooth mapping $F\colon \mathfrak{M} \times [0, 1] \to \mathfrak{N}$, such that $F(\cdot, i) = f_i$, $i = 0, 1$.

**Compatibility and contractibility.** The question of exactness is intimately related with the inner properties of the topological space $\mathcal{S}_R$. Suppose that $\mathcal{S}_R$ is a contractible space, *i.e.*; intuitively, it can be continuously shrunk to a point. Then, $H^1_{dR}(\mathcal{S}_R) = 0$ and the exactness can be replaced by the requirement of closeness: 1-forms $F^i$, $i = 1, 2, 3$ are compatible *iff* they all are closed. Thus, one needs to check that $dF^i = 0, i = 1, 2, 3$. In classical mechanics, this condition is written as curl $\boldsymbol{F} = \boldsymbol{0}$ or, using $\nabla$, as $\nabla \times \boldsymbol{F} = \boldsymbol{0}$.

**Compatibility in the general case.** The conditions $dF^i = 0, i = 1, 2, 3$ only become necessary and lose their sufficiency status when $\mathcal{S}_R$ has "holes" and other pathologies that break its contractibility. The following theorem defines the conditions of compatibility.

**Claim.** The necessary and sufficient conditions for compatibility of $F^i$ are:
(i) $dF^i = 0, i = 1, 2, 3$;
(ii) $\int_{c_k} F^i = 0, i = 1, 2, 3, k = 1, \dots, \beta_1(\mathcal{S}_R)$. Here, $\beta_1(\mathcal{S}_R)$ is a Betti number, calculated for $\mathcal{S}_R$, and $c_k$ are generators of[6] $H_1(\mathcal{S}_R; \mathbb{R})$.

Proof of this claim is given in [63].

**Remark 3.2.** A Betti number $\beta_k(\mathcal{S}_R)$ is a topological invariant that refers to the number of $k$-dimensional holes. Let us briefly give the formal description of Betti numbers [62, 64].

Let $s \in \mathbb{N}$, then denote by $\Delta_s \subset \mathbb{R}^s$ the standard $s$-simplex, formed by vectors $(e_i)_{i=0}^s$, where $e_0 = 0$ and for $i \in \{1, \dots, s\}$, $e_i = (0, \dots, 1, \dots 0)$, in which 1 stands in the $s$th place. That is, $\Delta_s$ has vertices in the zero vector and in vectors of the standard basis. Let $\mathfrak{M}$ be a topological space. A continuous map $\sigma\colon \Delta_s \to \mathfrak{M}$ is called a singular $s$-simplex in $\mathfrak{M}$. If $s = 0$, then the image, $\sigma(\Delta_0)$, is just a point in $\mathfrak{M}$. If $s = 1$, then $\Delta_1 = [0, 1]$ is a closed interval, and $\sigma$ is a path in $\mathfrak{M}$.

The set $S_s(\mathfrak{M})$ of all singular $s$-simplices in $\mathfrak{M}$ can serve as a base for constructing a group. Let $\sigma \in S_s(\mathfrak{M})$. Then, one can generate a group $F(\sigma)$ as follows: $F(\sigma) = \{\sigma\} \times \mathbb{Z}$, and the multiplication operation is defined as $(\sigma, l)(\sigma, k) := (\sigma, l + k)$. Thus, we arrive at the sequence $(F(\sigma))_{\sigma \in S_s(\mathfrak{M})}$ of groups. Any finite tuples $(g_1, \dots, g_l)$ and $(h_1, \dots, h_k)$, where $g_i$, $h_j$ belong to some $F(\sigma)$, can be multiplied just by concatenation: $(g_1, \dots, g_l)(h_1, \dots, h_k) := (g_1, \dots, g_l, h_1, \dots, h_k)$. This operation is associative, and the sequence of the length zero is its identity element. However, it does not satisfy the condition of inverses. Such an issue can be solved (see [64] for details) and we arrive, finally, at the group $C_s(\mathfrak{M})$ on the set $S_s(\mathfrak{M})$. This group is called the singular chain group in dimension $s$. Any of its elements that can be written as a formal linear combination of singular simplices with integer coefficients are called a singular $s$-chain in $\mathfrak{M}$.

---

6 Here $H_1(\mathcal{S}_R; \mathbb{R})$ is first homology group with real coefficients [62].

Let us comment on the intuitive view on singular chains. If chain "closes up" on itself and is not a "boundary" of any chain with higher dimension, then it must surround a "hole". Thus, in the framework of this view, a chain is an object that allows detection of "holes". The boundary of a chain is composed of faces of simplices that constitute the chain. From a formal point of view, for each $k = 0, 1, \ldots, s$, we define an affine map $F_{k,s}\colon \Delta_{s-1} \to \Delta_s$, which works as follows:

$$F_{k,s}(e_i) := \begin{cases} e_i, & i \in \{0, \ldots, i-1\}\,, \\ e_{i+1}, & i \in \{i, \ldots, s-1\}\,. \end{cases}$$

By the definition, $F_{k,s}$ maps $\Delta_{s-1}$ in the boundary face of $\Delta_s$ that is situated opposite to the vertex $e_i$. Now, let $\sigma\colon \Delta_s \to \mathfrak{M}$ be a singular simplex, which defines a $(s-1)$-chain $\partial_p \sigma$ by the relation

$$\partial_s \sigma := \sum_{k=0}^{s} (-1)^k \sigma \circ F_{k,s}\,.$$

The chain $\partial_s \sigma$ is called the boundary of $\sigma$, and the homomorphism

$$\partial_s\colon C_s(\mathfrak{M}) \to C_{s-1}(\mathfrak{M})$$

is the boundary operator. An $s$-chain $c$ is called a cycle if $\partial_s c = 0$ and a boundary if $c = \partial_{s+1} c_1$ for some $(s+1)$ chain $c_1$. Consider an example. If $\sigma\colon \Delta_1 \to \mathfrak{M}$, then $\partial\sigma = \sigma(1) - \sigma(0)$. If $\sigma$ is a cycle, then $\sigma(0) = \sigma(1)$, and $\sigma$ is a closed path.

Denote $Z_s(\mathfrak{M}) = \ker \partial_s$, that is, a group of all $s$-cycles, and $B_s(\mathfrak{M}) = \partial_{s+1}(C_{s+1}(\mathfrak{M}))$, that is, a group of images. Both $Z_s(\mathfrak{M})$ and $B_s(\mathfrak{M})$ are subgroups of $C_s(\mathfrak{M})$. The operation $\partial_s$ satisfies the property $\partial_s \circ \partial_{s+1} = 0$. Thus, $B_s(\mathfrak{M})$ is a subgroup of $Z_s(\mathfrak{M})$. We arrive at the following definition; the $s$th singular homology group of $\mathfrak{M}$ is the quotient group

$$H_s(\mathfrak{M}) := Z_s(\mathfrak{M})/B_s(\mathfrak{M})\,.$$

Appealing again to intuitive reasonings, one should say that $H_s(\mathfrak{M}) = 0$ iff no "$s$-dimensional holes". That is, every $s$-cycle is a boundary of some $(s+1)$-chain.

Finally, the integer $\beta_s(\mathfrak{M}) := \operatorname{rank}(H_s(\mathfrak{M}))$ is called the $s$th Betti number of $\mathfrak{M}$. In this definition, the item $\operatorname{rank}(H_s(\mathfrak{M}))$ is a number of elements in any finite basis for $H_s(\mathfrak{M})$. That is, any element of $H_s(\mathfrak{M})$ can be decomposed on a formal linear combination of $\beta_s(\mathfrak{M})$ elements.

**Local compatibility conditions.** The physical space that contains shapes is Euclidean. Thus, its Riemannian curvature tensor vanishes, $\mathfrak{R}(\boldsymbol{g}) = 0$, where $\boldsymbol{g}$ is a metric tensor on $\mathscr{E}$. The reference shape is also Euclidean. Thus, the pullback of the curvature vanishes as well: $y^*\mathfrak{R}(\boldsymbol{g}) = \mathbf{0}$. One has

$$y^*\mathfrak{R}(\boldsymbol{g}) = \mathfrak{R}(y^*\boldsymbol{g}) = \mathfrak{R}(\boldsymbol{C}^\flat)\,,$$

where $\boldsymbol{C}^\flat$ is a field of bilinear forms obtained by lowering index of $\boldsymbol{C}$. Therefore, a necessary condition for compatibility of $\boldsymbol{C}$ is $\mathfrak{R}(\boldsymbol{C}^\flat) = \mathbf{0}$. As was shown in [60], this condition is locally sufficient as well.

**Remark 3.3.** Applying the musical isomorphism $(\cdot)^\flat$ to the "first leg", we arrive at the following representation:

$$\boldsymbol{C}_X^\flat = C_{mn}|_X \boldsymbol{i}^m \otimes \boldsymbol{i}^n \,, \quad \text{where} \quad C_{mn}|_X = \sum_{k=1}^{3} \left.\frac{\partial x^k}{\partial X^m}\right|_X \left.\frac{\partial x^k}{\partial X^n}\right|_X \,.$$

In componentwise form, the equality $\mathfrak{R}(\boldsymbol{C}^\flat) = \boldsymbol{0}$ can be written in Cartesian coordinates as

$$\mathfrak{R}(\boldsymbol{C}^\flat) = \frac{1}{2}(\partial_k\partial_j C_{im} + \partial_i\partial_m C_{jk} - \partial_m\partial_j C_{ik} - \partial_i\partial_k C_{jm}) + (\boldsymbol{C}^{-1})^{rs}(A_{jkr}A_{ims} - A_{jmr}A_{iks}) = 0\,, \tag{3.18}$$

where

$$A_{ijk} = A_{jik} = \frac{1}{2}(\partial_j C_{ik} + \partial_i C_{jk} - \partial_k C_{ij})\,.$$

In infinitesimal strain approximation, the quadric terms in (3.18) are neglected, and this condition can be written in more simple form, namely the Saint-Venant equation:

$$\partial_k\partial_j \varepsilon_{im} + \partial_i\partial_m \varepsilon_{jk} - \partial_m\partial_j \varepsilon_{ik} - \partial_i\partial_k \varepsilon_{jm} = 0\,.$$

The version of this equation in curvilinear coordinates can be found in [65]. Using $\nabla$, one can write

$$\operatorname{Ink}(\boldsymbol{\varepsilon}) = \boldsymbol{0}\,, \quad \operatorname{Ink}(\boldsymbol{\varepsilon}) = \operatorname{curl}\operatorname{curl}^{\mathrm{T}} \boldsymbol{\varepsilon} = \nabla \times (\nabla \times \boldsymbol{\varepsilon})^{\mathrm{T}}\,.$$

Here, $\boldsymbol{\varepsilon}$ is an infinitesimal strain, $\boldsymbol{\varepsilon} := [\nabla \boldsymbol{u}]^{\mathrm{sym}}$, and $\boldsymbol{u}$ is a displacement vector field.

In classical works on conventional elasticity, the integrability conditions are formulated in terms of fields of infinitesimal strains, $\boldsymbol{\varepsilon} = [\nabla \boldsymbol{u}]^{\mathrm{sym}}$, and vortices, $\boldsymbol{\omega} = [\nabla \boldsymbol{u}]^{\mathrm{asym}}$, in the following manner:

$$\nabla \times \boldsymbol{\varepsilon} - (\nabla \vec{\boldsymbol{\omega}})^{\mathrm{T}} = \boldsymbol{0}\,,$$

where $\vec{\boldsymbol{\omega}}$ is a vector field, associated with $\boldsymbol{\omega}$. Raising the differentiation order in the left part of this equality, one obtains independent conditions for $\boldsymbol{\varepsilon}$ and $\boldsymbol{\omega}$. These are the Saint-Venant equation

$$\nabla \times (\nabla \times \boldsymbol{\varepsilon})^{\mathrm{T}} = \boldsymbol{0}\,,$$

and

$$\nabla \cdot \vec{\boldsymbol{\omega}}\,.$$

At the same time, these conditions are necessary, but insufficient; when these conditions are satisfied, the displacement field corresponding to the given $\boldsymbol{\varepsilon}$ and $\boldsymbol{\omega}$ is found up to a rigid motion and the gradient of some scalar function [66]. If this arbitrariness is leveled by additional conditions (for example, boundary conditions), then the displacements are uniquely determined by Cesaro's formulae. In terms of the bend-twist tensor[7] $\boldsymbol{\kappa} = \nabla\vec{\boldsymbol{\omega}}$ one can formulate more general compatibility conditions [66]:

$$\nabla \times \boldsymbol{\kappa} = \mathbf{0}\,, \qquad \boldsymbol{\kappa} = -(\nabla \times \boldsymbol{\varepsilon})^{\mathrm{T}}\,. \tag{3.19}$$

These conditions are necessary and sufficient. The latter condition explicitly shows the relation between infinitesimal strains and vortices.

**Remark 3.4.** It is possible to derive the compatibility conditions without reasoning from non-Euclidean geometry. In the case of infinitesimal strains, it is a trivial consequence of the symmetry of partial derivatives with respect to the order of differentiation (Schwartz's theorem). It is reproduced in all textbooks related to linear elasticity. For finite strains, this result may be obtained by direct differentiation of the Euclidean strains, see, for example, [68].

**Displacements in the reference shape with a given strain tensor.** For given 1-forms $F^i$, we can find the positions of reference points in the actual shape:

$$x(p) = \int_{p_0}^{p} \partial_i \otimes F^i \llcorner \dot{\chi}\,,$$

where $\chi$ is a curve on reference shape that starts in $p_0$ and ends in $p$.

If fields $\boldsymbol{\kappa}$ and $\boldsymbol{\varepsilon}$ are known and satisfy (3.19), then the sought vortices and displacements are determined by the Cesaro-type formulae [66]:

$$\vec{\boldsymbol{\omega}}(\boldsymbol{r}) = \vec{\boldsymbol{\omega}}(\boldsymbol{r}_0) + \int_{\boldsymbol{r}_0}^{\boldsymbol{r}} d\boldsymbol{r}' \cdot \boldsymbol{\kappa}(\boldsymbol{r}')\,,$$

$$\boldsymbol{u}(\boldsymbol{r}) = \boldsymbol{u}(\boldsymbol{r}_0) + \vec{\boldsymbol{\omega}}(\boldsymbol{r}_0) \times (\boldsymbol{r} - \boldsymbol{r}_0) + \int_{\boldsymbol{r}_0}^{\boldsymbol{r}} d\boldsymbol{r}' \cdot \{\boldsymbol{\varepsilon}(\boldsymbol{r}') + \boldsymbol{\kappa}(\boldsymbol{r}') \times (\boldsymbol{r} - \boldsymbol{r}')\}\,,$$

where $\boldsymbol{r}_0$ is the position vector of a fixed point. These relations determine single-valued fields that do not depend on the path of integration, as it follows from the Stokes' theorem and the compatibility conditions (3.19).

## 3.7 Stresses

Ideas that led to the mathematical formulation of stresses[8] are rooted in works of James Bernoulli, John Bernoulli, and L. Euler [71, 72]. The concept of stresses as a tensor field of the second rank, generalizing the concept of hydrostatic pressure in an ideal fluid,

7 In [67] the tensor $\kappa$ is called a curvature-twist dyadic.

8 A detailed exposition of the historical aspects of the theory of stresses is given in [69, 70].

was formulated in the works of Cauchy [73, 74] and developed in the studies of Poisson and Navier [75]. Note that the coordinate representation of such fields assumed a certain analytical freedom, since different coordinate systems associated with the reference and deformed shapes are used. In this case, there is a need for coordinate transformations, the general form of which was given by Piola [76].

The classical Cauchy theory is based on the *cut* and *freezing principles*[9]. According to the first one, when a *part* $\mathscr{U}$ is cut from the actual shape $\mathscr{S}$ with an arbitrary piecewise smooth boundary $\partial\mathscr{U}$, the action of the remaining part of this shape on $\mathscr{U}$ can be replaced by the contact force density $\boldsymbol{t_n}$ distributed over $\partial\mathscr{U}$ [78, 79]. Here, $\boldsymbol{n}$ is an outward-pointing unit normal vector field on $\partial\mathscr{U}$. It is postulated that the resultant force $\mathscr{F}_{\mathscr{U}}$, acting on the part $\mathscr{U}$, is represented as the sum of the contact and body forces:

$$\mathscr{F}_{\mathscr{U}} = \int\limits_{\partial\mathscr{U}} \boldsymbol{t_n}\, dS + \int\limits_{\mathscr{U}} \boldsymbol{b}\, dV\,,$$

where $\boldsymbol{b}$ is the body force density. To formalize the contact forces, Cauchy used the postulate on the dependence of their density $\boldsymbol{t_n}$ only on the normal $\boldsymbol{n}$ to the cut surface, *i.e.*,

$$\boldsymbol{t_n}(x) := \boldsymbol{t}(x, \boldsymbol{n})\,.$$

This allowed him (under certain agreements on smoothness and using the famous tetrahedron argument) to prove the "action and reaction equality":

$$-\boldsymbol{t_n} = \boldsymbol{t_{-n}}\,,$$

and to prove the existence of stress tensor field $\boldsymbol{T}\colon \mathscr{S} \to \mathscr{V} \otimes \mathscr{V}^*$, *i.e.*, the field of linear transformation of the outer unit normal $\boldsymbol{n}$ to the contact force density $\boldsymbol{t_n}$:

$$\boldsymbol{t_n} = \boldsymbol{Tn}\,.$$

**Remark 3.5.** In the middle of the twentieth century, the Cauchy theorem on the existence of a stress tensor was reformulated in Noll's works, and a more general proof was constructed, which used not three linearly independent vectors, but only two. The proof was divided into a proof of the homogeneity and additivity of the relations determining the relationship between the unit normal and the density of contact forces [80].

Note that in classical reasonings, the part of the shape was considered as a "nice" subset of Euclidean space. Later, in the framework of potential theory (O. Kellogg [58]), the notion of "nice" subset was clarified: it is open set that is bounded by a finite

9 It seems that the cut principle dates back to the works of Galileo, and the freezing principle was formulated by Simon Stevin [77] in 1586 (Stevin S. "De beghinselen des waterwichts", 1586). A copy of the corresponding page can be found in [70, p. 198]. Note that Stevin considered the equilibrium of a liquid, mentally transferring the equilibrium laws for a rigid body to an arbitrary part of the liquid. This is reflected in the term "freezing".

number of smooth regular elementary surfaces. Each of them can be represented as a graph of some smooth function (see Section 3.1). This requirement allows one to use Stokes' theorem in its Euclidean formulation. Then, applying the freezing principle, using Eulerian Dynamics laws, one can write

$$\int_{\partial\mathcal{U}} \boldsymbol{t}_{\boldsymbol{n}}\,dS + \int_{\mathcal{U}} \boldsymbol{b}\,dV = \frac{d}{dt}\int_{\mathcal{U}} \rho\ddot{\boldsymbol{u}}\,dV\,, \quad \int_{\partial\mathcal{U}} \boldsymbol{p}\times\boldsymbol{t}_{\boldsymbol{n}}\,dS + \int_{\mathcal{U}} \boldsymbol{p}\times\boldsymbol{b}\,dV = \frac{d}{dt}\int_{\mathcal{U}} \rho\boldsymbol{p}\times\dot{\boldsymbol{u}}\,dV\,.$$

Here, $\boldsymbol{u}$ is a displacement vector field, $\boldsymbol{p}$ is a position vector field, and $\rho$ is the mass density. Since $\mathcal{U}$ is arbitrary, one obtains the differential equations of motion:

$$\nabla\cdot\boldsymbol{T} - \rho\ddot{\boldsymbol{u}} + \boldsymbol{b} = \boldsymbol{0}\,, \quad \boldsymbol{T}^{\mathrm{T}} = \boldsymbol{T}\,.$$

Suppose that we are given reference, $\mathcal{S}_R$, and actual, $\mathcal{S}$, shapes that are related by a deformation $y\colon \mathcal{S}_R \to \mathcal{S}$. Let $\mathcal{U}_0$ be a part of $\mathcal{S}_R$, which is a shape itself, and $\mathcal{U} = y(\mathcal{U}_0)$. Let $\boldsymbol{N}$, $\boldsymbol{n}$ be outward-pointing unit normal vector fields on $\partial\mathcal{U}_0$ and $\partial\mathcal{U}$.

The contact force density $\boldsymbol{t}_{\boldsymbol{n}}$ and the Cauchy stress tensor $\boldsymbol{T}\colon \mathcal{S} \to \mathcal{V}\otimes\mathcal{V}^*$ take part in the spatial description, *i.e.*, they allow one to calculate contact force acting on $\partial\mathcal{U}$ in actual coordinates. Transforming the surface integral over $\partial\mathcal{U}$ into an integral over $\partial\mathcal{U}_0$, one obtains [59]

$$\mathcal{C} = \int_{\partial\mathcal{U}} \boldsymbol{T}\boldsymbol{n}\,dS = \int_{\partial\mathcal{U}_0} \boldsymbol{P}\boldsymbol{N}\,dS_0\,,$$

where $dS_0$ is the reference area element, $dS$ is the actual area element related by *Nanson's formula* [8]

$$\boldsymbol{n}\,dS = J\boldsymbol{F}^{-\mathrm{T}}\boldsymbol{N}dS_0\,,$$

and $\boldsymbol{P}\colon \mathcal{S}_R \to \mathcal{V}\otimes\mathcal{V}^*$ is the first Piola–Kirchhoff stress tensor related with $\boldsymbol{T}$ by *Piola transformation* [37]

$$J\boldsymbol{T} = \boldsymbol{P}\boldsymbol{F}^{\mathrm{T}}\,,$$

where $J$ is the Jacobian of $\boldsymbol{F}$ (3.8). Introducing the contact force density $\boldsymbol{t}_{R;\boldsymbol{N}}$ in the reference coordinates, one arrives at

$$\boldsymbol{t}_{R;\boldsymbol{N}} = \boldsymbol{P}\boldsymbol{N}\,.$$

**Remark 3.6.** In the framework of this book, we use Cauchy and first Piola–Kirchhoff stress tensors, although other stress measures exist. The second Piola–Kirchhoff stress tensor $\boldsymbol{S}$ and the Kirchhoff stress tensor $\boldsymbol{\tau}$ may serve as examples. They are related with Cauchy and first Piola–Kirchhoff stress tensors as follows [81]:

$$\boldsymbol{S} = \boldsymbol{F}^{-1}\boldsymbol{P}\,, \quad \boldsymbol{\tau} = J\boldsymbol{T}\,.$$

## 3.8 Non-Linear Elasticity as Field Theory

### 3.8.1 Action and Its Lagrangian

In this section, we consider the elements of field theory that allow us to derive equations of motion and conservation laws in spatial and material descriptions by means of the most "economical" reasoning (E. Mach), provided by variational calculus. Let $\mathcal{S}_R$ be any shape taken as a reference and $\mathfrak{M} \subset \mathcal{E}^{\mathcal{S}_R \times \mathbb{T}}$ be the set of all motions. The main object of consideration is represented by the *action* $\mathfrak{A}$, which is a mapping

$$\mathfrak{A} : \mathfrak{M} \to \mathbb{R} .$$

We apply the *principle of additivity of action*: there exists the smooth mapping

$$\mathfrak{L} : \mathcal{S}_R \times \mathbb{T} \times \mathcal{E} \times \mathcal{V} \times \mathrm{Lin}(\mathcal{V};\mathcal{V}) \times \cdots \times \mathrm{Lin}_r(\mathcal{V},\dots,\mathcal{V};\mathcal{V}) \to \mathbb{R} ,$$
$$\mathfrak{L} : (X, t, x, \boldsymbol{u}, \boldsymbol{f}_1, \dots, \boldsymbol{f}_r) \mapsto \mathfrak{L}(X, t; x, \boldsymbol{u}, \boldsymbol{f}_1, \dots, \boldsymbol{f}_r) ,$$

called *Lagrangian volume density*, such that

$$\mathfrak{A}(y) = \int_{t_1}^{t_2} \int_{\mathcal{R}} \mathfrak{L}(X, t, y(X,t), \boldsymbol{V}(X,t), \boldsymbol{F}(X,t), \dots, \boldsymbol{F}^{(r)}(X,t))\, dV\, dt , \tag{3.20}$$

for any $y \in \mathfrak{M}$, $t_1, t_2 \in \mathbb{T}$, and $\mathcal{R} \subset \mathcal{S}_R$, which is a shape itself. The value of $r$ in (3.20) represents the order of gradient theory.

**Remark 3.7.** The list of variables presented in Lagrangian density depends on the theory considered. For example, with thermoelasticity, one should add a scalar field of temperature to the list. In the more general case, the Lagrangian density may also contain mixed derivatives with respect to space and time and high derivatives with respect to time.

Further, we consider the case when $r = 1$. That is, formula (3.20) reduces to

$$\mathfrak{A}(y) = \int_{t_1}^{t_2} \int_{\mathcal{R}} \mathfrak{L}(X, t, y(X,t), \boldsymbol{V}(X,t), \boldsymbol{F}(X,t))\, dV\, dt . \tag{3.21}$$

We say that the theory considered is the *first-order gradient theory*.

Moreover, we restrict ourselves to the following form of Lagrangian density[10]:

$$\mathfrak{L}(X, t, y(\mathfrak{X}), \boldsymbol{V}(\mathfrak{X}), \boldsymbol{F}(\mathfrak{X})) = \frac{1}{2}\rho_R(X)\boldsymbol{V}(\mathfrak{X})^2 - W(X, \boldsymbol{F}(\mathfrak{X})) - \Phi(y(\mathfrak{X}), t) , \tag{3.22}$$

where $\rho_R$ is the volume density of the solid, relative to the reference configuration, $W$ is the elastic energy volume density, and $\Phi$ is the potential of physical body forces.

10 Here symbol $\mathfrak{X}$ stands for $(X, t)$.

**Remark 3.8.** Equivalently, the action (3.21) may be written in the form

$$\mathfrak{A}(y) = \int_{t_1}^{t_2} (\mathfrak{K} - \mathfrak{W} + \mathfrak{P})\, dt\,,$$

where $\mathfrak{K} = \frac{1}{2}\int_{\mathscr{R}} \rho \boldsymbol{V}^2\, dV$ is the kinetic energy, $\mathfrak{W} = \int_{\mathscr{R}} W\, dV$ is the stored elastic (potential) energy, and $\mathfrak{P} = -\int_{\mathscr{R}} \Phi\, dV$ is work, generated by physical body forces.

### 3.8.2 Partial and Full Variations

We introduce a one-parameter family $\{g(\varepsilon)\}_{\varepsilon\in\mathbb{R}}$ of transformations of both coordinates and fields. Elements of this family constitute a Lie group $G$ (Chapter 14) with the group properties

$$g(0) = \mathrm{Id}\,, \quad g(\varepsilon) \circ g(\mu) = g(\varepsilon + \mu)\,.$$

The action $(X, t, y(X, t)) \mapsto g(\varepsilon) \circ (X, t, y(X, t))$ of $G$ on a triple $(X, t, y(X, t))$ generates the following mappings:

$$\widetilde{X}\colon (X, t, \varepsilon) \mapsto \widetilde{X}(X, t, \varepsilon)\,, \quad \widetilde{t}\colon (X, t, \varepsilon) \mapsto \widetilde{t}(X, t, \varepsilon)\,,$$
$$\widetilde{y}\colon (X, t, \varepsilon) \mapsto \widetilde{y}(\widetilde{X}(X, t, \varepsilon), \widetilde{t}(X, t, \varepsilon); \varepsilon)\,.$$

Here, $\widetilde{X}(X, t, 0) = X$, $\widetilde{t}(X, t, 0) = t$, and $\widetilde{y}(\widetilde{X}(X, t, 0), \widetilde{t}(X, t, 0); 0) = y(X, t)$.

Under action of the group $G$, the integral (3.21) transforms into

$$\widetilde{\mathfrak{A}}(\varepsilon) = \int_{\widetilde{t}_1(\varepsilon)}^{\widetilde{t}_2(\varepsilon)} \int_{\widetilde{\mathscr{R}}(\varepsilon)} \mathfrak{L}(\widetilde{X}(X, t, \varepsilon), \widetilde{t}(X, t, \varepsilon), \dots)\, d\widetilde{V}(\varepsilon)\, d\widetilde{t}(\varepsilon)\,. \tag{3.23}$$

Variations of spatial and temporal variables have the form[11]:

$$\overrightarrow{\delta X} = \left.\frac{d\widetilde{X}}{d\varepsilon}\right|_{\varepsilon=0} \delta\varepsilon\,, \quad \delta t = \left.\frac{d\widetilde{t}}{d\varepsilon}\right|_{\varepsilon=0} \delta\varepsilon\,.$$

By definition, full variation of the deformation can be expressed as

$$\delta y = \left.\frac{d\widetilde{y}}{d\varepsilon}\right|_{\varepsilon=0} \delta\varepsilon = \left.\left(\frac{\partial\widetilde{y}}{\partial\widetilde{X}}\frac{d\widetilde{X}}{d\varepsilon} + \frac{\partial\widetilde{y}}{\partial\widetilde{t}}\frac{d\widetilde{t}}{d\varepsilon} + \frac{\partial\widetilde{y}}{\partial\varepsilon}\right)\right|_{\varepsilon=0} \delta\varepsilon\,,$$

and, consequently,

$$\delta y = \overline{\delta} y + \boldsymbol{F}[\overrightarrow{\delta X}] + \boldsymbol{V}\delta t\,, \tag{3.24}$$

11 Note that $\overrightarrow{\delta X}$ is a vector field and $\delta t$ is a scalar field.

where $\overline{\delta}y$ denote partial variation

$$\overline{\delta}y = \left.\frac{\partial\tilde{y}}{\partial\varepsilon}\right|_{\varepsilon=0}\delta\varepsilon\,.$$

We introduce the following mappings:

$$\mathfrak{f}\colon \mathcal{S}_R\times\mathbb{T}\to\mathcal{S}_R\times\mathbb{T}\times\mathcal{E}\times\mathcal{V}\times\mathrm{Lin}(\mathcal{V};\mathcal{V})\,,$$

such that $\mathfrak{f}(X,t) := (X,t,y(X,t),\boldsymbol{V}(X,t),\boldsymbol{F}(X,t))$, and

$$\tilde{\mathfrak{f}}\colon \mathcal{S}_R\times\mathbb{T}\times\mathbb{R}\to\mathcal{S}_R\times\mathbb{T}\times\mathcal{E}\times\mathcal{V}\times\mathrm{Lin}(\mathcal{V};\mathcal{V})\,,$$

such that $\mathfrak{f}(X,t,\varepsilon) := \mathfrak{L}(\tilde{X}(X,t,\varepsilon),\tilde{t}(X,t,\varepsilon),\dots)$. Compositions of these maps with the Lagrangian density,

$$\mathfrak{L}\colon (X,t,x,\boldsymbol{u},\boldsymbol{f})\mapsto\mathfrak{L}(X,t,x,\boldsymbol{u},\boldsymbol{f})$$

lead to the following expressions:

$$\begin{aligned}\widehat{\mathfrak{L}} &= \mathfrak{L}\circ\mathfrak{f}\colon (X,t)\mapsto\mathfrak{L}(X,t,y(X,t),\boldsymbol{V}(X,t),\boldsymbol{F}(X,t))\,,\\ \widetilde{\mathfrak{L}} &= \mathfrak{L}\circ\tilde{\mathfrak{f}}\colon (X,t,\varepsilon)\mapsto\mathfrak{L}(\tilde{X}(X,t,\varepsilon),\tilde{t}(X,t,\varepsilon),\dots)\,.\end{aligned}$$

In the present section we will use the following notation. If $h\colon \mathcal{S}_R\times\mathbb{T}\times\mathcal{E}\times\mathcal{V}\times\mathrm{Lin}(\mathcal{V};\mathcal{V})\to W$ is any mapping with values in some set $W$, then $h|_{\mathfrak{f}}$ denotes $h\circ\mathfrak{f}$. This is a mapping defined on $\mathcal{S}_R\times\mathbb{T}$ with values in $W$. A similar notation for $\tilde{\mathfrak{f}}$ is used.

**First form for $\boldsymbol{\delta\mathfrak{A}}$.** Now we consider the variation of the action (3.23). Taking into account that integration region depends on $\varepsilon$, one arrives at the formula

$$\delta\mathfrak{A} = \left.\frac{d\widetilde{\mathfrak{A}}}{d\varepsilon}\right|_{\varepsilon=0}\delta\varepsilon = \left.\left[\int\limits_{\tilde{t}_1(\varepsilon)}^{\tilde{t}_2(\varepsilon)}\int\limits_{\widetilde{\mathcal{R}}(\varepsilon)}\left(\frac{d\widetilde{\mathfrak{L}}}{d\varepsilon}+\widetilde{\mathfrak{L}}\,\mathrm{Div}\left(\frac{d\tilde{X}}{d\varepsilon}\right)+\widetilde{\mathfrak{L}}\frac{d}{dt}\frac{d\tilde{t}}{d\varepsilon}\right)d\tilde{V}\,d\tilde{t}\right]\right|_{\varepsilon=0}\delta\varepsilon\,,$$

or, equivalently,

$$\delta\mathfrak{A} = \int\limits_{t_1}^{t_2}\int\limits_{\mathcal{R}}\left(\delta\widehat{\mathfrak{L}}+\widehat{\mathfrak{L}}\,\mathrm{div}_R\,\overrightarrow{\delta X}+\widehat{\mathfrak{L}}\frac{d}{dt}\delta t\right)dV\,dt\,. \tag{3.25}$$

Here, $\delta\widehat{\mathfrak{L}} = \left.\frac{d\widetilde{\mathfrak{L}}}{d\varepsilon}\right|_{\varepsilon=0}\delta\varepsilon$ is a variation of the Lagrangian density that can be calculated by the chain rule as follows:

$$\begin{aligned}\delta\widehat{\mathfrak{L}} = \Bigg(&\left.\frac{\partial\mathfrak{L}}{\partial X}\right|_{\tilde{\mathfrak{f}}}\left[\frac{d\tilde{X}}{d\varepsilon}\right]+\left.\frac{\partial\mathfrak{L}}{\partial t}\right|_{\tilde{\mathfrak{f}}}\left[\frac{d\tilde{t}}{d\varepsilon}\right]+\left.\frac{\partial\mathfrak{L}}{\partial x}\right|_{\tilde{\mathfrak{f}}}\left[\frac{d\tilde{y}}{d\varepsilon}\right]\\ &\left.+\left.\frac{\partial\mathfrak{L}}{\partial\boldsymbol{u}}\right|_{\tilde{\mathfrak{f}}}\left[\frac{d\dot{\tilde{y}}}{d\varepsilon}\right]+\left.\frac{\partial\mathfrak{L}}{\partial\boldsymbol{f}}\right|_{\tilde{\mathfrak{f}}}\left[\frac{d}{d\varepsilon}\frac{\partial\tilde{y}}{\partial\tilde{X}}\right]\Bigg)\right|_{\varepsilon=0}\delta\varepsilon\,.\end{aligned}$$

Taking into account the equalities

$$\frac{\partial \widehat{\mathfrak{L}}}{\partial X} = \left.\frac{\partial \mathfrak{L}}{\partial X}\right|_{\mathfrak{f}} + \left.\frac{\partial \mathfrak{L}}{\partial x}\right|_{\mathfrak{f}} \frac{\partial y}{\partial X} + \left.\frac{\partial \mathfrak{L}}{\partial \boldsymbol{u}}\right|_{\mathfrak{f}} \frac{\partial \dot{y}}{\partial X} + \left.\frac{\partial \mathfrak{L}}{\partial \boldsymbol{f}}\right|_{\mathfrak{f}} \frac{\partial \boldsymbol{F}}{\partial X}, \tag{3.26}$$

$$\frac{\partial \widehat{\mathfrak{L}}}{\partial t} = \left.\frac{\partial \mathfrak{L}}{\partial t}\right|_{\mathfrak{f}} + \left.\frac{\partial \mathfrak{L}}{\partial x}\right|_{\mathfrak{f}} \frac{\partial y}{\partial t} + \left.\frac{\partial \mathfrak{L}}{\partial \boldsymbol{u}}\right|_{\mathfrak{f}} \frac{\partial \dot{y}}{\partial t} + \left.\frac{\partial \mathfrak{L}}{\partial \boldsymbol{f}}\right|_{\mathfrak{f}} \frac{\partial \boldsymbol{F}}{\partial t}, \tag{3.27}$$

together with (3.24), we arrive at the following result[12]:

$$\delta \widehat{\mathfrak{L}} = \frac{\partial \widehat{\mathfrak{L}}}{\partial X}\left[\overrightarrow{\delta X}\right] + \frac{\partial \widehat{\mathfrak{L}}}{\partial t}[\delta t] + \left.\frac{\partial \mathfrak{L}}{\partial x}\right|_{\mathfrak{f}}\left[\overline{\delta} y\right] + \left.\frac{\partial \mathfrak{L}}{\partial \boldsymbol{u}}\right|_{\mathfrak{f}}\left[\frac{d}{dt}\overline{\delta} y\right] + \left.\frac{\partial \mathfrak{L}}{\partial \boldsymbol{f}}\right|_{\mathfrak{f}}\left[\frac{\partial \overline{\delta} y}{\partial X}\right]. \tag{3.28}$$

Values of $\mathfrak{x} \mapsto \frac{\partial \mathfrak{L}}{\partial \boldsymbol{f}}(\mathfrak{x})$ are linear mappings from $\mathrm{Lin}(\mathscr{V};\mathscr{V})$ to $\mathbb{R}$. Since the space $\mathrm{Lin}(\mathscr{V};\mathscr{V})$ is endowed with Euclidean structure with scalar product (:), there exists a unique field $\mathfrak{L}_{\boldsymbol{f}} : \mathfrak{x} \mapsto \mathfrak{L}_{\boldsymbol{f}}(\mathfrak{x}) \in \mathrm{Lin}(\mathscr{V};\mathscr{V})$ of linear mappings, such that for any field $\boldsymbol{T} : \mathfrak{x} \mapsto \boldsymbol{T}(\mathfrak{x}) \in \mathrm{Lin}(\mathscr{V};\mathscr{V})$, the following equality holds:

$$\frac{\partial \mathfrak{L}}{\partial \boldsymbol{f}}[\boldsymbol{T}] = \mathfrak{L}_{\boldsymbol{f}} : \boldsymbol{T}\,.$$

Similar representations hold for fields $\frac{\partial \mathfrak{L}}{\partial x}$ and $\frac{\partial \mathfrak{L}}{\partial \boldsymbol{u}}$ of linear functionals. There exists a unique vector field $\mathfrak{L}_x : \mathfrak{x} \mapsto \mathfrak{L}_x(\mathfrak{x}) \in \mathscr{V}$ and $\mathfrak{L}_{\boldsymbol{u}} : \mathfrak{x} \mapsto \mathfrak{L}_{\boldsymbol{u}}(\mathfrak{x}) \in \mathscr{V}$, such that for any vector field $\boldsymbol{v} : \mathfrak{x} \mapsto \boldsymbol{v}(\mathfrak{x}) \in \mathscr{V}$, the following representations hold:

$$\frac{\partial \mathfrak{L}}{\partial x}[\boldsymbol{v}] = \mathfrak{L}_x \cdot \boldsymbol{v}\,, \quad \text{and} \quad \frac{\partial \mathfrak{L}}{\partial \boldsymbol{u}}[\boldsymbol{v}] = \mathfrak{L}_{\boldsymbol{u}} \cdot \boldsymbol{v}\,.$$

Hence, the expression (3.28) for $\delta \widehat{\mathfrak{L}}$ takes the form:

$$\delta \widehat{\mathfrak{L}} = \frac{\partial \widehat{\mathfrak{L}}}{\partial X}\left[\overrightarrow{\delta X}\right] + \frac{\partial \widehat{\mathfrak{L}}}{\partial t}[\delta t] + \mathfrak{L}_x|_{\mathfrak{f}} \cdot \overline{\delta} y + \mathfrak{L}_{\boldsymbol{u}}|_{\mathfrak{f}} \cdot \frac{d}{dt}\overline{\delta} y + \mathfrak{L}_{\boldsymbol{f}}|_{\mathfrak{f}} : \frac{\partial \overline{\delta} y}{\partial X}\,. \tag{3.29}$$

Substituting (3.29) into (3.25) taking into account (2.12) and (2.13), and the equality

$$\frac{d}{dt}\left(\mathfrak{L}_{\boldsymbol{u}}|_{\mathfrak{f}} \cdot \overline{\delta} y\right) = \frac{d}{dt}\,\mathfrak{L}_{\boldsymbol{u}}|_{\mathfrak{f}} \cdot \overline{\delta} y + \mathfrak{L}_{\boldsymbol{u}}|_{\mathfrak{f}} \cdot \frac{d}{dt}\overline{\delta} y\,,$$

one can obtain

$$\begin{aligned}\delta \mathfrak{A} = \int_{t_1}^{t_2}\int_{\mathscr{R}} \Big\{ &\mathrm{div}_R\left(\widehat{\mathfrak{L}}\overrightarrow{\delta X} + \mathfrak{L}_{\boldsymbol{f}}^{\mathrm{T}}\big|_{\mathfrak{f}}\left[\overline{\delta} y\right]\right) + \frac{d}{dt}\left(\widehat{\mathfrak{L}}\delta t + \mathfrak{L}_{\boldsymbol{u}}|_{\mathfrak{f}} \cdot \overline{\delta} y\right) \\ &+ \left(\mathfrak{L}_x|_{\mathfrak{f}} - \mathrm{div}_R\left(\mathfrak{L}_{\boldsymbol{f}}|_{\mathfrak{f}}\right) - \frac{d}{dt}\left(\mathfrak{L}_{\boldsymbol{u}}|_{\mathfrak{f}}\right)\right) \cdot \overline{\delta} y \Big\}\, dV\, dt\,. \end{aligned} \tag{3.30}$$

**12** Note that partial variation commutes with differential operators. That is, $\overline{\delta}\dot{y} = (d/dt)\overline{\delta} y$, and $\overline{\delta}\frac{\partial y}{\partial X} = \frac{\partial \overline{\delta} y}{\partial X}$.

Using the following notations for *Noether current* components

$$\delta \boldsymbol{J}_1 = \mathfrak{L}|_{\mathfrak{f}}\, \overrightarrow{\delta X} + \mathfrak{L}_{\boldsymbol{f}}^{\mathrm{T}}\big|_{\mathfrak{f}} \left[\overline{\delta y}\right], \tag{3.31}$$

$$\delta \boldsymbol{J}_2 = \mathfrak{L}|_{\mathfrak{f}}\, \delta t + \mathfrak{L}_{\boldsymbol{u}}|_{\mathfrak{f}} \cdot \overline{\delta y}, \tag{3.32}$$

and the *Euler–Lagrange operator*

$$\mathbb{E}_y = \mathfrak{L}_x|_{\mathfrak{f}} - \mathrm{div}_R \left( \mathfrak{L}_{\boldsymbol{f}}|_{\mathfrak{f}} \right) - \frac{d}{dt} \left( \mathfrak{L}_{\boldsymbol{u}}|_{\mathfrak{f}} \right), \tag{3.33}$$

one can rewrite (3.30) more concisely:

$$\delta \mathfrak{A} = \int_{t_1}^{t_2} \int_{\mathcal{R}} \left\{ \mathrm{div}_R\, \delta \boldsymbol{J}_1 + \frac{d}{dt} \delta \boldsymbol{J}_2 + \mathbb{E}_y \cdot \overline{\delta y} \right\} dV\, dt. \tag{3.34}$$

**Second form for $\delta\mathfrak{A}$.** Now we obtain the second form for $\delta\mathfrak{A}$. Keeping in mind the expressions (3.26), (3.27), and (3.29), as well as the relation (3.24) between partial and full variations, one can derive the equality

$$\delta \widehat{\mathfrak{L}} + \widehat{\mathfrak{L}}\, \mathrm{div}_R\, \overrightarrow{\delta X} + \widehat{\mathfrak{L}} \frac{d}{dt} \delta t$$
$$= \left.\frac{\partial \mathfrak{L}}{\partial X}\right|_{\mathfrak{f}} [\overrightarrow{\delta X}] + \left.\frac{\partial \mathfrak{L}}{\partial t}\right|_{\mathfrak{f}} \delta t + \mathfrak{L}_x|_{\mathfrak{f}} \cdot \delta y + \mathfrak{L}_{\boldsymbol{u}}|_{\mathfrak{f}} \cdot \delta \dot{y} + \mathfrak{L}_{\boldsymbol{f}}|_{\mathfrak{f}} : \delta \frac{\partial y}{\partial X} + \widehat{\mathfrak{L}}\, \mathrm{div}_R\, \overrightarrow{\delta X} + \widehat{\mathfrak{L}} \frac{d}{dt} \delta t.$$

Taking into account commutation relations

$$\delta \frac{d}{dt} y = \frac{d}{dt} \delta y - \frac{\partial y}{\partial X} \left[ \frac{d}{dt} \overrightarrow{\delta X} \right] - \dot{y} \frac{d}{dt} \delta t,$$

$$\delta \frac{\partial}{\partial X} y = \frac{\partial}{\partial X} \delta y - \frac{\partial y}{\partial X} \frac{\partial \overrightarrow{\delta X}}{\partial X} - \dot{y} \otimes \frac{\partial}{\partial X} \delta t,$$

and the following identities:

$$\mathrm{div}_R\, \overrightarrow{\delta X} = \boldsymbol{I} : \frac{\partial \overrightarrow{\delta X}}{\partial X},$$

$$\mathfrak{L}_{\boldsymbol{f}}|_{\mathfrak{f}} : \frac{\partial y}{\partial X} \frac{\partial \overrightarrow{\delta X}}{\partial X} = \left( \frac{\partial y}{\partial X} \right)^{\mathrm{T}} \mathfrak{L}_{\boldsymbol{f}}|_{\mathfrak{f}} : \frac{\partial \overrightarrow{\delta X}}{\partial X},$$

$$\mathfrak{L}_{\boldsymbol{f}}|_{\mathfrak{f}} : \left( \dot{y} \otimes \frac{\partial}{\partial X} \delta t \right) = \mathfrak{L}_{\boldsymbol{f}}^{\mathrm{T}}\big|_{\mathfrak{f}} [\dot{y}] \lrcorner \frac{\partial}{\partial X} \delta t,$$

one can obtain the following expression for $\delta\mathfrak{A}$:

$$\delta \mathfrak{A} = \int_{t_1}^{t_2} \int_{\mathcal{R}} \left\{ \left.\frac{\partial \mathfrak{L}}{\partial X}\right|_{\mathfrak{f}} [\overrightarrow{\delta X}] + \left.\frac{\partial \mathfrak{L}}{\partial t}\right|_{\mathfrak{f}} \delta t + \left( \mathfrak{L}|_{\mathfrak{f}} \boldsymbol{I} - \left( \frac{\partial y}{\partial X} \right)^{\mathrm{T}} \mathfrak{L}_{\boldsymbol{f}}|_{\mathfrak{f}} \right) : \frac{\partial}{\partial X} \overrightarrow{\delta X} \right.$$
$$+ \left( \mathfrak{L}|_{\mathfrak{f}} - \mathfrak{L}_{\boldsymbol{u}}|_{\mathfrak{f}} \cdot \dot{y} \right) \frac{d}{dt} \delta t + \mathfrak{L}_x|_{\mathfrak{f}} \cdot \delta y + \mathfrak{L}_{\boldsymbol{u}}|_{\mathfrak{f}} \cdot \frac{d}{dt} \delta y$$
$$\left. - \left( \frac{\partial y}{\partial X} \right)^{\mathrm{T}} \left[ \mathfrak{L}_{\boldsymbol{u}}|_{\mathfrak{f}} \right] \cdot \frac{d}{dt} \overrightarrow{\delta X} + \mathfrak{L}_{\boldsymbol{f}}|_{\mathfrak{f}} : \frac{\partial}{\partial X} \delta y - \mathfrak{L}_{\boldsymbol{f}}^{\mathrm{T}}\big|_{\mathfrak{f}} [\dot{y}] \lrcorner \frac{\partial}{\partial X} \delta t \right\} dV\, dt. \tag{3.35}$$

**Standard notations.** Now introduce notations for the fields:

- *Eshelby energy-momentum tensor*

$$\boldsymbol{e} = \mathfrak{L}|_{\mathfrak{f}} \boldsymbol{I} - \left(\frac{\partial y}{\partial X}\right)^{\mathrm{T}} \mathfrak{L}_{\boldsymbol{f}}|_{\mathfrak{f}} ,$$

- *Hamiltonian density*

$$\mathfrak{H} = \mathfrak{L}|_{\mathfrak{f}} - \mathfrak{L}_{\boldsymbol{u}}|_{\mathfrak{f}} \cdot \dot{y} ,$$

- *Energy flux (Umov–Poynting vector)*

$$\boldsymbol{U} = \mathfrak{L}_{\boldsymbol{f}}^{\mathrm{T}}\big|_{\mathfrak{f}} [\dot{y}] ,$$

- *Density of the canonical impulse*

$$\boldsymbol{K} = \left(\frac{\partial y}{\partial X}\right)^{\mathrm{T}} [\mathfrak{L}_{\boldsymbol{u}}|_{\mathfrak{f}}] ,$$

- *Density of the physical impulse*

$$\boldsymbol{p} = \mathfrak{L}_{\boldsymbol{u}}|_{\mathfrak{f}} ,$$

- *The first Piola–Kirchhoff stress tensor*

$$\boldsymbol{P} = - \mathfrak{L}_{\boldsymbol{f}}|_{\mathfrak{f}} .$$

With such notations, the expression (3.35) may be rewritten in the form:

$$\delta\mathfrak{A} = \int_{t_1}^{t_2}\int_{\mathcal{R}} \left\{ \left.\frac{\partial \mathfrak{L}}{\partial X}\right|_{\mathfrak{f}} [\overrightarrow{\delta X}] + \left.\frac{\partial \mathfrak{L}}{\partial t}\right|_{\mathfrak{f}} \delta t + \boldsymbol{e}\colon \frac{\partial}{\partial X}\overrightarrow{\delta X} + \mathfrak{H}\frac{d}{dt}\delta t + \mathfrak{L}_x|_{\mathfrak{f}} \cdot \delta y \right.$$
$$\left. + \boldsymbol{p} \cdot \frac{d}{dt}\delta y - \boldsymbol{K} \cdot \frac{d}{dt}\overrightarrow{\delta X} - \boldsymbol{P}\colon \frac{\partial}{\partial X}\delta y - \boldsymbol{U} \lrcorner \frac{\partial}{\partial X}\delta t \right\} dV\, dt . \qquad (3.36)$$

**Remark 3.9.** Customarily, the following agreements on notations are used in continuum mechanics. Variables of the Lagrangian density $\mathfrak{L}$ are denoted by the same symbols as the corresponding fields. That is, one writes

$$\mathfrak{L}\colon (X, t, y, \boldsymbol{V}, \boldsymbol{F}) \mapsto \mathfrak{L}(X, t, y, \boldsymbol{V}, \boldsymbol{F}) ,$$

and use notations $\frac{\partial \mathfrak{L}}{\partial y}, \frac{\partial \mathfrak{L}}{\partial \boldsymbol{V}}, \frac{\partial \mathfrak{L}}{\partial \boldsymbol{F}}$ for $\frac{\partial \mathfrak{L}}{\partial x}, \frac{\partial \mathfrak{L}}{\partial \boldsymbol{u}}, \frac{\partial \mathfrak{L}}{\partial \boldsymbol{f}}$. Although these notations are not strictly correct, they are convenient. Another customarily (but also strictly not correct) agreement is that $\mathfrak{L}_{\boldsymbol{f}}$ is denoted as $\frac{\partial \mathfrak{L}}{\partial \boldsymbol{F}}$. Thus, one have the same notation for the 3rd order tensor field and for the corresponding a unique 2nd order tensor field which is associated with the former by the operation (:). Similar remark holds for $\mathfrak{L}_x$ and $\mathfrak{L}_{\boldsymbol{u}}$. By virtue of such agreements (3.33) has the following representation:

$$\mathbb{E}_y = \frac{\partial \mathfrak{L}}{\partial y} - \mathrm{div}_R \frac{\partial \mathfrak{L}}{\partial \boldsymbol{F}} - \frac{d}{dt}\frac{\partial \mathfrak{L}}{\partial \boldsymbol{V}} .$$

### 3.8.3 Field Equations

Consider the particular case when only physical fields (*i.e.*, $y$) undergo variations; then formula (3.34) reduces to

$$\delta\mathfrak{A} = \int_{t_1}^{t_2}\int_{\mathcal{R}} \mathbb{E}_y \bar{\delta} y \, dV \, dt \, .$$

From this follows the field equation

$$\mathfrak{L}_x|_\mathfrak{f} - \mathrm{div}_R \left( \mathfrak{L}_f|_\mathfrak{f} \right) - \frac{d}{dt} \left( \mathfrak{L}_u|_\mathfrak{f} \right) = \mathbf{0} \, ,$$

or, with the notation introduced above,

$$\mathfrak{L}_x|_\mathfrak{f} + \mathrm{div}_R \, \boldsymbol{P} = \frac{d}{dt} \boldsymbol{p} \, .$$

Here, $\mathfrak{L}_x|_\mathfrak{f}$ denotes the mass forces.

Since $\mathfrak{L}$ is defined by (3.22), we arrive at the following field equation in the direct motion description[13]:

$$\mathrm{div}_R \frac{\partial W}{\partial \boldsymbol{F}} - \frac{\partial \Phi}{\partial y} = \rho_R \dot{\boldsymbol{V}} \, . \tag{3.37}$$

### 3.8.4 Action Invariance Conditions

Consider another particular case, when only coordinates undergo variations. Moreover, we suppose that these variations are defined by a Lie group.

**I. Spatial translations.** Let $\widetilde{X}(X, t, \varepsilon) = X + \varepsilon \boldsymbol{X}_0$, where $\boldsymbol{X}_0 \in \mathscr{V}$ is a constant vector, and let $\widetilde{t}(X, t, \varepsilon) = t$, $\widetilde{y}(X, t, \varepsilon) = y$. Since

$$\overrightarrow{\delta X} = \boldsymbol{X}_0 \delta\varepsilon \, , \quad \frac{d}{dt}\overrightarrow{\delta X} = \mathbf{0} \, , \quad \frac{\partial}{\partial X}\overrightarrow{\delta X} = \mathbf{0} \, ,$$

the variation (3.36) reduces to

$$\delta\mathfrak{A} = \int_{t_1}^{t_2}\int_{\mathcal{R}} \left\{ \left.\frac{\partial \mathfrak{L}}{\partial X}\right|_\mathfrak{f} [\boldsymbol{X}_0]\delta\varepsilon \right\} dV \, dt \, .$$

The equality $\delta\mathfrak{A} = 0$ must be valid for each spatial translation. Since $\boldsymbol{X}_0$ is arbitrary, we arrive at the following condition:

$$\left.\frac{\partial \mathfrak{L}}{\partial X}\right|_\mathfrak{f} = 0 \, .$$

13 Note that formally $\frac{\partial W}{\partial \boldsymbol{F}}$ denote the third-order tensor field. However, traditionally this symbol is used for a unique second-order tensor field, which is associated with the former by the operation (:). We mean the latter case.

**II. Temporal translations.** Let $\tilde{t}(X, t, \varepsilon) = t + \varepsilon t_0$, where $t_0 \in \mathbb{R}$ is a constant, and let $\tilde{X}(X, t, \varepsilon) = X$, $\tilde{y}(X, t, \varepsilon) = y$. Since

$$\delta t = t_0 \delta\varepsilon \,, \quad \frac{d}{dt}\delta t = 0 \,, \quad \frac{\partial}{\partial X}\delta t = \delta t \,,$$

the variation (3.36) reduces to

$$\delta\mathfrak{A} = \int_{t_1}^{t_2}\int_{\mathcal{R}} \left\{ \left.\frac{\partial \mathfrak{L}}{\partial t}\right|_{\mathrm{f}} t_0 \delta\varepsilon \right\} dV\, dt \,.$$

The equality $\delta\mathfrak{A} = 0$ must be valid for each temporal translation. Since $t_0$ is arbitrary, we arrive at the following condition:

$$\left.\frac{\partial \mathfrak{L}}{\partial t}\right|_{\mathrm{f}} = 0 \,.$$

**III. Spatial rotations.** Let $\tilde{X}(X, t, \varepsilon) = o + \mathbf{Q}(\varepsilon)[X - o]$, where $o \in \mathscr{E}$ is a fixed point and $\mathbb{R} \ni \varepsilon \mapsto \mathbf{Q}(\varepsilon) \in O(3)$ is smooth field of orthogonal mappings, such that $\mathbf{Q}(0) = \boldsymbol{I}$. Temporal coordinate and fields are fixed: $\tilde{t}(X, t, \varepsilon) = t$, $\tilde{y}(X, t, \varepsilon) = y$. Since $\mathbf{Q}$ is smooth, we may write

$$\mathbf{Q}(\varepsilon) = \mathbf{Q}(0) + \varepsilon\boldsymbol{\omega} + \boldsymbol{o}(\varepsilon) \quad \text{as} \quad \varepsilon \to 0 \,,$$

where $\boldsymbol{\omega} = \left.\dfrac{d\mathbf{Q}(\varepsilon)}{d\varepsilon}\right|_{\varepsilon=0} [1] \in \mathrm{Lin}(\mathscr{V};\mathscr{V})$.

**Remark 3.10.** Linear mapping $\boldsymbol{\omega}$ is antisymmetric *i.e.*, $\boldsymbol{\omega}^{\mathrm{T}} = -\boldsymbol{\omega}$. This expression may be obtained in the following way. One should differentiate the identity $\mathbf{Q}(\varepsilon)\mathbf{Q}^{\mathrm{T}}(\varepsilon) = \boldsymbol{I}$ with respect to $\varepsilon$ and put $\varepsilon = 0$.

According to the expressions

$$\overrightarrow{\delta X} = \boldsymbol{\omega}[X - o]\delta\varepsilon \,, \quad \frac{d}{dt}\overrightarrow{\delta X} = \mathbf{0} \,, \quad \frac{\partial}{\partial X}\overrightarrow{\delta X} = \boldsymbol{\omega}\delta\varepsilon \,,$$

the variation (3.36) reduces to

$$\delta\mathfrak{A} = \int_{t_1}^{t_2}\int_{\mathcal{R}} \left\{ \left.\frac{\partial \mathfrak{L}}{\partial X}\right|_{\mathrm{f}} \boldsymbol{\omega}[X - o] + \mathbf{e}{:}\boldsymbol{\omega} \right\} \delta\varepsilon\, dV\, dt \,.$$

The equality $\delta\mathfrak{A} = 0$ must be valid for each spatial rotation. Hence,

$$\left.\frac{\partial \mathfrak{L}}{\partial X}\right|_{\mathrm{f}} \boldsymbol{\omega}[X - o] + \mathbf{e}{:}\boldsymbol{\omega} = 0 \,.$$

By the identities

$$\left.\frac{\partial \mathfrak{L}}{\partial X}\right|_{\mathrm{f}} \boldsymbol{\omega}[X - o] = \left\{ (X - o) \otimes \left.\frac{\partial \mathfrak{L}}{\partial X}\right|_{\mathrm{f}} \right\} {:}\boldsymbol{\omega}^{\mathrm{T}} \,,$$

and $\boldsymbol{e}:\boldsymbol{\omega} = \boldsymbol{e}^{\mathrm{T}}:\boldsymbol{\omega}^{\mathrm{T}}$, we have

$$\left\{(X - o) \otimes \left.\frac{\partial \mathfrak{L}}{\partial X}\right|_{\mathfrak{f}} + \boldsymbol{e}^{\mathrm{T}}\right\} : \boldsymbol{\omega} = 0 \,.$$

Since $\boldsymbol{\omega}$ is arbitrary antisymmetric linear mapping, we finally arrive at

$$\operatorname{Asym}\left\{(X - o) \otimes \left.\frac{\partial \mathfrak{L}}{\partial X}\right|_{\mathfrak{f}} + \boldsymbol{e}^{\mathrm{T}}\right\} = \boldsymbol{0} \,,$$

where Asym[...] is the antisymmetric part of the linear mapping [...].

Now consider the particular case of variation, when coordinates are fixed, while physical fields are varied, but, in contrast with Section 3.8.3, their variation is defined by a Lie group.

**IV. Translation of the deformation.** Let $\tilde{y}(X, t, \varepsilon) = y(X, t) + \varepsilon \boldsymbol{y}_0$, where $\boldsymbol{y}_0 \in \mathscr{V}$ is a constant vector, and let $\tilde{X}(X, t, \varepsilon) = X$, $\tilde{t}(X, t, \varepsilon) = t$. Since

$$\delta y = \boldsymbol{y}_0 \delta\varepsilon \,, \quad \frac{d}{dt}\delta y = \boldsymbol{0} \,, \quad \frac{\partial}{\partial X}\delta y = \boldsymbol{0} \,,$$

the variation (3.36) reduces to

$$\delta\mathfrak{A} = \int_{t_1}^{t_2}\int_{\mathscr{R}} \{\mathfrak{L}_x|_{\mathfrak{f}} \cdot \boldsymbol{y}_0 \delta\varepsilon\} \, dV \, dt \,.$$

The equality $\delta\mathfrak{A} = 0$ must be valid for each such translation. Since $\boldsymbol{y}_0$ is arbitrary, we arrive at the following condition:

$$\mathfrak{L}_x|_{\mathfrak{f}} = \boldsymbol{0} \,.$$

**V. Rotations of deformation.** Let $\tilde{y}(X, t, \varepsilon) = o + \boldsymbol{Q}(\varepsilon)[y(X, t) - o]$, where $o \in \mathscr{E}$ is a fixed point, and $\mathbb{R} \ni \varepsilon \mapsto \boldsymbol{Q}(\varepsilon) \in O(3)$ is smooth field of orthogonal mappings such that $\boldsymbol{Q}(0) = \boldsymbol{I}$. Spatial and temporal points are fixed: $\tilde{X}(X, t, \varepsilon) = X$, $\tilde{t}(X, t, \varepsilon) = t$. Like in III, we obtain

$$\delta y = \boldsymbol{\omega}[y(X, t) - o]\delta\varepsilon \,, \quad \frac{d}{dt}\delta y = \boldsymbol{\omega}\left[\frac{\partial y}{\partial t}\right]\delta\varepsilon \,, \quad \frac{\partial}{\partial X}\delta y = \boldsymbol{\omega}\frac{\partial y}{\partial X}\delta\varepsilon \,,$$

where $\boldsymbol{\omega}$ is antisymmetric linear mapping as in III. In this case, the variation (3.36) reduces to

$$\delta\mathfrak{A} = \int_{t_1}^{t_2}\int_{\mathscr{R}} \left\{\mathfrak{L}_x|_{\mathfrak{f}} \cdot \boldsymbol{\omega}[y(X, t) - o] + \boldsymbol{p} \cdot \boldsymbol{\omega}\left[\frac{\partial y}{\partial t}\right] - \boldsymbol{P}:\left(\boldsymbol{\omega}\frac{\partial y}{\partial X}\right)\right\} \delta\varepsilon \, dV \, dt \,.$$

The invariance condition is as follows:

$$\mathfrak{L}_x|_{\mathfrak{f}} \cdot \boldsymbol{\omega}[y(X, t) - o] + \boldsymbol{p} \cdot \boldsymbol{\omega}\left[\frac{\partial y}{\partial t}\right] - \boldsymbol{P}:\left(\boldsymbol{\omega}\frac{\partial y}{\partial X}\right) = 0 \,.$$

By the identities

$$\boldsymbol{u} \cdot \boldsymbol{\omega}[\boldsymbol{v}] = (\boldsymbol{v} \otimes \boldsymbol{u}^\flat):\boldsymbol{\omega}^{\mathrm{T}} ,$$
$$\boldsymbol{L}:(\boldsymbol{\omega}\boldsymbol{M}) = (\boldsymbol{M}\boldsymbol{L}^{\mathrm{T}}):\boldsymbol{\omega}^{\mathrm{T}} ,$$

which are hold for all vectors $\boldsymbol{u}, \boldsymbol{v} \in \mathscr{V}$ and all linear mappings $\boldsymbol{L}, \boldsymbol{M}, \boldsymbol{\omega} \in \mathrm{Lin}(\mathscr{V};\mathscr{V})$, one has

$$\left\{(y(X,t) - o) \otimes \mathfrak{L}_x^\flat\Big|_{\mathfrak{f}} + \frac{\partial y}{\partial t} \otimes \boldsymbol{p}^\flat - \frac{\partial y}{\partial X}\boldsymbol{P}^{\mathrm{T}}\right\} :\boldsymbol{\omega} = 0 .$$

Hence,

$$\mathrm{Asym}\left\{(y(X,t) - o) \otimes \mathfrak{L}_x^\flat\Big|_{\mathfrak{f}} + \frac{\partial y}{\partial t} \otimes \boldsymbol{p}^\flat - \frac{\partial y}{\partial X}\boldsymbol{P}^{\mathrm{T}}\right\} = \boldsymbol{0} .$$

## 3.9 Constitutive Relations

### 3.9.1 Principle of Material-Frame Indifference

We will consider the changes of frame

$$(x,t) \mapsto \varphi_{12}(x,t) = (x^*, t^*) ,$$

defined by the relations (2.31) and (2.32)

$$x^* = a(t) + \boldsymbol{Q}(t)(x - b) ,$$
$$t^* = t - c ,$$

and will assume that mappings $a$ and $\boldsymbol{Q}$ are smooth. A change of frame defines one-parametric families of transformations on scalars, vectors, and second-rank tensors:

(i) Scalars remain unchanged.

(ii) Let $\boldsymbol{v} \in \mathscr{V}$. It can be represented by a point difference $\boldsymbol{v} = x - y$ in the frame $\varphi_1$. In the frame $\varphi_2$, the following relations hold:

$$x^* = a(t) + \boldsymbol{Q}(t)(x - b) , \quad y^* = a(t) + \boldsymbol{Q}(t)(y - b) ,$$

and for $\boldsymbol{v}^* = x^* - y^*$, we have

$$\boldsymbol{v}^* = \boldsymbol{Q}(t)\boldsymbol{v} . \tag{3.38}$$

The last equality defines the one-parametric family $\{f_t\}_{t\in\mathscr{T}}$ of vector transformations

$$f_t : \mathscr{V} \to \mathscr{V} , \quad \boldsymbol{v} \mapsto \boldsymbol{v}^* .$$

(iii) Let $\boldsymbol{T} \in \mathrm{Lin}(\mathscr{V};\mathscr{V})$. If $\boldsymbol{w} = \boldsymbol{T}[\boldsymbol{v}]$, then using relations $\boldsymbol{v}^* = \boldsymbol{Q}(t)\boldsymbol{v}$ and $\boldsymbol{w}^* = \boldsymbol{Q}(t)\boldsymbol{w}$ one arrives at the equality $\boldsymbol{w}^* = \boldsymbol{T}^*[\boldsymbol{v}^*]$, where

$$\boldsymbol{T}^* = \boldsymbol{Q}(t)\boldsymbol{T}^*\boldsymbol{Q}^{\mathrm{T}}(t) . \tag{3.39}$$

We obtained the family $\{g_t\}_{t\in\mathscr{T}}$ of tensor transformations

$$g_t : \mathrm{Lin}(\mathscr{V};\mathscr{V}) \to \mathrm{Lin}(\mathscr{V};\mathscr{V}) , \quad \boldsymbol{T} \mapsto \boldsymbol{T}^* .$$

Mappings whose values are scalars, vectors, or tensors are called *frame indifferent* or *objective* [37] if both dependent and independent variables transform according to the expressions (2.31), (2.32), (3.38), and (3.39) under a change of frame. That is, let $h$, $\boldsymbol{v}$ and $\boldsymbol{T}$ be scalar, vector, and tensor fields relative to $\varphi_1$, and $h^*$, $\boldsymbol{v}^*$ and $\boldsymbol{T}^*$ be scalar, vector, and tensor fields relative to $\varphi_2$. These fields are frame indifferent, if

$$\begin{aligned} h^*(x^*, t^*) &= h(x, t) , \\ \boldsymbol{v}^*(x^*, t^*) &= \boldsymbol{Q}(t)\, \boldsymbol{v}(x, t) , \\ \boldsymbol{T}^*(x^*, t^*) &= \boldsymbol{Q}(t)\, \boldsymbol{T}(x, t)\, \boldsymbol{Q}^{\mathrm{T}}(t) , \end{aligned} \tag{3.40}$$

where $(x, t)$ and $(x^*, t^*)$ are connected by (2.31) and (2.32).

Let us consider a motion in a frame $\varphi_1$, defined by the mapping

$$y : \mathcal{S}_R \times \mathbb{T} \to \mathscr{E} , \quad (X, t) \mapsto x .$$

To the reference shape $\mathcal{S}_R$ considered in $\varphi_1$ there corresponds a shape $\mathcal{S}_R^*$, considered in $\varphi_2$. According to (2.31) and (2.32), these shapes are related by the equalities $X^* = a(\tilde{t}) + \boldsymbol{Q}(\tilde{t})(X - b)$, and $\tilde{t}^* = \tilde{t} - c$, in which $\tilde{t}$ is fixed. The same motion is described in a frame $\varphi_2$ by

$$x^* = y^*(X^*, t^*) = a(t) + \boldsymbol{Q}(t)\,(y(X, t) - b) , \quad t^* = t - c . \tag{3.41}$$

The equality (3.41) allows us to obtain the relation between deformation gradients $\boldsymbol{F}$ and $\boldsymbol{F}^*$. The former one is calculated relatively to the frame $\varphi_1$, *i.e.*, $\boldsymbol{F} = \frac{\partial y}{\partial X}$, and the latter one is calculated relative to the frame $\varphi_2$, *i.e.*, $\boldsymbol{F}^* = \frac{\partial y^*}{\partial X^*}$. From the chain rule, it follows that

$$\boldsymbol{F}^* = \boldsymbol{Q}(t) \frac{\partial y}{\partial X} \frac{\partial X}{\partial X^*} ,$$

and, hence,

$$\boldsymbol{F}^* = \boldsymbol{Q}(t) \boldsymbol{F} \boldsymbol{Q}^{\mathrm{T}}(\tilde{t}) . \tag{3.42}$$

Now we give a definition to the physical processes, which are the “same”. By a *dynamical process*, we mean an ordered pair $(y, \boldsymbol{T})$, consisting of the time-dependent deformation $y$ and the Cauchy stress tensor $\boldsymbol{T}$. Two dynamical processes $(y, \boldsymbol{T})$ and $(y^*, \boldsymbol{T}^*)$ given in different rigid frames are called *equivalent processes*, if they are related to each other by a change of frame (3.40) and (3.41).

Experience shows that two solids with the same reference shape and the same mass distribution may react quite differently under the influence of the same forces. This difference is described in the following manner: we say that the two solids consist of different materials. The material properties of a solid are described as conditions that give a restriction on possible dynamical processes. These conditions are called *constitutive equations* or *constitutive assumptions*. In this book, we consider constitutive equations involving only motion and the Cauchy stress tensor.

We state the fundamental physical principle according to which material properties are indifferent, *i.e.*, independent of the frame of reference. In mathematical terms, this means that the constitutive equations are subject to the following[14]

**Principle of material frame indifference.** *Constitutive equations must be invariant under change of a frame of reference. If a constitutive equation is satisfied for a process with a motion and a Cauchy stress tensor given by*

$$x = y(X, t)\,, \quad \boldsymbol{T} = \boldsymbol{T}(X, t)\,,$$

*then it must be satisfied also for any equivalent process* $(y^*, \boldsymbol{T}^*)$.

### 3.9.2 The Cauchy Polar Decomposition Theorem

Further, we will use the important result obtained Cauchy: *polar decomposition theorem.*

**Theorem** (Polar decomposition theorem). *For any linear mapping* $\boldsymbol{T} \in \mathrm{Inv}(\mathscr{V};\mathscr{V})$, *there exists a unique orthogonal linear mapping* $\boldsymbol{R} \in \mathrm{O}(3)$ *and positive-definite*[15] *symmetric linear mappings* $\boldsymbol{U}, \boldsymbol{V} \in \mathrm{InvLin}(\mathscr{V};\mathscr{V})$, *such that the following equalities hold:*

$$\boldsymbol{T} = \boldsymbol{R}\boldsymbol{U}\,, \quad \boldsymbol{T} = \boldsymbol{V}\boldsymbol{R}\,.$$

*Here* $\boldsymbol{U}^2 = \boldsymbol{T}^{\mathrm{T}}\boldsymbol{T}$, *and* $\boldsymbol{V}^2 = \boldsymbol{T}\boldsymbol{T}^{\mathrm{T}}$.

**Remark 3.11.** Proof of this theorem may be found in [82] and is provided as follows. The transformation $\boldsymbol{T}^{\mathrm{T}}\boldsymbol{T}$ is positive[16] and one can define its square root $\boldsymbol{U} := \sqrt{\boldsymbol{T}^{\mathrm{T}}\boldsymbol{T}}$, which is invertible and symmetric. Now, take $\boldsymbol{R} = \boldsymbol{T}\boldsymbol{U}^{-1}$. Since

$$\boldsymbol{R}\boldsymbol{R}^{\mathrm{T}} = \boldsymbol{T}\boldsymbol{U}^{-1}\boldsymbol{U}^{-T}\boldsymbol{T}^{\mathrm{T}} = \boldsymbol{T}(\boldsymbol{U}^2)^{-1}\boldsymbol{T}^{\mathrm{T}} = \boldsymbol{I}\,,$$

one has $\boldsymbol{R} \in \mathrm{O}(3)$. Thus, we have obtained the desired decomposition $\boldsymbol{T} = \boldsymbol{R}\boldsymbol{U}$. The second case is considered analogously.

**Cauchy–Green measures.** With application to the deformation gradient $\boldsymbol{F}$, the polar decomposition theorem asserts that there exists a unique orthogonal linear mapping

14 We give a formulation similar to Truesdell and Noll in the book [37].

15 A linear map $\boldsymbol{L}$ is positive definite if $\boldsymbol{u} \cdot \boldsymbol{L}[\boldsymbol{u}] > 0$ for any non-zero vector $\boldsymbol{u}$. For a positive definite operator the notion of square root is defined (Chapter 12, Section 12.5.4).

16 Indeed, for any vector $\boldsymbol{u} \neq \boldsymbol{0}$, and one has

$$\boldsymbol{u} \cdot \boldsymbol{T}^{\mathrm{T}}\boldsymbol{T}[\boldsymbol{u}] = \boldsymbol{T}[\boldsymbol{u}] \cdot \boldsymbol{T}[\boldsymbol{u}] \geq 0\,.$$

Since $\boldsymbol{T}$ is invertible, $\boldsymbol{T}[\boldsymbol{u}] \neq \boldsymbol{0}$ and the sign $\geq$ can be changed by $>$.

$\boldsymbol{R} \in \mathrm{O}(3)$, such that[17] $\det \boldsymbol{R} = 1$, and positive-definite symmetric linear mappings $\boldsymbol{U}, \boldsymbol{V} \in \mathrm{Inv}(\mathscr{V};\mathscr{V})$ called *right* and *left stretch tensors*, respectively, such that

$$\boldsymbol{F} = \boldsymbol{R}\boldsymbol{U}\,, \quad \boldsymbol{F} = \boldsymbol{V}\boldsymbol{R}\,. \tag{3.43}$$

Right and left stretch tensors define *right* and *left Cauchy–Green tensors*:

$$\boldsymbol{C} := \boldsymbol{U}^2 = \boldsymbol{F}^{\mathrm{T}}\boldsymbol{F}\,, \quad \text{and} \quad \boldsymbol{B} := \boldsymbol{V}^2 = \boldsymbol{F}\boldsymbol{F}^{\mathrm{T}}\,.$$

Let us consider the mechanical sense of the equalities (3.43). According to the equality (3.6), the deformation gradient maps non-deformed infinitesimal fiber into a deformed one. The deformation of the fiber decomposes on rotation, defined by $\boldsymbol{R}$, and stretch, defined by $\boldsymbol{U}$ or $\boldsymbol{V}$. Rotation is "rigid motion", which can be seen by the observer associated with the space $\mathscr{E}$. Another observer, rotating with the considering fiber, will see stretches only. Thus, objective quantities, which can be set as measures of deformation, are represented by $\boldsymbol{U}$ and $\boldsymbol{V}$, or, equivalently, by right and left Cauchy–Green tensors $\boldsymbol{C}$ and $\boldsymbol{B}$.

### 3.9.3 Simple Material

Let $\mathrm{Sh}(\mathscr{E})$ be the set of all shapes in $\mathscr{E}$. Define a family $\{\mathfrak{g}_{\mathscr{S}_R}\}_{\mathscr{S}_R \in \mathrm{Sh}(\mathscr{E})}$ of mappings

$$\mathfrak{g}_{\mathscr{S}_R} : \mathscr{S}_R \times \mathrm{Lin}(\mathscr{V};\mathscr{V}) \to \mathrm{Sym}(\mathscr{V})\,,$$

which are called *response functionals*. Such a family represents the mechanical behavior of a solid. The constitutive relation of the solid with *simple material* relative to the reference shape $\mathscr{S}_R$ is the following relation:

$$\boldsymbol{T}_X = \mathfrak{g}_{\mathscr{S}_R}(X, \boldsymbol{F}_X)\,. \tag{3.44}$$

Let $\mathscr{S}_R$ and $\widetilde{\mathscr{S}}_R$ be any two reference shapes and $\mathscr{S}$ be some actual shape. Let $y\colon \mathscr{S}_R \to \mathscr{S}$ and $\tilde{y}\colon \widetilde{\mathscr{S}}_R \to \mathscr{S}$ be deformations and $\lambda = \tilde{y}^{-1} \circ y\colon \mathscr{S}_R \to \widetilde{\mathscr{S}}_R$. The following diagram illustrates the relations between $y$, $\tilde{y}$, $\lambda$, and their deformation gradients:

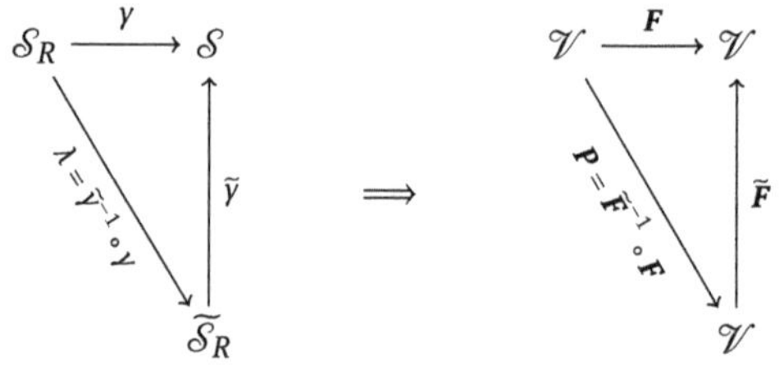


17 Since $\det \boldsymbol{F} > 0$.

Thus one has the following relation:

$$\mathfrak{g}_{\mathcal{S}_R}(X, \boldsymbol{F}_X) = \mathfrak{g}_{\mathcal{S}_R}(X, \widetilde{\boldsymbol{F}}_{\lambda(X)}\boldsymbol{P}_X)\,.$$

The right-hand side of this equality may be denoted by $\widetilde{\mathfrak{g}}_{\widetilde{\mathcal{S}}_R}(\widetilde{X}, \widetilde{\boldsymbol{F}}_{\widetilde{X}})$, *i.e.*,

$$\widetilde{\mathfrak{g}}_{\widetilde{\mathcal{S}}_R}(\widetilde{X}, \widetilde{\boldsymbol{F}}_{\widetilde{X}}) := \mathfrak{g}_{\mathcal{S}_R}\left(\lambda^{-1}(\widetilde{X}), \widetilde{\boldsymbol{F}}_{\widetilde{X}}\boldsymbol{P}_{\lambda^{-1}(\widetilde{X})}\right)\,.$$

Thus, we have the response functional $\widetilde{\mathfrak{g}}_{\widetilde{\mathcal{S}}_R}$. However, we need to reflect that we are dealing with the same solid. Hence, it should be $\widetilde{\mathfrak{g}}_{\widetilde{\mathcal{S}}_R} = \mathfrak{g}_{\widetilde{\mathcal{S}}_R}$. This motivates the following.

**Axiom.** *For any shapes $\mathcal{S}_R$ and $\widetilde{\mathcal{S}}_R$ response functionals $\mathfrak{g}_{\mathcal{S}_R}$ and $\mathfrak{g}_{\widetilde{\mathcal{S}}_R}$ are related by*

$$\mathfrak{g}_{\widetilde{\mathcal{S}}_R}(\widetilde{X}, \widetilde{\boldsymbol{F}}_{\widetilde{X}}) = \mathfrak{g}_{\mathcal{S}_R}(X, \boldsymbol{F}_X)\,, \quad \textit{where} \quad \widetilde{X} = \lambda(X)\,, \quad \widetilde{\boldsymbol{F}} = \boldsymbol{F}\boldsymbol{P}^{-1}\,. \tag{3.45}$$

**Remark 3.12.** We can consider $\mathcal{S}_R$ as a “test” reference shape. Relative to this shape response functional $\mathfrak{g}_{\mathcal{S}_R}$ was obtained from experiments. Then, according to the relation (3.45) response functionals for the same solid relative to another reference shapes can be built.

Not all mappings $\mathfrak{g}_{\mathcal{S}_R}$ satisfy the principle of material-frame indifference. To obtain the appropriate type of such mappings, let $\boldsymbol{Q}$ be an orthogonal tensor. Then, according to (3.40) and the principle of material-frame indifference[18] we have

$$\mathfrak{g}_{\mathcal{S}_R^*}(X^*, \boldsymbol{F}^*_{X^*}) = \boldsymbol{Q}(t)\mathfrak{g}_{\mathcal{S}_R}(X, \boldsymbol{F}_X)\boldsymbol{Q}^{\mathrm{T}}(t)\,.$$

The left-hand side of the equality may be considered as the value of the response functional that is relative to the reference shape $\mathcal{S}_R^*$. According to the axiom, taking $\boldsymbol{P} = \boldsymbol{Q}(\widetilde{t})$, we arrive at the following equality:

$$\mathfrak{g}_{\mathcal{S}_R}(X, \boldsymbol{F}^*_{X^*}\boldsymbol{Q}(\widetilde{t})) = \boldsymbol{Q}(t)\mathfrak{g}_{\mathcal{S}_R}(X, \boldsymbol{F}_X)\boldsymbol{Q}^{\mathrm{T}}(t)\,.$$

Taking into account the equality (3.42) and the identity $\boldsymbol{Q}^{\mathrm{T}}(\widetilde{t})\boldsymbol{Q}(\widetilde{t}) = \boldsymbol{I}$, we obtain the following expression:

$$\mathfrak{g}_{\mathcal{S}_R}(X, \boldsymbol{Q}(t)\boldsymbol{F}_X) = \boldsymbol{Q}(t)\mathfrak{g}_{\mathcal{S}_R}(X, \boldsymbol{F}_X)\boldsymbol{Q}^{\mathrm{T}}(t)\,,$$

which holds for each orthogonal tensor $\boldsymbol{Q}$. Now, let $\boldsymbol{F} = \boldsymbol{R}\boldsymbol{U}$ be a polar decomposition (see (3.43)). Taking $\boldsymbol{Q} = \boldsymbol{R}^{\mathrm{T}}$ leads to the final expression:

$$\mathfrak{g}_{\mathcal{S}_R}(X, \boldsymbol{U}_X) = \boldsymbol{R}^{\mathrm{T}}\mathfrak{g}_{\mathcal{S}_R}(X, \boldsymbol{F}_X)\boldsymbol{R}\,. \tag{3.46}$$

This expression defines the functional restriction on possible mappings $\mathfrak{g}_{\mathcal{S}_R}$ and reflects the requirement from the principle of material-frame indifference. From (3.46), it follows that

$$\boldsymbol{T}_X = \boldsymbol{R}\mathfrak{g}_{\mathcal{S}_R}(X, \boldsymbol{U}_X)\boldsymbol{R}^{\mathrm{T}}\,. \tag{3.47}$$

The reference shape $\mathcal{S}_R$ is *stress free*, if $\mathfrak{g}_{\mathcal{S}_R}(X, \boldsymbol{I}_X) = \boldsymbol{0}$ for each $X \in \mathcal{S}_R$. The assumption of the existence of a stress-free reference shape for any solid is traditional in conventional elasticity.

18 We consider equivalent processes. According to the principle of material-frame indifference since $\boldsymbol{T}_X = \mathfrak{g}_{\mathcal{S}_R}(X, \boldsymbol{F}_X)$, then $\boldsymbol{T}^*_X = \mathfrak{g}_{\mathcal{S}_R^*}(X^*, \boldsymbol{F}^*_X)$, and symbol $\mathfrak{g}$ is the same.

### 3.9.4 Representation Theorems

Suppose that $\mathscr{V}$ is an $n$-dimensional Euclidean vector space. We will consider mappings $\varepsilon\colon \mathrm{Lin}(\mathscr{V};\mathscr{V})^k \to \mathbb{R}$ and $\boldsymbol{f}\colon \mathrm{Lin}(\mathscr{V};\mathscr{V})^k \to \mathrm{Lin}(\mathscr{V};\mathscr{V})$, where $\mathrm{Lin}(\mathscr{V};\mathscr{V})^k = \mathrm{Lin}(\mathscr{V};\mathscr{V}) \times \cdots \times \mathrm{Lin}(\mathscr{V};\mathscr{V})$ ($k$ times). Our goal is to introduce the notion of isotropic mapping [37].

Mapping $\varepsilon\colon (\boldsymbol{A}_1,\dots,\boldsymbol{A}_k) \mapsto \varepsilon(\boldsymbol{A}_1,\dots,\boldsymbol{A}_k)$ is called *isotropic*, if the following equality

$$\varepsilon(\boldsymbol{A}_1,\dots,\boldsymbol{A}_k) = \varepsilon\left(\boldsymbol{Q}\boldsymbol{A}_1\boldsymbol{Q}^{\mathrm{T}},\dots,\boldsymbol{Q}\boldsymbol{A}_k\boldsymbol{Q}^{\mathrm{T}}\right)$$

holds for all orthogonal mappings $\boldsymbol{Q}$ and all tuples $(\boldsymbol{A}_1,\dots,\boldsymbol{A}_k)$ from the domain of $\varepsilon$.

Isotropic mapping $\varepsilon\colon \boldsymbol{A} \mapsto \varepsilon(\boldsymbol{A})$ of one variable is called the *orthogonal invariant* (or, simply, *invariant*) of $\boldsymbol{A}$. In the case of several variables, $\varepsilon$ is called a *simultaneous invariant*. The following theorem sets necessary and sufficient conditions for $\varepsilon$ to be invariant:

**Theorem.** *A mapping $\varepsilon\colon \mathrm{Sym}(\mathscr{V}) \to \mathbb{R}$ is isotropic (invariant) iff it can be represented as a function of the principal invariants of its argument. That is,*

$$\varepsilon(\boldsymbol{A}) = \tilde{\varepsilon}(I_1(\boldsymbol{A}), I_2(\boldsymbol{A}), I_n(\boldsymbol{A}))\,.$$

Mapping $\boldsymbol{f}\colon (\boldsymbol{A}_1,\dots,\boldsymbol{A}_k) \mapsto \boldsymbol{f}(\boldsymbol{A}_1,\dots,\boldsymbol{A}_k)$ is called *isotropic*, if the following equality:

$$\boldsymbol{Q}\boldsymbol{f}(\boldsymbol{A}_1,\dots,\boldsymbol{A}_k)\boldsymbol{Q}^{\mathrm{T}} = \boldsymbol{f}\left(\boldsymbol{Q}\boldsymbol{A}_1\boldsymbol{Q}^{\mathrm{T}},\dots,\boldsymbol{Q}\boldsymbol{A}_k\boldsymbol{Q}^{\mathrm{T}}\right)$$

holds for all orthogonal mappings $\boldsymbol{Q}$ and all tuples $(\boldsymbol{A}_1,\dots,\boldsymbol{A}_k)$ from the domain of $\boldsymbol{f}$.

Now, we formulate the classical representation theorem. It is formulated for the general case of $n$-dimensional real vector space $\mathscr{V}$.

**Theorem** (The Rivlin and Ericksen representation theorem). *A mapping $\boldsymbol{f}\colon \boldsymbol{A} \mapsto \boldsymbol{f}(\boldsymbol{A})$, defined on symmetric tensors, is isotropic iff it can be expressed as follows:*

$$\boldsymbol{D} = \boldsymbol{f}(\boldsymbol{A}) = \varphi_0\boldsymbol{I} + \varphi_1\boldsymbol{A} + \varphi_2\boldsymbol{A}^2 + \cdots + \varphi_{n-1}\boldsymbol{A}^{n-1}\,.$$

*Here $\varphi_k = \tilde{\varphi}_k(I_1(\boldsymbol{A}),\dots,I_n(\boldsymbol{A}))$.*

*Proof*[19]. Let $\boldsymbol{A}$ be a symmetric tensor. Consider an eigenvector $\boldsymbol{e}$ of the tensor $\boldsymbol{A}$. If $\mathscr{U} = \{x\boldsymbol{e} \mid x \in \mathbb{R}\}$ denotes the one-dimensional vector space generated by $\boldsymbol{e}$, then one has the direct sum $\mathscr{V} = \mathscr{U} \otimes \mathscr{U}^{\perp}$ of the line $\mathscr{U}$ and the plane $\mathscr{U}^{\perp}$. Define orthogonal transformation $\boldsymbol{Q}$ for which $\mathscr{U}^{\perp}$ is a fixed set[20]:

$$\boldsymbol{Q}\boldsymbol{u} = \begin{cases} -\boldsymbol{u}, & \boldsymbol{u} \in \mathscr{U}\,,\\ \boldsymbol{u}, & \boldsymbol{u} \in \mathscr{U}^{\perp}\,. \end{cases}$$

19 We follow [37].

20 $\boldsymbol{Q}$ is a reflection on the plane $\mathscr{U}^{\perp}$. If one choose a basis $(\boldsymbol{e}_i)_{i=1}^n$ for $\mathscr{V}$, where $\boldsymbol{e}_1 = \boldsymbol{e}$, and $(\boldsymbol{e}_i)_{i=2}^n$ is some basis for $\mathscr{U}^{\perp}$, then $[\boldsymbol{Q}^i{}_j] = \mathrm{diag}\,\{-1, 1, \dots, 1\}$.

According to the definition[21], $\boldsymbol{QAQ}^{\mathrm{T}} = \boldsymbol{A}$. Since $\boldsymbol{f}$ is an isotropic function, $\boldsymbol{QDQ}^{\mathrm{T}} = \boldsymbol{D}$ or $\boldsymbol{QD} = \boldsymbol{DQ}$. Thus,

$$\boldsymbol{Q}(\boldsymbol{De}) = \boldsymbol{D}(\boldsymbol{Qe}) = -\boldsymbol{De} \,.$$

Hence, $\boldsymbol{De} \in \mathcal{U}$. That is, $\boldsymbol{De} = d\boldsymbol{e}$ for some $d \in \mathbb{R}$. This means that $\boldsymbol{e}$ is an eigenvector for $\boldsymbol{D}$.

Another eigenvectors of $\boldsymbol{A}$ are considered in an analogous way. Thus, one has that all eigenvectors of $\boldsymbol{A}$ are eigenvectors of $\boldsymbol{D}$.

Denote all distinct eigenvalues of $\boldsymbol{A}$ by $a_1, \ldots, a_m$, where $m \leq n$, and the corresponding eigenvectors by $\boldsymbol{e}_1, \ldots, \boldsymbol{e}_m$. These vectors are similarly eigenvectors for $\boldsymbol{D}$, which correspond to eigenvalues $d_1, \ldots, d_m$. Note that these numbers may be not distinct. The system of $m$ equations

$$d_k = \varphi_0 + \varphi_1 a_k + \varphi_2 a_k^2 + \cdots + \varphi_{m-1} a_k^{m-1} \,, \quad k = 1, \ldots, m \,,$$

has the unique solution $\{\varphi_0, \varphi_1, \ldots, \varphi_{m-1}\}$, since its determinant is equal to

$$\prod_{j<k} (a_j - a_k) \neq 0 \,.$$

Thus,

$$\boldsymbol{D} = \boldsymbol{f}(\boldsymbol{A}) = \varphi_0 \boldsymbol{I} + \varphi_1 \boldsymbol{A} + \varphi_2 \boldsymbol{A}^2 + \cdots + \varphi_{m-1} \boldsymbol{A}^{m-1} \,, \quad m \leq n \,.$$

Since $\boldsymbol{Qf}(\boldsymbol{A})\boldsymbol{Q}^{\mathrm{T}} = \boldsymbol{f}(\boldsymbol{QAQ}^{\mathrm{T}})$, one has

$$\boldsymbol{f}(\boldsymbol{QAQ}^{\mathrm{T}}) = \varphi_0 \boldsymbol{I} + \varphi_1 (\boldsymbol{QAQ}^{\mathrm{T}}) + \cdots + \varphi_{m-1} (\boldsymbol{QAQ}^{\mathrm{T}})^{m-1} \,.$$

This implies that coefficients $\varphi_k$ remain unchanged if one changes $\boldsymbol{A}$ into $\boldsymbol{Q}(\boldsymbol{A})\boldsymbol{Q}^{\mathrm{T}}$. Thus, they are invariants of $\boldsymbol{A}$. □

**Remark 3.13.** Coefficients $\varphi_k$ are uniquely determined by $\boldsymbol{f}(\boldsymbol{A})$ if all eigenvalues are distinct, that is, $n = m$. If $m < n$, then, $n - m$ functions $\varphi_k$ may be chosen to be zero.

Now, we return to simple materials. Intuitively speaking, the notion of homogeneous solid isotropy means that such a solid can be brought into a shape, no rotations of which can be detected by the stress measurer. The following theorem takes place [37, 52]:

**Theorem.** *Simple material is isotropic iff its constitutive relation may be written in the form:*

$$\boldsymbol{T} = \boldsymbol{f}(\boldsymbol{B}) \,,$$

*relative to some undistorted shape. Here,* $\boldsymbol{B}$ *is a left Cauchy–Green tensor and* $\boldsymbol{f}$ *is isotropic mapping.*

---

21 Denote $\boldsymbol{B} = \boldsymbol{QAQ}^{\mathrm{T}}$. Then $\boldsymbol{QA} = \boldsymbol{BQ}$. If $\boldsymbol{u} \in \mathcal{U}$ then it directly follows that $\boldsymbol{Bu} = \boldsymbol{Au}$. Now suppose that $\boldsymbol{u} \in \mathcal{U}^{\perp}$. Since $\boldsymbol{e} \cdot \boldsymbol{Au} = \boldsymbol{u} \cdot \boldsymbol{Ae} = 0$, one has that $\boldsymbol{Au} \in \mathcal{U}^{\perp}$. Then we again have $\boldsymbol{Bu} = \boldsymbol{Au}$ and it proves the equality $\boldsymbol{B} = \boldsymbol{A}$.

From the theorem and the Rivlin and Ericksen representation theorem, it follows that relative to some undistorted shape,

$$\boldsymbol{T} = \varphi_0 \boldsymbol{I} + \varphi_1 \boldsymbol{B} + \varphi_2 \boldsymbol{B}^2 .$$

According to the Caley–Hamilton theorem, $\boldsymbol{B}^3 - I_1(\boldsymbol{B})\boldsymbol{B}^2 + I_2(\boldsymbol{B})\boldsymbol{B} - I_3(\boldsymbol{B})\boldsymbol{I} = \boldsymbol{0}$, and we can rewrite the considered constitutive relation in another form:

$$\begin{aligned}\boldsymbol{T} = {} & \beth_0(I_1(\boldsymbol{B}), I_2(\boldsymbol{B}), I_3(\boldsymbol{B}))\boldsymbol{I} + \beth_1(I_1(\boldsymbol{B}), I_2(\boldsymbol{B}), I_3(\boldsymbol{B}))\boldsymbol{B} \\ & + \beth_{-1}(I_1(\boldsymbol{B}), I_2(\boldsymbol{B}), I_3(\boldsymbol{B}))\boldsymbol{B}^{-1} ,\end{aligned} \tag{3.48}$$

## 3.10 Hyperelastic Solids

### 3.10.1 Expressions for Stresses

A hyperelastic solid is such an elastic solid for which the stress-strain relation is derived from a *strain energy density function* $W$. For simple solids, this function depends on the deformation gradient $\boldsymbol{F}$: $W = W(\boldsymbol{F})$. The first Piola–Kirchhoff stress tensor $\boldsymbol{P}$ can be derived by the relation

$$\boldsymbol{P} = \frac{\partial W}{\partial \boldsymbol{F}} .$$

Then, the Cauchy stress tensor $\boldsymbol{T}$ has the form

$$\boldsymbol{T} = J^{-1} \frac{\partial W}{\partial \boldsymbol{F}} \boldsymbol{F}^{\mathrm{T}} .$$

In terms of the right and left Cauchy–Green strains $\boldsymbol{C}$ and $\boldsymbol{B}$, the Cauchy stresses are expressed as follows:

$$\boldsymbol{T} = 2J^{-1} \boldsymbol{F} \frac{\partial W}{\partial \boldsymbol{C}} \boldsymbol{F}^{\mathrm{T}} , \quad \boldsymbol{T} = 2J^{-1} \boldsymbol{B} \frac{\partial W}{\partial \boldsymbol{B}} .$$

### 3.10.2 Universal Deformations

The non-linear equations of the elasticity theory are rather complicated, and the application of regular methods developed in linear theory do not allow the construction of analytic solutions of any non-trivial boundary value problem. At the same time, in the first half of the twentieth century, several analytical representations for the solution of particular non-linear problems were found. A characteristic feature of these solutions was that they could be produced in every incompressible isotropic elastic material by the application of surface traction only. For this reason, the deformations that are determined by such solutions are called *controllable*. Moreover, such deformations do not depend on a particular choice of elastic potential and, therefore, they are also called *universal*.

The general strategy for obtaining universal solutions was developed by J. L. Ericksen. He proved that for compressible materials, only homogeneous deformations are universal [83]. Studies of incompressible materials proved to be more fruitful. As a result of thorough research, he obtained four families of universal solutions for incompressible material [84]. Further development on this problem was achieved by Singh and Pipkin in their work on universal solutions for elastic dielectrics [85] and by Klingbeil and Shield, who described the fifth family for hyperelastic solids [86].

Here, we describe the procedure for constructing of universal deformations in more detail.

**Claim** (The Ericksen Theorem). *The universal deformations in compressible isotropic hyperelastic solid are necessarily homogeneous [83].*

We will prove this claim in the spirit of the original Ericksen argumentation, explicitly taking into account the Euclidean properties of physical space containing reference and deformed shapes. Note that there are other variants of the proof [87, 88].

Let $y$ be a deformation that brings the natural reference shape to a deformed (actual) shape and $\boldsymbol{F}$ be the corresponding deformation gradient. Let $\mathbf{c}$ be an Almansi strain, *i.e.*, $\mathbf{c} = \boldsymbol{B}^{-1} = \boldsymbol{F}^{-\mathrm{T}}\boldsymbol{F}^{-1}$. In the actual chart, the components of $\mathbf{c}$ can be written as follows:

$$c^a{}_b = g^{al}\left\{(\boldsymbol{F}^{-\mathrm{T}})^A{}_l(\boldsymbol{F}^{-1})^B{}_b G_{AB}\right\} \circ y^{-1} .$$

Suppose that the material is isotropic. Then, the strain energy function can be written as a dependence on principal invariants of $\mathbf{c}$:

$$W = W(I_1, I_2, I_3)\,; \quad I_1 = \operatorname{tr}\mathbf{c}\,, \quad I_2 = \frac{1}{2}((\operatorname{tr}\mathbf{c})^2 - \operatorname{tr}\mathbf{c}^2)\,, \quad I_3 = \det\mathbf{c}\,.$$

Cauchy stresses are defined as

$$\boldsymbol{T} = -2I_3^{1/2}\mathbf{c}\frac{\partial W}{\partial \mathbf{c}^{\mathrm{T}}} = -2I_3^{1/2}\mathbf{c}\sum_{N=1}^{3}\frac{\partial W}{\partial I_N}\frac{\partial I_N}{\partial \mathbf{c}^{\mathrm{T}}}\,. \tag{3.49}$$

Substituting (3.49) into the momentum equation, we obtain the following relation:

$$\operatorname{div}\boldsymbol{T} = -2\left\{\sum_{N=1}^{3}\frac{\partial W}{\partial I_N}\operatorname{div}\left(I_3^{1/2}\mathbf{c}\frac{\partial I_N}{\partial \mathbf{c}^{\mathrm{T}}}\right) + I_3^{1/2}\sum_{N=1}^{3}\sum_{M=1}^{3}\frac{\partial^2 W}{\partial I_N \partial I_M}(\nabla I_M)\mathbf{c}\frac{\partial I_N}{\partial \mathbf{c}^{\mathrm{T}}}\right\} = \mathbf{0}\,. \tag{3.50}$$

We are looking for solutions of (3.50) that do not depend on a particular choice of $W(I_1, I_2, I_3)$. Thus, $\mathbf{c}$ must be such that the multipliers for the derivatives with respect to the variables $I_1, I_2, I_3$ vanish, *i.e.*,

$$\operatorname{div}\left(I_3^{1/2}\mathbf{c}\frac{\partial I_N}{\partial \mathbf{c}^{\mathrm{T}}}\right) = \mathbf{0}\,,$$

$$I_3^{1/2}\left((\nabla I_M)\mathbf{c}\frac{\partial I_N}{\partial \mathbf{c}^{\mathrm{T}}} + (\nabla I_N)\mathbf{c}\frac{\partial I_M}{\partial \mathbf{c}^{\mathrm{T}}}\right) = \mathbf{0}\,.$$

Taking into account the expressions for derivatives of principal invariants,

$$\frac{\partial I_1}{\partial \boldsymbol{c}} = \boldsymbol{I}\,, \quad \frac{\partial I_2}{\partial \boldsymbol{c}} = I_1\boldsymbol{I} - \boldsymbol{c}\,, \quad \frac{\partial I_3}{\partial \boldsymbol{c}} = I_3\boldsymbol{c}^{-\mathrm{T}}\,,$$

with $N = 3$ and $M = 1$, we obtain

$$\operatorname{div}\left(I_3^{1/2}\boldsymbol{c}\frac{\partial I_3}{\partial \boldsymbol{c}^{\mathrm{T}}}\right) = \operatorname{div}\left(I_3^{1/2}\boldsymbol{c}\boldsymbol{c}^{-1}I_3\right) = \operatorname{div}\left(I_3^{3/2}\boldsymbol{I}\right) = \nabla I_3^{3/2} = \boldsymbol{0}\,,$$

$$(\nabla I_1)\boldsymbol{c}\frac{\partial I_3}{\partial \boldsymbol{c}^{\mathrm{T}}} + (\nabla I_3)\boldsymbol{c}\frac{\partial I_1}{\partial \boldsymbol{c}^{\mathrm{T}}} = (\nabla I_1)I_3 + \nabla I_3 = \boldsymbol{0}\,.$$

Thus, $I_3 = \text{const}$ and, consequently, $I_1 = \text{const}$. These equations imply that

$$\nabla(\operatorname{tr}\mathbf{c}) = \mathbf{0}\,, \quad \operatorname{div}\mathbf{c} = \mathbf{0}\,, \tag{3.51}$$

or, in components,

$$c^j{}_{j,i} = 0\,, \quad c^{ij}{}_{,j} = 0\,.$$

We assume that the actual shape, as well as the reference shape, are embedded into Euclidean space. Thus, the Riemann curvature tensor vanishes both on the metric tensor $\boldsymbol{g}$, induced by actual coordinates that can be considered with a chart on physical space, and on the metric tensor $\boldsymbol{G}$, induced by the Almansi strain field according to relation $\boldsymbol{c}^\flat = y_*\boldsymbol{G}$, *i.e.*,

$$\mathfrak{R}(\boldsymbol{c}^\flat) = \mathfrak{R}(y_*\boldsymbol{G}) = y_*\mathfrak{R}(\boldsymbol{G}) = \boldsymbol{0}\,.$$

In a chart, this equation can be written as

$$R_{ijkm} = \frac{1}{2}\left(c_{im,kj} + c_{jk,im} - c_{ik,mj} - c_{jm,ik}\right) + (\boldsymbol{c}^{-1})^{rs}(A_{jkr}A_{ims} - A_{jmr}A_{iks}) = 0\,,$$

where

$$A_{ijk} = A_{jik} = \frac{1}{2}(c_{ik,j} + c_{jk,i} - c_{ij,k})\,,$$

$c_{im}$ are components of $\boldsymbol{c}^\flat$ and $(\boldsymbol{c}^{-1})^{rs}$ are components of $(\boldsymbol{c}^\flat)^{-1}$. From (3.51), it follows that

$$A_{iik} = \frac{1}{2}(c_{ik,i} + c_{ik,i} - c_{ii,k}) = 0\,.$$

Hence,

$$R^{ij}{}_{ji} = -(\boldsymbol{c}^{-1})^{rs}A^j{}_{ir}A^i{}_{js} = 0 \tag{3.52}$$

Taking into account that all eigenvalues of $\boldsymbol{c}^{-1}$ are positive, and can be represented as $\mathbf{c}^{-1} = \mathbf{c}^{-1/2}\mathbf{c}^{-1/2}$, we rewrite the relation (3.52) in the form

$$A^{ij}{}_r(\boldsymbol{c}^{-1/2})^r{}_k A_{ijs}(\boldsymbol{c}^{-1/2})^{sk} = 0\,.$$

Since $\mathbf{c}^{-1/2}$ has an inverse,

$$A_{ijs} = 0\,,$$

and, consequently,

$$A_{ijk} + A_{ikj} = c_{jk,i} = 0\,,$$

so $c_{ij}$ is a constant tensor. It is a necessary and sufficient condition that $c_{ij}$ corresponds to a homogeneous deformation.

**Incompressible material.** If $J = \det \boldsymbol{F} = 1$, then the material is called *incompressible.* Cauchy stresses in incompressible material can be represented by the formula [45]

$$\boldsymbol{T} = -p\boldsymbol{I} + 2\frac{\partial W}{\partial I_1}\boldsymbol{c}^{-1} + \frac{\partial W}{\partial I_2}\boldsymbol{c}\,, \tag{3.53}$$

where $p$ is a hydrostatic component, $W$ is elastic potential, $\boldsymbol{c}$ the Almansi strain, while $I_1, I_2$ are the principal invariants of $\boldsymbol{c}$. As above, $\boldsymbol{I}$ denotes the identity tensor. By substituting (3.53) into the momentum equation $\nabla \cdot \boldsymbol{T} = \boldsymbol{0}$, one can obtain the following expression for gradient of $p$:

$$\nabla p = 2\left(\boldsymbol{c}^{-1}\nabla\frac{\partial W}{\partial I_1} - \boldsymbol{c}\nabla\frac{\partial W}{\partial I_2} + \frac{\partial W}{\partial I_1}\nabla\cdot\boldsymbol{c}^{-1} - \frac{\partial W}{\partial I_2}\nabla\cdot\boldsymbol{c}\right).$$

From the evident relation $\nabla \times \nabla p = \boldsymbol{0}$ it follows that

$$\nabla\times\left(\boldsymbol{c}^{-1}\nabla\frac{\partial W}{\partial I_1} - \boldsymbol{c}\nabla\frac{\partial W}{\partial I_2} + \frac{\partial W}{\partial I_1}\nabla\cdot\boldsymbol{c}^{-1} - \frac{\partial W}{\partial I_2}\nabla\cdot\boldsymbol{c}\right) = \boldsymbol{0}\,.$$

By their definition, universal deformations shall satisfy this equation for all $W(I_1, I_2)$. Thus, the multipliers at all derivatives of $W$ must be identical to zero. This consideration results in the following system of equations [45]:

$$\begin{aligned}
\nabla I_1 \times \left(\nabla I_1 \cdot \boldsymbol{c}^{-1}\right) &= \boldsymbol{0}\,,\\
\nabla I_2 \times (\nabla I_2 \cdot \boldsymbol{c}) &= \boldsymbol{0}\,,\\
\nabla I_2 \times \nabla\left(I_1\boldsymbol{c}^{-1}\right) + \nabla I_1 \times \left(\nabla I_2 \cdot \boldsymbol{c}^{-1}\right) - \nabla I_1 \times (\nabla I_1 \cdot \boldsymbol{c}) &= \boldsymbol{0}\,,\\
\nabla I_1 \times \nabla\left(I_2\boldsymbol{c}^{-1}\right) + \nabla I_2 \times (\nabla I_1 \cdot \boldsymbol{c}) - \nabla I_2 \times \left(\nabla I_2 \cdot \boldsymbol{c}^{-1}\right) &= \boldsymbol{0}\,,\\
\nabla \times \left(\nabla I_1 \cdot \boldsymbol{c}^{-1}\right) + \nabla I_1 \times \nabla \cdot \boldsymbol{c}^{-1} &= \boldsymbol{0}\,,\\
\nabla \times (\nabla I_2 \cdot \boldsymbol{c}) + \nabla I_2 \times \nabla \cdot \boldsymbol{c} &= \boldsymbol{0}\,,\\
\nabla \times \left(\nabla I_2 \cdot \boldsymbol{c}^{-1}\right) + \nabla I_2 \times \nabla \cdot \boldsymbol{c}^{-1} - \nabla \times (\nabla I_1 \cdot \boldsymbol{c}) - \nabla I_1 \times \nabla \cdot \boldsymbol{c} &= \boldsymbol{0}\,,\\
\nabla \times \nabla \cdot \boldsymbol{c}^{-1} &= \boldsymbol{0}\,,\\
\nabla \times \nabla \cdot \boldsymbol{c} &= \boldsymbol{0}\,,
\end{aligned}$$

which, together with incompressibility condition,

$$I_3(\boldsymbol{c}) = 1$$

and the compatibility equation (3.18) define the over-determined system. Despite the strong over-determination, this system allows non-trivial solutions. The solutions are listed below. Some of them are used as core solutions in the computations given in Chapter 11.

**Universal solutions for incompressible material.** The reference positions of material points that constitute the bodies, subjected to different families of universal deformations, are determined by appropriate coordinates. Since, by assumption, the physical space containing the shape of the bodies is Euclidean, the Cartesian coordinates can be assumed to be given. Cylindrical and spherical coordinates are related with Cartesians as follows.

- Family 1. Cartesian coordinates $(X, Y, Z)$.
- Families 2, 3, 5. Cylindrical coordinates $(R, \Phi, Z)$,

$$R = \sqrt{X^2 + Y^2}\,, \quad \Phi = \arctan\frac{Y}{X}\,; \qquad X = R\cos\Phi\,, \quad Y = R\sin\Phi\,.$$

- Family 4. Spherical coordinates $(\mathfrak{R}, \Theta, \Phi)$,

$$R = \sqrt{X^2 + Y^2 + Z^2}\,, \quad \Theta = \arctan\frac{\sqrt{X^2 + y^2}}{Z}\,, \quad \Phi = \arctan\frac{Y}{X}\,;$$
$$X = R\cos\Phi\sin\Theta\,, \quad Y = R\cos\Phi\sin\Theta\,, \quad Z = R\cos\Theta\,.$$

Deformation of the shape of the body can be associated with a transition to a new coordinate chart. Then, the coordinates of the material points (ordered triples of numbers) do not change, while the transition functions that determine the relation between reference and actual charts completely characterize the deformation. Such transition functions for different families of universal deformations are given in Tab. 3.1.

Positions of material points in actual and reference shapes can be expressed by elements of local bases, as is shown in Tab. 3.2.

For a given deformation, one can obtain the deformation gradient $\boldsymbol{F}$ and the Cauchy–Green strain $\boldsymbol{C}$. The results of the calculations are given in Tab. 3.3.

The principal invariants of Cauchy–Green strains, $I_1$ and $I_2$ are given in Tab. 3.4.

Substituting the expressions for the Cauchy–Green strains into the constitutive equation, we obtain the Cauchy stress tensor, whose physical components are listed below (only the upper triangular part of the symmetric matrix):

- Family 1.

$$\begin{bmatrix} -p(r) + \frac{a^2}{r^2}\overline{\psi}_1(r) + \frac{r^2}{a^2}\overline{\psi}_{-1}(r) & 0 & 0 \\ & -p(r) + \left(b^2 + c^2\right) r^2\overline{\psi}_1(r) + \frac{a^2}{r^2}\left(d^2 + e^2\right)\overline{\psi}_{-1}(r) & (bd + ce) \times \left(r\overline{\psi}_1(r) - \frac{a^2}{r}\overline{\psi}_{-1}(r)\right) \\ & & -p(r) + a^2\left(b^2 + c^2\right)\overline{\psi}_{-1}(r) + \left(d^2 + e^2\right)\overline{\psi}_1(r) \end{bmatrix}$$

**Tab. 3.1:** Transition functions for universal deformations.

| # | Reference chart | Actual chart | Transition function | Incomp. condition |
|---|---|---|---|---|
| 1 | $(X, Y, Z)$ | $(r, \phi, z)$ | $r = \sqrt{2aX}$,<br>$\phi = bY + cZ$,<br>$z = dY + eZ$ | $a(be - cd) = 1$ |
| 2 | $(R, \Phi, Z)$ | $(x, y, z)$ | $x = 1/2aR^2$,<br>$y = b\Phi + cZ$,<br>$z = d\Phi + eZ$ | $a(be - cd) = 1$ |
| 3 | $(R, \Phi, Z)$ | $(r, \phi, z)$ | $r = \sqrt{aR^2 + b}$,<br>$\phi = c\Phi + dZ$,<br>$z = e\Phi + fz$ | $a(cf - de) = 1$ |
| 4 | $(\mathfrak{R}, \Theta, \Phi)$ | $(\mathfrak{r}, \theta, \phi)$ | $\mathfrak{r} = (\pm\mathfrak{R}^3 + a)^{1/3}$,<br>$\theta = \pm\Theta$,<br>$\phi = \Phi$ | |
| 5 | $(R, \Phi, Z)$ | $(r, \phi, z)$ | $\phi = b \log R + c\Phi$,<br>$r = aR$,<br>$z = dZ$ | $a^2 cd = 1$ |

**Tab. 3.2:** Positions of material points in actual and reference shapes.

| # | Reference position | Actual position |
|---|---|---|
| 1 | $o + X\boldsymbol{i}_1 + Y\boldsymbol{i}_2 + Z\boldsymbol{i}_3$ | $o + \rho\,\boldsymbol{e}_r + z\,\boldsymbol{i}_3$ |
| 2 | $o + R\,\boldsymbol{e}_r + Z\,\boldsymbol{i}_3$ | $o + x\,\boldsymbol{i}_1 + y\,\boldsymbol{i}_2 + z\,\boldsymbol{i}_3$ |
| 3 | $o + R\,\boldsymbol{e}_r + Z\,\boldsymbol{i}_3$ | $o + r\,\boldsymbol{e}_r + z\,\boldsymbol{i}_3$ |
| 4 | $o + \mathfrak{R}\,\boldsymbol{e}_{\mathfrak{r}}$ | $o + \mathfrak{r}\,\boldsymbol{e}_{\mathfrak{r}}$ |
| 5 | $o + R\,\boldsymbol{e}_r$ | $o + x\,\boldsymbol{i}_1 + y\,\boldsymbol{i}_2 + z\,\boldsymbol{i}_3$ |

- Family 2.

$$\begin{bmatrix} -p(x) + 2ax\,\overline{\psi}_1(x) + \frac{1}{2ax}\overline{\psi}_{-1}(x) & 0 & 0 \\ & -p(x) + \frac{1}{2x}\left(ab^2 + 2c^2x\right)\overline{\psi}_1(x) + a\left(ad^2 + 2e^2x\right)\overline{\psi}_{-1}(x) & \frac{1}{2}(abd + 2cex) \times \left(\frac{1}{x}\overline{\psi}_1(x) - 2a\overline{\psi}_{-1}(x)\right) \\ & & -p(x) + \frac{1}{2x}\left(ad^2 + 2e^2x\right)\overline{\psi}_1(x) + a\left(ab^2 + 2c^2x\right)\overline{\psi}_{-1}(x) \end{bmatrix}$$

**Tab. 3.3:** Deformation gradients and left Cauchy–Green strains.

| # | Deformation gradients | Strains |
|---|---|---|
| 1 | $\begin{bmatrix} \frac{a}{r} & 0 & 0 \\ 0 & b & c \\ 0 & d & e \end{bmatrix}$ | $\begin{bmatrix} \frac{a^2}{r^2} & 0 & 0 \\ 0 & b^2+c^2 & b\,d+c\,e \\ 0 & b\,d+c\,e & d^2+e^2 \end{bmatrix}$ |
| 2 | $\begin{bmatrix} \sqrt{2\,a\,x} & 0 & 0 \\ 0 & b & c \\ 0 & d & e \end{bmatrix}$ | $\begin{bmatrix} 2\,a\,x & 0 & 0 \\ 0 & \frac{a\,b^2}{2x}+c^2 & \frac{a\,b\,d}{2x}+c\,e \\ 0 & \frac{a\,b\,d}{2x}+c\,e & \frac{a\,d^2}{2x}+e^2 \end{bmatrix}$ |
| 3 | $\begin{bmatrix} \frac{\sqrt{a(r^2-b)}}{r} & 0 & 0 \\ 0 & c & d \\ 0 & e & f \end{bmatrix}$ | $\begin{bmatrix} \frac{a(r^2-b)}{r^2} & 0 & 0 \\ 0 & \frac{ac^2}{r^2-b}+d^2 & \frac{ace}{r^2-b}+df \\ 0 & \frac{ace}{r^2-b}+df & \frac{ae^2}{r^2-b}+f^2 \end{bmatrix}$ |
| 4 | $\begin{bmatrix} \left(1-\frac{a}{\mathfrak{r}^3}\right)^{2/3} & 0 & 0 \\ 0 & 1 & 0 \\ 0 & 0 & 1 \end{bmatrix}$ | $\begin{bmatrix} \frac{(\mathfrak{r}^3-a)^{4/3}}{\mathfrak{r}^4} & 0 & 0 \\ 0 & \frac{1}{(\mathfrak{r}^3-a)^{2/3}} & 0 \\ 0 & 0 & \frac{\csc^2\theta}{(\mathfrak{r}^3-a)^{2/3}} \end{bmatrix}$ |
| 5 | $\begin{bmatrix} a & 0 & 0 \\ \frac{ab}{r} & c & 0 \\ 0 & 0 & d \end{bmatrix}$ | $\begin{bmatrix} \frac{1}{cd} & \frac{b}{cdr} & 0 \\ \frac{b}{cdr} & \frac{b^2+c^2}{cdr^2} & 0 \\ 0 & 0 & d^2 \end{bmatrix}$ |

**Tab. 3.4:** Principal invariants.

| # | $I_1$ | $I_2$ |
|---|---|---|
| 1 | $\frac{a^2}{r^2}+r^2\left(b^2+c^2\right)+d^2+e^2$ | $a^2\left(b^2+c^2\right)+\frac{a^2}{r^2}\left(d^2+e^2\right)+\frac{r^2}{a^2}$ |
| 2 | $\frac{a}{2x}\left(b^2+d^2\right)+2ax+c^2+e^2$ | $a^2\left(b^2+d^2\right)+\frac{1}{2a\,x}+2a\,x\left(c^2+e^2\right)$ |
| 3 | $a\left(\frac{bc^2+e^2}{r^2-b}-\frac{b}{r^2}+c^2+1\right)$ $+d^2r^2+f^2$ | $\frac{a}{r^2(r^2-b)}\left\{\frac{r^4}{a^2}-a\left(b-r^2\right)\left(c^2r^2+e^2\right)\right.$ $\left.+\left(b^2-2br^2+r^4\right)\left(d^2r^2+f^2\right)\right\}$ |
| 4 | $\frac{a^2-2a\mathfrak{r}^3+3\mathfrak{r}^6}{\mathfrak{r}^4(\mathfrak{r}^3-a)^{2/3}}$ | $\frac{2a^2-4a\mathfrak{r}^3+3\mathfrak{r}^6}{\mathfrak{r}^2(\mathfrak{r}^3-a)^{4/3}}$ |
| 5 | $\frac{1}{cd}\left(b^2+c\left(c+d^3\right)+1\right)$ | $\frac{d}{c}\left(b^2+c^2+1\right)+\frac{1}{d^2}$ |

- Family 3.

$$\left[\begin{array}{c:c:c} \begin{array}{l} -p(r)+\frac{r^2}{a(r^2-b)}\overline{\psi}_{-1}(r) \\ \quad +a\left(1-\frac{b}{r^2}\right)\overline{\psi}_1(r) \end{array} & 0 & 0 \\ \hdashline & \begin{array}{l} -p(r)+\left(\frac{ac^2}{r^2-b}+d^2\right)r^2\overline{\psi}_1(r) \\ \quad +\frac{a}{r^2}\left(ae^2+\left(r^2-b\right)f^2\right)\overline{\psi}_{-1}(r) \end{array} & \begin{array}{c} \left(ace+\left(r^2-b\right)df\right) \\ \times\frac{r^2\overline{\psi}_1(r)+a\left(b-r^2\right)\overline{\psi}_{-1}(r)}{r^3-br} \end{array} \\ \hdashline & & \begin{array}{l} -p(r) \\ \quad +a\left(ac^2+\left(r^2-b\right)d^2\right)\overline{\psi}_{-1}(r) \\ \qquad -\frac{ae^2+\left(r^2-b\right)f^2}{b-r^2}\overline{\psi}_1(r) \end{array} \end{array}\right]$$

- Family 4.

$$\begin{bmatrix} -p(\mathfrak{r}) + \dfrac{(\mathfrak{r}^3-a)^{4/3}}{\mathfrak{r}^4}\overline{\psi}_1(\mathfrak{r}) + \dfrac{\mathfrak{r}^4}{(\mathfrak{r}^3-a)^{4/3}}\overline{\psi}_{-1}(\mathfrak{r}) & 0 & 0 \\ & -p(\mathfrak{r}) + \dfrac{(\mathfrak{r}^3-a)^{2/3}}{\mathfrak{r}^2}\overline{\psi}_{-1}(\mathfrak{r}) + \dfrac{\mathfrak{r}^2}{(\mathfrak{r}^3-a)^{2/3}}\overline{\psi}_1(\mathfrak{r}) & 0 \\ & & -p(\mathfrak{r}) + \dfrac{(\mathfrak{r}^3-a)^{2/3}}{\mathfrak{r}^2}\overline{\psi}_{-1}(\mathfrak{r}) + \dfrac{\mathfrak{r}^2}{(\mathfrak{r}^3-a)^{2/3}}\overline{\psi}_1(\mathfrak{r}) \end{bmatrix}$$

- Family 5.

$$\begin{bmatrix} -p(r) + \frac{1}{cd}\{\left(b^2+c^2\right)d^2\overline{\psi}_{-1}(r) + \overline{\psi}_1(r)\} & \frac{1}{cd}\left(b\overline{\psi}_1(r) - d^2\overline{\psi}_{-1}(r)\right) & 0 \\ & -p(r) + \frac{1}{cd}\{d^2\overline{\psi}_{-1}(r) + \left(b^2+c^2\right)\overline{\psi}_1(r)\} & 0 \\ & & -p(r) + \frac{1}{d^2}\left(d^4\overline{\psi}_1(r) + \overline{\psi}_{-1}(r)\right) \end{bmatrix}$$

Here, we use the notation

$$\overline{\psi}_1(r) := \psi_1(I_1(r), I_2(r)), \quad \overline{\psi}_{-1}(r) := \psi_{-1}(I_1(r), I_2(r)).$$

In order to find hydrostatic component of stresses, one has to substitute these expressions into the momentum equations, written in corresponding coordinates. As the result, one can obtain

- Family 1.

$$-a^2r^3\,\partial_r p(r) - \left[a^4\left(d^2+e^2\right) - 3r^4\right]\overline{\psi}_{-1}(r) + a^4 r\,\partial_r\overline{\psi}_1(r) \\ - a^2\left[a^2 + r^4\left(b^2+c^2\right)\right]\overline{\psi}_1(r) + r^5\,\partial_r\overline{\psi}_{-1}(r) = 0$$

- Family 2.

$$\partial_x p(x) - 2a\,x\,\partial_x\overline{\psi}_1(x) - \frac{1}{2a\,x}\,\partial_x\overline{\psi}_{-1}(x) - 2a\,\overline{\psi}_1(x) - \frac{1}{2a\,x^2}\,\overline{\psi}_{-1}(x) = 0$$

- Family 3.

$$\partial_r p(r) - \left[\frac{a(b+r^2)}{r^3} + \frac{a\,r\,c^2}{b-r^2} - d^2 r\right]\overline{\psi}_1(r) \\ - \left[\frac{a\,b^2}{r^3}\left(f^2(b-r^2) - a\,e^2\right) + \frac{r(r^2-3b)}{a(b-r^2)^2}\right]\overline{\psi}_{-1}(r) \\ + \left(b-r^2\right)\frac{a}{r^2}\,\partial_r\overline{\psi}_1(r) + \frac{r^2}{a(b-r^2)}\,\partial_r\overline{\psi}_{-1}(r) = 0$$

- Family 4.

$$p(\mathfrak{r}) - \frac{2a^2}{\mathfrak{r}^3}\frac{a - 3\mathfrak{r}^3}{(\mathfrak{r}^3 - a)^{\frac{7}{3}}}\overline{\psi}_{-1}(\mathfrak{r}) + \frac{2a^2}{\mathfrak{r}^5(\mathfrak{r}^3 - a)^{\frac{2}{3}}}\overline{\psi}_1(\mathfrak{r})$$
$$- \frac{(\mathfrak{r}^3 - a)^{\frac{4}{3}}}{\mathfrak{r}^4}\partial_r\overline{\psi}_1(\mathfrak{r}) - \frac{\mathfrak{r}^4}{(\mathfrak{r}^3 - a)^{\frac{4}{3}}}\partial_r\overline{\psi}_{-1}(\mathfrak{r}) = 0$$

- Family 5.

$$\begin{cases} r\,\partial_r p(r,\phi) + (b^2 + c^2 - 1)(\frac{1}{c\,d}\overline{\psi}_1 - d\overline{\psi}_{-1}) = 0 \\ \partial_\phi p(r,\phi) + \frac{2b}{c}(\frac{1}{d}\overline{\psi}_1 - d\overline{\psi}_{-1}) = 0 \end{cases}$$

Integration of these equations gives the following expressions for the hydrostatic component:

- Family 1.

$$\begin{aligned} p = &\int \left\{ \left(\frac{a^2}{r^3} + r(b^2 + c^2)\right)\overline{\psi}_1(r) + \left(\frac{a^2}{r^3}(d^2 + e2) - 3\frac{r}{a^2}\right)\overline{\psi}_{-1}(r) \right. \\ &\left. - \frac{a^2}{r^2}\partial_r\overline{\psi}_1(r) - \frac{r^2}{a^2}\partial_r\overline{\psi}_{-1}(r)\right\} dr \\ = &-\frac{a^2}{r^2}\overline{\psi}_1(r) - \frac{r^2}{a^2}\overline{\psi}_{-1}(r) \\ &+ \int \left\{ \left(r^2(b^2 + c^2) - \frac{a^2}{r^2}\right)\overline{\psi}_1(r) + \left(\frac{a^2}{r^2}(d^2 + e2) - \frac{r^2}{a^2}\right)\overline{\psi}_{-1}(r)\right\} \frac{dr}{r} \\ = &-\frac{a^2}{r^2}\overline{\psi}_1(r) - \frac{r^2}{a^2}\overline{\psi}_{-1}(r) + \int \left\{\frac{1}{2}\frac{\partial I_1}{\partial r}\overline{\psi}_1(r) - \frac{1}{2}\frac{\partial I_2}{\partial r}\overline{\psi}_{-1}(r)\right\}\frac{dr}{r} \\ = &-\frac{a^2}{r^2}\overline{\psi}_1(r) - \frac{r^2}{a^2}\overline{\psi}_{-1}(r) + W + p_0\,. \end{aligned}$$

- Family 2.

$$\begin{aligned} p &= \int \left\{ 2a\,x\,\partial_x\overline{\psi}_1(x) + \frac{1}{2a\,x}\partial_x\psi_{-1}(x) + 2a\overline{\psi}_1(x) - \frac{1}{2a\,x^2}\overline{\psi}_{-1}(x)\right\} \\ &= 2a\,x\overline{\psi}_1(x) + \frac{1}{2a\,x}\overline{\psi}_{-1}(x) + p_0\,. \end{aligned}$$

- Family 3.

$$p = \int \left\{ \left[ \frac{a(b+r^2)}{r^3} + \frac{a\,r\,c^2}{b-r^2} - d^2 r \right] \overline{\psi}_1(r) \right.$$
$$+ \left[ \frac{a}{r^3}\left(f^2(b-r^2) - a\,e^2\right) + \frac{r(r^2-3b)}{a(b-r^2)^2} \right] \overline{\psi}_{-1}(r) - \left(b-r^2\right)\frac{a}{r^2}\,\partial_r \overline{\psi}_1(r)$$
$$\left. - \frac{r^2}{a(b-r^2)}\,\partial_r \overline{\psi}_{-1}(r) \right\}$$
$$= (r^2-b)\frac{a}{r^2}\,\overline{\psi}_1(r) - \frac{r^2}{a(r^2-b)}\,\overline{\psi}_{-1}(r)$$
$$+ \int \left\{ \left[ \frac{a(r^2-b)}{r^2} + \frac{a\,r^2\,c^2}{b-r^2} - d^2 r^2 \right] \overline{\psi}_1(r) \right.$$
$$\left. + \left[ \frac{a}{r^2}\left(f^2(b-r^2) - a\,e^2\right) + \frac{r^2}{a(r^2-b)} \right] \overline{\psi}_{-1}(r) \right\} \frac{dr}{r}\,.$$

- Family 4.

$$p = \int \left\{ \frac{2a^2}{\mathfrak{r}^3}\,\frac{a-3\mathfrak{r}^3}{(\mathfrak{r}^3-a)^{\frac{7}{3}}}\,\overline{\psi}_{-1}(\mathfrak{r}) - \frac{2a^2}{\mathfrak{r}^5(\mathfrak{r}^3-a)^{\frac{2}{3}}}\,\overline{\psi}_1(\mathfrak{r}) \right.$$
$$\left. + \frac{(\mathfrak{r}^3-a)^{\frac{4}{3}}}{\mathfrak{r}^4}\,\partial_r \overline{\psi}_1(\mathfrak{r}) + \frac{\mathfrak{r}^4}{(\mathfrak{r}^3-a)^{\frac{4}{3}}}\,\partial_r \overline{\psi}_{-1}(\mathfrak{r}) \right\} d\mathfrak{r}$$
$$= \frac{(\mathfrak{r}^3-a)^{\frac{4}{3}}}{\mathfrak{r}^4}\,\overline{\psi}_1(\mathfrak{r}) + \frac{\mathfrak{r}^4}{(\mathfrak{r}^3-a)^{\frac{4}{3}}}\,\overline{\psi}_{-1}(\mathfrak{r})$$
$$- 2a \int \left\{ \frac{a-2\mathfrak{r}^3}{\mathfrak{r}^2(\mathfrak{r}^3-a)^{\frac{4}{3}}}\,\overline{\psi}_{-1} + \frac{2\mathfrak{r}^3-a}{\mathfrak{r}^4(\mathfrak{r}^3-a)^{\frac{2}{3}}}\,\overline{\psi}_1 \right\} \frac{d\mathfrak{r}}{\mathfrak{r}}\,.$$

- Family 5.

$$p = \frac{1}{c}\left(\frac{1}{d}\overline{\psi}_1 - d\overline{\psi}_{-1}\right)\left[2b\,\phi + \left(1-b^2-c^2\right)\ln r\right] + p_0\,.$$

## 3.11 Linearized Elasticity

### 3.11.1 Linearized Kinematics

Let $\mathcal{S}_R$ be some reference shape. We consider its "infinitesimal" perturbations by introducing a one-parametric family $\{\mathcal{S}_\varepsilon\}_{\varepsilon\in]-a,a[}$ of shapes, $a > 0$, and a one-parametric family

$$\{y_\varepsilon\}_{\varepsilon\in]-a,a[}\,, \quad y_\varepsilon : \mathcal{S}_R \to \mathcal{S}_\varepsilon$$

of deformations. We suppose that this family satisfies the following properties:

(i) $\mathcal{S}_0 = \mathcal{S}_R$ and $y_0 = \mathrm{Id}_{\mathcal{S}_R} : \mathcal{S}_R \to \mathcal{S}_R$ is the identity map;

(ii) for each $X \in \mathcal{S}_R$ the mapping $\varepsilon \mapsto y_\varepsilon(X)$ from $]-a, a[$ to $\mathcal{E}$ is smooth.

First-order Taylor expansion of $\varepsilon \mapsto y_\varepsilon(X)$ in a neighborhood of $\varepsilon = 0$ has the form:

$$y_\varepsilon(X) = y_0(X) + \varepsilon \left.\frac{\partial y_\varepsilon(X)}{\partial \varepsilon}\right|_{\varepsilon=0} + \boldsymbol{O}(\varepsilon^2) \quad \text{as} \quad \varepsilon \to 0\,.$$

Since $y_0(X) = X$, one can denote $y_\varepsilon(X) - X$ by $\boldsymbol{u}_\varepsilon(X)$. This vector represents a displacement from the reference place $X$ to the actual one, $y_\varepsilon(X)$, which corresponds to the "infinitesimally" distorted shape $\mathcal{S}_\varepsilon$. One has the formula

$$\boldsymbol{u}_\varepsilon(X) = \varepsilon \boldsymbol{s}(X) + \boldsymbol{O}(\varepsilon^2) \quad \text{as} \quad \varepsilon \to 0\,,$$

where the vector $\boldsymbol{s}(X) := \left.\frac{\partial y_\varepsilon(X)}{\partial \varepsilon}\right|_{\varepsilon=0}$ represents a rate of "perturbation" at point $X$. Let us denote $\widetilde{\boldsymbol{u}}_\varepsilon := \varepsilon \boldsymbol{s}$. It is the principal linear part of $\boldsymbol{u}_\varepsilon$, an infinitesimal displacement.

The displacement gradient of $\boldsymbol{u}_\varepsilon$ at point $X \in \mathcal{S}_R$ has the form:

$$\boldsymbol{u}'_\varepsilon(X) = \boldsymbol{H}_\varepsilon(X) + \boldsymbol{O}(\varepsilon^2) \quad \text{as} \quad \varepsilon \to 0\,.$$

Here, $\boldsymbol{H}_\varepsilon(X) = \varepsilon \boldsymbol{s}'(X)$ is the gradient of $\widetilde{\boldsymbol{u}}_\varepsilon$:

$$\widetilde{\boldsymbol{u}}_\varepsilon(X + \boldsymbol{h}) = \widetilde{\boldsymbol{u}}_\varepsilon(X) + \boldsymbol{H}_\varepsilon(X)[\boldsymbol{h}] + \boldsymbol{o}(\|\boldsymbol{h}\|) \quad \text{as} \quad \boldsymbol{h} \to \boldsymbol{0}\,,$$

We introduce the infinitesimal strain and rotation measures (vorticity tensor):

$$\widetilde{\boldsymbol{E}}_\varepsilon = \frac{1}{2}\left(\boldsymbol{H}_\varepsilon + \boldsymbol{H}_\varepsilon^{\mathrm{T}}\right)\,, \quad \widetilde{\boldsymbol{R}}_\varepsilon = \frac{1}{2}\left(\boldsymbol{H}_\varepsilon - \boldsymbol{H}_\varepsilon^{\mathrm{T}}\right)\,.$$

Since $\widetilde{\boldsymbol{u}}'_\varepsilon = \widetilde{\boldsymbol{E}}_\varepsilon + \widetilde{\boldsymbol{R}}_\varepsilon$, it follows that

$$\boldsymbol{u}'_\varepsilon = \widetilde{\boldsymbol{E}}_\varepsilon + \widetilde{\boldsymbol{R}}_\varepsilon + \boldsymbol{O}(\varepsilon^2) \quad \text{as} \quad \varepsilon \to 0\,.$$

With accuracy up to $\boldsymbol{O}(\varepsilon^2)$, the exact polar decomposition $\boldsymbol{F}_\varepsilon = \boldsymbol{R}_\varepsilon \boldsymbol{U}_\varepsilon$ can be approximately replaced by the additive decomposition $\widetilde{\boldsymbol{u}}'_\varepsilon = \widetilde{\boldsymbol{E}}_\varepsilon + \widetilde{\boldsymbol{R}}_\varepsilon$.

We obtain the asymptotic expansions for the right Cauchy–Green strain $\boldsymbol{C}_\varepsilon$, the right and left stretch tensors $\boldsymbol{U}_\varepsilon$ and $\boldsymbol{V}_\varepsilon$, and the orthogonal tensor $\boldsymbol{R}_\varepsilon$. According to (3.17), one has the asymptotic expansions

$$\boldsymbol{C}_\varepsilon = \boldsymbol{I} + \left(\boldsymbol{H}_\varepsilon + \boldsymbol{H}_\varepsilon^{\mathrm{T}}\right) + \boldsymbol{O}(\varepsilon^2) \quad \text{as} \quad \varepsilon \to 0\,,$$
$$\boldsymbol{B}_\varepsilon = \boldsymbol{I} + \left(\boldsymbol{H}_\varepsilon + \boldsymbol{H}_\varepsilon^{\mathrm{T}}\right) + \boldsymbol{O}(\varepsilon^2) \quad \text{as} \quad \varepsilon \to 0\,,$$

for the right and left Cauchy–Green strain measures. From the theorem of polar decomposition, it follows that

$$\boldsymbol{F}_\varepsilon = \boldsymbol{I} + \boldsymbol{u}'_\varepsilon = \boldsymbol{R}_\varepsilon \boldsymbol{U}_\varepsilon\,,$$

where

$$\boldsymbol{U}_\varepsilon = \sqrt{\boldsymbol{C}_\varepsilon} = \boldsymbol{I} + \frac{1}{2}\left(\boldsymbol{H}_\varepsilon + \boldsymbol{H}_\varepsilon^{\mathrm{T}}\right) + \boldsymbol{O}(\varepsilon^2) \quad \text{as} \quad \varepsilon \to 0\,.$$

Thus, one has the following expansion for the orthogonal tensor $\boldsymbol{R}_\varepsilon$:

$$\boldsymbol{R}_\varepsilon = \boldsymbol{I} + \frac{1}{2}\left(\boldsymbol{H}_\varepsilon - \boldsymbol{H}_\varepsilon^{\mathrm{T}}\right) + \boldsymbol{O}(\varepsilon^2) \quad \text{as} \quad \varepsilon \to 0 .$$

In terms of $\widetilde{\boldsymbol{E}}_\varepsilon$ and $\widetilde{\boldsymbol{R}}_\varepsilon$, the previous formulae look as follows:

$$\boldsymbol{U}_\varepsilon = \boldsymbol{I} + \widetilde{\boldsymbol{E}}_\varepsilon + \boldsymbol{O}(\varepsilon^2)\,, \quad \boldsymbol{R}_\varepsilon = \boldsymbol{I} + \widetilde{\boldsymbol{R}}_\varepsilon + \boldsymbol{O}(\varepsilon^2) \quad \text{as} \quad \varepsilon \to 0 . \tag{3.54}$$

Finally, from the theorem of polar decomposition, one also has $\boldsymbol{F}_\varepsilon = \boldsymbol{V}_\varepsilon \boldsymbol{R}_\varepsilon$, and then

$$\boldsymbol{V}_\varepsilon = \boldsymbol{I} + \widetilde{\boldsymbol{E}}_\varepsilon + \boldsymbol{O}(\varepsilon^2) \quad \text{as} \quad \varepsilon \to 0 .$$

### 3.11.2 Linearized Constitutive Relations

Let the material be elastic, *i.e.*. Its constitutive relation has the form

$$\boldsymbol{T}_\varepsilon = \boldsymbol{R}_\varepsilon \mathfrak{g}_{\mathcal{S}_R}(\boldsymbol{U}_\varepsilon)\boldsymbol{R}_\varepsilon^{\mathrm{T}} ,$$

where mapping $\mathfrak{g}_{\mathcal{S}_R}$ : $\mathrm{SimLin}(\mathscr{V};\mathscr{V}) \to \mathrm{SimLin}(\mathscr{V};\mathscr{V})$ depends on the shape $\mathcal{S}_R$; $\boldsymbol{F}_\varepsilon = \boldsymbol{R}_\varepsilon \boldsymbol{U}_\varepsilon$. We will obtain asymptotic relations by virtue of (3.54). Using the first-order Taylor formula, one obtains

$$\mathfrak{g}_{\mathcal{S}_R}(\boldsymbol{U}_\varepsilon) = \mathfrak{g}_{\mathcal{S}_R}\left(\boldsymbol{I} + \widetilde{\boldsymbol{E}}_\varepsilon + \boldsymbol{O}(\varepsilon^2)\right) = \mathfrak{g}_{\mathcal{S}_R}(\boldsymbol{I}) + \boldsymbol{L}_{\mathcal{S}_R}[\widetilde{\boldsymbol{E}}_\varepsilon] + \boldsymbol{O}(\varepsilon^2) \quad \text{as} \quad \varepsilon \to 0 ,$$

where

$$\boldsymbol{L}_{\mathcal{S}_R} = \left.\frac{\partial \mathfrak{g}_{\mathcal{S}_R}(\boldsymbol{U})}{\partial \boldsymbol{U}}\right|_{\boldsymbol{U}=\boldsymbol{I}} : \mathrm{SimLin}(\mathscr{V};\mathscr{V}) \to \mathrm{SimLin}(\mathscr{V};\mathscr{V}) .$$

Denoting $\boldsymbol{T}_0 := \mathfrak{g}_{\mathcal{S}_R}(\boldsymbol{I})$ and performing calculations

$$\boldsymbol{T}_\varepsilon = \boldsymbol{R}_\varepsilon \mathfrak{g}_{\mathcal{S}_R}(\boldsymbol{U}_\varepsilon)\boldsymbol{R}_\varepsilon^{\mathrm{T}} = (\boldsymbol{I} + \widetilde{\boldsymbol{R}}_\varepsilon + \boldsymbol{O}(\varepsilon^2))(\boldsymbol{T}_0 + \boldsymbol{L}_{\mathcal{S}_R}[\widetilde{\boldsymbol{E}}_\varepsilon] + \boldsymbol{O}(\varepsilon^2))(\boldsymbol{I} - \widetilde{\boldsymbol{R}}_\varepsilon + \boldsymbol{O}(\varepsilon^2)) ,$$

one arrives at the formula

$$\boldsymbol{T}_\varepsilon = \boldsymbol{T}_0 + \widetilde{\boldsymbol{R}}_\varepsilon \boldsymbol{T}_0 - \boldsymbol{T}_0 \widetilde{\boldsymbol{R}}_\varepsilon + \boldsymbol{L}_{\mathcal{S}_R}[\widetilde{\boldsymbol{E}}_\varepsilon] + \boldsymbol{O}(\varepsilon^2) \quad \text{as} \quad \varepsilon \to 0 . \tag{3.55}$$

Here, $\boldsymbol{T}_0$ is the Cauchy stress tensor in the shape $\mathcal{S}_R$. The fourth-rank tensor $\boldsymbol{L}_{\mathcal{S}_R}$ is called the *elasticity tensor*.

If $\mathcal{S}_R$ is stress free, then $\boldsymbol{T}_0 = \boldsymbol{0}$, and

$$\boldsymbol{T}_\varepsilon = \boldsymbol{L}[\widetilde{\boldsymbol{E}}_\varepsilon] + \boldsymbol{O}(\varepsilon^2) \quad \text{as} \quad \varepsilon \to 0 ,$$

which represents the classical Hooke's law with accuracy up to $\boldsymbol{O}(\varepsilon^2)$. We suppressed the lower index of $\boldsymbol{L}$ in this case, since the latter relation is special: it is valid only for stress-free reference shapes.

### 3.11.3 Linearized Stresses

Let us denote

$$\widetilde{\boldsymbol{T}}_\varepsilon = \boldsymbol{T}_0 + \widetilde{\boldsymbol{R}}_\varepsilon \boldsymbol{T}_0 - \boldsymbol{T}_0 \widetilde{\boldsymbol{R}}_\varepsilon + \boldsymbol{L}_{\mathcal{S}_R}[\widetilde{\boldsymbol{E}}_\varepsilon] ,$$

then according to (3.55), one has

$$\boldsymbol{T}_\varepsilon = \widetilde{\boldsymbol{T}}_\varepsilon + \boldsymbol{O}(\varepsilon^2) \quad \text{as} \quad \varepsilon \to 0 .$$

The latter relation shows that in the case of $\boldsymbol{T}_0 = \boldsymbol{0}$, the tensor $\widetilde{\boldsymbol{T}}_\varepsilon$ may be considered as the principal linear part of the Cauchy stress tensor $\boldsymbol{T}_\varepsilon$.

Finally, one obtains that in the linear elastic case, respectively to a stress free reference shape, fields $\widetilde{\boldsymbol{u}}_\varepsilon$ and $\widetilde{\boldsymbol{T}}_\varepsilon$ play the role of "displacement" and "stress" fields, respectively. One can linearize the equation of motion and boundary conditions and express them in terms of the so-obtained fields.

**Remark 3.14.** After linearization, the equation of motion, $\operatorname{div} \boldsymbol{T}_\varepsilon + \boldsymbol{b} = \rho \ddot{\boldsymbol{u}}_\varepsilon$, takes the form

$$\frac{1}{2} \operatorname{div} \boldsymbol{L} \left[ \widetilde{\boldsymbol{u}}'_\varepsilon + \left[ \widetilde{\boldsymbol{u}}'_\varepsilon \right]^{\mathrm{T}} \right] + \boldsymbol{b} = \rho \ddot{\widetilde{\boldsymbol{u}}}_\varepsilon ,$$

with accuracy up to $\boldsymbol{O}(\varepsilon^2)$. In the isotropic case, one obtains the Lamé equation.

Now, we consider asymptotic relations for Piola–Kirchhoff stress tensors in the linear elastic case. First, we obtain the asymptotic expansion of $J_\varepsilon = \det \boldsymbol{F}_\varepsilon$. Since $\boldsymbol{F}_\varepsilon = \boldsymbol{I} + \boldsymbol{u}'_\varepsilon$, one has

$$J_\varepsilon = \det(\boldsymbol{I} + \boldsymbol{u}'_\varepsilon) = \det \boldsymbol{I} + \left. \frac{d \det \boldsymbol{A}}{d\boldsymbol{A}} \right|_{\boldsymbol{A}=\boldsymbol{I}} [\boldsymbol{u}'_\varepsilon] + o\left( \left\| \boldsymbol{u}'_\varepsilon \right\|^2 \right) ,$$

where $\left. \frac{d \det \boldsymbol{A}}{d\boldsymbol{A}} \right|_{\boldsymbol{A}=\boldsymbol{I}} \in \operatorname{Lin}(\operatorname{Lin}(\mathscr{V};\mathscr{V}); \mathbb{R})$. In terms of $\varepsilon$,

$$J_\varepsilon = \det \boldsymbol{I} + \left. \frac{d \det \boldsymbol{A}}{d\boldsymbol{A}} \right|_{\boldsymbol{A}=\boldsymbol{I}} \left[ \widetilde{\boldsymbol{u}}'_\varepsilon \right] + O(\varepsilon^2) \quad \text{as} \quad \varepsilon \to 0 .$$

**Remark 3.15.** Further, we use the following classical result. Let $\mathscr{U}$ be an $n$-dimensional vector space, $\boldsymbol{A}, \boldsymbol{B} \in \operatorname{Lin}(\mathscr{U}; \mathscr{U})$. Then,

$$\frac{d \det \boldsymbol{A}}{d\boldsymbol{A}} [\boldsymbol{B}] = \det \boldsymbol{A} \operatorname{tr}\{\boldsymbol{A}^{-1} \boldsymbol{B}\} .$$

Finally, one arrives at the relation

$$J_\varepsilon = 1 + \operatorname{tr} \widetilde{\boldsymbol{u}}'_\varepsilon + O(\varepsilon^2) \quad \text{as} \quad \varepsilon \to 0 ,$$

or, according to the additive decomposition of $\widetilde{\boldsymbol{u}}'_\varepsilon$,

$$J_\varepsilon = 1 + \operatorname{tr} \widetilde{\boldsymbol{E}}_\varepsilon + O(\varepsilon^2) \quad \text{as} \quad \varepsilon \to 0 .$$

Since $\boldsymbol{P}_\varepsilon = J_\varepsilon \boldsymbol{T}_\varepsilon \boldsymbol{F}_\varepsilon^{-\mathrm{T}}$, and taking into account that

$$J_\varepsilon = 1 + O(\varepsilon)\,, \quad \boldsymbol{F}_\varepsilon^{-\mathrm{T}} = \boldsymbol{I} + \boldsymbol{O}(\varepsilon)\,, \quad \boldsymbol{T}_\varepsilon = \widetilde{\boldsymbol{T}}_\varepsilon + \boldsymbol{O}(\varepsilon^2) \quad \text{as} \quad \varepsilon \to 0\,,$$

and $\widetilde{\boldsymbol{T}}_\varepsilon = \boldsymbol{L}[\widetilde{\boldsymbol{E}}_\varepsilon]$, one obtains the following relation for the first Piola–Kirchhoff stress tensor:

$$\boldsymbol{P}_\varepsilon = \widetilde{\boldsymbol{T}}_\varepsilon + \boldsymbol{O}(\varepsilon^2) \quad \text{as} \quad \varepsilon \to 0\,.$$

The relations obtained for the Cauchy and Piola–Kirchhoff stress tensors show that with accuracy up to $\boldsymbol{O}(\varepsilon^2)$, respectively to a stress-free reference shape, these fields are equal to $\widetilde{\boldsymbol{T}}_\varepsilon$.

### 3.11.4 The Green–Rivlin–Shield–Truesdell Formula

Suppose that $\mathcal{S}_R$ is a stress-free shape and $y\colon \mathcal{S}_R \to \mathcal{S}$ is a deformation onto some shape $\mathcal{S}$; its deformation gradient is $\boldsymbol{F}$. Consider "infinitesimal" perturbations of the latter shape. Analogously to 3.11.1, we introduce a one-parametric family $\{\mathcal{S}_\varepsilon\}_{\varepsilon\in]-a,a[}$ of shapes, $a > 0$, and the corresponding one-parametric family

$$\{y_\varepsilon\}_{\varepsilon\in]-a,a[}\,, \quad y_\varepsilon\colon \mathcal{S} \to \mathcal{S}_\varepsilon$$

of deformations, such that: 1) $\mathcal{S}_0 = \mathcal{S}$ and $y_0 = \mathrm{Id}_{\mathcal{S}}\colon \mathcal{S} \to \mathcal{S}$; 2) for each $x \in \mathcal{S}$ the mapping $\varepsilon \mapsto y_\varepsilon(x)$ from $]-a, a[$ to $\mathcal{E}$ is smooth. Relations between $y$, $y_\varepsilon$ and the corresponding shapes are shown in the diagram:

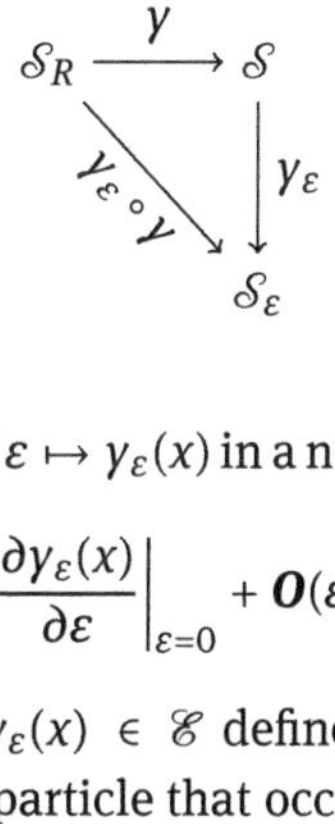

First-order Taylor expansion of $\varepsilon \mapsto y_\varepsilon(x)$ in a neighborhood of $\varepsilon = 0$ has the form:

$$y_\varepsilon(x) = y_0(x) + \varepsilon \left.\frac{\partial y_\varepsilon(x)}{\partial \varepsilon}\right|_{\varepsilon=0} + \boldsymbol{O}(\varepsilon^2) \quad \text{as} \quad \varepsilon \to 0\,.$$

The difference between points $x, y_\varepsilon(x) \in \mathcal{E}$ defines the translation vector $\boldsymbol{u}_\varepsilon(x) := y_\varepsilon(x) - x$, the displacement of the particle that occupied place $x$ in the shape $\mathcal{S}$, and the place $y_\varepsilon(x)$ in the shape $\mathcal{S}_\varepsilon$. The differential of the mapping $\varepsilon \mapsto \boldsymbol{u}_\varepsilon(x)$ is of the form

$$\widetilde{\boldsymbol{u}}_\varepsilon(x) := \varepsilon \boldsymbol{s}(x)\,, \quad \text{where} \quad \boldsymbol{s}(x) := \left.\frac{\partial y_\varepsilon(x)}{\partial \varepsilon}\right|_{\varepsilon=0}\,,$$

and represents the linear part of the displacement. The vector $\boldsymbol{s}(x)$ is the rate of change of this linear part with respect to $\varepsilon$.

For fixed $\varepsilon$, the mapping $\mathcal{S} \ni x \mapsto \tilde{\boldsymbol{u}}_\varepsilon(x)$ is the field of infinitesimal displacements; denote its gradient by $\boldsymbol{H}_\varepsilon : x \mapsto \boldsymbol{H}_\varepsilon(x)$, *i.e.*,

$$\tilde{\boldsymbol{u}}_\varepsilon(x+\boldsymbol{h}) = \tilde{\boldsymbol{u}}_\varepsilon(x) + \boldsymbol{H}_\varepsilon(x)[\boldsymbol{h}] + \boldsymbol{o}(\|\boldsymbol{h}\|) \quad \text{as} \quad \boldsymbol{h} \to \boldsymbol{0} \,,$$

and the corresponding infinitesimal strains and vortices by $\widetilde{\boldsymbol{E}}_\varepsilon$ and $\widetilde{\boldsymbol{R}}_\varepsilon$:

$$\widetilde{\boldsymbol{E}}_\varepsilon = \frac{1}{2}\left(\boldsymbol{H}_\varepsilon + \boldsymbol{H}_\varepsilon^{\mathrm{T}}\right)\,, \quad \widetilde{\boldsymbol{R}}_\varepsilon = \frac{1}{2}\left(\boldsymbol{H}_\varepsilon - \boldsymbol{H}_\varepsilon^{\mathrm{T}}\right)\,.$$

The total deformation is the composition $\chi_\varepsilon := y_\varepsilon \circ y$, and its gradient is the linear map $\boldsymbol{F}_\varepsilon = \boldsymbol{H}_\varepsilon \boldsymbol{F}$. The corresponding left Cauchy–Green strain tensor, $\boldsymbol{B}_\varepsilon = \boldsymbol{F}_\varepsilon \boldsymbol{F}_\varepsilon^{\mathrm{T}}$ and the related quantities can be represented for small $\varepsilon$ by the following asymptotic expansions:

$$\boldsymbol{B}_\varepsilon \doteq \boldsymbol{B} + \boldsymbol{H}_\varepsilon \boldsymbol{B} + \boldsymbol{B}\boldsymbol{H}_\varepsilon^{\mathrm{T}}\,, \quad \boldsymbol{B} = \boldsymbol{F}\boldsymbol{F}^{\mathrm{T}}\,,$$
$$\boldsymbol{B}_\varepsilon^{-1} \doteq \boldsymbol{B}^{-1} - \boldsymbol{B}^{-1}\boldsymbol{H}_\varepsilon - \boldsymbol{H}_\varepsilon^{\mathrm{T}}\boldsymbol{B}^{-1}\,.$$

Here, the symbol $\doteq$ indicates equality with up to $\boldsymbol{O}(\varepsilon^2)$.

The material is assumed to be isotropic, obeying the relation (3.48):

$$\boldsymbol{T} = \daleth_0 \boldsymbol{I} + \daleth_1 \boldsymbol{B} + \daleth_{-1}\boldsymbol{B}^{-1}\,,$$

where $\daleth_r$, $r = -1, 0, 1$, are functions on the principal invariants $I_k$, $k = 1, 2, 3$, of $\boldsymbol{B}$. This constitutive relation is written relatively to the stress-free shape $\mathcal{S}_R$.

In the case of small $\varepsilon$, the following asymptotic expansion holds:

$$\daleth_{r,\varepsilon} \doteq \daleth_r + \sum_{k=1}^{3} \frac{\partial \daleth_r}{\partial I_k}\left(\frac{dI_k}{d\boldsymbol{B}}\left\{\boldsymbol{H}_\varepsilon \boldsymbol{B} + \boldsymbol{B}\boldsymbol{H}_\varepsilon^{\mathrm{T}}\right\}\right)\,.$$

Using formulae for the principal invariants derivatives are [37]

$$\frac{dI_1(\boldsymbol{B})}{d\boldsymbol{B}}[\boldsymbol{A}] = \mathrm{tr}(\boldsymbol{A})\,, \quad \frac{dI_2(\boldsymbol{B})}{d\boldsymbol{B}}[\boldsymbol{A}] = \mathrm{tr}\left[\left(I_1(\boldsymbol{B})\boldsymbol{I} - \boldsymbol{B}^{\mathrm{T}}\right)\boldsymbol{A}^{\mathrm{T}}\right]\,,$$
$$\frac{dI_3(\boldsymbol{B})}{d\boldsymbol{B}}[\boldsymbol{A}] = \mathrm{tr}\left[\left(I_3(\boldsymbol{B})(\boldsymbol{B}^{-1})^{\mathrm{T}}\right)\boldsymbol{A}^{\mathrm{T}}\right]\,,$$

and substituting the relations obtained in (3.48), we obtain

$$\begin{aligned}\boldsymbol{T}_\varepsilon \doteq\ & \boldsymbol{T}_0 + \daleth_1\left(\boldsymbol{H}_\varepsilon\boldsymbol{B} + \boldsymbol{B}\boldsymbol{H}_\varepsilon^{\mathrm{T}}\right) - \daleth_{-1}\left(\boldsymbol{B}^{-1}\boldsymbol{H}_\varepsilon + \boldsymbol{H}_\varepsilon^{\mathrm{T}}\boldsymbol{B}^{-1}\right)\\ & + \sum_{r=-1}^{1}\left(\frac{\partial\daleth_r}{\partial I_1}\mathrm{tr}\left(\boldsymbol{H}_\varepsilon\boldsymbol{B} + \boldsymbol{B}\boldsymbol{H}_\varepsilon^{\mathrm{T}}\right) + \frac{\partial\daleth_r}{\partial I_2}\mathrm{tr}\left(I_1(\boldsymbol{B})\boldsymbol{H}_\varepsilon\boldsymbol{B} + I_1(\boldsymbol{B})\boldsymbol{B}\boldsymbol{H}_\varepsilon^{\mathrm{T}} - \boldsymbol{B}\boldsymbol{H}_\varepsilon\boldsymbol{B} - \boldsymbol{B}^2\boldsymbol{H}_\varepsilon^{\mathrm{T}}\right)\right.\\ & \left. + \frac{\partial\daleth_r}{\partial I_3}\mathrm{tr}\left(I_3(\boldsymbol{B})\boldsymbol{B}^{-1}\boldsymbol{H}_\varepsilon\boldsymbol{B} + I_3(\boldsymbol{B})\boldsymbol{H}_\varepsilon^{\mathrm{T}}\right)\right)\boldsymbol{B}^r\,,\end{aligned} \tag{3.56}$$

where $\boldsymbol{T}_0 = \daleth_0\boldsymbol{I} + \daleth_1\boldsymbol{B} + \daleth_{-1}\boldsymbol{B}^{-1}$.

The expression (3.56) can be simplified if one uses the identity

$$\beth_1\left(\boldsymbol{H}_\varepsilon\boldsymbol{B}+\boldsymbol{B}\boldsymbol{H}_\varepsilon^{\mathrm{T}}\right)-\beth_{-1}\left(\boldsymbol{B}^{-1}\boldsymbol{H}_\varepsilon+\boldsymbol{H}_\varepsilon^{\mathrm{T}}\boldsymbol{B}^{-1}\right)$$
$$=\boldsymbol{H}_\varepsilon\boldsymbol{T}_0+\boldsymbol{T}_0\boldsymbol{H}_\varepsilon^{\mathrm{T}}-2\left(\beth_0\widetilde{\boldsymbol{E}}_\varepsilon+\beth_{-1}\left\{\widetilde{\boldsymbol{E}}_\varepsilon\boldsymbol{B}^{-1}+\boldsymbol{B}^{-1}\widetilde{\boldsymbol{E}}_\varepsilon\right\}\right),$$

and the Cayley–Hamilton theorem, from which follows $\boldsymbol{B}^2 = I_1\boldsymbol{B} - I_2\boldsymbol{I} + I_3\boldsymbol{B}^{-1}$. As a final result, we obtain the *Green–Rivlin–Shield–Truesdell formula* [89]:

$$\boldsymbol{T}_\varepsilon \doteq \boldsymbol{T}_0+\widetilde{\boldsymbol{R}}_\varepsilon\boldsymbol{T}_0-\boldsymbol{T}_0\widetilde{\boldsymbol{R}}_\varepsilon+\boldsymbol{L}_\gamma[\widetilde{\boldsymbol{E}}_\varepsilon]\,, \tag{3.57}$$

where $\boldsymbol{L}_\gamma$ is the elasticity tensor

$$\boldsymbol{L}_\gamma[\widetilde{\boldsymbol{E}}_\varepsilon]=\widetilde{\boldsymbol{E}}_\varepsilon\boldsymbol{T}_0+\boldsymbol{T}_0\widetilde{\boldsymbol{E}}_\varepsilon-2\left(\beth_0\widetilde{\boldsymbol{E}}_\varepsilon+\beth_{-1}\left\{\widetilde{\boldsymbol{E}}_\varepsilon\boldsymbol{B}^{-1}+\boldsymbol{B}^{-1}\widetilde{\boldsymbol{E}}_\varepsilon\right\}\right)$$
$$+2\sum_{r=-1}^{1}\left\{\left(I_2\frac{\partial\beth_r}{\partial I_2}+I_3\frac{\partial\beth_r}{\partial I_3}\right)\mathrm{tr}\widetilde{\boldsymbol{E}}_\varepsilon+\frac{\partial\beth_r}{\partial I_1}\mathrm{tr}(\boldsymbol{B}\widetilde{\boldsymbol{E}}_\varepsilon)-I_3\frac{\partial\beth_r}{\partial I_2}\mathrm{tr}\left(\boldsymbol{B}^{-1}\widetilde{\boldsymbol{E}}_\varepsilon\right)\right\}\boldsymbol{B}^r.$$

## 3.12 Distributed Defects in Solids

### 3.12.1 Preliminary Remarks

The violation of the compatibility conditions characterizes the phenomenon of plastic deformation, which is a result of the development and redistribution of defects in the ideal body (crystal). A physical interpretation of the plastic deformation of crystals was formulated by Dehlinger, Taylor, Orowan, and Polanyi [90–94], who explained the divergence between the theoretical assessment of mechanical properties of crystals, obtained by Frenkel [95], and the experimental data. It should be mentioned that within the framework of these ideas, the plastic distortion $\boldsymbol{\beta}^{\mathrm{p}}$ was considered to be known, whereas its symmetric part $\boldsymbol{\varepsilon}^{\mathrm{p}}$ was believed to be incompatible. The measure of incompatibility was defined by the incompatibility tensor $\boldsymbol{\eta} = -\nabla\times(\nabla\times\boldsymbol{\varepsilon}^{\mathrm{p}})^{\mathrm{T}}$, or the dislocation density tensor $\boldsymbol{\alpha} = -\nabla\times\boldsymbol{\beta}^{\mathrm{p}}$ [66]. The latter characteristic has a clear geometrical interpretation as a subintegral expression for determining the Burgers vector [96]. In this case, the stressed state in the body is fully determined by elastic deformation, imposed on the plastic deformation such that the body does not split into parts [96]. This allows classification of the body as a local simple body. It should be noted that elastic deformation and rotation are interrelated as $(\nabla\times\boldsymbol{\varepsilon})^{\mathrm{T}}+\nabla\vec{\boldsymbol{\omega}}=\boldsymbol{K}$, where $\boldsymbol{K}$ is the Cosserat tensor characterizing the relative rotation of the crystalline lattice.

Thus, due to incompatible plastic deformation, the body is divided into a puzzle of infinitesimally small elements (fragments) that can be combined (assembled) as a solid body by means of elastic individual deformation of each such fragment. A more general situation occurs if we believe that, besides individual deformation, each fragment should be imparted an individual rotation not connected with deformation, thereby, defining the orientation of the directors (in the terminology of the theory of moment media). To this end, the bending-torsion tensor $\boldsymbol{\kappa}$ should be used, as proposed by De Wit; in the case of compatible deformation, this tensor is defined by the equality $\boldsymbol{\kappa} = \nabla\vec{\boldsymbol{\omega}}$, and in the general case, it is specified as an arbitrary smooth field [66]. In such a case, the stressed state of the body depends on two independent fields, *i.e.*, the elastic deformation and the elastic bending-torsion, which ensure integrability of the relations for full fields. The sources of internal stresses are determined by deformation incompatibility measures (the density of the dislocations $\boldsymbol{\alpha}$) and the measure of disclination density, determined by the equality $\boldsymbol{\theta} = -\nabla \times \boldsymbol{\kappa}^{\mathrm{p}}$. Speaking figuratively, owing to the continuous distribution of the two types of defects, *i.e.*, dislocations and disclinations, the body is divided into a puzzle of fragments that can be assembled as a solid body by means of elastic individual deformation and independent elastic orientation of the edges. In the framework of this approach, elastic distortion turns out to be indefinite, although its symmetric part is determined definitively. The stressed state of the body is characterized by the laws of state for media with moment stresses [97] or a microstructure [98]. Experimental identification of disclinations was done in the mid-twentieth century [99–101], in particular, in investigations of liquid crystal mechanics. It should be noted that pioneering works [66, 96] pointed to a close link between dislocations and disclination density tensors and the vectors of curvature and torsion of material connection. It also pointed to a link between the continuum theory of dislocations and the Einstein–Cartan theory of gravitation.

### 3.12.2 Total Distortion 1-Forms

The response of a conventional solid is determined with respect to a reference shape that is supposed to be embedded into three-dimensional Euclidean space in a relaxed state, *i.e.*, such a reference shape has to be free of stresses. In that case, the localization of deformation $y$ is defined by exact 1-forms $F^{\mathfrak{i}} = dy^{\mathfrak{i}}$, $\mathfrak{i} = 1, 2, 3$. The starting point in the construction of the theory of continuously distributed defects is the rejection of this proposal. The exact forms $F^{\mathfrak{i}} \in \Omega^1(\mathcal{S}_R)$ are replaced by some, not necessarily exact, forms $H^{\mathfrak{i}} \in \Omega^1(\mathcal{S}_R)$, *i.e.*, in general, $dH^{\mathfrak{i}} \neq 0$. The only condition that should be demanded is that $H^1 \wedge H^2 \wedge H^3 \neq 0$. We will refer to the 1-forms $H^{\mathfrak{i}} = H^{\mathfrak{i}}{}_a dx^a$ as *1-forms of total distortion* and consider them as being both spatial and time dependent.

**Remark 3.16.** We use fraktur notation $\mathfrak{i}$ for the index in $F^{\mathfrak{i}}$ and $H^{\mathfrak{i}}$ on purpose to designate that $F^{\mathfrak{i}}$ and $H^{\mathfrak{i}}$ are not components of a vector. Both $F^{\mathfrak{i}}$ and $H^{\mathfrak{i}}$ are 1-forms.

### 3.12.3 Four-Dimensional Space and the Homotopy Operator

**Frames.** Let $\mathscr{E}^3$ be a three-dimensional Euclidean affine space. Here, and until the end of this section, we use capital Latin letters to designate indices that vary from 1 to 3. The volume element of $\mathscr{E}^3$ is denoted by $\mu = dx^1 \wedge dx^2 \wedge dx^3$. That is, $\mu$ is a frame for $\Omega^3(\mathscr{E}^3)$. A coordinate frame $(\partial_A)_{A=1}^3$ on $\mathscr{E}^3$ defines the collection $(\mu_A)_{A=1}^3$ of the 2-forms $\mu_A := \partial_A \lrcorner \mu$, which serve as a frame for $\Omega^2(\mathscr{E}^3)$. Differential forms $\mu_A$ can be viewed as oriented surface elements.

We will use four-dimensional Euclidean affine space $\mathscr{E}^4 = \mathscr{E}^3 \times \mathbb{R}$. Lowercase Latin letters are used as indices that vary from 1 to 4. The coordinate frame $(\partial_A)_{A=1}^3$ on $\mathscr{E}^3$ induces the coordinate frame $(\partial_i) = ((\partial_I)_{I=1}^3, \partial_t)$ on $\mathscr{E}^4$. Here, $t$ is a natural coordinate on $\mathbb{R}$. In the following, we often will use the notation $\partial_4$ for $\partial_t$. The volume element on $\mathscr{E}^4$ is denoted by $\pi$ and can be represented as

$$\pi = \mu \wedge dt \,.$$

The differential form $\pi$ is a frame for $\Omega^4(\mathscr{E}^4)$. The frame $(\partial_i)_{i=1}^4$ generates the differential forms $\pi_a = \partial_a \lrcorner \pi$, $a = 1, 2, 3, 4$. Thus, the family $(\pi_a)_{a=1}^4$ forms a frame for $\Omega^3(\mathscr{E}^4)$.

We use two designations for the exterior derivative. The first one, $\overline{d}$, is defined as $\overline{d} = dx^I \wedge \partial_I$, while the second one, $d$, is defined via $d = dx^i \wedge \partial_i$. Operators $d$ and $\overline{d}$ are related by

$$d = \overline{d} + dt \wedge \partial_4 \,.$$

Note that the operator $\overline{d}$ can be considered as a differentiation with fixed $t$.

**Remark 3.17.** Note that $\mu_1 = dx^2 \wedge dx^3, \mu_2 = dx^3 \wedge dx^1, \mu_3 = dx^1 \wedge dx^2$ are 2-forms that can be defined equivalently as $\mu_I = *(dx^I)$, where $*$ is the Hodge star operator[22]. The index $I$ was chosen as lower by virtue of the definition $\mu_I = \partial_I \lrcorner \mu$. Such a designation was accepted in [102]. An analogous situation holds for $\pi_a$.

**Star-shaped regions.** Let $\mathscr{S}_R \subset \mathscr{E}^3$ be a shape. Assume that there exists a point $x_0 \in \mathscr{S}_R$, such that the line segment connecting $x_0$ and any other point $x \in \partial\mathscr{S}_R$ intersects $\partial\mathscr{S}_R$ only at $x$. In this case, the set $\mathscr{S}_R$ is called star shaped. In further reasoning, we consider only the star-shaped set $\mathscr{S}_R$.

**Homotopy operators.** In order to decompose total distortion on exact and non-exact parts, we introduce a special linear map. For each $p = 1, 2, 3, 4$, we define the linear map $\mathscr{H} : \Omega^p(\mathscr{S}_R \times \mathbb{R}) \to \Omega^{p-1}(\mathscr{S}_R \times \mathbb{R})$ such that for any $p$-form $\omega \in \Omega^p(\mathscr{S}_R \times \mathbb{R})$, one has [102]

$$\mathscr{H}\omega = \int_0^1 \lambda^{p-1} \mathscr{X} \lrcorner \overline{\omega}(\lambda) \, d\lambda \,,$$

22 See Chapter 13, Section 13.4.2.

where $\mathcal{X} = (x^i - x_0^i)\partial_i$, sum over $i = 1, 2, 3, 4$, and

$$\widetilde{\omega}(\lambda) = \omega_{i_1 \dots i_p}\left(x_0^b + \lambda\left(x^b - x_0^b\right)\right) dx^{i_1} \wedge \dots \wedge dx^{i_p} \,.$$

### 3.12.4 Decomposition of Total Distortion into Exact and Antiexact Parts

Let $\mathcal{H}$ be a homotopy operator. It has the property [102]

$$d\mathcal{H} + \mathcal{H} d = \mathrm{Id}_{\Omega^p(\mathcal{S}_R \times \mathbb{R})} \,.$$

Thus, total distortion $H^{\mathrm{i}}$ can be written as

$$H^{\mathrm{i}} = d\mathcal{H} H^{\mathrm{i}} + \mathcal{H} dH^{\mathrm{i}} \,. \tag{3.58}$$

By the construction, $d\mathcal{H} H^{\mathrm{i}}$ is an exact part of $H^{\mathrm{i}}$ that can be represented as exterior differential of some smooth function $y^{\mathrm{i}}$, *i.e.*, $d\mathcal{H} H^{\mathrm{i}} = dy^{\mathrm{i}}$. Thus, $y^{\mathrm{i}} = \mathcal{H} H^{\mathrm{i}} + h^{\mathrm{i}}$, where $dh^{\mathrm{i}} = 0$. The second item in (3.58) gives antiexact part of $H^{\mathrm{i}}$ that characterizes the distribution of defects.

If we denote the space of all exact 1-forms by $E^1(\mathcal{S}_R \times \mathbb{R})$ and the space of all antiexact 1-forms by $A^1(\mathcal{S}_R \times \mathbb{R})$, we have the following direct sum [102]:

$$\Omega^1(\mathcal{S}_R \times \mathbb{R}) = E^1(\mathcal{S}_R \times \mathbb{R}) \oplus A^1(\mathcal{S}_R \times \mathbb{R}) \,,$$

and the following result: if $v \in A^1(\mathcal{S}_R \times \mathbb{R})$ then $v = \mathcal{H} dv$. In other words, $\mathcal{H} d = \mathrm{Id}_{A^1(\mathcal{S}_R \times \mathbb{R})}$, and the homotopy operator $\mathcal{H}$ is inverse to $d$.

### 3.12.5 Continuity Equations of Defect Dynamics

We introduce the following differential forms, describing the density of defects and their currents [103]:

- 2-Forms of dislocation density:

$$\alpha^{\mathrm{i}} = \alpha^{A\mathrm{i}} \mu_A \,.$$

- 1-Form of dislocation current:

$$J^{\mathrm{i}} = J^{\mathrm{i}}_A dx^A \,.$$

- 2-Form of disclination current:

$$S^{\mathrm{i}} = S^{A\mathrm{i}} \mu_A \,.$$

- 3-Form of disclination density:

$$Q^{\mathrm{i}} = q^{\mathrm{i}} \mu \,.$$

In general, these differential forms depend on time as well as on spatial variables. Mappings $\alpha^i$, $J^i$, $S^i$ and $Q^i$ satisfy the following equations [103]:

$$\partial_t \alpha^i = -\overline{d}J^i - S^i\,, \quad \partial_t Q^i = -\overline{d}S^i\,, \quad \overline{d}\alpha^i = Q^i\,, \quad \overline{d}Q^i = 0\,. \tag{3.59}$$

In a chart, these equations can be rewritten as (here $e^{ABC}$ is the alternator (2.4))

$$\partial_t \alpha^{Ai} = -e^{ABC}\partial_B J^i_C - S^{Ai}\,, \quad \partial_A \alpha^{Ai} = q^i\,, \quad \partial_t q^i = -\partial_A S^{Ai}\,,$$

while the last equation in (3.59), $\overline{d}Q^i = 0$, is identically satisfied, since $Q^i$ is a 3-form.

The first integrals of the system (3.59) define the kinematic field equations [103]:

$$\partial_t k^i = \overline{d}w^i - S^i\,, \quad \overline{d}k^i = Q^i\,, \quad \partial_t \beta^i = \overline{d}V^i - J^i - w^i\,, \quad \overline{d}\beta^i = \alpha^i - k^i\,,$$

where

- $k^i = k^{Ai}\mu_A$ is a bend-twist 2-form;
- $w^i = w^i_A dx^A$ is a spin 1-form;
- $\beta^i = \beta^i_A dx^A$ is a distortion 1-form;
- $V^i$ is a velocity zero-form.

In a chart, these equations take the form

$$\begin{aligned} &\partial_t k^A = -S^{Ai} + e^{ABC}\partial_B w^i_C\,, && \partial_A k^{Ai} = q^i\,, \\ &\partial_t \beta^i_A = \partial_A V^i - J^i_A - w^i_A\,, && e^{ABC}\partial_B \beta^i_C = \alpha^{Ai} - k^{Ai}\,. \end{aligned}$$

### 3.12.6 4D Representation

Following [102, 103], we define disclination 3-forms $\Omega^i$ and dislocation 2-form $D^i$ in 4D-Euclidean space as

$$\begin{aligned} \Omega^i &= -S^i \wedge dt + Q^i = S^{Ai}\mu_A \wedge dt + q^i\mu\,, \\ D^i &= J^i \wedge dt + \alpha^i = J^i_A dx^A \wedge dt + \alpha^{Ai}\mu_A \end{aligned} \tag{3.60}$$

Then, the equations (3.59) can be rewritten in a more concise form:

$$d\Omega^i = 0\,, \quad dD^i = \Omega^i\,, \tag{3.61}$$

and their first integrals can be represented as

$$D^i = dH^i + K^i\,, \quad H^i = V^i dt + \beta^i\,, \quad K^i = -w^i \wedge dt + k^i\,.$$

### 3.12.7 The Momentum Equation

Let $\mathcal{T}$ be a kinetic energy density and $\Psi$ be a potential energy density. We define 3-forms in $\mathscr{E}^4$:

$$z_i = \frac{\partial(\mathcal{T} - \Psi)}{\partial H^i{}_a}\pi_a = \frac{\partial \mathcal{T}}{\partial H^i{}_4}\pi_4 - \frac{\partial \Psi}{\partial H^i{}_A}\pi_A\,.$$

Following conventional terminology, we will refer to

$$p_i := \frac{\partial \mathcal{T}}{\partial H^i{}_4} ,$$

as components of momentum, and to

$$\sigma_i{}^A := \frac{\partial \Psi}{\partial H^i{}_A} ,$$

as components of Piola–Kirchhoff stresses. Then,

$$z_i = -\sigma_i{}^A \pi_A + p_i \pi_4 = -\sigma_i{}^A \mu_A \wedge dt + p_i \mu .$$

The balance of momentum is tantamount to affirming that $z_i$ are exact 3-forms, *i.e.*,

$$dz_i = \left(\partial_4 p_i - \partial_A \sigma_i{}^A\right) \pi = 0 . \tag{3.62}$$

### 3.12.8 Equations in Matrix Form

Equations (3.59) and (3.62) can be rewritten in elegant matrix form. If we define the matrices

$$H = \begin{pmatrix} H^1 \\ H^2 \\ H^3 \end{pmatrix} , \quad K = \begin{pmatrix} k^1 \\ k^2 \\ k^3 \end{pmatrix} , \quad D = \begin{pmatrix} D^1 \\ D^2 \\ D^3 \end{pmatrix} , \quad \Omega = \begin{pmatrix} \Omega^1 \\ \Omega^2 \\ \Omega^3 \end{pmatrix} ,$$
$$Z = \begin{pmatrix} z_1 & z_2 & z_3 \end{pmatrix} ,$$

then equations (3.59) and (3.62) take the form

$$d\Omega = 0 , \quad \Omega = dD = dK , \quad D = dH + K , \quad dZ = 0 .$$

### 3.12.9 Relations of Dislocation and Disclination Forms with Connection, Curvature, and Torsion Forms

Dislocation 2-forms $D^i$ and disclination 3-forms $\Omega^i$ are closely related with the connection 1-form $\Gamma$ and the corresponding curvature 2-form $\Theta$, torsion 2-form $\Sigma$, and soldering 1-form $\nu$. In order to show this, recall the Cartan structural equations

$$\begin{aligned} d\nu &= -\Gamma \wedge \nu + \Sigma , \\ d\Sigma &= -\Gamma \wedge \Sigma + \Theta \wedge \nu , \\ d\Gamma &= -\Gamma \wedge \Gamma + \Theta , \\ d\Theta &= \Theta \wedge \Gamma - \Gamma \wedge \Theta . \end{aligned} \tag{3.63}$$

These equations can be formally resolved with respect to $v$, $\Sigma$, $\Gamma$, and $\Theta$ by applying the homotopy operator $\mathscr{H}$:

$$\begin{aligned}
v &= A[dy + \eta - \mathscr{H}(\Gamma_A \wedge dy)]\,,\\
\Sigma &= A[d\eta + \Gamma_A \wedge \eta + \mathscr{H}(d\Gamma_A \wedge dy) - \Gamma_A \wedge \mathscr{H}(\Gamma_A \wedge dy)]\,,\\
\Gamma &= A\Gamma_A A^{-1} - dAA^{-1}\,,\\
\Theta &= A(d\Gamma_A + \Gamma_A \wedge \Gamma_A)A^{-1}\,,
\end{aligned}$$

where $A$ is a solution of the matrix integral Riemann–Graves equation

$$A = I - \mathscr{H}(\Gamma_A)$$

while $y$, $\eta$ and $\Gamma_A$ are determined from the relations

$$\begin{aligned}
y &= y_0 + \mathscr{H}(A^{-1}v)\,,\\
\eta &= \mathscr{H}(A^{-1}\Sigma)\,,\\
\Gamma_A &= \mathscr{H}(A^{-1}\Theta A)\,.
\end{aligned}$$

Let us identify total distortion 1-form $H^i$ with soldering 1-form $v$ and dislocation 3-form with torsion 2-form $\Sigma$. Then, from (3.63) and (3.60), it follows that

$$\begin{aligned}
\Omega &= \Theta \wedge v - \Gamma \wedge \Sigma\,,\\
K &= \Gamma \wedge v\,.
\end{aligned} \tag{3.64}$$

Applying the homotopy operator $\mathscr{H}$ to the first two equations (3.63), we obtain explicit representations for $H^i$ and $D^i$:

$$\begin{aligned}
H &= A(dy + \eta - \mathscr{H}(\Gamma_A \wedge dy))\,,\\
D &= A(d\eta + \Gamma_A \wedge \eta + \mathscr{H}(d\Gamma_A \wedge dy) - \Gamma_A \wedge \mathscr{H}(\Gamma_A \wedge dy))\,.
\end{aligned}$$

Consequently, substituting these relations into (3.64), we obtain

$$\begin{aligned}
K &= A\left(\Gamma_A - A^{-1}dA\right) \wedge (dy + \eta - \mathscr{H}(\Gamma_A \wedge dy))\,,\\
\Omega &= A[(d\Gamma_A - \Gamma_A \wedge \Gamma_A) \wedge (dy + \eta - \mathscr{H}(\Gamma_A \wedge dy))\\
&\quad - (\Gamma_A - A^{-1}dA) \wedge (d\eta + \Gamma_A \wedge \eta + \mathscr{H}(\Gamma_A \wedge dy) - \Gamma_A \wedge \mathscr{H}(\Gamma_A \wedge dy))]\,.
\end{aligned}$$

Thus, all quantities are expressed in terms of $\eta$, $\Gamma_A$, and $A$. If disclinations are neglected, then the expressions obtained above can be sufficiently simplified. In this case, $\Omega = 0$, $K = 0$ and $A = I$. Then, we have

$$H = dy + \eta\,, \quad D = d\eta\,, \quad \text{or} \quad \eta = \mathscr{H}(D)\,.$$

If, additionally, dislocations are also neglected, *i.e.*, there are no defects at all, then we arrive at the conventional kinematic relations:

$$D = K = 0\,, \quad \Omega = 0\,, \quad \Gamma_A = 0\,, \quad \eta = 0\,, \quad A = I\,, \quad H = d\eta\,.$$

Note that the relations between differential forms, which describe the distribution of defects, and purely geometric objects – differential forms, related with connection on the principal bundle, yield from very abstract and general relations, which determine the properties of differentially closed systems, to which the Cartan structural equations belong.

### 3.12.10 Application of Yang–Mills Coupling Theory

Consider a Lagrangian $\mathscr{L}_0(dy)$ that describes the stress-strain state of a simple solid in conventional elasticity. Let $A \in G_0$ be an element of the Lie group $G_0$ that acts on $y$ from the left[23], *i.e.*, $y' = A \triangleright y$. If $A$ performs a homogeneous action, *i.e.*, it does not depend on the spatial variables and time, then $d$ and $A$ commute:

$$dy' = d(A \triangleright y) = A \triangleright dy \, .$$

In this case, the Lagrangian (constructed with account of frame-indifference principle) is invariant with respect to $A$:

$$\mathscr{L}_0(dy') = \mathscr{L}_0(A \triangleright dy) = \mathscr{L}_0(dy) \, .$$

If $A$ performs a non-homogeneous action, then the relation between $y'$ and $y$ becomes more complex:

$$dy' = (dA) \triangleright y + A \triangleright dy \, ,$$

and the Lagrangian is no longer invariant under the action of $A$:

$$\mathscr{L}_0(dy') = \mathscr{L}_0((dA) \triangleright y + A \triangleright dy) \neq \mathscr{L}_0(dy) \, .$$

The concept of minimal replacement, which is adopted in Yang–Mills theory, restores the invariance of the Lagrangian by replacing

$$d \mapsto D: \quad dy \mapsto Dy := dy + B[y] \, .$$

The fields $B[y]$ are referred to as *calibration fields*. The suitable choice of calibration fields allows preserving the invariance of the Lagrangian under simultaneous replacing of $y \mapsto A \triangleright y$ and $d \mapsto D$. In [102], it is shown that these calibration fields can be determined by some connection field $\Gamma$ and torsion $\varphi$, *i.e.*,

$$B[y] := \Gamma y + \varphi \, .$$

This means that for given dislocation and disclination forms, one can uniquely determine the Cartan structural equations and, consequently, the connection form and

23 See Chapter 14, Section 14.1.

the corresponding torsion, curvature, and solder form, and, after this, determine the calibration fields that compensate the alternation of the Lagrangian under non-homogeneous action $A$. This action locally defines the transformation of elementary volume into the unstressed state, which is impossible for solids with distributed defects in Euclidean space. However, together with the action $A$, the replacement $d \mapsto D$ is performed, which means the passage to non-Euclidean geometry (internal geometry) of the solid which does the transformations of elementary volumes into unstressed state is possible "in large", *i.e.*, for the whole solid.

The concept of minimal replacement involves the definition of the extended Lagrangian according to the concept of minimal coupling [102]. The extended Lagrangian for defected simple solids can be written as

$$\mathscr{L} = \mathscr{L}_0(D\gamma) + s_1\mathscr{L}_1(\Gamma) + s_2\mathscr{L}_2(\varphi) ,$$

where $\mathscr{L}_0(D\gamma)$ describes the action of the elastic properties of the material, $\mathscr{L}_1(\Gamma)$ describes the action of disclinations, and $\mathscr{L}_2(\varphi)$ describes the action of dislocations.

Stationarity conditions of the action with respect to variation of $\gamma$ give the equation

$$DZ = -2R \wedge \Theta , \tag{3.65}$$

under variation of $\Gamma$,

$$DG = \frac{1}{2}J , \tag{3.66}$$

and with respect to variation of $\varphi$,

$$Z = 2DR , \tag{3.67}$$

where $R = \frac{\partial \mathscr{L}}{\partial D}$ (for more details, see [102]). The equations (3.65), (3.66), and (3.67) are referred to as evolution equations.

# 4 Geometric Formalization of the Body and Its Representation in Physical Space

## 4.1 Geometric Motivation

In the conventional theory of elasticity, the reference shape is assumed to be free from stresses, whereas the configuration result in such a shape is called *natural*. The notion of natural configuration is often overlooked, yet it is intimately connected with the thermodynamical concept of the representative elementary volume (REV). In order to define macroscopic fields such as traction and stresses, one has to imagine shape of the body being subdivided on sufficiently small REVs. It is assumed that these REVs are sufficiently small in comparison with the whole shape, so that it can be considered an infinitesimal neighborhood of the position material point and sufficiently large in comparison with the material of microstructure, so that the axiom on the local thermodynamic equilibrium is valid. Then, one has to apply some fundamental principles to the arbitrary collection of REVs, such as the cut principle of Euler and Cauchy [52, p. 154], which, together with the freezing principle and Eulerian laws of dynamics, yields the definition of the macroscopic stress tensor field. Obviously, this field can be defined only up to some gauge, and therefore, the notion of "zero stresses" is also relative to this gauge. In this regard, it is seems reasonable to replace the notion of stress free REV by the notion of *uniform* REV, which means that the physical state of this REV, in particular, its stress state, is undistinguishable from some the standard state.

Assume now that we have found the required compromise, and the body can be presented as a collection of REVs. Global deformation of the body causes local deformation of REVs. Choosing the standard body of a simple geometric shape, for example, a cube, and implementing in it homogeneous deformation, one can determine response of a standard REV on the deformation and fit the law of response, which is also referred to as the constitutive law. It is emphasized that constitutive laws are local, *i.e.*, each of them relates the distortion of the shapes of the REV with the energy density or stresses added to it or reduced in it. Note that it is about increments of energy density (stresses), rather than about its absolute values.

Now, the constitutive law fittingly obtained for a standard REV is applied to all REVs constituting the shapes of the body. For inner boundary-value problem formulation we should assume that the body "is assembled" from the same elementary volumes as the standard body, and that all these volumes are in the same physical state as the volumes of the standard body before the experiment. Then, solving the relevant boundary value problem, we can determine the distribution of the energy (stresses) density in the modeled body. Therefore, we conclude that the classical approach assumes the hypothesis: REVs constituting the standard body prior to experiment and REVs constituting the modeled body, are in the same physical state.

https://doi.org/10.1515/9783110563214-004

According to the terminology of [104], the shape of the standard body and the reference shape of modeled body are uniform (or, equivalently, the shapes of these bodies are the images of the uniform configurations). Therefore, we shall hereinafter use the term “uniform reference configuration” instead of the term “natural reference configuration”. However, a uniform reference configuration does not exist in every case! The body may have a variety of reasons abolishing its uniformity, like defects, inhomogeneous temperature fields, inhomogeneous shrinkage caused by a chemical reaction, *etc.*

Experimentally the amount of inhomogeneity can be estimated by dissection of the body’s shape into a number of small parts, allowing them to deform independently. Many experimental techniques, like the drilling method, the milling method, and others were developed around this idea [105, 106]. Theoretically, such a technique is equivalent to elimination of couplings between parts such that they can be free and become uniform; however, taken as a whole, they are no longer a solid body. One can imagine these “relaxed” parts as a mosaic that cannot be folded without gaps and overlaps in the Euclidean space. The modeling of such discontinuous shapes within the framework of conventional continuum mechanics represents a substantial problem. There are, at least, two ways to address this problem:

(i) To reject the idea of uniform reference configuration and describe all fields relative to a non-uniform shape.
(ii) To assemble geometrically incompatible elementary volumes into a connected region by immersing them into a space with extra features, characterizing the non-Euclidean properties.

In this book, we consider only simple materials [104]. Their response is fully defined by linear transformation, determined by local deformation. Thus, each REV can be transformed to a uniform state by means of a linear transformation; these transformations taken together define a continuous field. At the same time, the geometry of the space with affine connection can be determined by Cartans’ moving frame method, which also uses the field of linear transformations of coordinate basis vectors. Thus, the construction of a region containing arbitrary distorted elementary volumes with mutually compatible shapes, gives use to the mathematical construction of a space with a general affine connection. The specific geometric properties of the thus-built space, *i.e.*, the inner geometry of the body, characterize the measure of incompatibility and sources of residual stresses.

In order to explain such argumentation, we shall give the simple example of a deformable membrane, assuming that the images of the reference and the actual shapes are known, such that one of them, or both, cannot be immersed into a Euclidean plane.

We assume that the observers, describing the “physical reality,” are fictional “two-dimensional creatures” populating the membrane, a two-dimensional surface, like inhabitants of Sphereland in *Sphereland: a fantasy about curved spaces*, written by D. J. Burger [107]. They cannot directly observe the deformation of the membrane, but feel changes in the physical properties of their two-dimensional world, caused by deformation. In addition to "two-dimensional observers," deformation of the membrane is followed by powerful "three-dimensional observers," who can see the shape of the membrane in its possible embedding in the enveloping three-dimensional space. Let us consider two situations.

An image of a uniform configuration of the membrane cannot be embedded into a Euclidean plane, but it can be embedded in an enveloping Euclidean space as part of the sphere (Figure 4.1). In such an embedding, small neighborhoods are in a uniform, or physically indistinguishable, state. Some of them are shown as circumferences in Figure 4.1. "Two-dimensional observers" drawing on measurements of infinitesimal geometrical objects cannot say anything about the "spherical" shape of their world and may be convinced that their world is flat. Suppose now that the physical space is flat. Any embedding of the membrane in the physical space has to be accompanied by a deformation. In this case, two-dimensional observers will find out that their uniform world is no longer uniform: circumferences, characterizing the neighborhoods in Figure 4.1, are distorted, and such distortion occurs without any visible reasons for the two-dimensional observers, because no external force fields have been applied. The distorted circumferences are shown in Figure 4.1 by the solid lines. For comparison, undistorted circumferences are shown nearby by the dotted lines. According to [8], the membrane acquired material inhomogeneity, remaining materially uniform, *i.e.*, made from one and the same material. It is clear that changes of geometric images can be observed only by the powerful "three-dimensional observers." Images in Figure 4.1 are shown from their point of view. Two-dimensional observers will notice only changes in the physical properties of their world. For example, the rigidity of the membrane changes ununiformally, while if the membrane possesses optical properties, it becomes optically inhomogeneous; a sort of a rainbow will appear in the world of two-dimensional observers.

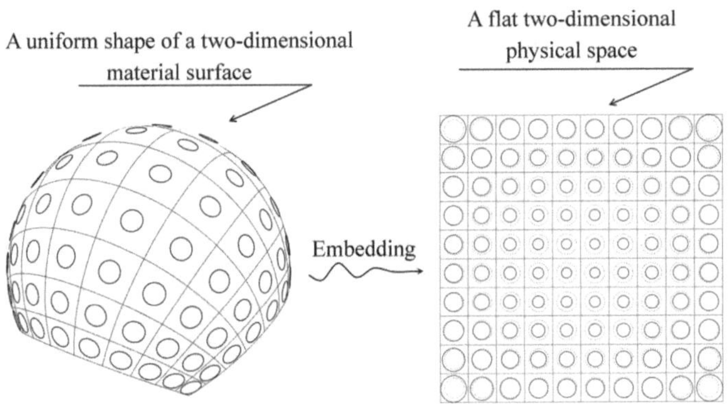


**Fig. 4.1:** Embedding of a two-dimensional body with a non-Euclidean inner geometry in a flat physical space.

**Remark 4.1.** The embedding of a two-dimensional body $\mathcal{S}$ into an ambient three-dimensional space $\mathcal{E}$, shown in Figure 4.1, is defined by using the formulae for a stereographic projection:

$$X = \frac{2x}{1 + x^2 + y^2}, \quad Y = \frac{2y}{1 + x^2 + y^2}, \quad Z = -\frac{x^2 + y^2 - 1}{x^2 + y^2 + 1}. \tag{4.1}$$

Here, $x, y$ are the surface coordinates that can be perceived by a two-dimensional observer, and $X, Y, Z$ are the Cartesian coordinates of the enveloping Euclidean space, defining the geometrical positions for the points of a two-dimensional embedded body that can be perceived by powerful three-dimensional creatures. The circumferences, symbolizing uniform neighborhoods of the points on $\mathcal{S}$ for three-dimensional observers are represented by the parametric formulae

$$X = X_0 + \frac{r_0}{\omega} n_3 \cos\varphi, \quad Y = Y_0 + \frac{r_0}{\omega} n_3 \sin\varphi, \quad Z = Z_0 + \frac{r_0}{\omega}(n_1 \cos\varphi + n_2 \sin\varphi),$$
$$\omega = \sqrt{n_3^2 + (n_1 \cos\varphi + n_2 \sin\varphi)^2}, \quad \varphi \in [0, 2\pi[,$$

where $X_0, Y_0, Z_0$ are three-dimensional coordinates of the centers of circumferences, calculated by (4.1) for the specified $x_0, y_0$, and $r_0$ is their radius (identical for all), and $(n_1, n_2, n_3)$ are the components of the normal vector to the body image.

The embedding of the body $\mathcal{S}$ in the two-dimensional physical space $\mathcal{P}$ is given by the transformation

$$\tilde{x} = \frac{X}{1 + Z}, \quad \tilde{y} = \frac{Y}{1 + Z}.$$

As a result of this transformation, the circumferences are transformed into closed curves, parameterized in $\mathcal{P}$ as follows: $(x, y) \mapsto (\tilde{x}, \tilde{y})$,

$$\tilde{x} = \frac{\omega X_0 + r_0 n_3 \cos\varphi}{\omega + \omega Z_0 - r_0(n_1 \cos\varphi + n_2 \sin\varphi)},$$
$$\tilde{y} = \frac{\omega Y_0 + r_0 n_3 \sin\varphi}{\omega + \omega Z_0 - r_0(n_1 \cos\varphi + n_2 \sin\varphi)},$$

where $\varphi \in [0, 2\pi[$.

The circumferences, symbolizing uniform neighborhoods, are transformed into distorted objects from the point of view of two-dimensional observers for reasons that are not related to the visible deformation of their two-dimensional world, whereas a three-dimensional observer visualizes this deformation explicitly. However, two-dimensional observers may assume that, for some reason, their two-dimensional world changes its geometric properties: its spherical geometry has been transformed to a flat geometry.

Both points of view can be used to describe the embedding of the body in the physical space; however, appealing to a powerful "multidimensional observer" is often cumbersome and related to difficult-to-interpret geometric images. In the present

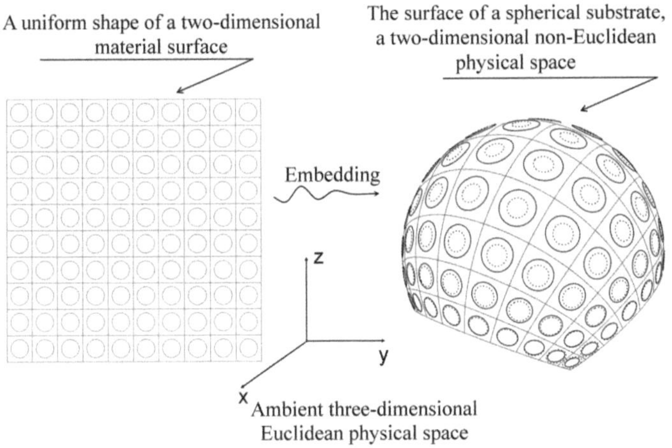


**Fig. 4.2:** Embedding of a two-dimensional body with a Euclidean inner geometry in a non-Euclidean physical space.

book, we develop methods of the "internal observer," who can use two notions: deformation of the body and change of the geometry of the space, containing this body.

In the example discussed, the space containing a uniform reference of the body shape has a non-Euclidian structure, while the physical space is a Euclidean plane. It is easy to imagine an inverted situation: a uniform shape is defined by flat geometry and the physical space is a Riemannian space with non-trivial curvature. Figure 4.2 illustrates this argument: a flat membrane is deposited such that it can slip onto the non-deformable spherical foundation. The images were also built by using transformation (4.1). Thus, incompatible deformations of the body are characterized by two geometrical concepts: the embedding in the physical space and the structure of the space containing its reference uniform configuration.

It is possible to obtain another geometric interpretation of a non-Euclidean structure of the reference shape. Let the body $\mathfrak{B}$ and the physical space $\mathscr{P}$ be presented by two-dimensional manifolds, where, by the condition, $\mathscr{P}$ is Euclidean. It can be shown by two orthogonal axes $(x_1, x_2)$ as in Figure 4.3. Assuming that the physical space of $\mathscr{P}$ is associated with the observer $A$, who "sees" everything happening in $\mathscr{P}$, but is unable to see what is beyond it, the observer lives in a flat world like the characters of *Flatland*, written by E. A. Abbott [108]. We assume that the physical space of $\mathscr{P}$ represents a typical layer of the three-dimensional space of events $\mathfrak{S} = \mathbb{T} \times \mathscr{P}$, in which the added dimension corresponds to time; motions in $\mathscr{P}$ are characterized by changes of the relative positions of points in $\mathscr{P}$ when the layer $\mathscr{P}$ is transferred along the added chronometric axis $\mathbb{T}$. In the space $\mathfrak{S}$, there exists a more powerful observer $B$, who sees all images of configurations of the body $\mathfrak{B}$ in total, *i.e.*, observes the world tube of the body $\mathfrak{B}$ (Figure 4.3). Each section of the global tube is an image of the configuration, corresponding to a certain time instant. If the body has a natural configuration,

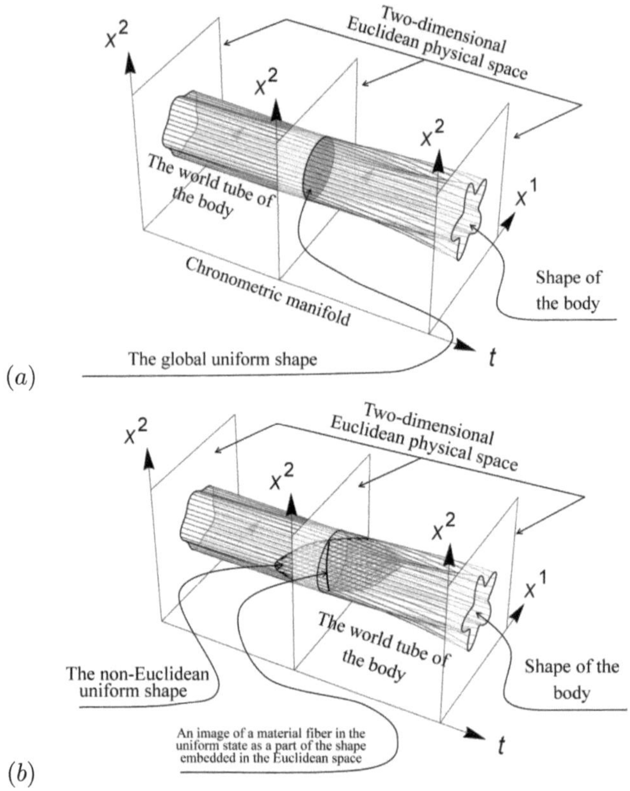


**Fig. 4.3:** A uniform reference configuration fully (a) and partially embedded (b) into the physical space of the observer.

the observer $A$ can see this configuration in one of the sections. However, in a general case, such a section may not be present; the observer $A$ concludes that deformations of the body $\mathfrak{B}$ are incompatible. At the same time, the observer $A$ may detect the sequence of sections (from his perspective, the sequence of time instants), in which the material points in specific curvilinear segments are in the natural state.

Unlike observer $A$, a more powerful observer $B$ sees all these segments in total and observes some two-dimensional surface in $\mathfrak{S}$, the intersection of the general surface and the global tube of the body. All material points of this surface are in a uniform state and, therefore, the surface represents an image of the natural configuration of the body. Thus, the observer $A$ sees the image of a uniform configuration part by part, whereas the observer $B$ sees the entire image. It is important that the image of the configuration for compatible and incompatible deformations, a two-dimensional surface and its "visualization," depends on the geometrical fantasy of the observer. The fact that the segments are "glued together" in the section of the global tube, which can be formalized by a smooth manifold, is based on the additional hypothesis on the smoothness of a uniform configuration. We can imagine a more general situation, when at every time instant the neighborhood of one point transforms to a uniform state. In this case, the observer $B$ sees a "thread" passing along the global tube.

## 4.2 Comparison between Conventional and Non-Euclidean Continuum Mechanics

The most significant aspects that need to be changed or clarified in the transition from the classical Cauchy theory of deformations (Euclidean deformations) to generalizations to manifolds with non-Euclidean connection are given in Tab. 4.1.

Here are the physical and geometric explanations:

(i) The affine structure of the physical space $\mathscr{E}$ allows one to determine the positions of points by means of the field of the placement vector $P \mapsto \boldsymbol{r} = P - O$. In the case of arbitrary smooth manifolds, the notion of the placement vector becomes meaningless; the position of the points is determined by local coordinates (an analogous situation arises in theoretical physics in the transition from the special theory of relativity to the general theory of relativity).

(ii) In Euclidean continuum mechanics, the infinitesimal material fiber is represented by the difference in the vectors of the places of its beginning and end. In the non-Euclidean case, the infinitesimal material fiber associated with the point $\mathfrak{X}$ of the body $\mathfrak{B}$ is represented by the class of curves equivalent at this point, *i.e.*, the tangent vector to the smooth manifold $\mathfrak{B}$ at the point $\mathfrak{X}$. The set of vectors tangent to the smooth manifold $\mathfrak{B}$ at the point $\mathfrak{X}$ is endowed with the

**Tab. 4.1:** The correspondence between notions of Euclidean and non-Euclidean theories of deformation.

| | Notions | Euclidean continuum mechanics | Non-Euclidean continuum mechanics |
|---|---|---|---|
| (i) | Place | Vector field of places $\boldsymbol{r}$ (in Euclidean sense) | Local coordinates $(x^k)$ |
| (ii) | Infinitesimal material fibers | $\Delta\boldsymbol{r} = \boldsymbol{r}_2 - \boldsymbol{r}_1$, where $\boldsymbol{r}_1, \boldsymbol{r}_2$ are placement vectors | Equivalence class of curves |
| (iii) | Deformation | $\gamma\colon \mathscr{E} \supset S_R \to S_t \subset \mathscr{E}$, or, since $\mathscr{E} \cong \mathscr{V}, \mathscr{V} \supset U_R \to U_t \subset \mathscr{V}$ | $\mathscr{P} \supset \varkappa_R(\mathfrak{B}) \to \varkappa_t(\mathfrak{B}) \subset \mathscr{P}$ |
| (iv) | Configuration | Not used | $\varkappa\colon \mathfrak{B} \to \mathscr{P}$ |
| (v) | Displacement vector | $\boldsymbol{u} = \boldsymbol{x} - \boldsymbol{X}$ | Not used |
| (vi) | Deformation gradient | $\boldsymbol{F} = \mathbf{1} + (\nabla \boldsymbol{u})^{\mathsf{T}}$, where $\nabla = \boldsymbol{e}^k \partial_k$ | $\mathscr{F} = T\varkappa$ |
| (vii) | Forces | Vectors | Covectors |
| (viii) | Tensors | $\mathscr{V} \otimes \mathscr{V}, \mathscr{V} \otimes \mathscr{V} \otimes \mathscr{V}, \ldots$ | $T_{\mathfrak{X}}\mathfrak{B} \otimes T_{\mathfrak{X}}\mathfrak{B}$, $T_{\mathfrak{X}}\mathfrak{B} \otimes T^*_{\mathfrak{X}}\mathfrak{B}, \ldots$ |
| (ix) | Fields | $\mathscr{E} \to \mathscr{V} \otimes \mathscr{V}$ | Sections |

structure of a vector space of dimension $n = \dim \mathfrak{B}$ and is denoted by $T_{\mathfrak{X}}\mathfrak{B}$. This is the tangent space to the manifold $\mathfrak{B}$ at the point $\mathfrak{X}$.

**Remark 4.2.** To distinguish the tangent vectors from the manifolds from the vectors of the translation space $\mathscr{V}$, the latter are denoted by lowercase bold Latin characters $\boldsymbol{u}, \boldsymbol{v}, \ldots$, and the tangent vectors are denoted by lowercase Latin characters $u, v, \ldots$.

(iii) In the Euclidean continuum mechanics, it is assumed that there is a global natural (stress free) *reference shape* $S_R \subset \mathscr{E}$, and the concept of the body is not used. Deformations are represented by mappings of the form $\mathscr{E} \supset S_R \to S_t \subset \mathscr{E}$, where $S_t$ is the actual shape. By virtue of the isomorphism $\mathscr{E} \cong \mathscr{V}$, deformations can be represented in an equivalent way as mappings of subsets of $\mathscr{V}$, *i.e.*, as the maps $\mathscr{V} \supset U_R \to U_t \subset \mathscr{V}$. In the non-Euclidean case, the deformation is the mapping[1] $\mathscr{P} \supset \varkappa_R(\mathfrak{B}) \to \varkappa_t(\mathfrak{B}) \subset \mathscr{P}$. Here, $\varkappa_R(\mathfrak{B})$, unlike in the classical case, can be a stressed shape.

(iv) The concept of configuration is absent in the Euclidean theory, since the body is identified with the natural reference shape. At the same time, the role of the "deformation" from the imaginary space of affine connection, that is the superstructure over the body, in which an unstressed (or, more generally, uniform) shape is considered, to the actual shape of the physical space is played by configuration.

(v) The affine structure of space in Euclidean theory allows one to introduce the displacement vector field $\boldsymbol{u} = \boldsymbol{x} - \boldsymbol{X}$, the value of which represents the difference of the places of the particle occupied by it in the actual and reference shapes. In the case when the physical space is an abstract Riemannian manifold, the concept of a vector displacement field is absent.

(vi) In Euclidean theory, the deformation gradient is determined by the field of linear transformations (the mappings $X \mapsto \boldsymbol{F}_X = y'(X)$, each of which, $\boldsymbol{F}_X$, linearizes the deformation $y$ in the neighborhood of the point $X$). Equivalently, the deformation gradient is represented in the form $\boldsymbol{F} = \mathbf{1} + (\nabla \boldsymbol{u})^{\mathrm{T}}$, where $\nabla \boldsymbol{u}$ denotes the gradient of the vector field (in Gibbs notation [46]). In the approach developed, the deformation gradient is determined from the configuration $\varkappa \colon \mathfrak{B} \to \mathscr{P}$ and is represented by the tangent mapping $\mathscr{F} = T\varkappa \colon T\mathfrak{B} \to T\mathscr{P}$ (see Chapter 13, Section 13.2.2 and Section 13.3.6).

(vii) In contrast to classical representations, it seems expedient to distinguish between vectors and covectors on a smooth manifold. We denote by $T_p^*\mathfrak{M}$ the cotangent space to the smooth manifold $\mathfrak{M}$ (representing the body or physical space), that is, the space of linear functionals over $T_p\mathfrak{M}$. From the physical point of view, the elements of the dual space are the forces [35].

1 Here, $\varkappa_R$ is the reference configuration, and $\varkappa_t$ is the actual one.

(viii) In classical mechanics, tensors and tensor fields are determined in the framework of the axioms of Euclidean space. At the same time, to work with fields in a non-Euclidean space, one needs a definition of a tensor that does not use the axioms of Euclidean space.

(ix) A tensor field on $\mathfrak{M}$ (including vector and covector) is considered as a section of the vector bundle constructed over $\mathfrak{M}$ (see Chapter 13, Section 13.3). The coordinate frame on $\dim \mathfrak{M} = s$ is denoted by the symbol $(\partial_i)_{i=1}^{s}$, and the coframe corresponding to it is denoted by $(dx^i)_{i=1}^{s}$ (see Chapter 13, Section 13.3.8). Thus, for an arbitrary smooth vector $u\colon \mathfrak{M} \to T\mathfrak{M}$ and covector $v\colon \mathfrak{M} \to T^*\mathfrak{M}$ fields one has decompositions: $u = u^i \partial_i$, $v = v_i dx^i$.

**Remark 4.3.** The use of vector bundles allows one to consider sections as mappings of one smooth manifold into another, which, in turn, allows one to determine the smoothness of the sections in the standard way (see Chapter 13, Section 13.1.3).

**Remark 4.4.** Tensors and tensor fields in Euclidean space are denoted by bold uppercase Latin symbols $\boldsymbol{T}, \boldsymbol{F}, \dots$, and tensor fields on smooth manifolds are handwritten Latin symbols $\mathscr{T}, \mathscr{F}, \dots$. The exception is $\mathscr{E}$ and $\mathscr{V}$, which are reserved for affine Euclidean space and the associated vector space.

(x) The Euclidean structure of space provides the observer with an abstract "instrument", which makes it possible to measure the lengths of vectors (infinitesimal fibers) and the angles between them. The presence of such a "device" is necessary for the construction of the theory of deformations. In the case of smooth manifolds, the role of the instrument is played by the Riemannian metric (see Chapter 13, Section 13.3.10). The latter exists on any smooth manifold. Using the Riemannian metric, one can introduce musical isomorphisms that associate a vector with a covector (operation $(\cdot)^\flat$) and vice versa (operation $(\cdot)^\sharp$). Note that from a physical point of view, musical isomorphisms allow measuring physical quantities in units of measurement unusual for them, for example, force (covector) in meters, and displacement (vector) in Newtons. In index notation, these operations correspond to the lowering and raising of indices.

(xi) Often, there is a need for the expression of tensor fields given at points of the actual shape through the points of the reference shape, and vice versa. In classical mechanics, these operations correspond to the Piola transformation. In the case of smooth manifolds, the analog of the Piola transform is pullback and pushforward operations. If $\mathscr{T}$ is a tensor field on $\mathscr{P}$, and $\mathscr{Q}$ is a tensor field on $\mathfrak{B}$, then the pullback of $\mathscr{T}$ is denoted by $\varkappa^*\mathscr{T}$, and through $\varkappa_*\mathscr{Q}$ the pushforward of $\mathscr{Q}$ is denoted. Here, $\varkappa\colon \mathfrak{B} \to \mathscr{P}$ is a smooth map. In the case of a scalar function, its pullback and pushforward are reduced to replacing its domain. For an arbitrary tensor field, this is not so; except for the replacement of the domain, it is necessary to replace the tangent and cotangent bundles with special recalculation formulae (see Chapter 13, Section 13.3.7).

## 4.3 A Body

A *body* $\mathfrak{B}$ is a set, whose elements are denoted by fraktur characters, $\mathfrak{X}, \mathfrak{Y}, \mathfrak{Z}, \dots$, and are referred to as *material points*. Its cardinal number, $\mathrm{Card}(\mathfrak{B})$, defines the type of mechanical theory, which describes the kinematics and dynamics of material points that constitute the body. In the framework of analytical mechanics, finite or countable sets of material points are taken into account, *i.e.*, $\mathrm{Card}(\mathfrak{B}) \leq \mathrm{Card}(\mathbb{N}) = \aleph_0$. The subject of research in continuum mechanics is bodies with cardinal number $\mathrm{Card}(\mathfrak{B}) = \mathrm{Card}(\mathbb{R}) = \aleph_1$. Therefore, we will treat a body as a set with continuum cardinality.

In order to describe interaction between the subsets of material points one needs to formalize the notion of a neighborhood. Such a notion can be formalized in the framework of topology. That is, one needs to assume that the body is a topological space. From the physical standpoint, different points should have disjoint neighborhoods, such as the space should be Hausdorff.

For an analytical description of physical fields that are acting on material points, one requires to cover the body with charts $(U, \varphi)$. Such charts locally introduce coordinates on the body $\mathfrak{B}$; each chart $(U, \varphi)$ consists of an open set $U \subset \mathfrak{B}$ and a homeomorphism $\varphi : U \to O$ between domain $U$ of the chart and an open set $O \subset \mathbb{R}^n$. The number $n$ is fixed. In the general case, such a covering can be provided by not less than a continuum of charts. The assumption that $\mathfrak{B}$ has a countable base allows one to deal with finite or countable numbers of charts. To deal with smooth fields only, an additional requirement is needed: if domains $U$ and $V$ of charts $(U, \varphi)$ and $(V, \psi)$ intersect $(U \cap V \neq \emptyset)$, then the composition, $\psi \circ \varphi^{-1}|_{\varphi(U \cap V)}$, that acts between open sets in $\mathbb{R}^n$, is a $C^\infty$-diffeomorphism. Thus, a smooth atlas on $\mathfrak{B}$ should be chosen.

One considers subsets (*parts*) of the body and provides integration over them, or their boundaries, with the purpose to formulate balance laws. The integration can be performed correctly in the case of oriented manifolds. Thus, one needs to introduce an orientation on $\mathfrak{B}$. This orientation is induced on all parts of the body. All above reasonings lead to the following properties, which the set $\mathfrak{B}$ of material points is required to satisfy:

($\mathfrak{B}_1$) $\mathrm{Card}(\mathfrak{B}) = \aleph_1$;

($\mathfrak{B}_2$) $\mathfrak{B}$ is an $n$-dimensional $C^\infty$-manifold[2].

($\mathfrak{B}_3$) $\mathfrak{B}$ is orientable manifold, *i.e.*, a volume form $\mu_0 \in \Omega^n(\mathfrak{B})$ exists for it.

Suppose that an orientation $[\mu_0]$ was chosen. By a *part of the body* $\mathfrak{B}$ we mean an $n$-dimensional embedded submanifold $\mathfrak{P} \subset \mathfrak{B}$ with boundary and orientation $[\iota_{\mathfrak{P}}^* \mu_0]$. The set of all parts of the body is denoted by $\mathrm{Part}(\mathfrak{B})$.

The representation of a body as a smooth manifold was used for the first time by W. Noll, C.-C. Wang, M. Gurtin, and A. Murdoch in [104, 109, 110]. For instance,

2 Brief information about the smooth manifold notion is given in Chapter 13, Section 13.1.

Noll introduced an axiomatic definition of a smooth body as a set, equipped with a family of its mappings in the physical space called configurations and satisfying some special system of axioms. Gurtin and Murdoch gave a self-contained definition of a two-dimensional material surface based on the Noll axioms.

**Remark 4.5.** The Noll system of axioms is based on the notion of the configuration $\varkappa$ as a mapping of the set $\mathfrak{B}$, representing a collection of material points constituting the body, to the absolute (Euclidean) physical space $\mathscr{E}$, *i.e.*, $\varkappa\colon \mathfrak{B} \to \mathscr{E}$. Let us assume that a class of configurations $\mathfrak{C}$ is given. The body $\mathfrak{B}$ is called a continuous body of the class $C^p$, if the class of configurations $\mathfrak{C}$ satisfies the following conditions [104]:

($\mathfrak{C}_1$) Each configuration $\varkappa \in \mathfrak{C}$ is a homeomorphism, and the image $\varkappa(\mathfrak{B})$ is an open set in $\mathscr{E}$.

($\mathfrak{C}_2$) If $\gamma, \varkappa \in \mathfrak{C}$, then the composition[3] $\lambda = \gamma \circ \hat{\varkappa}^{-1}\colon \varkappa(\mathfrak{B}) \to \gamma(\mathfrak{B})$ represents a mapping of the class $C^p$, which is called deformation of the body $\mathfrak{B}$ from the configuration $\varkappa$ to the configuration $\gamma$.

($\mathfrak{C}_3$) If $\varkappa \in \mathfrak{C}$ and if $\lambda\colon \varkappa(\mathfrak{B}) \to \mathscr{E}$ is a mapping of the class $C^p$, then $\lambda \circ \varkappa \in \mathfrak{C}$.

Thus, the order of body smoothness is defined by the smoothness of the compositions of its configurations. According to such a definition, the smooth body is a smooth manifold with trivial topological structure, *i.e.*, a trivial manifold, whose atlas contains only one chart [3]. Such manifolds do not cover the possible variants of bodies (the need for using non-trivial manifolds, representing the body, is also described in [109]).

M. Gurtin and A. Murdoch in [110] represent the material surface (a two-dimensional body) as a body, such that each image of a configuration into Euclidean affine space is a surface. Each surface $\Omega$ in $\mathscr{E}$ is a subset in $\mathscr{E}$, such that a pair $(T_p\Omega, \pi_p)$ is assigned to each point $p \in \Omega$. Here, $T_p\Omega$ is a two-dimensional vector subspace of $\mathscr{V}$, and $\pi_p$ is a mapping. The family $\{(T_p\Omega, \pi_p)\}_{p\in\Omega}$ satisfies the following axioms:

($\Omega_1$) For each point $p \in \Omega$, the mapping $\pi_p\colon \mathscr{N}_p \to \Omega$ is of class $C^2$. Its domain, $\mathscr{N}_p$, is an open neighborhood of $\mathbf{0}$ in $T_p\Omega$.

($\Omega_2$) For each $p \in \Omega$, there exists a neighborhood $U(p)$ of the point $p \in \Omega$ in $\mathscr{E}$, such that $\pi_p(\mathscr{N}_p) \cap U(p) = \Omega \cap U(p)$.

($\Omega_3$) For each $p \in \Omega$, the vector $\pi_p(\boldsymbol{h}) - (p + \boldsymbol{h})$ belongs to $(T_p\Omega)^\perp$ for each $\boldsymbol{h} \in \mathscr{N}_p$ and is of order $\boldsymbol{o}(\|\boldsymbol{h}\|)$ as $\|\boldsymbol{h}\| \to 0$;

($\Omega_4$) There exists a continuous field $\boldsymbol{m}\colon \Omega \to \mathscr{V}$, which never vanishes and has $m(P) \in (T_P\Omega)^\perp$ for each $p \in \Omega$.

Here, $(T_p\Omega)^\perp$ is the subspace of $\mathscr{V}$, which is orthogonal to $T_p\Omega$:

$$(T_p\Omega)^\perp = \{\boldsymbol{v} \in \mathscr{V} \mid \forall \boldsymbol{u} \in T_p\Omega\colon (\boldsymbol{u} \cdot \boldsymbol{v} = 0)\}\,.$$

In [110], the set $T_p\Omega$ is referred as a tangent space at point $p \in \Omega$. Such a definition of the tangent space essentially uses the Euclidean and affine structure of $\mathscr{E}$. Since

3 The mapping $\hat{\varkappa}\colon \mathfrak{B} \to \varkappa(\mathfrak{B})$ is obtained from $\varkappa\colon \mathfrak{B} \to \mathscr{E}$ by restricting its codomain to range.

we intend to consider bodies in Riemannian physical space, it is preferable to use the intrinsic definition of tangent space.

Consider an example. Let $\mathfrak{B}$ be a two-dimensional body that can be presented by a single sphere in a three-dimensional Euclidean space; its image is described by the equation $x_1^2+x_2^2+x_3^2 = 1$, where $x_1, x_2, x_3$ are the Cartesian coordinates. On this body, we can define an atlas that consists of two charts $(U_1, \varphi_1)$, $(U_2, \varphi_2)$, whose coordinate homeomorphisms are defined as stereographic projections. Thus, stereographic projections of the parts of the sphere on the plane, the points of which have a zero coordinate $x_3$ when this plane is identified with $\mathbb{R}^2$, are presented by the following mappings:

$$h_1 : \mathfrak{B} \setminus (0,0,-1) \ni (x_1, x_2, x_3) \mapsto \left( \frac{x_1}{1+x_3}, \frac{x_2}{1+x_3} \right),$$
$$h_2 : \mathfrak{B} \setminus (0,0,1) \ni (x_1, x_2, x_3) \mapsto \left( \frac{x_1}{1-x_3}, \frac{x_2}{1-x_3} \right).$$

the first of these establishes a one-to-one correspondence between the points of the sphere and the plane, "pierced" by a beam beginning in the "southern pole" point with the coordinates $(0, 0, -1)$, and the second establishes a one-to-one correspondence between the points of the sphere and the plane, "pierced" by a beam beginning in the "northern pole" point with the coordinates $(0, 0, 1)$.

In Figure 4.4 the dashed line with the sign of embedding designates the relation between sets and subsets, and the arrow designates the direction of action of the mapping.

The pre-images $h_1^{-1}(D)$, $h_2^{-1}(D)$ of the open disk

$$D = \{(\xi^1, \xi^2) \mid (\xi^1)^2 + (\xi^2)^2 < \tfrac{7}{4}\}$$

are used as $U_1$ and $U_2$. The restriction of the mappings $h_1^{-1}$ and $h_2^{-1}$ on $D$ are taken as $\varphi_1 : D \to U_1$ and $\varphi_2 : D \to U_2$:

$$\varphi_1 : (\xi^1, \xi^2) \mapsto \left( \frac{2\xi^1}{1+(\xi^1)^2+(\xi^2)^2}, \frac{2\xi^2}{1+(\xi^1)^2+(\xi^2)^2}, \frac{\xi^1+\xi^2-1}{1+(\xi^1)^2+(\xi^2)^2} \right);$$
$$\varphi_2 : (\eta^1, \eta^2) \mapsto \left( \frac{2\eta^1}{1+(\eta^1)^2+(\eta^2)^2}, \frac{2\eta^2}{1+(\eta^1)^2+(\eta^2)^2}, -\frac{\eta^1+\eta^2-1}{1+(\eta^1)^2+(\eta^2)^2} \right).$$

To the region $U_{12} = U_1 \cap U_2$ in $\mathbb{R}^2$ there corresponds a set of ordered pairs $(\xi^1, \xi^2)$, such that $\frac{4}{7} < (\xi^1)^2 + (\xi^2)^2 < \frac{7}{4}$. In this case, the transition homeomorphism is as follows:

$$\varphi_{12} : U_{12} \to U_{12}, \quad (\eta^1, \eta^2) \mapsto (\xi^1, \xi^2) = \left( \frac{(\eta^1)^2}{(\eta^1)^2+(\eta^2)^2}, \frac{(\eta^2)^2}{(\eta^1)^2+(\eta^2)^2} \right),$$

and is a $C^\infty$-diffeomorphism. Thus, the sphere $\mathfrak{B}$ is a two-dimensional manifold with the $C^\infty$-structure induced by the constructed atlas (Figure 4.4).

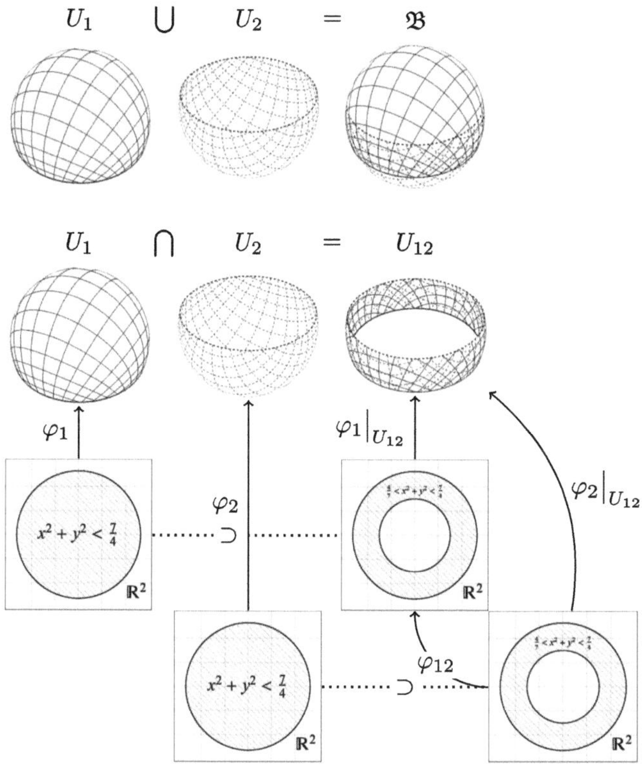


**Fig. 4.4:** Charting of the sphere and the transition function between the charts.

Since the body $\mathfrak{B}$ is a smooth manifold, there exists a Riemannian metric $\mathcal{G} \in \mathrm{Sec}(T^*\mathfrak{B} \otimes T^*\mathfrak{B})$. We refer to $\mathcal{G}$ as the *material metric*. Such a metric is not arbitrary; it is specific to considering physical theory. In Chapter 8, Section 8.4, the example of constructing the material metric for materially uniform simple bodies is shown. In another chapters where the material metric is involved, we assume that it is obtained from reasonings that are not dependent on those of such chapters.

Note that unlike the case of the body, the *spatial* Riemannian metric $g$ of the physical space $\mathcal{P}$ is pre-defined in our considerations. In a more general situation, both $\mathcal{G}$ and $g$ should be obtained from certain reasonings. In this case, the body $\mathfrak{B}$ and the physical space $\mathcal{P}$ are equal in rights. Such a consideration is beyond the scope of our book. The dimensions of the body and the physical space satisfy the following inequalities:

$$n = \dim \mathfrak{B} \le \dim \mathcal{P} = m \le 3\,.$$

## 4.4 Configurations

Let $\mathfrak{B}$ be a body and $\mathscr{P}$ be a physical space with the Riemannian metric $g$. The relation between the continual set of material points that constitute the body and the set of their places in the physical space is established by means of a *configuration*, that is, a map $\varkappa \colon \mathfrak{B} \to \mathscr{P}$. The image of a configuration $\varkappa$, *i.e.*, the set $\varkappa(\mathfrak{B})$, is referred to as a *shape* and is denoted by $\mathcal{S}_\varkappa$, if the body is fixed. The relations between a body, a configuration, and the shape are illustrated in the following diagram:

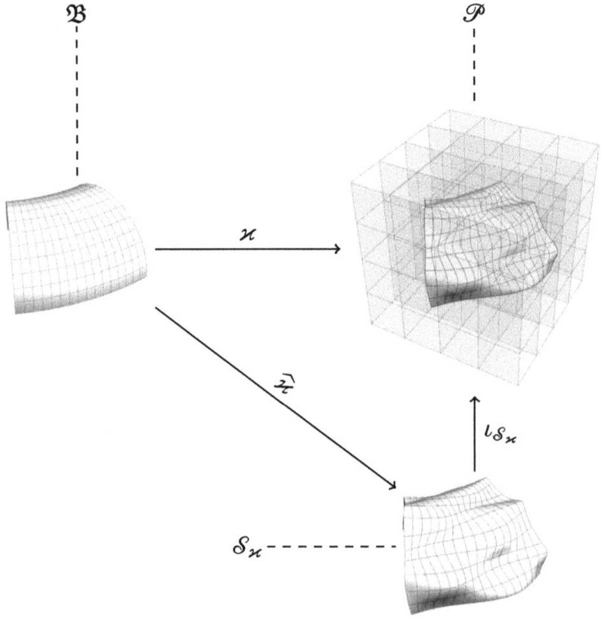


The diagram uses the following designations:
- $\iota_{\mathcal{S}_\varkappa} \colon \mathcal{S}_\varkappa \to \mathscr{P}$ is the inclusion map.
- $\hat{\varkappa} \colon \mathfrak{B} \to \mathcal{S}_\varkappa$ is the map that is obtained from $\varkappa \colon \mathfrak{B} \to \mathscr{P}$ by restricting its codomain to $\mathcal{S}_\varkappa$.
- $\varkappa = \iota_{\mathcal{S}_\varkappa} \circ \hat{\varkappa}$.

Since both the body and the physical space are smooth manifolds, we take from the set of all configurations $\varkappa$ only those that are of class $C^\infty$, that is, $\varkappa \in C^\infty(\mathfrak{B};\mathscr{P})$. This allows one to use differential calculus for further derivations. Additionally, $\varkappa$ is assumed to be a *smooth immersion*; at each material point $\mathfrak{X} \in \mathfrak{B}$ the tangent map $T_\mathfrak{X}\varkappa \colon T_\mathfrak{X}\mathfrak{B} \to T_{\varkappa(\mathfrak{X})}\mathscr{P}$ is injective. As we will discuss in Section 5.5, the map $T_\mathfrak{X}\varkappa$ represents the configuration gradient at $\mathfrak{X}$, and thus the assumption states that spatial images of infinitesimal material fibers do not coincide. This condition is also referred to as the *local impenetrability condition*. Such an assumption does not guarantee that

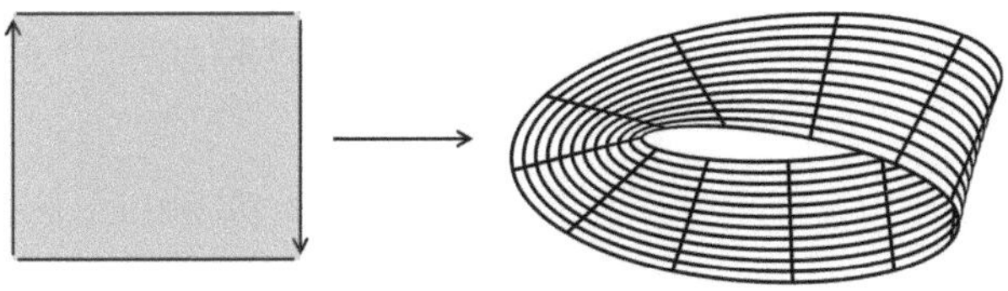

**Fig. 4.5:** Möbius strip.

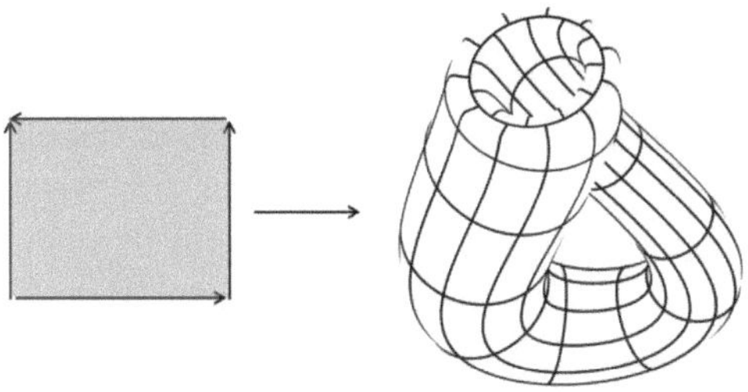

**Fig. 4.6:** Klein bottle.

this impenetrability condition is *global*. The class of all smooth immersions $\varkappa$ must be restricted by the provision that $\varkappa$ is a homeomorphism onto its image $\varkappa(\mathfrak{B}) \subset \mathscr{P}$ in the subspace topology[4], or, equivalently, $\hat{\varkappa}$ is a homeomorphism. Thus, to satisfy this requirement $\varkappa$ needs to be a *smooth embedding*.

Consider two examples to illustrate the difference between smooth immersion and smooth embedding, represented by the Möbius strip (Figure 4.5) and the Klein bottle (Figure 4.6), which can be considered as two-dimensional bodies. For the Möbius strip, one can define the configuration with the image in three-dimensional Euclidean space $\mathscr{E}$:

$$\boldsymbol{p}(r,\varphi) = \left(3 + r\cos\frac{\varphi}{2}\right)\cos\varphi\boldsymbol{i}_1 + \left(3 + r\cos\frac{\varphi}{2}\right)\sin\varphi\boldsymbol{i}_2 + r\sin\frac{\varphi}{2}\boldsymbol{i}_3\,,$$

$(r,\varphi) \in \,]-1,1[\, \times [0,2\pi[$. Here $(\boldsymbol{i}_1, \boldsymbol{i}_2, \boldsymbol{i}_3)$ is an orthonormal frame, and $\boldsymbol{p}$ is the position vector field. This set has no self-intersections and thus the corresponding configuration is a smooth embedding. At the same time, the projection of such a set on any plane, for example, $X_1OX_2$, is not a smooth embedding on the two-dimensional Euclidean space. There is a possibility to construct an embedding of the Möbius strip in a two-dimensional space that is topologically equivalent to the Klein bottle. If the physical space $\mathscr{P}$ is represented by a smooth manifold with such a structure, then for bodies that are topologically equivalent to the Möbius strip, there exists a continual family of configurations as smooth embeddings. This example provides an analogy to geometric continuum mechanics. If a body has internal stresses in Euclidean physical space, then the idea is to construct its embedding on some curved (non-Euclidean) space, where it will have a stress-free shape.

4 If $\mathfrak{S}_{\mathscr{P}}$ is a topology on $\mathscr{P}$, then the set $\mathfrak{S}_{\mathcal{S}_\varkappa} = \{O \cap \mathcal{S}_\varkappa \mid O \in \mathfrak{S}_{\mathscr{P}}\}$ is the subspace topology of $\mathcal{S}_\varkappa$.

The Klein bottle can be represented in three-dimensional Euclidean space $\mathscr{E}$ as the image of the mapping $(t, \theta) \mapsto \boldsymbol{p}(t, \theta)$,

$$\boldsymbol{p}(t, \theta) = \boldsymbol{\alpha}(t) + r(t)\left(\cos\theta\, \boldsymbol{G}(\boldsymbol{\tau}(t)) + \sin\theta \boldsymbol{i}_3\right) , \quad (t, \theta) \in ]0, 2\pi[ \times [0, 2\pi[ ,$$

where $\boldsymbol{\alpha}(t) = a(1-\cos t)\boldsymbol{i}_1 + b\sin t(1-\cos t)\boldsymbol{i}_2$, $r(t) = c - d(t-\pi)\sqrt{t(2\pi - t)}$, $\boldsymbol{G}(x\boldsymbol{i}_1 + y\boldsymbol{i}_2) = -y\boldsymbol{i}_1 + x\boldsymbol{i}_2$, $\boldsymbol{\tau}(t) = \|\boldsymbol{\alpha}'(t)\|^{-1}\boldsymbol{\alpha}'(t)$. Such a mapping is a smooth immersion but with self-intersections. The minimal (relative to a dimension) space, into which the Klein bottle can be embedded, is $\mathbb{R}^4$. This corresponds to the results of the theory of smooth manifolds, in which theorems on embeddings are proved. One of them, the "weak Whitney theorem," asserts that for any compact $n$-dimensional $C^r$-manifold $\mathfrak{M}$, where $r \geq 2$, there exists a $C^r$-embedding of $\mathfrak{M}$ into $\mathbb{R}^{2n+1}$. The proof can be found in [111, 112]. To strengthen the theorem, see the remark in [112, p. 27]. In particular, the "strong Whitney theorem" asserts that any smooth $n$-dimensional $(n > 1)$ manifold can be immersed into $\mathbb{R}^{2n-1}$ and can be embedded into $\mathbb{R}^{2n}$. In this case, the values of $2n - 1$ and $2n$ are unimprovable.

The different ways of Klein bottle parametrization are represented in [113]. The surface, which was illustrated in Figure 4.6, was obtained by the following parameter values: $(a, b, c, d) = (20, 8, 11/2, 2/5)$.

**Remark 4.6.** The necessity of carrying out research on configurations that are immersions is obvious for bodies whose dimension is less than the dimension of the physical space. However, such configurations can be of interest also in the case when the dimensions of the body and physical space coincide. The "Möbius crystal", investigated by Wang [109], is an example. Let us briefly consider the example.

The body is materially uniform [104, 109] if all its points are mutually materially isomorphic. Physically, this means that all its material points consist of the same material. Let the physical space be a three-dimensional Euclidean space $\mathscr{E}$ and a stress-free shape of a body $\mathfrak{B}$ represented by a long circular cylinder. Assume that the material is anisotropic and has a rhombic symmetry. This means that three crystallographic axes $(\boldsymbol{e}_1, \boldsymbol{e}_2, \boldsymbol{e}_3)$ are associated with each point of the body. These axes are such that the mechanical properties of the material do not change as the shape rotates through an angle $\pi$ around each of these axes. There are no other material symmetries. Assume that in the stress-free shape, all material points are oriented in such a way that the corresponding crystallographic axes are parallel, and one of them, for example $\boldsymbol{e}_1$, is parallel to the axis of the cylinder. In such a shape, there exists a unique smooth material isomorphism [109], represented by a translation in Euclidean space. We make the following transformations: first we tighten the cylinder so that its lower base rotates relative to the upper one by an angle $\pi$, then we bend the cylinder into the ring, and "glue" the bases. C.-C. Wang called the resulting solid of the toroidal form the "Möbius crystal". It is clear that any simply-connected part of the body thus obtained has a smooth field of material isomorphisms, but it cannot be extended to the whole body.

In the rest of the book, by a *configuration* of a body $\mathfrak{B}$ in the physical space $\mathscr{P}$ we mean a $C^\infty$-embedding $\varkappa\colon \mathfrak{B} \to \mathscr{P}$. The set of all such mappings is denoted by $\mathfrak{C}(\mathfrak{B}; \mathscr{P})$.

## 4.5 A Shape of a Body as a Submanifold of the Physical Space

### 4.5.1 A Shape as a Submanifold

Let $\mathfrak{B}$ be an $n$-dimensional body, $\mathscr{P}$ be a physical space with the Riemannian metric $g$, and $\varkappa \in \mathfrak{C}(\mathfrak{B}; \mathscr{P})$ be some configuration. The shape $\mathscr{S}_\varkappa \subset \mathscr{P}$ is endowed with subspace topology. Since $\varkappa$ is a smooth embedding, $\mathscr{S}_\varkappa$ is an *embedded n-dimensional submanifold*[5] *of* $\mathscr{P}$. That is, by means of $\varkappa$, $\mathscr{S}_\varkappa$ is endowed with a $C^\infty$-structure such that the inclusion map $\iota_{\mathscr{S}_\varkappa}\colon \mathscr{S}_\varkappa \to \mathscr{P}$ is a $C^\infty$-embedding. Such a structure is constructed as follows. If $A_{\mathfrak{B}} = \{(U_\alpha, \varphi_\alpha)\}_{\alpha \in I}$ is some smooth atlas from the smooth structure of the body $\mathfrak{B}$, then the set

$$A_\varkappa = \left\{\left(\varkappa(U_\alpha), \varphi_\alpha \circ \widehat{\varkappa}^{-1}|_{\varkappa(U_\alpha)}\right)\right\}_{\alpha \in I}$$

is a smooth atlas on $\mathscr{S}_\varkappa$. It generates the desired smooth structure of $\mathscr{S}_\varkappa$. The construction of a chart $(\varkappa(U_\alpha), \varphi_\alpha \circ \widehat{\varkappa}^{-1}|_{\varkappa(U_\alpha)})$ is illustrated in the following diagram:

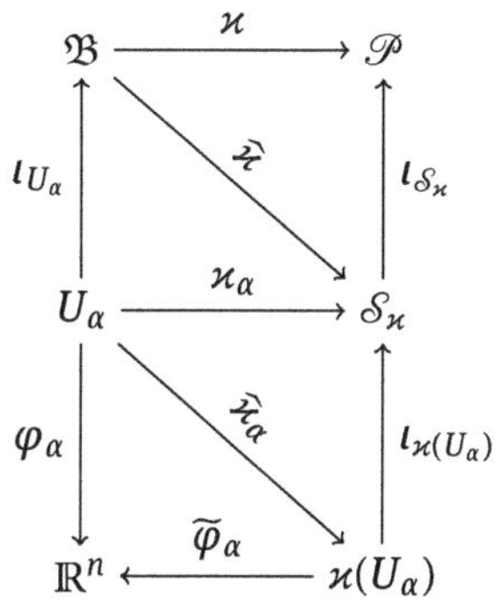


The diagram uses the following designations:

- $\varkappa_\alpha$ is the restriction of $\widehat{\varkappa}$ on $U_\alpha \subset \mathfrak{B}$, i.e., $\varkappa_\alpha = \widehat{\varkappa} \circ \iota_{U_\alpha}$.
- $\widehat{\varkappa}_\alpha$ obtained from $\widehat{\varkappa}$ by restricting of its domain and codomain on $U_\alpha$ and $\varkappa(U_\alpha)$, i.e., $\iota_{\varkappa(U_\alpha)} \circ \widehat{\varkappa}_\alpha = \widehat{\varkappa} \circ \iota_{U_\alpha}$.
- $\widetilde{\varphi}_\alpha = \varphi_\alpha \circ \widehat{\varkappa}_\alpha^{-1}$ is the coordinate homeomorphism.

5 See Chapter 13, Section 13.1.4 for more detail.

### 4.5.2 The Local $k$-slice Condition

One can define a smooth atlas on the shape $\mathscr{S}_\varkappa$ by means of a smooth atlas of the body $\mathfrak{B}$. Meanwhile, as was stated above, only objects that belong to the physical space $\mathscr{P}$ (curves, tangent vectors, scalar functions, *etc.*) are observable. If an observer introduces some coordinate system on $\mathscr{P}$, then he intends to use such a system for constructing a smooth atlas of $\mathscr{S}_\varkappa$. This can be formalized by the *local k-slice condition* [3]. Let us consider this formalization in more detail.

Let $\mathfrak{M}$ be a smooth $n$-dimensional manifold and $(U, \varphi)$ a smooth chart on $\mathfrak{M}$. A subset $S \subset U$ is a $k$-slice of $U$, $k \in \{0, \dots, n\}$, if $\varphi(S) \subset \mathbb{R}^n$ has the form

$$\varphi(S) = \{(x^1, \dots, x^k, x^{k+1}, \dots, x^n) \in \mathbb{R}^n \mid x^{k+1} = \dots = x^n = 0\}\,.$$

Here, $(x^i)$ are local coordinates that are generated by the coordinate homeomorphism $\varphi$. By the definition, the last $n - k$ coordinates of each point from $S$ are equal to zero.

In accordance with [3], $S$ satisfies the *local k-slice condition* if for each point $p$ of $S$ there exists a smooth chart $(U, \varphi)$ on $\mathfrak{M}$, such that $p \in U$ and $U \cap S$ is a $k$-slice of $U$. The theorem [3] asserts that: *if $S \subset \mathfrak{M}$ is an embedded k-dimensional submanifold of $\mathfrak{M}$, then S satisfies the local k-slice condition.* Therefore, there exists a family $\{(U_p, \varphi_p)\}_{p\in S}$ of charts on $\mathfrak{M}$ that cover $S$ and such that for any $p \in S$, the chart domain of $(U_p, \varphi_p)$ contains $p$, and $S \cap U_p$ is a $k$-slice of $U_p$. The family[6] $A_S = \{(S \cap U_p, \pi \circ \varphi|_{S\cap U_p})\}_{p\in S}$ constitutes a smooth atlas of $S$. These reasonings are illustrated in the diagram

$$\begin{array}{ccc}
S \cap U_p & \xrightarrow{\iota_{U_p \cap S}} & U_p \\
{\scriptstyle (x^1, \dots, x^k, 0, \dots, 0)}\big\downarrow & & \big\downarrow{\scriptstyle (x^1, \dots, x^n)} \\
\mathbb{R}^k & \xleftarrow[\pi]{} & \mathbb{R}^n
\end{array}$$

Here, $\pi\colon \mathbb{R}^n \to \mathbb{R}^k$ is the projection onto the first $k$ coordinates, that is,

$$\pi\colon (x^1, \dots, x^k, x^{k+1}, \dots, x^n) \mapsto (x^1, \dots, x^k)\,.$$

6 In [3] it is proven that such atlas generates a smooth structure that is the unique one in which $S$ can be considered as a submanifold.

### 4.5.3 The Induced Riemannian Space Structure

Let $\mathfrak{B}$ be a $n$-dimensional smooth body and $\mathscr{P}$ be an $m$-dimensional physical space with a Riemannian metric $\mathscr{g}$. The Riemannian space structure of $\mathscr{P}$ can be induced on each shape $\mathcal{S}_\varkappa$ by means of the pullback $\iota^*_{\mathcal{S}_\varkappa}$. The diagram

$$
\begin{array}{ccc}
\mathscr{P} & \xleftarrow{\;\pi_1\;} & T^*\mathscr{P} \otimes T^*\mathscr{P} \\
{\scriptstyle \iota_{\mathcal{S}_\varkappa}}\uparrow & & \downarrow{\scriptstyle \iota^*_{\mathcal{S}_\varkappa}} \\
\mathcal{S}_\varkappa & \xleftarrow[\;\pi_2\;]{} & T^*\mathcal{S}_\varkappa \otimes T^*\mathcal{S}_\varkappa
\end{array}
$$

illustrates the action of $\iota^*_{\mathcal{S}_\varkappa}$; here $\pi_1$ and $\pi_2$ are the projections from the corresponding total spaces of vector bundles onto their bases. Thus, since $\mathscr{g} \in \operatorname{Sec}(T^*\mathscr{P} \otimes T^*\mathscr{P})$ is the Riemannian metric on $\mathscr{P}$, the section

$$\iota^*_{\mathcal{S}_\varkappa}\mathscr{g} \in \operatorname{Sec}\left(T^*\mathcal{S}_\varkappa \otimes T^*\mathcal{S}_\varkappa\right)$$

is the Riemannian metric on $\mathcal{S}_\varkappa$. It acts in the following way:

$$\forall u, v \in T\mathcal{S}_\varkappa :\ \iota^*_{\mathcal{S}_\varkappa}\mathscr{g}(u, v) := \mathscr{g}(T\iota_{\mathcal{S}_\varkappa} \llcorner u, T\iota_{\mathcal{S}_\varkappa} \llcorner v)\,. \tag{4.2}$$

Here, $T\iota_{\mathcal{S}_\varkappa} : T\mathcal{S}_\varkappa \to T\mathscr{P}$ is the tangent map of the inclusion map. If $(s^\alpha)_{\alpha=1}^n$ are local coordinates on $\mathcal{S}_\varkappa$ that are induced from the local $k$-slice condition, and $(x^i)_{i=1}^m$ are local coordinates on $\mathscr{P}$, then $T_x\iota_{\mathcal{S}_\varkappa}$ is represented by the dyadic decomposition

$$T_x\iota_{\mathcal{S}_\varkappa} = \left.\frac{\partial x^i}{\partial s^\alpha}\right|_{\varphi(x)} \partial_{x^i}|_x \otimes ds^\alpha|_x\,,$$

where $x^i : (s^1, \dots, s^n) \mapsto x^i(s^1, \dots, s^n)$, $i = 1, \dots, m$, is the coordinate representation of $\iota_{\mathcal{S}_\varkappa}$ in a neighborhood of $x$, and $\varphi$ is the coordinate homeomorphism that generates $(s^\alpha)_{\alpha=1}^n$. Thus, by (4.2) one has

$$\iota^*_{\mathcal{S}_\varkappa}\mathscr{g} = g_{ij}\frac{\partial x^i}{\partial s^\alpha}\frac{\partial x^j}{\partial s^\beta} ds^\alpha \otimes ds^\beta\,. \tag{4.3}$$

Here, $g_{ij} = \mathscr{g}(\partial_{x^i}, \partial_{x^j})$.

### 4.5.4 A Shape and the Physical Space. Intrinsic versus Spatial

Let the physical space be the classical three-dimensional Euclidean space $\mathscr{E}$. Imagine that there are two bodies, $\mathfrak{B}_1$ and $\mathfrak{B}_2$, where the first one is three-dimensional and the second one is two-dimensional. Suppose that these bodies are in contact, that is, in some configurations $\varkappa_1$ and $\varkappa_2$ one has that the surface $\varkappa_2(\mathfrak{B}_2)$ is a part of the boundary $\partial\varkappa_1(\mathfrak{B}_1)$ for an open set $\varkappa_1(\mathfrak{B}_1)$. The stress and strain states in $\mathfrak{B}_1$ are defined via three-dimensional tensor fields that have values in the tensor product $\mathscr{V} \otimes \mathscr{V}^*$ of the associated vector space $\mathscr{V}$. Since $\mathfrak{B}_2$ is two-dimensional, one requires clarifying the notion of stress and strain on such a body. This can be done in the framework of the theory developed by Gurtin and Murdoch in [110]. One obtains two-dimensional tensor fields that cannot be set into any correspondence with three-dimensional ones without additional geometrical considerations. In particular, this consideration defines the operation In[...] that allows one to transform the surface divergence $\mathrm{div}_s$ of the two-dimensional Cauchy stress tensor field $\boldsymbol{\sigma}_s$ into a three-dimensional vector field and thus to write the balance equality

$$\mathrm{div}\,\boldsymbol{T} + \boldsymbol{b} = \mathrm{In}[\mathrm{div}_s\,\boldsymbol{\sigma}_s]\,.$$

Here, $\boldsymbol{T}$ is the Cauchy stress tensor in $\mathfrak{B}_1$ and $\boldsymbol{b}$ is a body force. Let us consider this problem in more detail.

Let $\mathscr{P}$ be a physical space and $\mathcal{S}_\varkappa$ be some shape. Vector bundle structures over $\mathcal{S}_\varkappa$ are defined such that they are compatible with features of pure geometrical objects, that "populate" the manifold, figuratively speaking, its "inhabitants", such as smooth curves $\gamma\colon \mathbb{I} \to \mathcal{S}_\varkappa$ and smooth functions $f\colon \mathcal{S}_\varkappa \to \mathbb{R}$. Similarly, vector bundle structures over $\mathscr{P}$ can also be defined by smooth curves $\gamma\colon \mathbb{I} \to \mathscr{P}$ and smooth functions $f\colon \mathscr{P} \to \mathbb{R}$. These structures are not related directly. For example, a smooth section $u \in \mathrm{Vec}(\mathcal{S}_\varkappa)$ takes values in $T\mathcal{S}_\varkappa$, but not in $T\mathscr{P}$. This means that the observer in $\mathscr{P}$ cannot measure values of $u$ without using extra data. This data is provided by the tangent map $T\iota_{\mathcal{S}_\varkappa}$. Denote this map by $\mathrm{In}_{\mathcal{S}_\varkappa}$. That is,

$$\mathrm{In}_{\mathcal{S}_\varkappa} := T\iota_{\mathcal{S}_\varkappa}\colon T\mathcal{S}_\varkappa \to T\mathscr{P}\,.$$

Applied to the aforementioned section $u$, such a map gives a new one, $v = \mathrm{In}_{\mathcal{S}_\varkappa} \llcorner u : \mathcal{S}_\varkappa \to T\mathscr{P}$. The new field $v$ takes values in $T\mathscr{P}$ and, thus, these values can be measured by the observer. We see that $\mathrm{In}_{\mathcal{S}_\varkappa}$ plays the role of the additional geometric data, which aligns tangent vectors from $T\mathcal{S}_\varkappa$ with *spatial* tangent vectors from $T\mathscr{P}$. Note that such a property was implicitly used in (4.2).

Let us consider the correspondence between tangent vectors on shape and spatial ones in more detail. The restrictions of $\mathrm{In}_{\mathcal{S}_\varkappa}$ on tangent fibers, *i.e.*, the maps

$$\mathrm{In}_{\mathcal{S}_\varkappa;x} := T_x\iota_{\mathcal{S}_\varkappa}\colon T_x\mathcal{S}_\varkappa \to T_x\mathscr{P}\,, \tag{4.4}$$

$x \in \mathcal{S}_\varkappa$, are injective and linear, with $\mathrm{rank}(\mathrm{In}_{\mathcal{S}_\varkappa;x}) = \dim\mathfrak{B} \le \dim\mathscr{P}$. This means that for any $x \in \mathcal{S}_\varkappa$, the map $\mathrm{In}_{\mathcal{S}_\varkappa;x}$ is an isomorphism onto its image. Thus, one has the opportunity to identify vectors from $T_x\mathcal{S}_\varkappa$ with the corresponding vectors from the $n$-dimensional subspace of $T_x\mathscr{P}$. Neglecting such an opportunity results in more detailed, but more complicated, theory. As a rule, we use such identification only in the case when the body and the physical space have the same dimensions, *i.e.*,

$\dim \mathfrak{B} = \dim \mathscr{P}$. That is, if $\dim \mathfrak{B} = \dim \mathscr{P}$, then we identify linear spaces $T_x\mathcal{S}_\varkappa$ and $T_x\mathscr{P}$. Each tangent vector $u \in T_x\mathcal{S}_\varkappa$ is identified with the corresponding spatial tangent vector $\mathrm{In}_{\mathcal{S}_\varkappa;x}(u) \in T_x\mathscr{P}$.

Note that when $\dim \mathfrak{B} = \dim \mathscr{P}$, the shape $\mathcal{S}_\varkappa$ is an open subset of $\mathscr{P}$. This corresponds to the classical considerations of continuum mechanics in which shapes are defined as open sets with boundaries that are regular in the sense of Kellogg.

# 5 Strain Measures

## 5.1 Review on Cauchy Theory

The geometric ideas for describing the small change in the shape and volume of bodies under the action of external fields applied to them go back to the works of Bernoulli and Euler, but the geometric relations of the theory of finite deformations were obtained by Augustin-Louis Cauchy and are given in most modern monographs on continuum mechanics without significant changes [59]. Despite the fact that Cauchy's arguments are inextricably linked with the postulates of Euclidean space, the Cauchy method is easily carried over to spaces of a more general structure, in particular, Riemannian ones. In this regard, it seems appropriate to comment on the provisions it adopts (the authors used the works [73, 78, 114]).

i) *The existence of a natural shape.* The concept of the body is not used by Cauchy to describe deformations, it is sufficient to consider the body shapes, among which, as expected, there is a special, natural stress-free one. It is in this shape that there is a body in the absence of external fields, and the application leads to distortion of the shape and deformation.
ii) *Euclidean structure of the physical space.* Appealing to the classical analytic geometry of Descartes, Cauchy identifies points in a distorted shape by their Cartesian coordinates $x$, $y$, $z$. Note that Cauchy used a distorted shape to identify points, because it is observable.
iii) *The existence of displacements.* The Cauchy deformation is considered as a process of "returning" a shape to an undistorted state (in modern terminology this is an inverse deformation [115]), which is realized for the shape as a whole and is characterized by a continuous displacement field with components

$$\tilde{\xi} : (x, y, z) \to \tilde{\xi}(x, y, z) , \quad \tilde{\eta} : (x, y, z) \to \tilde{\eta}(x, y, z) , \quad \tilde{\zeta} : (x, y, z) \to \tilde{\zeta}(x, y, z)$$

in the Cartesian frame [78]. Thus, "incompatible deformations" are excluded.
iv) *The physical state of the body depends on the distortions of elementary material fibers.* Cauchy postulates this implicitly, considering as arguments of all field values the ratio of the lengths of infinitesimal material segments with their distortion in the composition of the body shape. To calculate the lengths, the Cartesian metric generated by the relations $\boldsymbol{i}_p \cdot \boldsymbol{i}_s = \delta_{ps}$, where $(\boldsymbol{i}_s)_{s=1}^3$ is Cartesian frame, is used.

**Remark 5.1.** An infinitely small material segment was determined by the coordinates of the particles (molecules, in the interpretation of the physics of the nineteenth century) located at its ends. In the actual form, these material particles correspond to the Cartesian coordinates $x$, $y$, $z$ and $x + \Delta x$, $y + \Delta y$, $z + \Delta z$. The coordinates of the particles

https://doi.org/10.1515/9783110563214-005

in the reference shape were represented by expressions [78]

$$x-\xi, y-\eta, z-\zeta\,, \quad \text{and} \quad x+\Delta x-\xi-\Delta\xi, y+\Delta y-\eta-\Delta\eta, z+\Delta z-\zeta-\Delta\zeta\,,$$

where $\xi=\tilde{\xi}(x,y,z)$, $\eta=\tilde{\eta}(x,y,z)$, $\zeta=\tilde{\zeta}(x,y,z)$.

To quantify the distortion of material fibers, Cauchy calculates their lengths $r$ and $r/(1+\varepsilon)$ in distorted and undistorted shapes:

$$r^2=(\Delta x)^2+(\Delta y)^2+(\Delta z)^2\,, \quad \left(\frac{r}{1+\varepsilon}\right)^2=(\Delta x-\Delta\xi)^2+(\Delta y-\Delta\eta)^2+(\Delta z-\Delta\zeta)^2\,,$$

and, taking into account their infinitesimality, decomposes the last relation by the Taylor formula of the first order [78]:

$$\begin{aligned}\left(\frac{r}{1+\varepsilon}\right)^2 &= \left(\Delta x-\frac{\partial\xi}{\partial x}\Delta x-\frac{\partial\xi}{\partial y}\Delta y-\frac{\partial\xi}{\partial z}\Delta z-\cdots\right)^2\\ &+\left(\Delta y-\frac{\partial\eta}{\partial x}\Delta x-\frac{\partial\eta}{\partial y}\Delta y-\frac{\partial\eta}{\partial z}\Delta z-\cdots\right)^2\\ &+\left(\Delta z-\frac{\partial\zeta}{\partial x}\Delta x-\frac{\partial\zeta}{\partial y}\Delta y-\frac{\partial\zeta}{\partial z}\Delta z-\cdots\right)^2.\end{aligned} \tag{5.1}$$

In the context of modern terminology, this means that the study material is *simple* [37, p. 60]. Expression (5.1) is characterized by a quadratic form corresponding to an ellipsoid, a triple of mutually orthogonal axes and three numbers corresponding to the main semi-axes. This Cauchy ellipsoid assumes the main characteristic of the deformed state at a point. Since these constructions implicitly exclude rotations accompanying deformations, one can see in them a principle that, within the framework of modern terminology, is called the principle of objectivity or the principle of material indifference[1].

For the sake of justice, it should be noted that Cauchy did not complete the construction of the theory of finite deformations, and the famous Cauchy polar decomposition, which explicitly distinguishes the rotations accompanying deformations was not strictly proved [59], despite the fact that he carried out this research for about 10 years. Here, however, it should be taken into account that Cauchy did not have a modern theory of operators in finite-dimensional spaces and carried out all constructions in terms of semiaxes of ellipsoids.

v) *Duality of distorted and undistorted shapes.* In the work [114], unlike in [78], to identify the Cauchy points, their coordinates are used in an undistorted shape, thereby emphasizing the duality of distorted and undistorted shapes. Historically, earlier references to this duality are available in the works of J. Lagrange [117].

1 Note that the need to use the principle of material indifference was pointed out by Poisson, who first drew attention to the fact that rigid stresses in the body should not change in tension [116]. From the point of view of the general theory of relativity, the principle of material indifference is a special case of the principle of general covariance [11].

In fact, there is no reason to prefer one of the shapes over another, and the equations characterizing the deformation can be written both relative to the reference and relative to the actual shapes. Mathematically, the transition from one description to another is carried out on the basis of the formulae for the transformation of areas and volumes in Euclidean space (Nanson's formula [37, 124]) and leads to the well-known relations of Piola [37, p. 124]. The duality of the relations obtained in this case is shown in [115].

The generalization of the positions i)–v) and the transfer of the generalization to structurally inhomogeneous bodies, *i.e.*, manifolds with non-Euclidean geometry, reduces to a modification of the procedure for comparing infinitesimal material fibers associated with different points of the body with a uniform (reference) fiber. Geometrically, this means the definition of the *non-Euclidean material connection* on the body [8]. For simple materials, it is an affine connection[2], and the basic relationships needed for such a generalization are similar in shape to the relations arising when using common curvilinear coordinates[3], although, of course, their geometric meaning acquires a new content.

## 5.2 Configurations and Deformations

Let $\mathfrak{B}$ be an $n$-dimensional smooth manifold ($\dim \mathfrak{B} = n$) that represents a body and let $\mathscr{P}$ be an $m$-dimensional smooth manifold ($\dim \mathscr{P} = m$) that represents a physical space. Suppose that $\mathscr{P}$ is endowed with the Riemannian metric $g$. Changing the shape of the body $\mathfrak{B}$, embedded in $\mathscr{P}$, is quantified by deformation map, which is the subject of present section.

In conventional continuum mechanics the pivot role in deformation quantification belongs to the analysis of shapes of material fibers, which change in the course of deformation. In the framework of the geometrical approach, material fibers can be defined as smooth curves $\chi\colon \mathbb{I} \to \mathfrak{B}$ on the body. Such curves are observable only if some configuration $\varkappa \in \mathfrak{C}(\mathfrak{B}; \mathscr{P})$ is chosen. That is, one observes not the set $\chi(\mathbb{I})$ of material points, but the set $\varkappa \circ \chi(\mathbb{I})$ of their places.

Let $\varkappa_R$ be a configuration that the observer considers as a *reference*. Thus, $\varkappa_R \circ \chi(\mathbb{I})$ is the reference image of the material fiber. If $\varkappa$ is a configuration, which image, $\varkappa(\mathfrak{B})$, the observer considers as being *now*, as *actual*, the set $\varkappa \circ \chi(\mathbb{I})$ is the actual image

2 By the method of the moving frame in space one can introduce the affine connection. To do this, one uses a smooth field of non-degenerate linear transformations. Taking into account that the response of simple materials is completely determined by the smooth field of linear transformations, W. Noll constructed an affine connection on the body with the help of the moving frame method, transforming it into a uniform shape [104].

3 Apparently, the brothers E. Cosserat and F. Cosserrat [118] began to use systematically curvilinear coordinates in the theory of elasticity.

of the same material fiber. Using the measurement tool provided by the Riemannian metric $g$, one can calculate the lengths of reference and actual shapes of the material fiber. Therefore, it is possible to compare metric properties of the reference and actual images of the fiber. To this end, define the map

$$\gamma\colon \varkappa_R(\mathfrak{B}) \to \varkappa(\mathfrak{B})\,, \quad \gamma := \hat{\varkappa} \circ \hat{\varkappa}_R^{-1}\,,$$

which is referred to as a *deformation of* $\mathfrak{B}$ *from the configuration* $\varkappa_R$ *into the configuration* $\varkappa$ [104]. Relations between the configurations and deformations are illustrated in the following diagram:

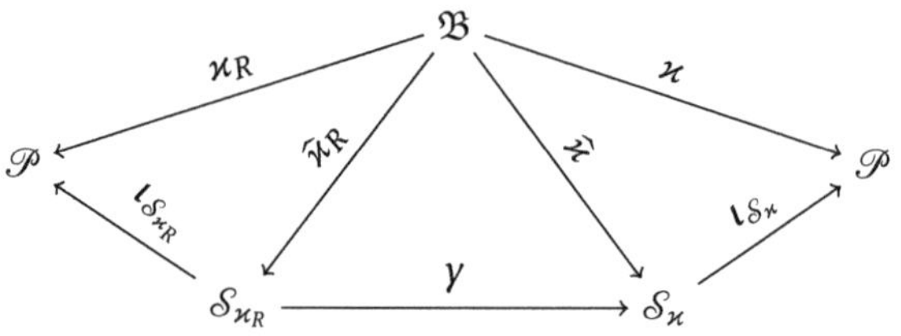


In the diagram, $\mathscr{S}_{\varkappa_R} = \varkappa_R(\mathfrak{B})$ and $\mathscr{S}_{\varkappa} = \varkappa(\mathfrak{B})$. The artistic form of this diagram is shown in Figure 5.1.

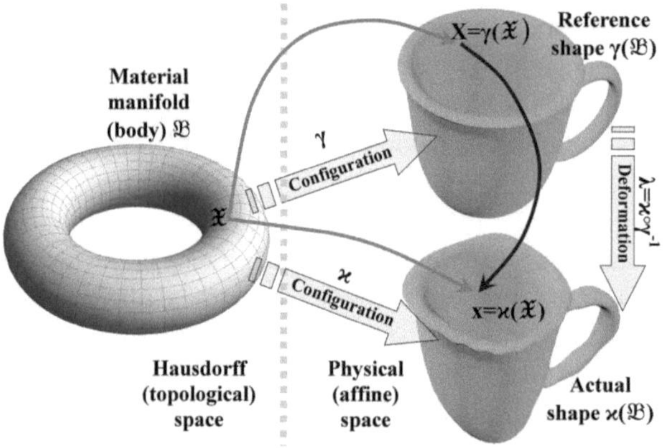


**Fig. 5.1:** Configurations and deformations.

## 5.3 Coordinate Representations of Configurations and Deformations

### 5.3.1 The General Case

Since the body $\mathfrak{B}$ and the physical space $\mathscr{P}$ are represented by smooth manifolds with $\dim \mathfrak{B} = n$ and $\dim \mathscr{P} = m$, one can take some smooth atlases $A_{\mathfrak{B}} = \{(U_\alpha, \varphi_\alpha)\}_{\alpha\in I}$, and

$A_{\mathscr{P}} = \{(V_\beta, \psi_\beta)\}_{\beta\in J}$ from their $C^\infty$-structures. Let $\varkappa, \varkappa_R \in \mathfrak{C}(\mathfrak{B};\mathscr{P})$ be configurations. Since $\mathcal{S}_{\varkappa_R}$ and $\mathcal{S}_\varkappa$ are embedded $n$-dimensional submanifolds of $\mathscr{P}$, one can choose smooth atlases $A_{\mathcal{S}_{\varkappa_R}} = \{(W^R_\delta, \sigma_{R;\delta})\}_{\delta\in K_R}$ and $A_{\mathcal{S}_\varkappa} = \{(W_\tau, \sigma_\tau)\}_{\tau\in K}$ from their smooth structures.

Fix some material point $\mathfrak{X} \in \mathfrak{B}$. Denote $X = \varkappa_R(\mathfrak{X})$ and $x = \varkappa(\mathfrak{X})$. The points $X$ and $x$ determine the spatial places of the material point $\mathfrak{X}$. Obtain coordinate representations of the configurations $\varkappa_R$ and $\varkappa$ in some chart domain $U_\alpha$, $\alpha \in I$, such that $\mathfrak{X} \in U_\alpha$. Choose indices $\beta_1, \beta_2 \in J$, such that $X \in V_{\beta_1}$ and $x \in V_{\beta_2}$. Then, denoting $U_{\alpha\beta_1\beta_2} := U_\alpha \cap \varkappa_R^{-1}(V_{\beta_1}) \cap \varkappa^{-1}(V_{\beta_2})$, one arrives at the following coordinate representations for the configurations $\varkappa_R$ and $\varkappa$:

$$\begin{aligned}\widetilde{\varkappa}_{R;\alpha,\beta_1,\beta_2} &= \psi_{\beta_1} \circ \varkappa_R \circ \varphi_\alpha^{-1} : \varphi_\alpha(U_{\alpha\beta_1\beta_2}) \to \psi_{\beta_1}(V_{\beta_1}) ,\\ \widetilde{\varkappa}_{\alpha,\beta_1,\beta_2} &= \psi_{\beta_2} \circ \varkappa \circ \varphi_\alpha^{-1} : \varphi_\alpha(U_{\alpha\beta_1\beta_2}) \to \psi_{\beta_2}(V_{\beta_2}) .\end{aligned}$$

These mappings act on open subsets of $\mathbb{R}^n$ and take values in $\mathbb{R}^m$. Obtain coordinate representations for their restrictions on shapes, $\widehat{\varkappa}_R$ and $\widehat{\varkappa}$. Choose such indices $\delta \in K_R$ and $\tau \in K$ that $X \in W^R_\delta$ and $x \in W_\tau$. Denoting $U_{\alpha\delta\tau} := U_\alpha \cap \widehat{\varkappa}_R^{-1}(W^R_\delta) \cap \widehat{\varkappa}^{-1}(W_\tau)$, one obtains the following coordinate representations:

$$\begin{aligned}\widetilde{\widehat{\varkappa}}_{R;\alpha,\delta,\tau} &= \sigma_{R;\delta} \circ \widehat{\varkappa}_R \circ \varphi_\alpha^{-1} : \varphi_\alpha(U_{\alpha\delta\tau}) \to \sigma_{R;\delta}\left(W^R_\delta\right) ,\\ \widetilde{\widehat{\varkappa}}_{\alpha,\delta,\tau} &= \sigma_\tau \circ \widehat{\varkappa} \circ \varphi_\alpha^{-1} : \varphi_\alpha(U_{\alpha\delta\tau}) \to \sigma_\tau(W_\tau) .\end{aligned}$$

Such mappings act on open subsets of $\mathbb{R}^n$ into $\mathbb{R}^n$. Finally, one arrives at the coordinate representations for deformation $y = \widehat{\varkappa} \circ \widehat{\varkappa}_R^{-1}$ and its inverse, $y^{-1}$, in neighborhoods of $X$ and $x$:

$$\begin{aligned}\widetilde{y}_{\tau,\delta} &= \sigma_\tau \circ y \circ \sigma_{R;\delta}^{-1} : \sigma_{R;\delta}(S_{R;\alpha\delta\tau}) \to \sigma_\tau(W_\tau) ,\\ \widetilde{y^{-1}}_{\delta,\tau} &= \sigma_{R;\delta} \circ y^{-1} \circ \sigma_\tau^{-1} : \sigma_\tau(S_{\alpha\delta\tau}) \to \sigma_{R;\delta}\left(W^R_\delta\right) .\end{aligned}$$

Here, $S_{R;\alpha\delta\tau} = \varkappa_R(U_{\alpha\delta\tau})$ and $S_{\alpha\delta\tau} = \varkappa(U_{\alpha\delta\tau})$. Such mappings act between open sets in $\mathbb{R}^n$. Relations between restricted configurations, deformations, and their coordinate representations are illustrated in the diagram

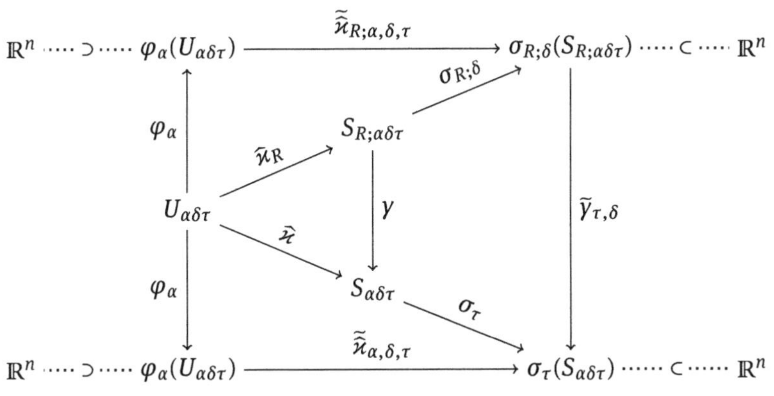

### 5.3.2 The Euclidean Case

In the particular case when $\mathscr{P}$ is represented by $m$-dimensional affine-Euclidean space $\mathscr{E}$ with associated vector space $\mathscr{V}$, one can choose the minimal atlas $A_{\mathscr{E}}^{\min}$. Such an atlas consists of one chart $(\mathscr{E}, \mathscr{D})$, in which $\mathscr{D}$ is the mapping (2.2), *i.e.*,

$$\forall a \in \mathscr{E}: \quad \mathscr{D}(a) := ((a - o) \cdot \boldsymbol{i}_k)_{k=1}^{m} ,$$

where $(o; (\boldsymbol{i}_k)_{k=1}^{m})$ is a fixed orthonormal coordinate system on $\mathscr{E}$. One may add other charts to the minimal atlas $A_{\mathscr{E}}^{\min}$, which are $C^{\infty}$-compatible with the original chart $(\mathscr{E}, \mathscr{D})$. This way is equivalent to introducing curvilinear coordinates on $\mathscr{E}$.

Consider coordinate representations of configurations and deformations in the Euclidean space $\mathscr{E}$. We use two copies for $\mathbb{R}^m$: the space $\mathbb{R}_d^m$, which contains Cartesian coordinates, and the space $\mathbb{R}_c^m$, which contains curvilinear ones. Choose some point $\mathfrak{X} \in \mathfrak{B}$ and pick up a chart $(U_\alpha, \varphi_\alpha)$ from the atlas $A_{\mathfrak{B}}$, such that $U_\alpha \ni \mathfrak{X}$. Let $\varkappa \in \mathfrak{C}(\mathfrak{B}; \mathscr{E})$ be a configuration. The affine structure of the physical space allows us to define the following mappings:

$$\begin{aligned}
\overline{\varkappa}_\alpha &= \varkappa \circ \varphi_\alpha^{-1} \colon O_\alpha \to \mathscr{E} ; \\
\overline{\overline{\varkappa}}_\alpha &= \boldsymbol{p} \circ \overline{\varkappa} \colon O_\alpha \to \mathscr{V} ; \\
\varkappa_{d;\alpha} &= \mathscr{D} \circ \overline{\varkappa} \colon O_\alpha \to \mathbb{R}_d^m .
\end{aligned}$$

Here, $O_\alpha \subset \mathbb{R}^n$ is an open set, the codomain of $\varphi_\alpha$. The mapping $\overline{\varkappa}_\alpha$ provides spatial representation of material points; their places in the physical space $\mathscr{E}$ are assigned to their local coordinates in the chart $(U_\alpha, \varphi_\alpha)$. The mapping $\overline{\overline{\varkappa}}_\alpha$ provides the "vectorization" of the configuration; to each material point, which is represented by local coordinates in $(U_\alpha, \varphi_\alpha)$, its position relative to the fixed origin $o$ is assigned by means of the position vector field. Finally, the mapping $\varkappa_{d;\alpha}$ provides the "arithmetization" of the configuration; a $m$-tuple of the Cartesian coordinates is assigned to a place on each material point.

Let us introduce curvilinear coordinates in $\mathscr{E}$ by means of some smooth atlas $A_c = \{(V_\beta, \psi_\beta)\}_{\beta \in J}$, which is equivalent to $A_{\mathscr{E}}^{\min}$. Let $(V_\beta, \psi_\beta)$ be any chart from such an atlas, for which $\mathfrak{X} \in V_\beta$. The mapping

$$\varkappa_{c;\alpha,\beta} = \psi_\beta \circ \varkappa \circ \varphi_\alpha^{-1} \colon \mathbb{R}^n \supset \varphi_\alpha \left( U_\alpha \cap \varkappa^{-1}(V_\beta) \right) \to \psi_\beta(V_\beta \cap \varkappa(U_\alpha)) \subset \mathbb{R}_c^m ,$$

assigns to each point from a portion of $U_\alpha$ the curvilinear coordinates of its place in $\mathscr{E}$. The equality

$$h_{c;\beta} = \psi_\beta \circ \mathscr{D}^{-1} \colon \mathbb{R}_d^m \to \mathbb{R}_c^m$$

defines a transformation of Cartesian coordinates into curvilinear ones. By means of such a mapping, one can obtain the following expression for $\varkappa_{c;\alpha,\beta}$:

$$\varkappa_{c;\alpha,\beta} = h_{c;\beta} \circ \mathscr{D} \circ \varkappa \circ \varphi_\alpha^{-1} ,$$

and the formula that links $\varkappa_{c;\alpha,\beta}$ and $\varkappa_{d;\alpha}$:

$$\varkappa_{d;\alpha} = h_{c;\beta}^{-1} \circ \varkappa_{c;\alpha,\beta} .$$

To such a relation there corresponds the diagram ($W_{\alpha\beta} = U_\alpha \cap \varkappa^{-1}(V_\beta)$):

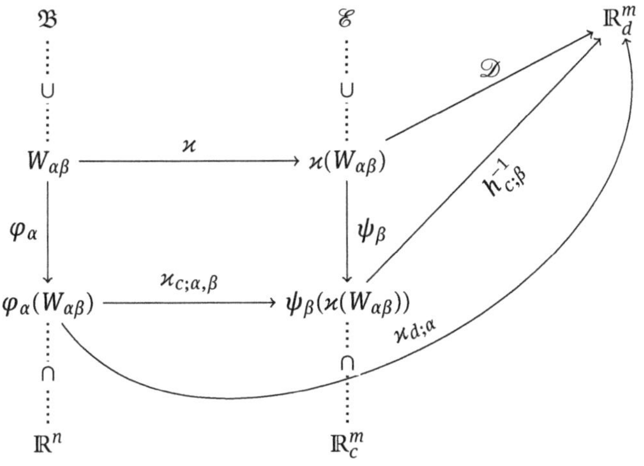


Let $\varkappa_R, \varkappa \in \mathfrak{C}(\mathfrak{B};\mathscr{E})$ be two configurations. Then, $y = \hat{\varkappa} \circ \hat{\varkappa}_R^{-1}$ is the corresponding deformation. Since $y$ acts between submanifolds, one can apply the local $k$-slice condition, which allows one to obtain the corresponding atlases $A_{\mathcal{S}_{\varkappa_R}} = \{(W_\delta^R, \ell_{R;\delta})\}_{\delta \in K_R}$ and $A_{\mathcal{S}_\varkappa} = \{(W_\tau, \ell_\tau)\}_{\tau \in K}$ for $\mathcal{S}_{\varkappa_R}$ and $\mathcal{S}_\varkappa$. Thus, the deformation has the following representation:

$$\tilde{y}_{\tau,\delta} = \ell_\tau \circ y \circ \ell_{R;\delta}^{-1} \colon \mathbb{R}^n \supset Q_{\alpha\delta\tau} \to R_{\alpha\delta\tau} \subset \mathbb{R}^n ,$$

where $Q_{\alpha\delta\tau}$ and $R_{\alpha\delta\tau}$ are open sets, which are defined similarly as in the preceding section.

The attempt to represent the deformation $y$ in terms of $\mathscr{D}$ directly can fail in the case $n < m$, since one obtains a mapping that is not defined on some open subset of[4] $\mathbb{R}_d^m$. Such a mapping $y_d$ cannot be differentiated by itself; for each point $y$ of its domain $D_d$, one needs to find an open neighborhood $O_{d;y}$ and such a smooth mapping $\Gamma_{d;y}$ defined on this neighborhood that its restriction to $D_d$ coincides with the coordinate representation of $y$, *i.e.*, $\Gamma_{d;y}|_{D_d} = y_d$.

Consider the case $n = m$. Denote by $\mathscr{D}_{\varkappa_R} \colon \mathcal{S}_{\varkappa_R} \to \mathbb{R}_d^m$ and $\mathscr{D}_\varkappa \colon \mathcal{S}_\varkappa \to \mathbb{R}_d^m$ the restrictions of $\mathscr{D}$ to the shapes $\mathcal{S}_{\varkappa_R}$ and $\mathcal{S}_\varkappa$. Thus, $\{(\mathcal{S}_{\varkappa_R}, \mathscr{D}_{\varkappa_R})\}$ and $\{(\mathcal{S}_\varkappa, \mathscr{D}_\varkappa)\}$ are the corresponding atlases. The coordinate representation of $y$ is the following:

$$y_d = \mathscr{D}_\varkappa \circ y \circ \mathscr{D}_{\varkappa_R}^{-1} \colon \mathbb{R}_d^m \supset \mathscr{D} \circ \varkappa_R(U_\alpha) \to \mathbb{R}_d^m .$$

4 The set $\varkappa_R(U_\alpha)$ is open in $\varkappa_R(\mathfrak{B})$ but not open in $\mathscr{E}$. Therefore, $\mathscr{D} \circ \varkappa_R(U_\alpha)$ is not open in $\mathbb{R}_d^m$.

Here, $\mathscr{D} \circ \varkappa_R(U_\alpha)$ is an open subset of $\mathbb{R}^m_d$. The relation between $y$ and $y_d$ is illustrated by the diagram

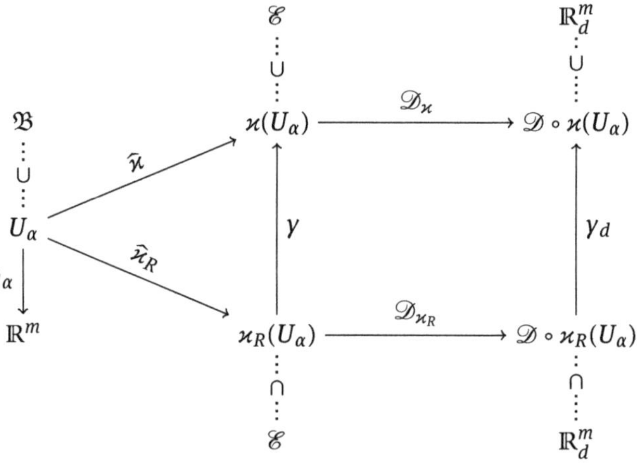


Now consider the representation of the deformation $y$ in the curvilinear coordinates. The atlas $A_c = \{(V_\beta, \psi_\beta)\}_{\beta \in J}$ gives two atlases for the shapes $\mathcal{S}_{\varkappa_R}$ and $\mathcal{S}_\varkappa$, which are denoted by $A_{c;\varkappa_R} = \{(\mathcal{S}_{\varkappa_R} \cap V_\delta, \widetilde{\psi}_{R;\delta})\}_{\delta \in J_{\varkappa_R}}$ and $A_{c;\varkappa} = \{(\mathcal{S}_\varkappa \cap V_\beta, \widetilde{\psi}_\beta)\}_{\beta \in J_{\varkappa_R}}$. In these atlases, coordinate mappings $\widetilde{\psi}_{R;\delta}$ and $\widetilde{\psi}_\beta$ are restrictions of the original coordinate mappings from the atlas $A_c$. Choose $\delta \in J_{\varkappa_R}$, $\beta \in J_\varkappa$, such that $\varkappa_R(\mathfrak{X}) \in V_\delta$ and $\varkappa(\mathfrak{X}) \in V_\beta$. Denoting $U_{\alpha\beta\delta} = U_\alpha \cap \varkappa^{-1}(V_\beta) \cap \varkappa_R^{-1}(V_\delta)$ one arrives at the coordinate representation

$$y_c = \widetilde{\psi}_\beta \circ y \circ \widetilde{\psi}_{R;\delta}^{-1} \colon \mathbb{R}^m_c \supset \psi_\delta \circ \varkappa_R(U_{\alpha\beta\delta}) \to \psi_\beta(\mathcal{S}_\varkappa \cap V_\beta) \subset \mathbb{R}^m_c \,.$$

The relation between $y$ and $y_c$ is shown in the diagram

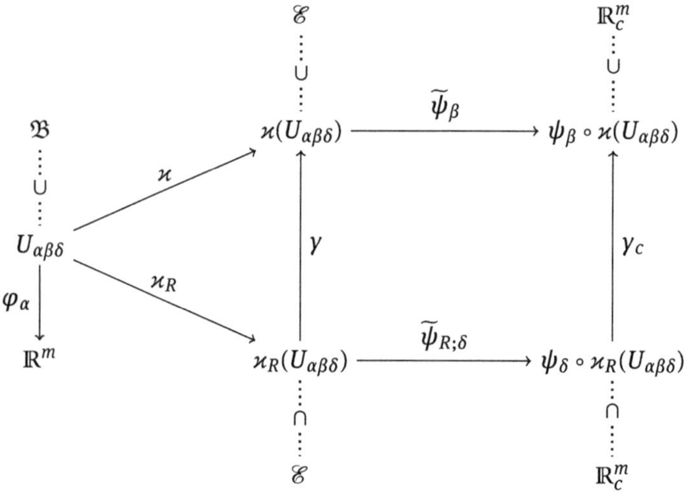

## 5.4 Two-Point Tensors

### 5.4.1 Two-Point Tensor Bundle

The deformation gradient $\boldsymbol{F}$ and the first Piola–Kirchhoff stress tensor $\boldsymbol{P}$ that arose in conventional elasticity (Chapter 3) are examples of two-point tensors. To generalize these notions on smooth manifolds one requires some generalization for the notion of the tensor bundle. Such a generalization comes down to the notion of a two-point tensor bundle, which is introduced in the present section.

Let $(\mathfrak{E}, \mathfrak{N}, \pi)$ be a smooth real vector bundle, $\mathfrak{M}$ be a smooth manifold, and $f \in C^\infty(\mathfrak{M}; \mathfrak{N})$ be some smooth mapping. The mapping $f$ induces the new vector bundle $(f^*\mathfrak{E}, \mathfrak{M}, \pi^*)$ upon the original one, $(\mathfrak{E}, \mathfrak{N}, \pi)$. The vector bundle $(f^*\mathfrak{E}, \mathfrak{M}, \pi^*)$ is referred to as the *pullback bundle* [119]. The total space of such a bundle is represented by the set

$$f^*\mathfrak{E} := \{(p, x) \in \mathfrak{M} \times \mathfrak{E} \mid f(p) = \pi(x)\} = \bigcup_{p \in \mathfrak{M}} \{p\} \times \mathfrak{E}_{f(p)} ,$$

and the projection $\pi^* : f^*\mathfrak{E} \to \mathfrak{M}$ is defined via $\pi^*(p, x) := p$. Thus, fiber over $p$ has the form

$$(f^*\mathfrak{E})_p = \{p\} \times \mathfrak{E}_{f(p)} .$$

The notion of the pullback bundle allows one to define a *two-point tensor bundle.* If $\mathfrak{M}$ and $\mathfrak{N}$ are smooth manifolds, $f \in C^\infty(\mathfrak{M}; \mathfrak{N})$, then[5]

$$T^{(r,s)} f^* T\mathfrak{N} \otimes T^{(k,l)} T\mathfrak{M} ,$$

is the total space of such a bundle. Its base is $\mathfrak{M}$ and fiber over $p$ is the following:

$$\{p\} \times \left( T^{(r,s)}(T_{f(p)}\mathfrak{N}) \otimes T^{(k,l)}(T_p\mathfrak{M}) \right) .$$

Each element of $T^{(r,s)} f^* T\mathfrak{N} \otimes T^{(k,l)} T\mathfrak{M}$ is referred to as a *two-point tensor over the map* $f$ [60, 120, 121].

Choose some open set $V \subset \mathfrak{N}$ and let $U = f^{-1}(V)$. Assume that $(E_\alpha)$, $(E^\alpha)$ are the local frame and coframe for $T\mathfrak{M}$, defined on $U$, while $(e_i)$, $(e^i)$ are the local frame and coframe for $T\mathfrak{N}$, defined on $V$. By

$$e_{i_1} \otimes \cdots \otimes e_{i_r} \otimes e^{j_1} \otimes \cdots \otimes e^{j_s} \otimes E_{\alpha_1} \otimes \cdots \otimes E_{\alpha_k} \otimes E^{\beta_1} \otimes \cdots \otimes E^{\beta_l} ,$$

we denote local section, which to each point $p \in U$ assigns the mixed tensor

$$p \mapsto e_{i_1}|_q \otimes \cdots \otimes e_{i_r}|_q \otimes e^{j_1}|_q \otimes \cdots \otimes e^{j_s}|_q \otimes E_{\alpha_1}|_p \otimes \cdots \otimes E_{\alpha_k}|_p \otimes E^{\beta_1}|_p \otimes \cdots \otimes E^{\beta_l}|_p ,$$

5 One can define bundles of another form, like $T^{(k,l)} T\mathfrak{M} \otimes T^{(r,s)} f^* T\mathfrak{N}$, etc.

where $q = f(p)$. The family of all such sections constitutes the local frame for $T^{(r,s)} f^* T\mathfrak{N} \otimes T^{(k,l)} T\mathfrak{M}$.

In this book, we use second-order two-point tensors. All possible types of such tensors are listed below:

$$
\begin{aligned}
\mathcal{T} &\in \operatorname{Sec}(f^* T\mathfrak{N} \otimes T^*\mathfrak{M}) , & \mathcal{T} &: p \mapsto T^i{}_\alpha(p) e_i|_{f(p)} \otimes E^\alpha|_p , \\
\mathcal{T} &\in \operatorname{Sec}(f^* T\mathfrak{N} \otimes T\mathfrak{M}) , & \mathcal{T} &: p \mapsto T^{i\alpha}(p) e_i|_{f(p)} \otimes E_\alpha|_p , \\
\mathcal{T} &\in \operatorname{Sec}(f^* T^*\mathfrak{N} \otimes T^*\mathfrak{M}) , & \mathcal{T} &: p \mapsto T_{i\alpha}(p) e^i|_{f(p)} \otimes E^\alpha|_p , \\
\mathcal{T} &\in \operatorname{Sec}(f^* T^*\mathfrak{N} \otimes T\mathfrak{M}) , & \mathcal{T} &: p \mapsto T_i{}^\alpha(p) e^i|_{f(p)} \otimes E_\alpha|_p .
\end{aligned}
\tag{5.2}
$$

Consider the transformation laws of such tensors. Assume that primed frames $(E'_\alpha)$, $(e'_i)$ and not primed frames $(E_\alpha)$, $(e_i)$ are related via equalities,

$$
E'_\alpha = \Omega^\beta{}_\alpha E_\beta , \quad e'_i = \omega^j{}_i e_j ,
$$

where $\Omega : p \mapsto \Omega(p) \in \mathrm{GL}(\dim \mathfrak{M}; \mathbb{R})$ and $\omega : p \mapsto \omega(p) \in \mathrm{GL}(\dim \mathfrak{N}; \mathbb{R})$ are smooth local fields of invertible matrices defined on some open subsets $\mathfrak{M}$ and $\mathfrak{N}$, respectively. Then, for coframes one has

$$
E'^\alpha = (\Omega^{-1})^\alpha{}_\beta E^\beta , \quad e'^i = (\omega^{-1})^i{}_j e^j .
$$

Thus, the following transformation laws for the tensors (5.2) take place:

$$
\begin{aligned}
T'^i{}_\alpha &= (\omega^{-1})^i{}_j \Omega^\beta{}_\alpha T^j{}_\beta , & T'^{i\alpha} &= (\omega^{-1})^i{}_j (\Omega^{-1})^\alpha{}_\beta T^{j\beta} , \\
T'_{i\alpha} &= \omega^j{}_i \Omega^\beta{}_\alpha T_{j\beta} , & T'_i{}^\alpha &= \omega^j{}_i (\Omega^{-1})^\alpha{}_\beta T_j{}^\beta .
\end{aligned}
$$

There is a close relationship between second-order two-point tensors and vector bundle homomorphisms [3]. The following commutative diagrams illustrate actions of such homomorphisms:

$$
\begin{array}{ccc}
T\mathfrak{M} & \xrightarrow{\mathcal{T}} & T\mathfrak{N} \\
{\scriptstyle \pi_{T\mathfrak{M}}}\downarrow & & \downarrow{\scriptstyle \pi_{T\mathfrak{N}}} \\
\mathfrak{M} & \xrightarrow[f]{} & \mathfrak{N}
\end{array}
\qquad\qquad
\begin{array}{ccc}
T^*\mathfrak{M} & \xrightarrow{\mathcal{T}} & T\mathfrak{N} \\
{\scriptstyle \pi_{T^*\mathfrak{M}}}\downarrow & & \downarrow{\scriptstyle \pi_{T\mathfrak{N}}} \\
\mathfrak{M} & \xrightarrow[f]{} & \mathfrak{N}
\end{array}
$$

$$
\begin{array}{ccc}
T\mathfrak{M} & \xrightarrow{\mathcal{T}} & T^*\mathfrak{N} \\
{\scriptstyle \pi_{T\mathfrak{M}}}\downarrow & & \downarrow{\scriptstyle \pi_{T^*\mathfrak{N}}} \\
\mathfrak{M} & \xrightarrow[f]{} & \mathfrak{N}
\end{array}
\qquad\qquad
\begin{array}{ccc}
T^*\mathfrak{M} & \xrightarrow{\mathcal{T}} & T^*\mathfrak{N} \\
{\scriptstyle \pi_{T^*\mathfrak{M}}}\downarrow & & \downarrow{\scriptstyle \pi_{T^*\mathfrak{N}}} \\
\mathfrak{M} & \xrightarrow[f]{} & \mathfrak{N}
\end{array}
$$

By the definition, for each mapping $\mathcal{T}$ from such diagrams the restriction to each fiber is linear. There is the following one-to-one correspondence between such mappings and second-order two-point tensors:

$$\begin{aligned}
\mathscr{T} : T\mathfrak{M} \to T\mathfrak{N} \quad &\leftrightarrow \quad \mathscr{T} \in \mathrm{Sec}(f^* T\mathfrak{N} \otimes T^*\mathfrak{M}) ,\\
\mathscr{T} : T^*\mathfrak{M} \to T\mathfrak{N} \quad &\leftrightarrow \quad \mathscr{T} \in \mathrm{Sec}(f^* T\mathfrak{N} \otimes T\mathfrak{M}) ,\\
\mathscr{T} : T\mathfrak{M} \to T^*\mathfrak{N} \quad &\leftrightarrow \quad \mathscr{T} \in \mathrm{Sec}(f^* T^*\mathfrak{N} \otimes T^*\mathfrak{M}) ,\\
\mathscr{T} : T^*\mathfrak{M} \to T^*\mathfrak{N} \quad &\leftrightarrow \quad \mathscr{T} \in \mathrm{Sec}(f^* T^*\mathfrak{N} \otimes T\mathfrak{M}) .
\end{aligned}$$

### 5.4.2 The Transpose and Orthogonal Tensors

The transpose and orthogonal tensors are heavily used in continuum mechanics. One of the many examples is the Cauchy polar decomposition results in Cauchy–Green strain measures.

In the conventional finite-strain theory, the transpose $\boldsymbol{F}^{\mathrm{T}}$ to the deformation gradient $\boldsymbol{F} : \mathscr{S}_R \to \mathrm{Lin}(\mathscr{V};\mathscr{V})$ is defined as a unique field $\boldsymbol{F}^{\mathrm{T}} : \mathscr{S}_R \to \mathrm{Lin}(\mathscr{V};\mathscr{V})$ that satisfies the property (3.11),

$$\forall \boldsymbol{u}, \boldsymbol{v} \in \mathscr{V} : \boldsymbol{v} \cdot \boldsymbol{F}_X[\boldsymbol{u}] = \boldsymbol{u} \cdot \boldsymbol{F}_X^{\mathrm{T}}[\boldsymbol{v}] ,$$

for all $X \in \mathscr{S}_R$. The orthogonal tensor in Euclidean space is defined by (12.5), *i.e.*, it is a linear mapping $\boldsymbol{Q} \in \mathrm{Lin}(\mathscr{V};\mathscr{V})$ that preserves the inner product:

$$\forall \boldsymbol{u}, \boldsymbol{v} \in \mathscr{V} : \boldsymbol{Q}[\boldsymbol{u}] \cdot \boldsymbol{Q}[\boldsymbol{v}] = \boldsymbol{u} \cdot \boldsymbol{v} .$$

Each of these definitions essentially uses Euclidean properties of the associated vector space $\mathscr{V}$. In the case of smooth manifolds, instead of Cartesian inner product one can use either the contraction $\langle \cdot, \cdot \rangle$ or the Riemannian metric $g$ (if it is fixed). Since no affine structure is presumed, one needs to consider four possible two-point tensors (5.2). This leads to several definitions of the transpose and orthogonal tensors [60, 121].

In the book, we use the definitions of the transpose tensor for the first one in (5.2). These definitions agree with those in [121] but have some technical changes. In the article referred to, the mapping $f : \mathfrak{M} \to \mathfrak{N}$ (or, in terms of the article, $\varphi : \mathscr{B} \to \mathscr{S}$) is assumed to be a bijective. Such an assumption is strong enough, since we apply the theory developed to the case when $f$ is a configuration, *i.e.*, $f$ is bijective onto its image only.

Let $f \in C^\infty(\mathfrak{M};\mathfrak{N})$. Consider the pullback bundle $\iota^*_{f(\mathfrak{M})} T\mathfrak{N}$. Its base is the set $f(\mathfrak{M})$ and the fiber over each point $p \in f(\mathfrak{M})$ is represented by the set $\{p\} \times T_p\mathfrak{N}$. Such a vector bundle represents the restriction of $T\mathfrak{N}$ to $f(\mathfrak{M})$.

Assume that smooth manifolds $\mathfrak{M}$ and $\mathfrak{N}$ are equipped with Riemannian metrics $\mathscr{G}$ and $g$. For $\mathscr{T} \in \mathrm{Sec}(f^* T\mathfrak{N} \otimes T^*\mathfrak{M})$ we define its *transpose* as the two-point tensor

$$\mathscr{T}^{\mathrm{T}} \in \mathrm{Sec}\left( (\widehat{f}^{-1})^* T\mathfrak{M} \otimes \iota^*_{f(\mathfrak{M})} T^*\mathfrak{N} \right) ,$$

such that for each $x \in f(\mathfrak{M})$ the mapping $\mathcal{T}_x^{\mathrm{T}} \in \mathrm{Lin}(T_x\mathfrak{N}; T_y\mathfrak{M})$, where $y = \widehat{f}^{-1}(x)$, satisfies the condition

$$\forall u \in T_y\mathfrak{M}\ \forall v \in T_x\mathfrak{N}: \quad g_x(\mathcal{T}_y u, v) = \mathcal{G}_y\left(u, \mathcal{T}_x^{\mathrm{T}} v\right) . \tag{5.3}$$

The two-point tensor $\mathcal{T}^{\mathrm{T}}$ can be considered as the vector bundle homomorphism

$$\mathcal{T}^{\mathrm{T}} : \iota^*_{f(\mathfrak{M})} T\mathfrak{N} \to T\mathfrak{M} .$$

Note that such a mapping is defined not on the total space of the tangent bundle $(Tf(\mathfrak{M}), f(\mathfrak{M}), \pi)$, but on the total space of the vector bundle $(\iota^*_{f(\mathfrak{M})} T\mathfrak{N}, f(\mathfrak{M}), \pi^*)$. The relation between the mappings $\mathcal{T}^{\mathrm{T}}$ and $\widehat{f}$ is illustrated on the diagram

$$\begin{array}{ccc}
\iota^*_{f(\mathfrak{M})} T\mathfrak{N} & \xrightarrow{\ \mathcal{T}^{\mathrm{T}}\ } & T\mathfrak{M} \\
{\scriptstyle \pi^*}\downarrow & & \downarrow{\scriptstyle \pi_{T\mathfrak{M}}} \\
f(\mathfrak{M}) & \xrightarrow[\ \widehat{f}^{-1}\ ]{} & \mathfrak{M}
\end{array}$$

The formula for the transpose tensor components follows directly from (5.3). If $\mathcal{T} = T^i{}_\alpha e_i \otimes E^\alpha$, then $\mathcal{T}^{\mathrm{T}}$ has the following dyadic decomposition:

$$\mathcal{T}^{\mathrm{T}} = (\mathcal{T}^{\mathrm{T}})^\alpha{}_i E_\alpha \otimes e^i ,$$

where $(\mathcal{T}^{\mathrm{T}})^\alpha{}_i = \mathcal{T}^{\mathrm{T}}(E^\alpha, e_i)$. According to (5.3), in which $u = E_\alpha$, $v = e_j$, one obtains,

$$(G)_{\alpha\beta}(\mathcal{T}^{\mathrm{T}})^\beta{}_j = g_{ij} T^i{}_\alpha .$$

Finally, one arrives at the formula

$$(\mathcal{T}^{\mathrm{T}})^\alpha{}_i = g_{ij} G^{\alpha\beta} T^j{}_\beta , \tag{5.4}$$

where $G_{\alpha\beta} = \mathcal{G}(E_\alpha, E_\beta)$ and $g_{ij} = g(e_i, e_j)$. Note that this formula corresponds to the classical one, (3.15), obtained in Chapter 3.

Another definition of the transpose does not require the metrics. We use the pullback bundle $\iota^*_{f(\mathfrak{M})} T^*\mathfrak{N}$. Its base is $f(\mathfrak{M})$ and the fiber over each point $p \in f(\mathfrak{M})$ is represented by the set $\{p\} \times T^*_p\mathfrak{N}$. For $\mathcal{T} \in \mathrm{Sec}(f^* T\mathfrak{N} \otimes T^*\mathfrak{M})$, we define its *conjugate* as the two-point tensor

$$\mathcal{T}^* \in \mathrm{Sec}\left((\widehat{f}^{-1})^* T^*\mathfrak{M} \otimes \iota^*_{f(\mathfrak{M})} T\mathfrak{N}\right) ,$$

such that for each $x \in f(\mathfrak{M})$ the mapping $\mathcal{T}_x^* \in \mathrm{Lin}(T^*_x\mathfrak{N}; T^*_y\mathfrak{M})$, where $y = \widehat{f}^{-1}(x)$, satisfies the condition

$$\forall u \in T_y\mathfrak{M}\ \forall v \in T^*_x\mathfrak{N}: \quad \langle \mathcal{T}_y u, v\rangle_x = \left\langle u, \mathcal{T}_x^* v\right\rangle_y . \tag{5.5}$$

The physical interpretation of the definition (5.5) is as follows. Such a definition determines the energy conjugacy: the power generated by the spatial density of forces $v$ on the physical field of the velocity $\mathscr{T} u$ is equal to the power developed by the material density of forces $\mathscr{T}^* v$ on the material velocity $u$.

The two-point tensor $\mathscr{T}^*$ can be considered as the following vector bundle homomorphism:

$$\mathscr{T}^* : \iota^*_{f(\mathfrak{M})} T^*\mathfrak{N} \to T^*\mathfrak{M} .$$

The relation between the mappings $\mathscr{T}^*$ and $\widehat{f}$ is shown on the diagram

$$\begin{array}{ccc} \iota^*_{f(\mathfrak{M})} T^*\mathfrak{N} & \xrightarrow{\;\mathscr{T}^*\;} & T^*\mathfrak{M} \\ \pi^* \big\downarrow & & \big\downarrow \pi_{T^*\mathfrak{M}} \\ f(\mathfrak{M}) & \xrightarrow[\;\widehat{f}^{-1}\;]{} & \mathfrak{M} \end{array}$$

If $\mathscr{T} = T^i{}_\alpha e_i \otimes E^\alpha$, then the formula (5.5) implies that $\mathscr{T}^*$ can be written as the following dyadic decomposition:

$$\mathscr{T}^* = T^i{}_\alpha E^\alpha \otimes e_i .$$

This means that the dyadic decomposition of $\mathscr{T}^*$ can be obtained from the original of $\mathscr{T}$, just by interchanging its basis vectors in the dyad.

The *orthogonal tensor* $\mathcal{Q}$ is a two-point tensor $\mathcal{Q} \in \mathrm{Sec}(f^* T\mathfrak{N} \otimes T^*\mathfrak{M})$, such that

$$\forall x \in \mathfrak{M}\ \forall u, v \in T_x\mathfrak{M}: \quad \mathscr{G}_x(u, v) = g_{f(x)}(\mathcal{Q}_x u, \mathcal{Q}_x v) . \tag{5.6}$$

Using (5.3) one obtains $\mathscr{G}_x(u, v) = \mathscr{G}_x(u, \mathcal{Q}^{\mathrm{T}}_{f(x)} \mathcal{Q}_x v)$, for all $u, v \in T_x\mathfrak{M}$. This leads to the relation $\mathscr{G}_x(u, (\mathcal{Q}^{\mathrm{T}}_{f(x)} \mathcal{Q}_x - \mathrm{Id}_{T_x\mathfrak{M}})v) = 0$, which holds for each $u, v \in T_x\mathfrak{M}$. Thus, one has $\mathcal{Q}^{\mathrm{T}}_{f(x)} \mathcal{Q}_x = \mathrm{Id}_{T_x\mathfrak{M}}$, and we arrive at the conventional relation for the orthogonal tensor.

## 5.5 Configuration Gradient

Any section $u \in \mathrm{Vec}(\mathfrak{B})$ of the tangent bundle $T\mathfrak{B}$ generates some material fiber $\chi : \mathbb{I} \to \mathfrak{B}$, such that $u_{\chi(t)} = \chi'(t)$, for all $t \in \mathbb{I}$. Figuratively speaking, the material fiber is "assembled" from the values of such a field. Conversely, for each material fiber there exists a smooth vector field with such a property. Thus, tangent vectors or, equivalently, derivations of the space of germs, can be viewed as *infinitesimal elements of material fibers*. In a fixed material point $\mathfrak{X} \in \mathfrak{B}$ all such objects form the *infinitesimal neighborhood* of $\mathfrak{X}$, which is formalized by the tangent space $T_{\mathfrak{X}}\mathfrak{B}$. Each element from $T_{\mathfrak{X}}\mathfrak{B}$ is the velocity at $\mathfrak{X}$ for some material fiber that starts at $\mathfrak{X}$.

A configuration $\varkappa \in \mathfrak{C}(\mathfrak{B};\mathscr{P})$ transforms material fibers into observable spatial fibers. The same role for infinitesimal objects is played by the *configuration gradient* $\mathscr{F} = T\varkappa$, which is the tangent bundle homomorphism, or, equivalently, the second-order two-point tensor. That is,

$$\mathscr{F} : T\mathfrak{B} \to T\mathscr{P}\,, \quad \text{or}\,, \quad \mathscr{F} \in \operatorname{Sec}(\varkappa^* T\mathscr{P} \otimes T^*\mathfrak{B})\,.$$

As follows from the first row of (5.2), $\mathscr{F}$ has the dyadic decomposition $\mathscr{F} = F^i{}_\alpha e_i \otimes E^\alpha$. Here, $(E^\alpha)$ is a local coframe on $\mathfrak{B}$, and $(e_i)$ is a local frame on $\mathscr{P}$. In particular, choosing coordinate frames, one obtains

$$\mathscr{F} = \frac{\partial \varkappa^i(\mathfrak{X}^1, \ldots, \mathfrak{X}^n)}{\partial \mathfrak{X}_\alpha} \partial_i \otimes d\mathfrak{X}^\alpha\,,$$

where $\varkappa^i : (\mathfrak{X}^1, \ldots, \mathfrak{X}^n) \mapsto \varkappa^i(\mathfrak{X}^1, \ldots, \mathfrak{X}^n)$, $i = 1, \ldots, m$, is the coordinate representation of the configuration $\varkappa$.

If $\mathfrak{X} \in \mathfrak{B}$, then by $\mathscr{F}_\mathfrak{X}$ we denote the corresponding linear map that acts between tangent fibers. That is, $\mathscr{F}_\mathfrak{X} = T_\mathfrak{X}\varkappa \in \operatorname{Lin}(T_\mathfrak{X}\mathfrak{B}; T_{\varkappa(\mathfrak{X})}\mathscr{P})$.

Let $\varkappa_R, \varkappa \in \mathfrak{C}(\mathfrak{B};\mathscr{P})$ be two configurations and $y = \hat{\varkappa} \circ \hat{\varkappa}_R^{-1}$ the corresponding deformation, then $Ty$ corresponds to the *deformation gradient*. Consider the classical case when $\mathscr{P} = \mathscr{E}$, *i.e.*, when the physical space is represented by the affine-Euclidean space $\mathscr{E}$ with the translation vector space $\mathscr{V}$. Let $\dim \mathfrak{B} = \dim \mathscr{E} = m$. Compare the classical and smooth manifold approach for the notion of the deformation gradient. For a moment we will forget our convention about identifying of tangent spaces $T_X\mathscr{S}_\varkappa$ and $T_X\mathscr{P}$, which was stated in Section 4.5.4.

According to conventional elasticity, the deformation gradient, $\boldsymbol{F}$, is the total derivative of the deformation $y : \mathscr{S}_{\varkappa_R} \to \mathscr{S}_\varkappa$, which is defined on the open set $\mathscr{S}_{\varkappa_R}$, *i.e.*, $\boldsymbol{F} = y'$. In an orthonormal frame $(\boldsymbol{i}_k)_{k=1}^m$, the linear map $\boldsymbol{F}_X = y'(X)$ has the following dyadic representation:

$$\boldsymbol{F}_X = \left.\frac{\partial x^k}{\partial X^j}\right|_{(X^1,\ldots,X^m)} \boldsymbol{i}_k \oslash \boldsymbol{i}^j\,, \tag{5.7}$$

where $(X^i)_{i=1}^m$, $(x^i)_{i=1}^m$ are Cartesian coordinates of points in the reference and actual shapes, $(X^1, \ldots, X^m) \mapsto (x^1, \ldots, x^m) = \mathscr{D} \circ y \circ \mathscr{D}^{-1}(X^1, \ldots, X^m)$ is the coordinate representation of $y$, and vectors $\boldsymbol{i}^k$ are defined via relations $\boldsymbol{i}^k \cdot \boldsymbol{i}_s = \delta^k_s$.

**Remark 5.2.** In the framework of the present reasonings, for the dyadic representation in Euclidean space, we use the symbol $\oslash$ with the purpose of having a notation that is different from the general tensor product $\otimes$. In classical mechanics, the dyad $\boldsymbol{u} \oslash \boldsymbol{v}$ (linear map from $\mathscr{V}$ to $\mathscr{V}$) is defined as follows:

$$(\boldsymbol{u} \oslash \boldsymbol{v}) \cdot \boldsymbol{w} := (\boldsymbol{v} \cdot \boldsymbol{w})\boldsymbol{u}\,.$$

The Cartesian coordinate map $\mathscr{D}$ can be restricted to each of the shapes $\mathscr{S}_{\varkappa_R}$, $\mathscr{S}_\varkappa$ and to obtain atlases that are consist on one chart. One arrives at the following representation:

$$T_X\gamma = \left.\frac{\partial x^k}{\partial X^j}\right|_{(X^1,\dots,X^m)} \partial_{x^k}|_x \otimes dX^j|_X \,.$$

In this case, $\mathrm{In}_{\mathcal{S}_{\varkappa_R};X} = \boldsymbol{i}_k \otimes dX^k|_X$. Hence,

$$\mathrm{In}_{\mathcal{S}_\varkappa;x} \circ T_X\gamma \circ \mathrm{In}^{-1}_{\mathcal{S}_{\varkappa_R};X} = \left.\frac{\partial x^k}{\partial X^j}\right|_{(X^1,\dots,X^m)} \boldsymbol{i}_k \otimes i^j \approx \left.\frac{\partial x^k}{\partial X^j}\right|_{(X^1,\dots,X^m)} \boldsymbol{i}_k \oslash \boldsymbol{i}^j = \boldsymbol{F}_X \,. \tag{5.8}$$

On the left-hand side from the sign $\approx$, the element of the coframe $(i^k)_{k=1}^m$ appears, and on the right-hand side, stands the element of the dual *vectorial* frame $(\boldsymbol{i}^k)_{k=1}^m$. Such a distinction is the consequence of the ambient space Euclidean structure. When $T_X\gamma$, considered as a bilinear form, acts on a vector, then such an action is realized via the operation $\llcorner$: $T_X\gamma \llcorner u$, which leads to contraction of such a vector with the corresponding *covector* in the dyadic product. The Euclidean structure is not used. On the contrary, according to classical continuum mechanics, the bilinear form $\boldsymbol{F}_X$, acts on a vector by means of $(\cdot)$ : $\boldsymbol{F}_X \cdot \boldsymbol{u}$. This leads to the scalar product of such a vector with the corresponding *vector* in the dyadic product.

Choose local coordinates $(Z^i)_{i=1}^m$ for the shape $\mathcal{S}_{\varkappa_R}$ and "freeze" them into the shape. In a formal sense, this means that between coordinate mappings on $\mathcal{S}_{\varkappa_R}$ and $\mathcal{S}_\varkappa$, the equality $\sigma = \sigma_R \circ \gamma^{-1}$ holds. In this case,

$$\mathrm{In}_{\mathcal{S}_{\varkappa_R};X} = \left.\frac{\partial X^k}{\partial Z^j}\right|_{(Z^1,\dots,Z^m)} \boldsymbol{i}_k \otimes dZ^j|_X \,,$$

$$\mathrm{In}_{\mathcal{S}_\varkappa;x} = \left.\frac{\partial x^k}{\partial Z^j}\right|_{(Z^1,\dots,Z^m)} \boldsymbol{i}_k \otimes dZ^j|_x \,.$$

Denote local frames on $\mathcal{S}_{\varkappa_R}$ and $\mathcal{S}_\varkappa$ by $\boldsymbol{E}_k$ and $\boldsymbol{e}_k$:

$$\boldsymbol{E}_k = \frac{\partial X^j}{\partial Z^k}\boldsymbol{i}_j \,, \quad \boldsymbol{E}^k = \frac{\partial Z^k}{\partial X^j}\boldsymbol{i}^j \,, \quad \boldsymbol{e}_k = \frac{\partial x^j}{\partial Z^k}\boldsymbol{i}_j \,, \quad \boldsymbol{e}^k = \frac{\partial Z^k}{\partial x^j}\boldsymbol{i}^j \,.$$

Thus, from the formula (5.7) and the chain rule, one obtains the classical formula for $\boldsymbol{F}_X$ [45]: $\boldsymbol{F}_X = \boldsymbol{e}_k|_x \oslash \boldsymbol{E}^k|_X$. The expression for $T_X\gamma$ takes the form $T_X\gamma = \partial_{Z^k}|_x \otimes dZ^k|_X$, since the coordinate representation $\tilde{\gamma} = \sigma \circ \gamma \circ \sigma_R^{-1}$ is the identity map. In such a case, we obtain

$$\mathrm{In}_{\mathcal{S}_\varkappa;x} \circ T_X\gamma \circ \mathrm{In}^{-1}_{\mathcal{S}_{\varkappa_R};X} = \left.\frac{\partial Z^k}{\partial X^j}\right|_{(X^1,\dots,X^m)} \boldsymbol{e}_k|_x \otimes i^j = \boldsymbol{e}_k|_x \otimes E^k|_X \approx \boldsymbol{e}_k|_x \oslash \boldsymbol{E}^k|_X = \boldsymbol{F}_X \,. \tag{5.9}$$

Here, $i^j \in \mathcal{V}^*$.

**Remark 5.3.** In terms of the operation $\otimes$, the relations (5.8) and (5.9) between $T_X\gamma$ and $\boldsymbol{F}_X$ are replaced by the formula

$$\boldsymbol{F}_X = \mathrm{In}_{\mathcal{S}_\varkappa;x} \circ T_X\gamma \circ \mathrm{In}^{-1}_{\mathcal{S}_{\varkappa_R};X} \,.$$

## 5.6 Left and Right Cauchy–Green Strain Tensors

### 5.6.1 Spatial Measurements in Material Description

The configuration gradient $\mathcal{F}$ provides the spatial description for an infinitesimal material fiber. Using the metrical data given by the Riemannian metric $g$, one can calculate the length of a material fiber in the spatial description and the angle between spatial realizations of two material fibers. At the same time, the length and angle can be considered in the *material description*. Assume that some material metric $\mathcal{G}$ in the body $\mathfrak{B}$ is chosen.

Let $\varkappa \in \mathfrak{C}(\mathfrak{B};\mathcal{P})$ be some configuration and $\mathcal{F} = T\varkappa$ its configuration gradient. A vector $u \in T\mathfrak{B}$ represents some infinitesimal material fiber, which spatial realization is $v = \mathcal{F} \llcorner u$. The square of its length is given by $g(v, v)$:

$$g(v, v) = g(\mathcal{F} \llcorner u, \mathcal{F} \llcorner u) = \mathcal{G}\left(u, \mathcal{F}^{\mathrm{T}} \circ \mathcal{F} \llcorner u\right),$$

where (5.3) was used. That is,

$$\|v\|_g^2 = \mathcal{G}(u, \mathcal{C} \llcorner u), \quad \text{where}, \quad \mathcal{C} = \mathcal{F}^{\mathrm{T}} \circ \mathcal{F} : T\mathfrak{B} \to T\mathfrak{B}. \tag{5.10}$$

Here, $\mathcal{C}$ is a vector bundle homomorphism over $\mathfrak{B}$; for each $\mathfrak{X} \in \mathfrak{B}$ one has the symmetric linear mapping

$$\mathcal{C}_{\mathfrak{X}} = \mathcal{F}^{\mathrm{T}}_{\varkappa(\mathfrak{X})} \circ \mathcal{F}_{\mathfrak{X}} \in \operatorname{Lin}(T_{\mathfrak{X}}\mathfrak{B}; T_{\mathfrak{X}}\mathfrak{B}).$$

Assume that $\chi\colon ]a, b[ \to \mathfrak{B}$ is some material fiber. Its spatial realization is represented by the smooth curve $\varkappa \circ \chi\colon ]a, b[ \to \mathfrak{B}$. By the spatial length of material fiber $\chi$ we mean the integral

$$l_g(\chi) = \int_a^b \sqrt{g_{\varkappa\circ\chi(t)}(\mathcal{F}_{\chi(t)}u(t), \mathcal{F}_{\chi(t)}u(t))}\, dt,$$

where $u(t) = \chi'(t)$ is the velocity vector. Using the formula (5.10) one obtains

$$l_g(\chi) = \int_a^b \sqrt{\mathcal{G}_{\chi(t)}(u(t), \mathcal{C}_{\chi(t)}u(t))}\, dt. \tag{5.11}$$

Finally, let $\chi_1\colon ]a, b[\to \mathfrak{B}$, $\chi_2\colon ]c, d[\to \mathfrak{B}$ be two material fibers, which intersect at $\mathfrak{X} = \chi_1(t_1) = \chi_2(t_2)$. Let $u_i\colon t \mapsto u_i(t) = \chi_i'(t)$ be the velocity vector of $\chi_i$, $i = 1, 2$. The observable angle $\angle(\varkappa \circ \chi_1, \varkappa \circ \chi_2)_{\varkappa(\mathfrak{X})}$ between the fibers $\chi_1$ and $\chi_2$ at the observable point $\varkappa(\mathfrak{X})$ is the angle between their velocity vectors. That is,

$$\cos \angle(\varkappa \circ \chi_2, \varkappa \circ \chi_1)_X = \frac{g_X(\mathcal{F}_{\mathfrak{X}}u_1(t_1), \mathcal{F}_{\mathfrak{X}}u_2(t_2))}{\sqrt{g_X(\mathcal{F}_{\mathfrak{X}}u_1(t_1), \mathcal{F}_{\mathfrak{X}}u_1(t_1))}\sqrt{g_X(\mathcal{F}_{\mathfrak{X}}u_2(t_2), \mathcal{F}_{\mathfrak{X}}u_2(t_2))}}.$$

where $X = \varkappa(\mathfrak{X})$. Again using the relation (5.3) and the definition of $\mathscr{C}$ leads to the formula

$$\cos \angle(\varkappa \circ \chi_2, \varkappa \circ \chi_1)_X = \frac{\mathscr{G}_{\mathfrak{X}}(u_1(t_1), \mathscr{C}_{\mathfrak{X}} u_2(t_2))}{\sqrt{\mathscr{G}_{\mathfrak{X}}(u_1(t_1), \mathscr{C}_{\mathfrak{X}} u_1(t_1))}\sqrt{\mathscr{G}_{\mathfrak{X}}(u_2(t_2), \mathscr{C}_{\mathfrak{X}} u_2(t_2))}} . \tag{5.12}$$

Formulae (5.10), (5.11), and (5.12) provide the *material description* for the length and angle of the observed fibers. In the spatial description, one needs to know the configuration gradient $\mathscr{F}$ and the spatial metric $g$, while in the material description, one requires the metric $\mathscr{G}$ and the bundle homomorphism $\mathscr{C}$.

Consider the classical case of the Euclidean physical space. If the body is identified with some stress-free shape, then configurations are treated as deformations, and to the tensor $\mathscr{C} = \mathscr{F}^{\mathrm{T}} \circ \mathscr{F}$ there corresponds a right Cauchy–Green strain tensor $\boldsymbol{C} = \boldsymbol{F}^{\mathrm{T}} \circ \boldsymbol{F}$. Thus, one can treat $\mathscr{C}$ as the generalization of the classical right Cauchy–Green strain tensor. For this reason, we refer to $\mathscr{C}$ as the right Cauchy–Green strain tensor as well.

### 5.6.2 The Cauchy Polar Decomposition Theorem

In conventional considerations, right and left Cauchy–Green tensors appear by virtue of the Cauchy polar decomposition theorem. Such reasonings can be generalized on smooth manifolds [60]. For each $\mathfrak{X} \in \mathfrak{B}$, there exist orthogonal tensors[6] $\mathscr{R}_{\mathfrak{X}} \colon T_{\mathfrak{X}}\mathfrak{B} \to T_{\varkappa(\mathfrak{X})}\mathscr{P}$, such that

$$\mathscr{F} = \mathscr{R} \circ \mathscr{U} , \qquad \mathscr{F}_{\mathfrak{X}} = \mathscr{R}_{\mathfrak{X}} \circ \mathscr{U}_{\mathfrak{X}} ,$$
$$\mathscr{F} = \mathscr{V} \circ \mathscr{R} , \qquad \mathscr{F}_{\mathfrak{X}} = \mathscr{V}_{\varkappa(\mathfrak{X})} \circ \mathscr{R}_{\mathfrak{X}} ,$$

where the symmetric[7] positive-definite tensors $\mathscr{U}_{\mathfrak{X}} \colon T_{\mathfrak{X}}\mathfrak{B} \to T_{\mathfrak{X}}\mathfrak{B}$ and $\mathscr{V}_X \colon T_X\mathscr{P} \to T_X\mathscr{P}$ are, in conventional terminology, *right and left stretch tensors.*

For simple bodies, the following combinations of the elements from $(\mathscr{F}, \mathscr{F}^{\mathrm{T}})$ are used: the *right $\mathscr{C}$ and left $\mathscr{B}$ Cauchy–Green strain tensors*, which are vector bundle homomorphisms

$$\mathscr{C} = \mathscr{F}^{\mathrm{T}} \circ \mathscr{F} \colon T\mathfrak{B} \to T\mathfrak{B} ,$$
$$\mathscr{B} = \mathscr{F} \circ \mathscr{F}^{\mathrm{T}} \colon \iota^*_{\mathscr{S}_\varkappa} T\mathscr{P} \to T\mathscr{P} .$$

In the corresponding tangent fibers,

$$\mathscr{C}_{\mathfrak{X}} \colon T_{\mathfrak{X}}\mathfrak{B} \to T_{\mathfrak{X}}\mathfrak{B} , \qquad \mathscr{C}_{\mathfrak{X}} = \mathscr{F}^{\mathrm{T}}_{\mathfrak{X}} \circ \mathscr{F}_{\mathfrak{X}} ;$$
$$\mathscr{B}_X \colon T_X\mathscr{P} \to T_X\mathscr{P} , \qquad \mathscr{B}_X = \mathscr{F}_X \circ \mathscr{F}^{\mathrm{T}}_X , \qquad X = \varkappa(\mathfrak{X}) .$$

6 Orthogonal tensor was defined by (5.6).
7 $\mathscr{U}_{\mathfrak{X}} = \mathscr{U}^{\mathrm{T}}_{\mathfrak{X}}$ and $\mathscr{V}_{\mathfrak{X}} = \mathscr{V}^{\mathrm{T}}_{\mathfrak{X}}$.

The right and left stretch tensors are expressed via the Cauchy–Green strain tensors as follows:

$$\mathscr{U}_{\mathfrak{X}} = \sqrt{\mathscr{F}_X^{\mathrm{T}} \circ \mathscr{F}_{\mathfrak{X}}} = \sqrt{\mathscr{C}_{\mathfrak{X}}}\,,$$
$$\mathscr{V}_X = \sqrt{\mathscr{F}_{\mathfrak{X}} \circ \mathscr{F}_X^{\mathrm{T}}} = \sqrt{\mathscr{B}_X}\,.$$

### 5.6.3 Cauchy–Green Strain Measures as Pullback and Pushforward of Metrics

Let $\mathfrak{X} \in \mathfrak{B}$, $X = \varkappa(\mathfrak{X})$. For any vector $u \in T_{\mathfrak{X}}\mathfrak{B}$, the following relation holds:

$$\mathscr{C}_{\mathfrak{X}}(u) = \mathscr{F}_X^{\mathrm{T}}(\mathscr{F}_{\mathfrak{X}}u) = (\mathscr{F}_X^{\mathrm{T}})^{\alpha}{}_{i}(\mathscr{F}_{\mathfrak{X}})^{i}{}_{\beta}\partial_{\mathfrak{X}^\alpha}|_{\mathfrak{X}} \otimes d\mathfrak{X}^{\beta}|_{\mathfrak{X}} \llcorner u\,.$$

The equality (5.4) implies that

$$\mathscr{C}_{\mathfrak{X}} = (\mathscr{G}_{\mathfrak{X}})^{\alpha\gamma}(g_X)_{ij}(\mathscr{F}_{\mathfrak{X}})^{i}{}_{\gamma}(\mathscr{F}_{\mathfrak{X}})^{j}{}_{\beta}\partial_{\mathfrak{X}^\alpha}|_{\mathfrak{X}} \otimes d\mathfrak{X}^{\beta}|_{\mathfrak{X}}\,.$$

Finally, omitting the argument signs, one obtains

$$\mathscr{C} = G^{\alpha\gamma} g_{ij}\frac{\partial\varkappa^i}{\partial\mathfrak{X}^\gamma}\frac{\partial\varkappa^j}{\partial\mathfrak{X}^\beta}\partial_{\mathfrak{X}^\alpha} \otimes d\mathfrak{X}^\beta\,. \tag{5.13}$$

Similarly, for any $u \in T_X\mathscr{P}$, one has

$$\mathscr{B}_X(u) = \mathscr{F}_{\mathfrak{X}}\left(\mathscr{F}_X^{\mathrm{T}}u\right) = (\mathscr{F}_{\mathfrak{X}})^{i}{}_{\alpha}\left(\mathscr{F}_X^{\mathrm{T}}\right)^{\alpha}{}_{j}\,\partial_{X^i}|_X \otimes dX^j|_X \llcorner u\,.$$

Using similar reasoning, one can obtain

$$\mathscr{B} = G^{\alpha\beta} g_{lj}\frac{\partial\varkappa^l}{\partial\mathfrak{X}^\beta}\frac{\partial\varkappa^k}{\partial\mathfrak{X}^\alpha}\partial_{X^k} \otimes dX^j\,. \tag{5.14}$$

Denote $\mathscr{C}^{\flat_b} := \mathscr{C}^{1\flat_b}$. By the definition of $(\cdot)^\flat$, such a tensor is represented by the following dyadic representation:

$$\mathscr{C}^{\flat_b} = g_{ij}\frac{\partial\varkappa^i}{\partial\mathfrak{X}^\alpha}\frac{\partial\varkappa^j}{\partial\mathfrak{X}^\beta}d\mathfrak{X}^\alpha \otimes d\mathfrak{X}^\beta\,.$$

Thus, $\mathscr{C}$ is the pullback of the spatial metric:

$$\mathscr{C}^{\flat_b} = \varkappa^* g\,.$$

An analogous formula holds for the pushforward of the material metric. One needs to use the mapping $\hat{\varkappa}\colon \mathfrak{B} \to \varkappa(\mathfrak{B})$. The tangent map $\widehat{\mathscr{F}} = T\hat{\varkappa}\colon T\mathfrak{B} \to T\varkappa(\mathfrak{B})$ serves as the "internal" strain measure and is related with $\mathscr{F} = T\varkappa$ by the equality $\mathscr{F} = \mathrm{In}_{\mathcal{S}_\varkappa} \circ \widehat{\mathscr{F}}$. The use of $\iota^*_{\mathcal{S}_\varkappa} g$ as the metric on $\mathcal{S}_\varkappa$ and the relation (5.3), allows one to define $\widehat{\mathscr{F}}^{\mathrm{T}}\colon T\mathcal{S}_\varkappa \to T\mathfrak{B}$. Thus, one can obtain the "internal" Cauchy–Green tensor $\widehat{\mathscr{B}} = \widehat{\mathscr{F}} \circ \widehat{\mathscr{F}}^{\mathrm{T}}$.

Set $\widehat{\mathscr{B}}^{\sharp_p} := \widehat{\mathscr{B}}^{2\sharp_p}$. By the definition of $(\cdot)^\sharp$, one has

$$\widehat{\mathscr{B}}^{\sharp_p} = G^{\alpha\beta}\frac{\partial\hat{\varkappa}^k}{\partial\mathfrak{X}^\alpha}\frac{\partial\hat{\varkappa}^l}{\partial\mathfrak{X}^\beta}\partial_{X^k} \otimes \partial_{X^l}\,.$$

The last formula implies

$$\widehat{\mathscr{B}}^{\sharp_p} = \widehat{\mathfrak{X}}_*(\overset{*}{\mathscr{G}}) ,$$

where $\overset{*}{\mathscr{G}} := \mathscr{G}^{1\sharp_b 2\sharp_b}$.

*Thus, the right Cauchy–Green tensor defines the pullback of the spatial metric, and the left one, the pushforward of the material metric.* These reasonings are illustrated by the diagram

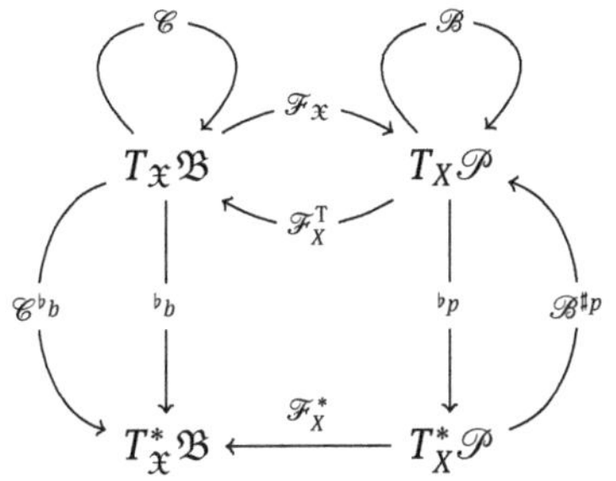

# 6 Motion

## 6.1 Motion as a Curve

A motion of the body $\mathfrak{B}$ in the physical space $\mathscr{P}$ can be described by a continual sequence of configurations. To such continual sequence there corresponds a continual family of shapes. Such a notion can be formalized as follows [60]. Let $\mathbb{T} = ]t_1, t_2[ \subset \mathbb{R}$ be an interval of instants ($t_1$ and $t_2$ may be finite or infinite). The mapping $M\colon \mathbb{T} \to \mathfrak{C}(\mathfrak{B}; \mathscr{P})$, which assigns to each instant $t \in \mathbb{T}$ some configuration $\varkappa_t$, i.e., $M\colon t \mapsto \varkappa_t$, represents a *motion of the body* $\mathfrak{B}$ *over the time interval* $\mathbb{T}$. That is, motion is a curve in $\mathfrak{C}(\mathfrak{B}; \mathscr{P})$. Equivalently, the motion $M$ can be represented as the family $M = \{\varkappa_t\}_{t\in\mathbb{T}}$ of configurations.

Any motion $M = \{\varkappa_t\}_{t\in\mathbb{T}}$ defines two types of mappings. Let $\mathfrak{X} \in \mathfrak{B}$ be a material point. Then, the mapping

$$\chi_{M;\mathfrak{X}}\colon \mathbb{T} \to \mathscr{P}\,, \quad \chi_{M;\mathfrak{X}}(t) := \varkappa_t(\mathfrak{X})\,,$$

represents a *spatial trajectory* of $\mathfrak{X}$. Another mapping is obtained by fixation of a time $t \in \mathbb{T}$. That is,

$$\chi_{M;t}\colon \mathfrak{B} \to \mathscr{P}\,, \quad \chi_{M;t}(\mathfrak{X}) := \varkappa_t(\mathfrak{X})\,.$$

Such a mapping is just the configuration $\varkappa_t$.

If for any $\mathfrak{X} \in \mathfrak{B}$, the mapping $\chi_{M;\mathfrak{X}}$ is of class $C^\infty$, then we refer to $M$ as a $C^\infty$- *(or smooth) motion*. In this book, we consider only smooth motions.

Fix a point $\mathfrak{X} \in \mathfrak{B}$. A smooth motion $M = \{\varkappa_t\}_{t\in\mathbb{T}}$ defines continual families of smooth charts[1] $\{(U_t, \varphi_t)\}_{t\in\mathbb{T}}$ and $\{(V_t, \psi_t)\}_{t\in\mathbb{T}}$, such that

$$\forall t \in \mathbb{T}\colon (\mathfrak{X} \in U_t) \wedge (\varkappa_t(U_t) \subset V_t)\,.$$

Thus, for the point $\mathfrak{X}$ one has $\mathbb{T} \ni t \mapsto \varkappa_t(\mathfrak{X}) \in \bigcup_{t\in\mathbb{T}} V_t$. The mappings $\chi_{M;\mathfrak{X}}$ and $\chi_{M;t}$ have the following coordinate representations:

$$\begin{aligned}\tilde{\chi}_{M;\mathfrak{X}} &= \psi_t \circ \chi_{M;\mathfrak{X}} \in C^\infty(\mathbb{T}; \mathbb{R}^m)\,,\\ \tilde{\chi}_{M;t} &= \psi_t \circ \chi_{M;t} \circ \varphi_t^{-1} \in C^\infty(\mathbb{R}^n; \mathbb{R}^m)\,.\end{aligned}$$

---

1 Since one can pick up countable sets of charts that cover $\mathfrak{B}$ and $\mathscr{P}$, some elements of the families $\{(U_t, \varphi_t)\}_{t\in\mathbb{T}}$ and $\{(V_t, \psi_t)\}_{t\in\mathbb{T}}$ may coincide. Nevertheless, in our reasonings it is convenient to use index set $\mathbb{T}$.

https://doi.org/10.1515/9783110563214-006

The action of these mappings is illustrated in the diagram

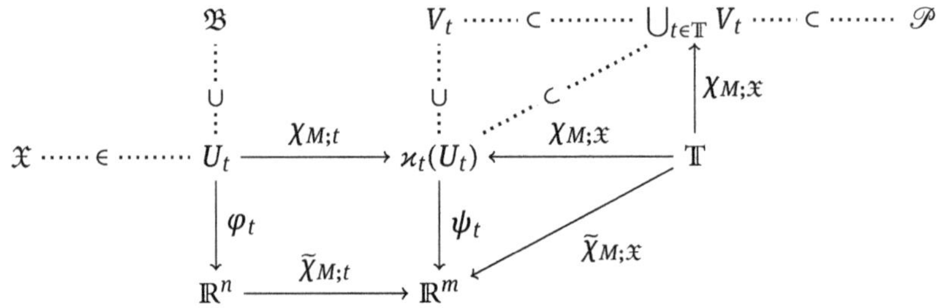


Note that in the framework of conventional continuum mechanics, the motion is considered as a family of deformations. Let $\varkappa_R$ be some chosen reference configuration. Thus, the motion is a family of transformations of the reference shape $\varkappa_R(\mathfrak{B})$ into actuals, $\varkappa_t(\mathfrak{B})$. Such transformations are represented by the family $\{\gamma_t\}_{t\in\mathbb{T}}$ of deformations $\gamma_t = \widehat{\varkappa}_t \circ \widehat{\varkappa}_R^{-1}$.

Relations between time-dependent configurations and deformations are illustrated in the diagram

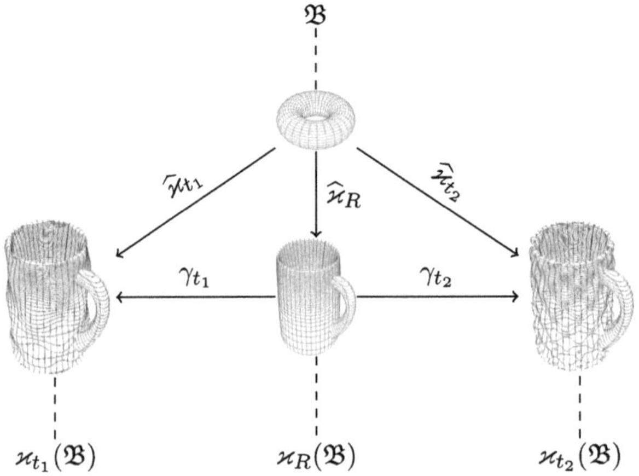


It is noteworthy that in this pictorial diagram, the body is shown as a solid torus, while its shapes are depicted as "mugs" that are topologically equivalent to each other. As such, we wish to emphasize that the body and its shapes can be significantly different metric manifolds.

## 6.2 Velocity

Let $M = \{\varkappa_t\}_{t\in\mathbb{T}}$ be a $C^\infty$-motion. Typically, there are two equivalent descriptions for kinematical and physical fields in continuum mechanics. The first one provides their definition as mappings with respect to positions of material points in some fixed shape. This is the *material description*. In this case, one observes the motion of individual particles. The second one involves the field determination as mappings with respect to fixed spatial points. This is the *spatial description*. The velocity fields in material and spatial descriptions are introduced as follows.

**Material description.** The notion of the velocity vector to a curve, defined by the formula (13.18), being applied to the curve $\chi_{M;\mathfrak{X}}$ gives two types of mappings. Fix a material point $\mathfrak{X} \in \mathfrak{B}$. Then, one has the mapping

$$V_{M;\mathfrak{X}} \colon \mathbb{T} \to T\mathscr{P}\,, \quad V_{M;\mathfrak{X}}(t) := \chi'_{M;\mathfrak{X}}(t) \in T_{\varkappa_t(\mathfrak{X})}\mathscr{P}\,.$$

Since the image of $\mathbb{T}$ under mapping $\chi_{M;\mathfrak{X}}$ is the spatial trajectory of the point $\mathfrak{X}$, which was observed at the instant interval $\mathbb{T}$, the vector $V_{M;\mathfrak{X}}(t_0)$, for $t_0 \in \mathbb{T}$ represents the *velocity of the point $\mathfrak{X}$ at some instant $t_0$*. Thus, one has the field of velocities of a fixed material point. In contrast, one may be interested in velocities of all material points in a fixed instant $t \in \mathbb{T}$. Let us define the mapping

$$V_{M;t} \colon \mathfrak{B} \to T\mathscr{P}\,, \quad V_{M;t}(\mathfrak{X}) := \chi'_{M;\mathfrak{X}}(t) \in T_{\varkappa_t(\mathfrak{X})}\mathscr{P}\,.$$

The value $V_{M;t}(\mathfrak{X}_0)$ at $\mathfrak{X}_0 \in \mathfrak{B}$ represents the *velocity of some point $\mathfrak{X}_0$ at the instant $t$*. Finally, one arrives at the assignment

$$V \colon \mathfrak{B} \times \mathbb{T} \to T\mathscr{P}\,, \quad (\mathfrak{X}, t) \mapsto \chi'_{M;\mathfrak{X}}(t)\,.$$

which is called the *material velocity field*.

For a material point $\mathfrak{X} \in \mathfrak{B}$ and instant $t_0 \in \mathbb{T}$, one has the decomposition $V_{M;\mathfrak{X}}(t_0) = V^i_{M;\mathfrak{X}}(t_0)\partial_i|_X$, where $X = \varkappa_{t_0}(\mathfrak{X})$,

$$V^i_{M;\mathfrak{X}}(t_0) = \left.\frac{d[\pi^i_m \circ \psi \circ \chi_{M;\mathfrak{X}}(\tau)]}{d\tau}\right|_{\tau = t_0}\,, \qquad i = 1, \dots, m\,.$$

Here, $\psi$ is a coordinate homeomorphism with domain on $\mathscr{P}$ containing $X$.

The material velocity $V$ defines the family $\{V_{M;t}\}_{t\in\mathbb{T}}$ of mappings $V_{M;t} \colon \mathfrak{B} \to T\mathscr{P}$. Each mapping $V_{M;t}$ assigns to every material point $\mathfrak{X} \in \mathfrak{B}$ a tangent vector $V_{M;t}(\mathfrak{X}) \in T_{\varkappa_t(\mathfrak{X})}\mathscr{P}$. In this regard, $V_{M;t}$ is exactly a smooth section of the pullback bundle $\varkappa_t^* T\mathscr{P}$, i.e., $V_{M;t} \in \operatorname{Sec}(\varkappa_t^* T\mathscr{P})$.

**Spatial description.** Let $t \in \mathbb{T}$. The *spatial velocity* is defined as the section of the pullback bundle $\iota^*_{\mathcal{S}_{\varkappa_t}} T\mathscr{P}$:

$$v_t = V(\cdot, t) \circ \widehat{\varkappa}_t^{-1} \colon \mathcal{S}_{\varkappa_t} \to T\mathscr{P}\,.$$

Finally, we define *spatial velocity field* as follows:

$$v \colon \bigcup_{t\in\mathbb{T}} (\mathcal{S}_{\varkappa_t} \times \{t\}) \to T\mathscr{P}\,, \quad (X, t) \mapsto v_t(X)\,.$$

## 6.3 $(m+1)$-Formalism and Acceleration

Acceleration fields are introduced in the framework of Newtonian space-time[2] $(\mathbb{T}\times\mathscr{P}, \mathfrak{G}, D, \nabla^{(m+1)}, \mathrm{pr}_1)$. The space-time manifold is represented by the $(m+1)$-dimensional product manifold $\mathbb{T}\times\mathscr{P}$. The absolute time function is the projection $\mathrm{pr}_1 : \mathbb{T}\times\mathscr{P}\to\mathbb{T}$. The affine connection $\nabla^{(m+1)}$ arises from the Levi–Civita connection $\nabla$ on $\mathscr{P}$ with coefficients $\gamma^i{}_{jk}$. If we denote coefficients of $\nabla^{(m+1)}$ by $\Gamma^i{}_{jk}$, then we have

$$\Gamma^i{}_{jk} = \begin{cases} \gamma^i{}_{jk}, & i,j,k\in\{1,\dots,m\}\,,\\ 0, & (k=0)\wedge(i,j\in\{0,1,\dots,m\})\,,\\ 0, & (j=0)\wedge(i,k\in\{0,1,\dots,m\})\,. \end{cases}$$

Let $M=\{\varkappa_t\}_{t\in\mathbb{T}}$ be a smooth motion. For a material point $\mathfrak{X}\in\mathfrak{B}$, its world line is represented by the curve[3]

$$\chi^{(m+1)}_{M;\mathfrak{X}} : \mathbb{T}\to\mathbb{T}\times\mathscr{P}\,,\quad \chi^{(m+1)}_{M;\mathfrak{X}}(t) := (t, \chi_{M;\mathfrak{X}}(t))\,.$$

The $(m+1)$-velocity $V^{(m+1)}_{M;\mathfrak{X}}$ of the point $\mathfrak{X}$ is the field

$$V^{(m+1)}_{M;\mathfrak{X}} : \mathbb{T}\to T(\mathbb{T}\times\mathscr{P})\,,\quad V^{(m+1)}_{M;\mathfrak{X}}(t) := (\chi^{(m+1)}_{M;\mathfrak{X}})'(t)\,.$$

Taking the natural isomorphism $T_{(t,X)}(\mathbb{T}\times\mathscr{P})\cong\mathbb{T}\oplus T_X\mathscr{P}$ into account gives

$$\forall t\in\mathbb{T}:\ V^{(m+1)}_{M;\mathfrak{X}}(t) = (1, V_{M;\mathfrak{X}}(t))\,.$$

The $(m+1)$-acceleration of the material point $\mathfrak{X}$ is defined as follows:

$$A^{(m+1)}_{M;\mathfrak{X}} : \mathbb{T}\to T(\mathbb{T}\times\mathscr{P})\,,\quad A^{(m+1)}_{M;\mathfrak{X}} = \nabla^{(m+1)}_{V^{(m+1)}_{M;\mathfrak{X}}} V^{(m+1)}_{M;\mathfrak{X}}\,.$$

Like the $(m+1)$-velocity, the $(m+1)$-acceleration vector field splits into two fields as follows:

$$\forall t\in\mathbb{T}:\ A^{(m+1)}_{M;\mathfrak{X}}(t) = (0, A_{M;\mathfrak{X}}(t))\,,$$

where $A_{M;\mathfrak{X}} : \mathbb{T}\to T\mathscr{P}$ is the desired acceleration of the point $\mathfrak{X}$. Taking into account the expressions for $\Gamma^i{}_{jk}$ one obtains the following representation of $A^{(m+1)}_{M;\mathfrak{X}}$ in components:

$$\left(A^{(m+1)}_{M;\mathfrak{X}}\right)^0 = 0\,,$$

$$\left(A^{(m+1)}_{M;\mathfrak{X}}\right)^i = \dot V^i_{M;\mathfrak{X}} + \left(\gamma^i{}_{jk}\circ\varkappa_t\right) V^j_{M;\mathfrak{X}} V^k_{M;\mathfrak{X}}\,,\quad i=1,\dots,m\,.$$

2 The main aspects of Newtonian space-time are given in Chapter 2, Section 2.5.

3 We use upper index $m+1$ to indicate that the corresponding mapping takes values in space-time.

Finally, one obtains the sought acceleration of the point $\mathfrak{X}$:

$$A_{M;\mathfrak{X}} : \mathbb{T} \to T\mathscr{P} \,, \quad A^i_{M;\mathfrak{X}} = \dot{V}^i_{M;\mathfrak{X}} + \gamma^i{}_{jk} V^j_{M;\mathfrak{X}} V^k_{M;\mathfrak{X}} \,, \quad i = 1, \ldots, m \,.$$

The field

$$A : \mathfrak{B} \times \mathbb{T} \to T\mathscr{P} \,, \quad A(\mathfrak{X}, t) := A_{M;\mathfrak{X}}(t) \,,$$

is called the *material acceleration*. The *spatial acceleration* is the map

$$a_t = A(\cdot, t) \circ \widehat{\varkappa}_t^{-1} : \mathcal{S}_{\varkappa_t} \to T\mathscr{P} \,.$$

Since $V(\mathfrak{X}, t) = v(\chi_{M;\mathfrak{X}}(t), t)$, one has

$$\dot{V}^i = \dot{v}^i + V^j \partial_j v^i \,.$$

This relation implies that

$$a^i_t = \dot{v}^i + v^j \partial_j v^i + \gamma^i{}_{jk} v^j v^k \,.$$

If $\dim \mathfrak{B} = \dim \mathscr{P}$, then any shape is an open set. In this case, the latter expression can be represented in the component-free form

$$a_t = \frac{\partial v_t}{\partial t} + \nabla_{v_t} v_t \,.$$

Here, $\partial v_t / \partial t$ is calculated with respect to a fixed spatial point.

## 6.4 Flows and Lie Derivatives

The preceding sections show the close relation between a motion and a continual set of curves, which represent spatial trajectories of material points. Such trajectories are integral curves for the material velocity field, and the latter is its infinitesimal generator. At the same time, the notion of flow is similarly based on the continual set of curves. The theory of flows on smooth manifolds gives a tool that is necessary for formulating the transport theorem and balance equations on smooth manifolds, which represent the body and the physical space. Such a tool is a Lie derivative. The main aspects of flows and Lie derivatives are briefly considered below.

### 6.4.1 Vector Fields and Integral Curves

Suppose that $\mathfrak{M}$ is a smooth $s$-dimensional manifold. A vector field on $\mathfrak{M}$ is a section $u : \mathfrak{M} \to T\mathfrak{M}$ of the tangent bundle $T\mathfrak{M}$. The material and spatial velocities give the examples of mappings that have $T\mathfrak{M}$ as a codomain, but their domains are contained in the Cartesian product $\mathfrak{M} \times \mathbb{R}$. Thus, such mappings do not satisfy the definition of a vector field. To this end, we introduce the following definition. Suppose that $\mathfrak{U} \subset \mathfrak{M}\times\mathbb{R}$ is an open subset. A *time-dependent vector field on* $\mathfrak{M}$ [3] is a continuous mapping $u\colon \mathfrak{U} \to T\mathfrak{M}$, such that

$$\forall (p,t) \in \mathfrak{U}\colon u(p,t) \in T_p\mathfrak{M} .$$

If $t \in \mathbb{R}$ is such that the set $\mathfrak{U}_t = \{p \in \mathfrak{M} \mid (p,t) \in \mathfrak{U}\}$ is not empty, then by definition the continuous mapping $u_t = u(\cdot, t)\colon \mathfrak{U}_t \to T\mathfrak{M}$ is a vector field in the conventional sense.

Any smooth curve $\chi\colon \mathbb{I} \to \mathfrak{M}$ generates the velocity vector field $t \mapsto \chi'(t)$. Vice versa, a vector field is given and one needs to obtain a curve for which such a field is the velocity vector field. Suppose that $u\colon \mathfrak{M} \to T\mathfrak{M}$ is a vector field. An *integral curve of* $u$ [3] is a smooth curve $\chi\colon \mathbb{I} \to \mathfrak{M}$ with the property

$$\forall t \in \mathbb{I}\colon \quad \chi'(t) = u_{\chi(t)} .$$

For the sake of simplicity, assume that $0 \in \mathbb{I}$. The point $\chi(0)$ is called the *starting point of* $\chi$. The question of determining the integral curve $\chi$, which has a point $p \in \mathfrak{M}$ as its starting point, reduces to the solution of the system of ordinary differential equations (ODE) system with an initial condition. Choose a smooth chart $(U, \varphi)$, such that $p \in U$. Then, in local coordinates one has $\varphi \circ \chi(t) = (\chi^1(t), \dots, \chi^s(t))$ and $u = u^i \partial_i$. One obtains the following *autonomous* ODE system:

$$\frac{d\chi^i}{dt} = u^i(\chi^1, \dots, \chi^s) , \quad i = 1, \dots, s ,$$

with the initial condition $(\chi^1(0), \dots, \chi^s(0)) = \varphi(p)$. The solution for this system is defined in an open interval $]-\varepsilon, \varepsilon[$, for some $\varepsilon > 0$. The curve $\chi$ is the sought integral curve of $u$ that starts at $p$.

Let $u\colon \mathfrak{U} \to T\mathfrak{M}$ be a time-dependent vector field. An *integral curve of* $u$ is a smooth curve $\chi\colon \mathbb{I} \to \mathfrak{M}$ with the property

$$\forall t \in \mathbb{I}\colon ((t, \chi(t)) \in \mathfrak{U}) \wedge (\chi'(t) = u(\chi(t), t)) .$$

Choose a smooth chart $(U, \varphi)$. Then, in local coordinates one has $\varphi \circ \chi(t) = (\chi^1(t), \dots, \chi^s(t))$ and $u = u^i \partial_i$. One obtains the following *non-autonomous* ODE system:

$$\frac{d\chi^i}{dt} = u^i(\chi^1, \dots, \chi^s, t) , \quad i = 1, \dots, s .$$

### 6.4.2 Flow

A *global flow* on a smooth manifold $\mathfrak{M}$ is a continuous left action of the additive group $\mathbb{R}$ on $\mathfrak{M}$. In other words, it is a continuous map $F\colon \mathbb{R} \times \mathfrak{M} \to \mathfrak{M}$, which satisfies the properties

$$\forall p \in \mathfrak{M}\ \forall t, s \in \mathbb{R}\colon \quad F(t, F(s,p)) = F(t+s, p) ,$$

$$\forall p \in \mathfrak{M}\colon \quad F(0,p) = p .$$

A global flow $F\colon \mathbb{R}\times\mathfrak{M} \to \mathfrak{M}$, $(t,p) \mapsto F(t,p)$ defines two types of maps, whether $t$ or $p$ is fixed. For any $t \in \mathbb{R}$, one has the continuous mapping $F_t\colon \mathfrak{M} \to \mathfrak{M}$, such that $F_t = F(t,\cdot)$. It satisfies the group property

$$F_t \circ F_s = F_{t+s}\,, \quad F_0 = \mathrm{Id}_{\mathfrak{M}}\,.$$

For each $p \in \mathfrak{M}$, one has the curve $F^{(p)}\colon \mathbb{R} \to \mathfrak{M}$, $F^{(p)} = F(\cdot, p)$. For any $p \in \mathfrak{M}$, define $u_p = (F^{(p)})'(0) \in T_p\mathfrak{M}$. If $F$ is a smooth global flow, then the assignment $u\colon p \mapsto u_p$ is a smooth vector field on $\mathfrak{M}$. Each curve $F^{(p)}$ is an integral curve of $u$ [3]. Due to this fact, the vector field $u$ is called the *infinitesimal generator of F*.

Suppose that $u \in \mathrm{Vec}(\mathfrak{M})$. The integral curve of $u$ is defined from the ODE system with an initial condition. This initial value problem has the unique solution defined on some open interval, which does not coincide with $\mathbb{R}$, in general. This means that $u$ cannot be an infinitesimal generator of a global flow $F$, since every curve $F^{(p)}$ must be defined on $\mathbb{R}$. For this reason, we introduce the notion of local flow [3].

Let $\mathfrak{M}$ be a smooth manifold. A *flow domain* is an open subset $G \subset \mathbb{R}\times\mathfrak{M}$, such that for all $p \in \mathfrak{M}$, the set $G^{(p)} = \{t \in \mathbb{R} \mid (t,p) \in G\}$ is an open interval containing[4] 0. A *local flow* on $\mathfrak{M}$ is a continuous mapping $F\colon G \to \mathfrak{M}$ that satisfies the following properties: for all $p \in \mathfrak{M}$ one has

$$F(0,p) = p\,,$$

and for all $s \in G^{(p)}$ and $t \in G^{(F(s,p))}$, such that $s+t \in G^{(p)}$,

$$F(t, F(s,p)) = F(t+s, p)\,.$$

For all $t \in \mathbb{R}$, define the set

$$\mathfrak{M}_t = \{p \in \mathfrak{M} \mid (t,p) \in G\}\,.$$

As for global flow, define the mapping $F_t\colon \mathfrak{M}_t \to \mathfrak{M}$ as $F_t(p) := F(t,p)$ and the curve $F^{(p)}\colon G^{(p)} \to \mathfrak{M}$ as $F^{(p)}(t) = F(t,p)$.

If $F$ is a smooth local flow, the infinitesimal generator of $F$ is defined similarly as for global flow: $u_p = (F^{(p)})'(0) \in T_p\mathfrak{M}$. Then, one has that $u\colon p \mapsto u_p$ is a smooth vector field on $\mathfrak{M}$, and each curve $F^{(p)}$ is an integral curve of $u$ [3]. On the contrary, suppose that we are given a vector field $u \in \mathrm{Vec}(\mathfrak{M})$. There is a unique smooth local flow $F\colon G \to \mathfrak{M}$ whose infinitesimal generator is $u$. This flow has the following properties[5]:

(i) $F$ cannot be extended to a flow on a larger flow domain.
(ii) For any $p \in \mathfrak{M}$, the curve $F^{(p)}\colon G^{(p)} \to \mathfrak{M}$ is the unique integral curve of $u$ that starts at $p$. This curve cannot be extended to an integral curve on any larger open interval.

4 It follows that $\{0\}\times\mathfrak{M} \subset G$.

5 These properties constitute the theorem that is proven in [3].

(iii) For any $p \in \mathfrak{M}$, if $s \in G^{(p)}$, then $G^{(F(s,p))} = \{t - s \mid t \in G^{(p)}\}$.
(iv) For each $t \in \mathbb{R}$, the set $\mathfrak{M}_t$ is open in $\mathfrak{M}$. The mapping $F_t\colon \mathfrak{M}_t \to \mathfrak{M}_{-t}$ (the codomain of the original map was restricted to $\mathfrak{M}_t$) is a diffeomorphism with inverse $F_{-t}$.

### 6.4.3 Lie Derivatives

Consider a pair of vector fields $u, v \in \mathrm{Vec}(\mathfrak{M})$ on a smooth manifold $\mathfrak{M}$. The vector field $u$ generates the local flow $F\colon G \to \mathfrak{M}$. Let $p \in \mathfrak{M}$. The sets $\mathfrak{M}_t$, $\mathfrak{M}_{-t}$ contain $p$ for values of $t$ that belong to a sufficiently small interval $]-\varepsilon, \varepsilon[$ for $\varepsilon > 0$. Since[6] $T_pF_t\colon T_p\mathfrak{M} \to T_{F_t(p)}\mathfrak{M}$ and $T_{F_t(p)}F_{-t}\colon T_{F_t(p)}\mathfrak{M} \to T_p\mathfrak{M}$, one obtains that $(T_pF_t)v_p \in T_{F_t(p)}\mathfrak{M}$ and $(T_{F_t(p)}F_{-t})v_{F_t(p)} \in T_p\mathfrak{M}$. The *Lie derivative of v with respect to u at point p* has two equivalent definitions [3, 122, 123]:

$$(\mathcal{L}_u v)_p = \lim_{t\to 0} \frac{(v \circ F_t)_p - (T_pF_t)v_p}{t}\,, \tag{6.1}$$

$$(\mathcal{L}_u v)_p = \lim_{t\to 0} \frac{(T_{F_t(p)}F_{-t})v_{F_t(p)} - v_p}{t}\,, \tag{6.2}$$

and if these limits exist, (6.1) and (6.2) represent a vector from $T_p\mathfrak{M}$. Formula (6.1) defines the Lie derivative by means of a pushforward, while (6.2) defines the Lie derivative as a pullback. The latter definition can be rewritten as follows[7]:

$$(\mathcal{L}_u v)_p = \lim_{t\to 0} \frac{(F_t^* v)_p - v_p}{t}\,. \tag{6.3}$$

Here, $(F_t^* v)_p$ denotes the value of $F_t^* v$ at point $p$. The mapping $p \mapsto (\mathcal{L}_u v)_p$ is well defined and represents a smooth vector field on $\mathfrak{M}$ [3]. This field is denoted by $\mathcal{L}_u v$.

Definitions (6.1) and (6.2) (or (6.3)) require finding local flows that are generated by $u$. Such an operation may be cumbersome in the computational aspect, but, since the Lie derivative can be defined as a limit at $t \to 0$, its calculation can be carried out in a simpler way. In order to do this, one needs to take into account the principal parts of $u$, $v$ only, when the expansion with respect to $t$ is performed (see [122, p. 126]):

$$\mathcal{L}_u v = [u, v]\,,$$

where $[u, v]$ are Lie brackets (13.20).

As an example, calculate Lie derivatives directly, by (6.1), (6.2), and with the definition of Lie brackets. Suppose that $\mathcal{E}$ is a two-dimensional affine Euclidean space

6 The sets $\mathfrak{M}_t$ and $\mathfrak{M}_t$ are open submanifolds of $\mathfrak{M}$. Since $p \in \mathfrak{M}_t$ and $p \in \mathfrak{M}_{-t}$, one has $T_p\mathfrak{M}_t = T_p\mathfrak{M}_{-t} = T_p\mathfrak{M}$.

7 Recall that $F_t^{-1} = F_{-t}$. Thus, $F_t^* v = TF_{-t} \circ v \circ F_t$ (for the definition of pullback, see Chapter 13, Section 13.3.7).

with a fixed Cartesian coordinate system $(o, (\boldsymbol{i}_1, \boldsymbol{i}_2))$. Let $\mathscr{O}$ be a set

$$\mathscr{O} = \mathscr{E} \setminus \left\{x \in \mathscr{E} \mid \boldsymbol{p}(x) = x^1 \boldsymbol{i}_1, x^1 \geq 0\right\} .$$

It is open, because it was obtained from $\mathscr{E}$ by puncturing points from the axis $ox^1$ that are in the positive direction. Induce the topological structure from $\mathscr{E}$ on $\mathscr{O}$. Consider a smooth atlas on $\mathscr{O}$, which consists of two charts. The first one is standard[8], $(\mathscr{O}, \mathscr{D})$. The second one is $(\mathscr{O}, H)$, where $H = h^{-1} \circ \mathscr{D}$, and the mapping

$$h\colon ]0, +\infty[ \times ]0, 2\pi[ \to \mathbb{R}^2 , \quad (r, \varphi) \mapsto (x^1, x^2)$$

is defined by the rule $x^1 = r \cos \varphi, \quad x^2 = r \sin \varphi$.

Each such chart covers the whole $\mathscr{O}$, and this means that one has the transformation of variables on $\mathscr{O}$. The symbols $\partial_1, \partial_2$ denote the coordinate frame that corresponds to the chart $(\mathscr{O}, \mathscr{D})$, and the symbols $\partial_r, \partial_\varphi$ denote the coordinate frame in the chart $(\mathscr{O}, H)$. Using the formula $\partial_{i'} = \frac{\partial x^j}{\partial x^{i'}} \partial_j$, which relates "old" frame $(\partial_j)$ to the "new" one, $(\partial_{i'})$, with respect to a transformation $x^{i'} = x^{i'}(x^j)$, one arrives at the relations

$$\partial_r = \cos \varphi \partial_1 + \sin \varphi \partial_2 , \quad \partial_\varphi = -r \sin \varphi \partial_1 + r \cos \varphi \partial_2 .$$

In addition, introduce the field $e_{\langle\varphi\rangle} = \frac{1}{r} \partial_\varphi$.

For a geometric interpretation, one needs to take into account the correspondences $\boldsymbol{i}_1 \leftrightarrow \partial_1, \boldsymbol{i}_2 \leftrightarrow \partial_2$. To this end, define the mappings

$$\Phi_t\colon (r, \varphi) \mapsto (R, \Psi) ,$$

which are coordinate representations of flows in the chart $(\mathscr{O}, H)$, calculate Lie derivatives at point $p = H^{-1}(r, \varphi)$ according to (6.1), (6.2), and compare them with similar derivations with Lie brackets. The map $f_t = H^{-1} \circ \Phi_t \circ H$ represents a flow that meets $\Phi_t$.

**(i)** Define the flow that is generated by $\partial_r$. Its coordinate representation can be obtained from the ODE system

$$\frac{dr}{dt} = 1 , \quad \frac{d\varphi}{dt} = 0 ,$$

with initial conditions $r = r_0, \varphi = \varphi_0$. The solution for this system has the form

$$\Phi_t(r_0, \varphi_0) = (r_0 + t, \varphi_0) .$$

Hence, one obtains the mapping $(r, \varphi) \mapsto \Phi_t(r, \varphi)$, which has the following coordinate representation in $(x^1, x^2)$:

$$(r, \varphi) \mapsto \Xi_t(r, \varphi) = ((r + t) \cos \varphi, (r + t) \sin \varphi) .$$

8 Here the symbol $\mathscr{D}$ is used for the restriction of the mapping $\mathscr{D}_{(o,(\boldsymbol{i}_1,\boldsymbol{i}_2))}$ (2.2) to $\mathscr{O}$.

Calculate the Lie derivatives $\mathcal{L}_{\partial_r}\partial_r$, $\mathcal{L}_{\partial_r}\partial_\varphi$ and $\mathcal{L}_{\partial_r}e_{\langle\varphi\rangle}$ at point $p$ using formula (6.1). The tangent map $Tf_t\colon T\mathcal{O} \to T\mathcal{O}$ has the form

$$Tf_t = \partial_R \otimes dr + \partial_\Psi \otimes d\varphi\,,$$

where $(R, \Psi) = (r + t, \varphi)$. Hence,

$$Tf_t\partial_r = \partial_R\,, \quad Tf_t\partial_\varphi = \partial_\Psi\,, \quad Tf_t e_{\langle\varphi\rangle} = \tfrac{1}{r}\partial_\Psi\,.$$

Since $\partial_r \circ f_t = \partial_R$, $\partial_\varphi \circ f_t = \partial_\Psi$, $e_{\langle\varphi\rangle} \circ f_t = \frac{1}{R}\partial_\Psi$, one has

$$[\mathcal{L}_{\partial_r}\partial_r]_p = \lim_{t\to 0}\frac{\partial_R - Tf_t\partial_r}{t} = 0\,;$$

$$[\mathcal{L}_{\partial_r}\partial_\varphi]_p = \lim_{t\to 0}\frac{\partial_\Psi - Tf_t\partial_\varphi}{t} = 0\,;$$

$$[\mathcal{L}_{\partial_r}e_{\langle\varphi\rangle}]_p = \lim_{t\to 0}\frac{\frac{1}{R}\partial_\Psi - Tf_t e_{\langle\varphi\rangle}}{t} = \lim_{t\to 0}\frac{\frac{1}{r+t}\partial_\Psi - \frac{1}{r}\partial_\Psi}{t} = -\frac{1}{r}e_{\langle\varphi\rangle}\,.$$

A similar result can be obtained if one calculates the Lie brackets:

$$[\partial_r, \partial_r] = \partial_r\partial_r - \partial_r\partial_r = 0\,,$$

$$[\partial_r, \partial_\varphi] = \partial_r\partial_\varphi - \partial_\varphi\partial_r = 0\,,$$

$$[\partial_r, e_{\langle\varphi\rangle}] = -\frac{1}{r}e_{\langle\varphi\rangle}\,.$$

Now calculate the Lie derivatives via formula (6.2). The inverse to the coordinate representation of the flow $\Phi_t$ has the form

$$\Phi_t^{-1}(R, \Psi) = (R - t, \Psi)\,, \quad R = r + t\,, \quad \Psi = \varphi\,,$$

at a point with polar coordinates $(R, \Psi)$. The tangent map to $f_t^{-1}$ is

$$Tf_t^{-1} = \partial_r \otimes dR + \partial_\varphi \otimes d\Psi\,.$$

For the vector field $\partial_\varphi$, one has $\{\partial_\varphi \circ f_t\}_p = \partial_\Psi$, and hence,

$$\{(f_t)^*\partial_\varphi\}_q = \{\partial_r \otimes dR + \partial_\varphi \otimes d\Psi\}\partial_\Psi = \partial_\varphi\,.$$

Thus, the Lie derivative is determined by the expressions

$$[\mathcal{L}_{\partial_r}\partial_\varphi]_p = \lim_{t\to 0}\frac{\{(f_t)^*\partial_\varphi\}_p - \partial_\varphi}{t} = \lim_{t\to 0}\frac{\partial_\varphi - \partial_\varphi}{t} = 0\,.$$

For the vector field $e_{\langle\varphi\rangle}$, one has $\{e_{\langle\varphi\rangle} \circ f_t\}_p = \frac{1}{R}\partial_\Psi$. Its pullback is equal to

$$\{(f_t)^* e_{\langle\varphi\rangle}\}_p = \{\partial_r \otimes dR + \partial_\varphi \otimes d\Psi\}\tfrac{1}{R}\partial_\Psi = \frac{r}{r+t}e_{\langle\varphi\rangle}\,,$$

and, hence, the Lie derivative is equal to

$$[\mathcal{L}_{\partial_r}e_{\langle\varphi\rangle}]_p = \lim_{t\to 0}\frac{\{(f_t)^* e_{\langle\varphi\rangle}\}_p - e_{\langle\varphi\rangle}}{t} = -\lim_{t\to 0}\frac{1}{r+t}e_{\langle\varphi\rangle} = -\frac{1}{r}e_{\langle\varphi\rangle}\,.$$

**(ii)** For the flow generated by $\partial_\varphi$, one has the ODE system

$$\frac{dr}{dt} = 0\,, \quad \frac{d\varphi}{dt} = 1\,,$$

with initial conditions $r = r_0$, $\varphi = \varphi_0$. The solution for this system gives the coordinate representation for the flow:

$$\Phi_t(r_0, \varphi_0) = (r_0, t + \varphi_0)\,.$$

In coordinates $(x^1, x^2)$ one has another representation:

$$(r, \varphi) \mapsto \Xi_t(r, \varphi) = (r\cos(t + \varphi), r\sin(t + \varphi))\,.$$

Calculate $\mathscr{L}_{\partial_\varphi}\partial_r$ at point $p$ according to formula (6.1). The tangent map $Tf_t\colon T\mathscr{O} \to T\mathscr{O}$ has the form

$$Tf_t = \partial_R \otimes dr + \partial_\Psi \otimes d\varphi\,,$$

where $(R, \Psi) = (r, t + \varphi)$. Thus, one obtains

$$Tf_t\partial_r = \partial_R\,.$$

Since $\{\partial_r \circ f_t\}_p = \partial_R$, one arrives at the equalities

$$[\mathscr{L}_{\partial_\varphi}\partial_r]_p = \lim_{t\to 0}\frac{\partial_R - Tf_t\partial_r}{t} = 0\,.$$

Calculation of the Lie brackets gives a similar result:

$$[\partial_\varphi, \partial_r] = \partial_\varphi\partial_r - \partial_\varphi\partial_r = 0\,.$$

Now, use the formula (6.2) to calculate the Lie derivatives. The mapping, which is inverse to the coordinate representation of the flow $\Phi_t$, has the following expression at point with polar coordinates $(R, \Psi)$:

$$\Phi_t^{-1}(t) = (R, \Psi - t)\,, \quad r = R\,, \quad \Psi = \varphi + t\,.$$

The corresponding tangent map is represented by the expression

$$Tf_t^{-1} = \partial_r \otimes dR + \partial_\varphi \otimes d\Psi\,,$$

which implies the expression for values of pullback for $\partial_r$ at point $p$:

$$\{f_t^*\partial_r\}_p = \{\partial_r \otimes dR + \partial_\varphi \otimes d\Psi\}\partial_R = \partial_r\,,$$

and the expression for the Lie derivative of $\partial_r$ among $\partial_\varphi$:

$$[\mathscr{L}_{\partial_\varphi}\partial_r]_p = \lim_{t\to 0}\frac{\{f_t^*\partial_r\}_p - \partial_r}{t} = \lim_{t\to 0}\frac{\partial_r - \partial_r}{t} = 0\,.$$

**(iii)** Finally, for the coordinate representation of the flow generated by $e_{\langle\varphi\rangle}$, we have the ODE system

$$\frac{dr}{dt} = 0\,, \qquad \frac{d\varphi}{dt} = \frac{1}{r}\,,$$

with the initial conditions $r = r_0$, $\varphi = \varphi_0$. Its solution has the form

$$\Phi_t(r_0, \varphi_0) = \left(r_0, \tfrac{t}{r_0} + \varphi_0\right)\,.$$

Correspondingly, the flow, which is generated by $e_{\langle\varphi\rangle}$, can be represented by the expression $(r, \varphi) \mapsto \Phi_t(r, \varphi)$.

Calculate the Lie derivative for $\partial_r$ with respect to $e_{\langle\varphi\rangle}$, *i.e.*, $\mathscr{L}_{e_{\langle\varphi\rangle}} \partial_r$, at point $p$ using (6.1). One has the following expression for the tangent map $Tf_t \colon T\mathcal{O} \to T\mathcal{O}$:

$$Tf_t = \partial_R \otimes dr - \frac{t}{r^2} \partial_\Psi \otimes dr + \partial_\Psi \otimes d\varphi\,,$$

where $(R, \Psi) = (r, \frac{t}{r} + \varphi)$. Thus, one obtains

$$Tf_t \partial_r = \partial_R - \frac{t}{r^2} \partial_\Psi\,.$$

Since $\{\partial_r \circ f_t\}_p = \partial_R$, and due to the continuity of $\partial_\varphi$, one has

$$[\mathscr{L}_{e_{\langle\varphi\rangle}} \partial_r]_p = \lim_{t \to 0} \frac{\partial_R - Tf_t \partial_r}{t} = \frac{1}{r^2} \lim_{t \to 0} \partial_{\frac{t}{r} + \varphi} = \frac{1}{r} e_{\langle\varphi\rangle}\,.$$

As was expected, the Lie brackets from the vector fields $e_{\langle\varphi\rangle}$ and $\partial_r$ give the same result:

$$[e_{\langle\varphi\rangle}, \partial_r] = \frac{1}{r} \partial_\varphi \partial_r - \partial_r \left(\frac{1}{r} \partial_\varphi\right) = \frac{1}{r} e_{\langle\varphi\rangle}\,.$$

Let us do the final calculation of the Lie derivative using (6.2). The inverse mapping to the coordinate representation of the flow $\Phi_t$ at point with polar coordinates $(R, \Psi)$ has the form

$$\Phi_t^{-1}(R, \Psi) = \left(R, -\frac{t}{R} + \Psi\right)\,, \quad R = r\,, \quad \Psi = \varphi + t\,.$$

Taking this into account, one obtains the expression for the tangent map:

$$Tf_t^{-1} = \partial_r \otimes dR + \frac{t}{R^2} \partial_\varphi \otimes dR + \partial_\varphi \otimes d\Psi\,.$$

The relations obtained allow one to represent pullback of the field $\partial_r$ in the form

$$\left\{f_t^* \partial_r\right\}_p = \partial_r + \frac{t}{(r+t)^2} \partial_\varphi\,,$$

hence,

$$[\mathscr{L}_{e_{\langle\varphi\rangle}} \partial_r]_p = \lim_{t \to 0} \frac{\left\{f_t^* \partial_r\right\}_p - \partial_r}{t} = \lim_{t \to 0} \frac{r}{(r+t)^2} e_{\langle\varphi\rangle} = \frac{1}{r} e_{\langle\varphi\rangle}\,.$$

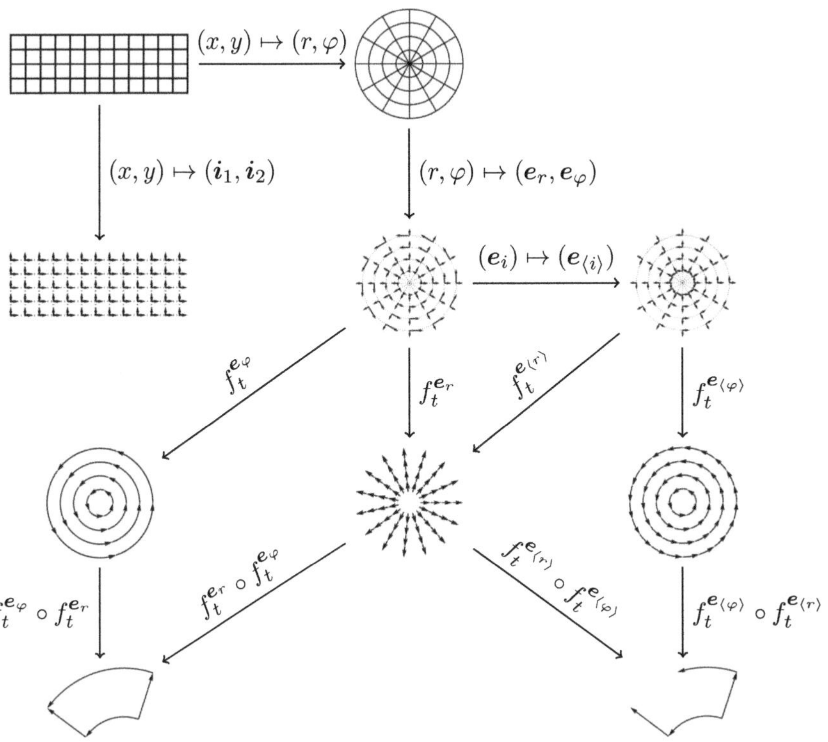


**Fig. 6.1:** Flows and Lie derivatives.

The geometric interpretation of the relations obtained is shown in Figure 6.1. It contains the space $\mathscr{E}$, with two coordinate systems in its part: the Cartesian $(x, y)$, with the basis $(\boldsymbol{i}_1, \boldsymbol{i}_2)$, and a polar one, $(r, \varphi)$, with the frame $(\boldsymbol{e}_r, \boldsymbol{e}_\varphi)$, which is the result of embedding $(\partial_r, \partial_\varphi)$ into $\mathscr{E}$. Coordinate grids of both systems are shown: rectangular and a system of concentric circles and rays. The pair $(\boldsymbol{e}_{\langle r \rangle}, \boldsymbol{e}_{\langle \varphi \rangle})$ is the frame, obtained from the corresponding polar frame by normalizing its vectors.

The lower half of the figure is devoted to the construction of flows along the corresponding vector fields. In the case when the flows were considered along the basis vectors of the polar system, a closed parallelogram was constructed. A similar case for a normalized frame leads to an unclosed parallelogram. In the second case, the Lie derivative is different from zero. This illustrates the argument that the Lie derivative is a measure of the non-closure of a parallelogram constructed on curves generated by flows.

The Lie derivative is generalized to arbitrary tensor fields by a formula that is analogous to (6.3). Suppose that $u \in \mathrm{Vec}(\mathfrak{M})$ and $F\colon G \to \mathfrak{M}$ is the local flow generated by $u$. Let $\mathscr{T} \in \mathrm{Sec}(T^{(k,l)}(T\mathfrak{M}))$. Then, the Lie derivative of $\mathscr{T}$ with respect to $u$ at a point

$p \in \mathfrak{M}$ is the limit [3, 123]

$$(\mathscr{L}_u \mathscr{T})_p = \lim_{t \to 0} \frac{(F_t^* \mathscr{T})_p - \mathscr{T}_p}{t} .$$

In the framework of calculus on the manifolds, it is shown that $(\mathscr{L}_u \mathscr{T})_p$ exists for all $p \in \mathfrak{M}$, and the assignment $p \mapsto (\mathscr{L}_u \mathscr{T})_p$ is an element of $\mathrm{Sec}(T^{(k,l)}(T\mathfrak{M}))$. Such an assignment is denoted by $\mathscr{L}_u \mathscr{T}$. The operation $\mathscr{L}_u$ satisfies the following properties [3]:
(i) $\forall f \in C^\infty(\mathfrak{M})$: $\mathscr{L}_u f = u(f)$;
(ii) $\forall f \in C^\infty(\mathfrak{M})\ \forall \mathscr{T} \in \mathrm{Sec}(T^{(k,l)}(T\mathfrak{M}))$: $\mathscr{L}_u(f\mathscr{T}) = (\mathscr{L}_u f)\mathscr{T} + f\mathscr{L}_u \mathscr{T}$;
(iii) $\forall \mathscr{T}, \mathscr{Q} \in \mathrm{Sec}(T^{(k,l)}(T\mathfrak{M}))$: $\mathscr{L}_u(\mathscr{T} \otimes \mathscr{Q}) = (\mathscr{L}_u \mathscr{T}) \otimes \mathscr{Q} + \mathscr{T} \otimes \mathscr{L}_u \mathscr{Q}$;
(iv) $\forall f \in C^\infty(\mathfrak{M})$: $\mathscr{L}_u(df) = d(\mathscr{L}_u f)$.

In particular, these properties imply that if $\mathscr{g} = g_{ij} dx^i \otimes dx^j$ is a Riemannian metric on $\mathfrak{M}$, then its Lie derivative is represented by the expression

$$\mathscr{L}_u \mathscr{g} = \left\{ u^k \partial_k g_{ij} + g_{kj} \partial_i u^k + g_{ik} \partial_j u^k \right\} dx^i \otimes dx^j . \tag{6.4}$$

Consider the case of differential forms, which are the elements of $\Omega^k(\mathfrak{M})$. The Lie derivative is defined in the similar way as for arbitrary smooth tensor fields. Suppose that $u \in \mathrm{Vec}(\mathfrak{M})$. The operation $\mathscr{L}_u$ has the following properties [3]:
(i) $\forall \omega \in \Omega^k(\mathfrak{M})\ \forall \eta \in \Omega^l(\mathfrak{M})$: $\mathscr{L}_u(\omega \wedge \eta) = (\mathscr{L}_u \omega) \wedge \eta + \omega \wedge (\mathscr{L}_u \eta)$;
(ii) for all $\omega \in \Omega^k(\mathfrak{M})$ *Cartan's magic formula* holds:

$$\mathscr{L}_u \omega = i_u(d\omega) + d(i_u \omega) . \tag{6.5}$$

### 6.4.4 Time-Dependent Flow

A flow, in some sense, is a family of integral curves of a vector field on a smooth manifold. Thus, a family of autonomous ODE systems corresponds to a flow. A time-dependent flow is similarly a family of integral curves, but for time-dependent vector fields. In this case, one deals with a family of non-autonomous ODE systems.

Let $\mathbb{I} \subset \mathbb{R}$ be an open interval and $G \subset \mathbb{I} \times \mathbb{I} \times \mathfrak{M}$ an open subset. Suppose that for each $t_0 \in \mathbb{I}$ and $p \in \mathfrak{M}$ the set

$$G^{(t_0,p)} = \{t \in \mathbb{I} \mid (t, t_0, p) \in G\}$$

is an open interval that contains[9] $t_0$. Also define the set

$$\mathfrak{M}_{t_0,t_1} = \{p \in \mathfrak{M} \mid (t_0, t_1, p) \in G\} ,$$

9 In particular, this implies that $\{t_0\} \times \{t_0\} \times \mathfrak{M} \subset G$.

for each $t_0, t_1 \in \mathbb{I}$. A *time-dependent flow* [3, 123] is a continuous mapping $F\colon G \to \mathfrak{M}$ with the properties: for all $p \in \mathfrak{M}$ and $t_0 \in \mathbb{I}$, one has

$$F(t_0, t_0, p) = p\,,$$

and if $p \in \mathfrak{M}_{t_1,t_0}$ and $F(t_1, t_0, p) \in \mathfrak{M}_{t_2,t_1}$, then

$$F(t_2, t_1, F(t_1, t_0, p)) = F(t_2, t_0, p)\,.$$

For each $t_0, t_1 \in \mathbb{I}$, let us introduce the map $F_{t_0,t_1}\colon \mathfrak{M}_{t_0,t_1} \to \mathfrak{M}$ by the relation $F_{t_0,t_1}(p) := F(t_0, t_1, p)$. Then, the properties of $F$ take the form

$$F_{t_2,t_1} \circ F_{t_1,t_0}(p) = F_{t_2,t_0}(p)\,, \quad F_{t_0,t_0}(p) = p\,.$$

Thus, a time-dependent flow can be considered as a two-parametric family $\{F_{t_0,t_1}\}_{t_0,t_1\in\mathbb{I}}$ of mappings $F_{t_0,t_1}$. This family is called a time-dependent flow itself in [123].

A flow $F$ generates for every $t_0 \in \mathbb{I}$ and $p \in \mathfrak{M}$ a curve

$$F^{(t_0,p)}\colon G^{(t_0,p)} \to \mathfrak{M}$$

by the rule $F^{(t_0,p)}(t) := F(t, t_0, p)$. One has $F^{(t_0,p)}(t_0) = p$. If $F$ is smooth, then one can consider the velocity vector $u_{p,t_0} = (F^{(t_0,p)})'(t_0)$. The assignment $(p, t) \mapsto u_{p,t}$ is a time-dependent vector field on $\mathfrak{M} \times \mathbb{I}$. Conversely, suppose that $u\colon \mathfrak{M} \times \mathbb{I} \to \mathfrak{M}$ is a smooth time-dependent vector field. There exists an open subset $G \subset \mathbb{I} \times \mathbb{I} \times \mathfrak{M}$ and a smooth map $F\colon G \to \mathfrak{M}$ that satisfies the following properties[10]:

(i) For any $t_0 \in \mathbb{I}$ and $p \in \mathfrak{M}$ the set $G^{(t_0,p)}$ is open interval that contains $t_0$. The curve $F^{(t_0,p)}\colon G^{(t_0,p)} \to \mathfrak{M}$ is smooth and represents the unique integral curve of $u$ with the initial condition $F^{(t_0,p)}(t_0) = p$. This curve cannot be extended to an integral curve on any larger open interval.

(ii) If $t_1 \in G^{(t_0,p)}$ and $q = F^{(t_0,p)}(t_1)$, then

$$G^{(t_1,q)} = G^{(t_0,p)}\,, \quad F^{(t_1,q)} = F^{(t_0,p)}\,.$$

(iii) The set $\mathfrak{M}_{t_0,t_1}$ is an open subset in $\mathfrak{M}$ for all $t_0, t_1 \in \mathbb{I}$. The mapping $F_{t_0,t_1}\colon \mathfrak{M}_{t_0,t_1} \to \mathfrak{M}$ is a diffeomorphism from $\mathfrak{M}_{t_0,t_1}$ onto $\mathfrak{M}_{t_1,t_0}$. Its inverse is $F_{t_1,t_0}$.

(iv) If $p \in \mathfrak{M}_{t_1,t_0}$ and $F(t_1, t_0, p) \in \mathfrak{M}_{t_2,t_1}$ then $p \in \mathfrak{M}_{t_2,t_0}$ and

$$F_{t_2,t_1} \circ F_{t_1,t_0}(p) = F_{t_2,t_0}(p)\,.$$

This map, $F$, is the time-dependent flow of $u$.

Let $F$ be a time-dependent flow, which is generated by a smooth time-dependent vector field $u\colon \mathfrak{M} \times \mathbb{I} \to T\mathfrak{M}$, and $\mathcal{T}_t$ a smooth tensor field, which depends on points

10 These properties constitute the theorem proved in [3].

from $\mathfrak{M}$ and the numerical parameter $t \in \mathbb{I}$. The *autonomous Lie derivative* [60] is defined via relation

$$\mathcal{L}_u \mathcal{T}_t = \frac{d}{dt}\{F^*_{t;s}\mathcal{T}_s\}\bigg|_{t=s}, \tag{6.6}$$

and the consequence of such a definition is the equality [123]:

$$\frac{d}{dt}F^*_{t,s}\mathcal{T}_t = F^*_{t,s}\mathcal{L}_u\mathcal{T}_t + F^*_{t,s}\frac{\partial \mathcal{T}_t}{\partial t}, \tag{6.7}$$

which is used in this book. In the equality (6.7), one needs clarification for the symbol $\frac{\partial \mathcal{T}_t}{\partial t}$. We give such a clarification for the case $\frac{\partial u}{\partial t}$, where $u : \mathfrak{M}\times\mathbb{I} \to T\mathfrak{M}$; the general case is considered in an analogous way. In calculating the partial derivative of some function, one means that other variables of such a function are fixed. In our considerations, the variable $p$ is fixed. This means that the mapping $u(p,\cdot) : \mathbb{I} \to T\mathfrak{M}$ takes values on the same vector space, $T_p\mathfrak{M}$, and the expression $\frac{u(p,t+h)-u(p,t)}{h}$ makes sense. Choose some Riemannian metric $g$ on $\mathfrak{M}$. At a point $p$ it represents a scalar product. The convergence of the function $\frac{u(p,t+h)-u(p,t)}{h}$ to some vector from $T_p\mathfrak{M}$ as $h \to 0$ is understood as the norm convergence, induced by scalar product. The symbol $\frac{\partial u}{\partial t}$ denotes such a vector. Its coordinate representation has the form $(\frac{\partial u^i}{\partial t})_{i=1}^{s}$.

The Lie derivative from a time-dependent tensor field $\mathcal{T}$ is the operation $\mathbb{L}$ that is defined by the formula [60]:

$$\mathbb{L}_u \mathcal{T}_t = \frac{d}{dt}\left\{F^*_{t;s}\mathcal{T}_t\right\}\bigg|_{t=s}. \tag{6.8}$$

The autonomous Lie derivative is related with $\mathbb{L}$ as follows [60]:

$$\mathbb{L}_u \mathcal{T}_t = \frac{\partial \mathcal{T}_t}{\partial t} + \mathcal{L}_u\mathcal{T}_t . \tag{6.9}$$

## 6.5 Motion as a Time-Dependent Flow

Suppose that the body and the physical space have the same dimensions. Consider a smooth motion $M = \{\varkappa_t\}_{t\in\mathbb{T}}$. It generates the family $\{\hat{\varkappa}_t\}_{t\in\mathbb{T}}$ of diffeomorphisms. Such a family, in turn, defines a two-parametric family $\{\psi_{t,s}\}_{t,s\in\mathbb{T}}$ of diffeomorphisms

$$\psi_{t,s} = \hat{\varkappa}_t \circ \hat{\varkappa}_s^{-1} : \mathcal{S}_{\varkappa_s} \to \mathcal{S}_{\varkappa_t},$$

which satisfies the following relations:

$$\begin{aligned} \forall t,s,q \in \mathbb{T}: &\quad \psi_{t,s}\circ\psi_{s,q} = \psi_{t,q}, \\ \forall t \in \mathbb{T}: &\quad \psi_{t,t} = \mathrm{Id}_{\mathcal{S}_{\varkappa_t}} . \end{aligned}$$

Thus, $\{\psi_{t,s}\}_{t,s\in\mathbb{T}}$ represents a time-dependent flow.

Recall that the spatial velocity field is the mapping

$$v : \bigcup_{t\in\mathbb{T}} (\mathcal{S}_{\varkappa_t} \times \{t\}) \to T\mathcal{P} .$$

Fix $s \in \mathbb{T}$ and $x \in \mathcal{S}_{\varkappa_s}$. Denote $\mathfrak{X} = \widehat{\varkappa}_s^{-1}(x)$. Consider the smooth curve

$$\gamma_{s;x} : \mathbb{T} \to \mathscr{P} , \quad \gamma_{s;x} : t \mapsto \psi_{t,s}(x) .$$

Such a curve coincides with the curve $\chi_{M;\mathfrak{X}}$. Thus, one obtains $\gamma'_{s;x}(t) = V(\mathfrak{X}, t)$ for all $t \in \mathbb{T}$, or,

$$\forall t \in \mathbb{T} : \quad \gamma'_{s;x}(t) = v(\gamma_{s;x}(t), t) ,$$

with the initial condition $\gamma_{s;x}(s) = x$. Thus, $\gamma_{s;x}$ is an integral curve of $v$.

# 7 Stress Measures

The conventional representations of stresses (Chapter 3) substantially use the Euclidean-affine structure of physical space, which contains both the reference and the actual shapes of the body. Below we list some characteristic features of this approach.

- The existence of the Cauchy stress tensor $\boldsymbol{T}$ is stated via the classical "Cauchy tetrahedron argument" that is based on the Euclidean structure.
- Integrals that represent the resulting contact forces essentially use the Euclidean parallel translation rule. Since contact force densities take values in $\mathscr{V}$, one can sum these values taken at different points. Such summation is required by the definition of the integration in Euclidean space.
- Let $S$ denote $\partial\varkappa_R(\mathfrak{P})$ or $\partial\varkappa(\mathfrak{P})$, where $\mathfrak{P} \in \operatorname{Part}(\mathfrak{B})$. For a point $x \in S$, one has the decomposition $T_xS \oplus (T_xS)^\perp = \mathscr{V}$. Such a direct sum involves the Euclidean structure of $\mathscr{V}$. Thus, $\boldsymbol{n}_x$ denotes the unit normal vector, the element of $(T_xS)^\perp$.

In this chapter, the notion of stresses is generalized on smooth manifolds. We follow the ideas set out in [124].

## 7.1 Concentrated Forces and Force Densities

**Concentrated force.** Let $M = \{\varkappa_t\}_{t\in\mathbb{T}}$ be a smooth motion and $\mathfrak{X} \in \mathfrak{B}$ be a material point. A *concentrated force*, which acts on the material point $\mathfrak{X}$ at instant $t \in \mathbb{T}$, is a covector $f_t \in T^*_{\varkappa_t(\mathfrak{X})}\mathscr{P}$. Suppose that a family $\{f_t\}_{t\in\mathbb{T}}$ of concentrated forces, which act on the material point $\mathfrak{X}$, corresponds to the given motion $M$. The motion $M$, in turn, generates the field $\{V_{M;t}\}_{t\in\mathbb{T}}$ of velocities $V_{M;t} \in \operatorname{Sec}(\varkappa_t^* T\mathscr{P})$. For each $t \in \mathbb{T}$, the scalar $f_t \llcorner V_{M;t}(\mathfrak{X})$ can be interpreted as the power that the force $f$ expends at instant $t$.

Finally, for each instant $t \in \mathbb{T}$, a concentrated force that acts on some material point at that instant is represented by an element of the pullback bundle $\varkappa_t^* T\mathscr{P}$.

**Force density.** As is usual in continuum mechanics, we assume that forces that act on the body $\mathfrak{B}$ are divided into two groups. The first one is represented by *body forces*, which are distributed over the volume of any part of the body. The second group is represented by *surface forces*, which are distributed over the boundary of any part. Thus, both of these groups consist of force densities, which act on the velocity to produce power densities (volume or surface). The latter are represented by ordinary differential forms and can be integrated (over the part of the body or its boundary) to obtain the resulting powers.

Suppose that $M = \{\varkappa_t\}_{t\in\mathbb{T}}$ is a smooth motion. Let $\mathfrak{P} \in \operatorname{Part}(\mathfrak{B})$ be a part of the body $\mathfrak{B}$. Recall that by a part we mean an $n$-dimensional embedded submanifold with boundary. Consider force densities in material and spatial descriptions.

https://doi.org/10.1515/9783110563214-007

A *body force density in material description* is given by a family $\{\beta^{\mathrm{mat}}_{\mathfrak{P};t}\}_{t\in\mathbb{T}}$ of sections

$$\forall t \in \mathbb{T}: \quad \beta^{\mathrm{mat}}_{\mathfrak{P};t} \in \mathrm{Sec}\left(\varkappa_t^* T^* \mathscr{P} \otimes \Lambda^n(T^*\mathfrak{P})\right) .$$

A *surface force density in material description* is a family $\{\tau^{\mathrm{mat}}_{\mathfrak{P};t}\}_{t\in\mathbb{T}}$ of sections

$$\forall t \in \mathbb{T}: \quad \tau^{\mathrm{mat}}_{\mathfrak{P};t} \in \mathrm{Sec}\left(\varkappa_t^* T^* \mathscr{P} \otimes \Lambda^{n-1}(T^*\partial\mathfrak{P})\right) .$$

We also postulate that in the material description, the resulting power expended by body and surface forces at instant $t \in \mathbb{T}$ is of the form

$$p_t^{\mathrm{mat}} := \int_{\mathfrak{P}} V_{M;t} \lrcorner \beta^{\mathrm{mat}}_{\mathfrak{P};t} + \int_{\partial\mathfrak{P}} V_{M;t} \lrcorner \tau^{\mathrm{mat}}_{\mathfrak{P};t} .$$

A *body force density in spatial description* is a family $\{\beta^{\mathrm{spat}}_{\mathfrak{P};t}\}_{t\in\mathbb{T}}$ of sections

$$\forall t \in \mathbb{T}: \quad \beta^{\mathrm{spat}}_{\mathfrak{P};t} \in \mathrm{Sec}\left(\iota^*_{\varkappa_t(\mathfrak{P})} T^* \mathscr{P} \otimes \Lambda^n\left(T^*\varkappa_t(\mathfrak{P})\right)\right) .$$

A *surface force density in spatial description* is a family $\{\tau^{\mathrm{spat}}_{\mathfrak{P};t}\}_{t\in\mathbb{T}}$ of sections

$$\forall t \in \mathbb{T}: \quad \tau^{\mathrm{spat}}_{\mathfrak{P};t} \in \mathrm{Sec}\left(\iota^*_{\varkappa_t(\mathfrak{P})} T^* \mathscr{P} \otimes \Lambda^{n-1}\left(T^*\partial\varkappa_t(\mathfrak{P})\right)\right) .$$

We postulate that in the spatial description, the resulting power expended by body and surface forces at instant $t \in \mathbb{T}$ is of the form

$$p_t^{\mathrm{spat}} := \int_{\varkappa_t(\mathfrak{P})} v_t \lrcorner \beta^{\mathrm{spat}}_{\mathfrak{P};t} + \int_{\partial\varkappa_t(\mathfrak{P})} v_t \lrcorner \tau^{\mathrm{spat}}_{\mathfrak{P};t} .$$

Usually, it is assumed that the value of a body force at a point is independent from any part that contains this point:

$$\forall t \in \mathbb{T}: \quad \beta_t^{\mathrm{mat}} \in \mathrm{Sec}\left(\varkappa_t^* T^* \mathscr{P} \otimes \Lambda^n\left(T^*\mathfrak{B}\right)\right) .$$
$$\forall t \in \mathbb{T}: \quad \beta_t^{\mathrm{spat}} \in \mathrm{Sec}\left(\iota^*_{\mathcal{S}_{\varkappa_t}} T^* \mathscr{P} \otimes \Lambda^n\left(T^*\mathcal{S}_{\varkappa_t}\right)\right) .$$

In this regard, factually, only a surface force depends on a part. The collection $\{(\beta_t^{\mathrm{mat}}, \tau^{\mathrm{mat}}_{\mathfrak{P};t})\}_{\mathfrak{P}\in\mathrm{Part}(\mathfrak{B}),t\in\mathbb{T}}$ represents a force system on $\mathfrak{B}$ in material description, and the collection $\{(\beta_t^{\mathrm{spat}}, \tau^{\mathrm{spat}}_{\mathfrak{P};t})\}_{\mathfrak{P}\in\mathrm{Part}(\mathfrak{B}),t\in\mathbb{T}}$ represents a force system on $\mathfrak{B}$ in spatial description. Thus, the total force $F_t$, which acts on the part $\mathfrak{P}$ at instant $t \in \mathbb{T}$, is a functional over velocities. In material and spatial descriptions, it is represented by the values

$$F_t^{\mathrm{mat}}(V) := \int_{\mathfrak{P}} V_{M;t} \lrcorner \beta_t^{\mathrm{mat}} + \int_{\partial\mathfrak{P}} V_{M;t} \lrcorner \tau^{\mathrm{mat}}_{\mathfrak{P};t} , \tag{7.1}$$

$$F_t^{\mathrm{spat}}(v) := \int_{\varkappa_t(\mathfrak{P})} v_t \lrcorner \beta_t^{\mathrm{spat}} + \int_{\partial\varkappa_t(\mathfrak{P})} v_t \lrcorner \tau^{\mathrm{spat}}_{\mathfrak{P};t} . \tag{7.2}$$

## 7.2 Inclined Hyperplanes

The notion of the stress tensor presupposes the possibility to choose a hyperplane (surface element) with orientation. For this purpose, in Riemannian-oriented manifolds one can use a unit normal vector field. At the same time, the Riemannian metric $\mathcal{G}$ on the body $\mathfrak{B}$ is obtained from certain physical reasonings. These reasonings may refer to the concept of stresses. In this regard, one needs a way to construct stresses on $\mathfrak{B}$ without appealing to the metric. Particularly it means that the notion of the oriented hyperplane needs to be introduced in the metric-independent way. The required possible definition of the oriented hyperplane is based on the notion of the inclined hyperplane [124].

Let $\mathfrak{X} \in \mathfrak{B}$. Consider the tangent space $T_{\mathfrak{X}}\mathfrak{B}$. By a *hyperplane* $H_{\mathfrak{X}}$ at the point $\mathfrak{X}$ we mean an $(n-1)$-dimensional vector subspace of $T_{\mathfrak{X}}\mathfrak{B}$. Denote the collection of all hyperplanes at $\mathfrak{X}$ by $G_{n-1}(T_{\mathfrak{X}}\mathfrak{B})$. The set $G_{n-1}(T_{\mathfrak{X}}\mathfrak{B})$ can be endowed by a smooth $(n-1)$-dimensional manifold structure [3, 125]. This manifold is called the *Grassmann manifold*. The union

$$G_{n-1}(T\mathfrak{B}) = \bigsqcup_{\mathfrak{X}\in\mathfrak{B}} G_{n-1}(T_{\mathfrak{X}}\mathfrak{B}),$$

is the *Grassmann bundle of hyperplanes*.

**Remark 7.1.** Let us briefly describe charts on $G_{n-1}(T_{\mathfrak{X}}\mathfrak{B})$ [3]. For any one-dimensional subspace $I \subset T_{\mathfrak{X}}\mathfrak{B}$, one can find a hyperplane $H_{\mathfrak{X}}$ such that $T_{\mathfrak{X}}\mathfrak{B} = H_{\mathfrak{X}} \oplus I$. Let $\mathcal{L} \in \mathrm{Lin}(H_{\mathfrak{X}}; I)$ be a linear map. Its graph, the set

$$\Gamma_{\mathcal{L}} = \{(u, \mathcal{L}(u)) \in H_{\mathfrak{X}} \times I \mid u \in H_{\mathfrak{X}}\},$$

can be identified with the following $(n-1)$-dimensional subspace of $T_{\mathfrak{X}}\mathfrak{B}$:

$$\Phi(\mathcal{L}) = \{u + \mathcal{L}(u) \in H_{\mathfrak{X}} \oplus I \mid u \in H_{\mathfrak{X}}\}.$$

Such an identification is provided due to the fact that $T_{\mathfrak{X}}\mathfrak{B}$ can be considered as a result for both external and internal direct sums. One has that $\Phi(\mathcal{L}) \cap I = \{0\}$. Conversely, let $\Delta \subset T_{\mathfrak{X}}\mathfrak{B}$ be an $(n-1)$-dimensional subspace such that $\Delta \cap I = \{0\}$. Denote by $\mathrm{pr}_1 : T_{\mathfrak{X}}\mathfrak{B} \to H_{\mathfrak{X}}$ and $\mathrm{pr}_2 : T_{\mathfrak{X}}\mathfrak{B} \to I$ the projections that are determined by external direct sum decomposition. The map $\mathrm{pr}_1|_\Delta : \Delta \to H_{\mathfrak{X}}$ is injective by virtue of the condition $\Delta \cap I = \{0\}$. Thus, $\mathrm{pr}_1|_\Delta$ is a vector space isomorphism. The mapping $\mathcal{L} := \mathrm{pr}_2|_\Delta \circ (\mathrm{pr}_1|_\Delta)^{-1} : H_{\mathfrak{X}} \to I$ is linear and $\Phi(\mathcal{L}) = \Delta$.

Let $U_I$ be a subset of $G_{n-1}(T_{\mathfrak{X}}\mathfrak{B})$, which consists of all $(n-1)$-dimensional subspaces whose intersections with $I$ consists of the zero vector only. The map

$$\Phi : \mathrm{Lin}(H_{\mathfrak{X}}; I) \to U_I, \quad u \mapsto \Phi(\mathcal{L}),$$

is a bijection. Denote $\varphi := \Phi^{-1} : U_I \to \mathrm{Lin}(H_{\mathfrak{X}}; I)$. The space $\mathrm{Lin}(H_{\mathfrak{X}}; I)$ can be identified with $\mathbb{R}^{n-1}$ by choosing bases for $H_{\mathfrak{X}}$ and $I$. Thus, the ordered pair $(U_I, \varphi)$, where $\varphi : U_I \to \mathbb{R}^{n-1}$, can be considered as a chart on $G_{n-1}(T_{\mathfrak{X}}\mathfrak{B})$.

Let $H_{\mathfrak{X}}$ be a hyperplane. Denote

$$\mathrm{Ann}(H_{\mathfrak{X}}) := \{\nu \in T^*_{\mathfrak{X}}\mathfrak{B} \mid \forall u \in H_{\mathfrak{X}} : \nu(u) = 0\} .$$

This is the vector space of all annihilators of $H_{\mathfrak{X}}$. Since $\dim \mathrm{Ann}(H_{\mathfrak{X}}) = 1$, one has

$$\forall \nu \in \mathrm{Ann}(H_{\mathfrak{X}}) : \ker \nu = \begin{cases} T_{\mathfrak{X}}\mathfrak{B}, & \nu = 0 , \\ H_{\mathfrak{X}}, & \nu \neq 0 . \end{cases}$$

Using the space $\mathrm{Ann}(H_{\mathfrak{X}})$ of annihilators one can introduce the notion of a *side* of $H_{\mathfrak{X}}$. Define the following equivalence relation $\sim_{H_{\mathfrak{X}}}$ on $T_{\mathfrak{X}}\mathfrak{B} \setminus H_{\mathfrak{X}}$:

$$\forall u_1, u_2 \in T_{\mathfrak{X}}\mathfrak{B} \setminus H_{\mathfrak{X}} : \quad (u_1 \sim_{H_{\mathfrak{X}}} u_2) \Leftrightarrow (\exists \nu \in \mathrm{Ann}(H_{\mathfrak{X}}) : \quad \nu(u_1)\nu(u_2) > 0) .$$

If $\nu(u_1)\nu(u_2) > 0$ for some $\nu \in \mathrm{Ann}(H_{\mathfrak{X}})$ then it follows $\nu \neq 0$. Any other non-zero $\tilde{\nu} \in \mathrm{Ann}(H_{\mathfrak{X}})$ can be expressed through $\nu$: $\tilde{\nu} = \alpha\nu$, for some $\alpha \in \mathbb{R}$. Thus, $\tilde{\nu}(u_1)\tilde{\nu}(u_2) > 0$. The relation $\sim_{H_{\mathfrak{X}}}$ is then well defined. If $u_1 \sim_{H_{\mathfrak{X}}} u_2$, then we say that *vectors $u_1$ and $u_2$ are on the same side of $H_{\mathfrak{X}}$*. If $u_1 \nsim_{H_{\mathfrak{X}}} u_2$, then we say that *vectors $u_1$ and $u_2$ are on the opposite sides of $H_{\mathfrak{X}}$*. Note that if $u \in T_{\mathfrak{X}}\mathfrak{B} \setminus H_{\mathfrak{X}}$, then vectors $u$ and $-u$ are on the opposite sides. Thus, the quotient set consists of two classes – two *sides* of $H_{\mathfrak{X}}$:

$$(T_{\mathfrak{X}}\mathfrak{B} \setminus H_{\mathfrak{X}})/ \sim_{H_{\mathfrak{X}}} = \{[u], [-u]\} .$$

An *inclined hyperplane* is any of the pairs $h_{\mathfrak{X}} = (H_{\mathfrak{X}}, [u])$ and $-h_{\mathfrak{X}} = (H_{\mathfrak{X}}, [-u])$. The collection of all inclined hyperplanes at $\mathfrak{X}$ is denoted by $G^{\perp}_{n-1}(T_{\mathfrak{X}}\mathfrak{B})$. One has the following natural map:

$$p_{\mathfrak{X}} : G^{\perp}_{n-1}(T_{\mathfrak{X}}\mathfrak{B}) \to G_{n-1}(T_{\mathfrak{X}}\mathfrak{B}) , \quad p_{\mathfrak{X}}(H_{\mathfrak{X}}, [u]) := H_{\mathfrak{X}} ,$$

which ignores the inclination.

Again, let $H_{\mathfrak{X}}$ be a hyperplane. Choosing a *positive inclination* means choosing exactly one of classes from $(T_{\mathfrak{X}}\mathfrak{B} \setminus H_{\mathfrak{X}})/ \sim_{H_{\mathfrak{X}}}$. A vector $u$ is *positively inclined* if it belongs to the equivalence class that corresponds to the positive inclination. Let $h_{\mathfrak{X}} = (H_{\mathfrak{X}}, [v])$ be the hyperplane $H_{\mathfrak{X}}$, together with chosen positive inclination. Define the set

$$\mathrm{Ann}^+(H_{\mathfrak{X}}) := \{\nu \in \mathrm{Ann}(H_{\mathfrak{X}}) \mid \forall u \in [v] : \nu(u) > 0\} .$$

Introduce the equivalence relation $\sim_{\mathfrak{X},+}$ on $T^*_{\mathfrak{X}}\mathfrak{B} \setminus \{0\}$:

$$\forall \nu_1, \nu_2 \in T^*_{\mathfrak{X}}\mathfrak{B} \setminus \{0\} : (\nu_1 \sim_{\mathfrak{X},+} \nu_2) \Leftrightarrow (\exists a > 0 : \nu_2 = a\nu_1) .$$

By the definition, each pair of elements from $\mathrm{Ann}^+(H_{\mathfrak{X}})$ is $\sim_{\mathfrak{X},+}$-related. This implies $\mathrm{Ann}^+(H_{\mathfrak{X}}) \in (T^*_{\mathfrak{X}}\mathfrak{B} \setminus \{0\})/ \sim_{\mathfrak{X},+}$. Since any non-zero covector $\nu \in T^*_{\mathfrak{X}}\mathfrak{B}$ defines the unique hyperplane, $\ker \nu$, one can identify spaces $G^{\perp}_{n-1}(T_{\mathfrak{X}}\mathfrak{B})$ and $(T^*_{\mathfrak{X}}\mathfrak{B} \setminus \{0\})/ \sim_{\mathfrak{X},+}$.

Choosing any inner product[1] on $T_{\mathfrak{X}}\mathfrak{B}$, one can identify $\mathrm{Ann}^+(H_{\mathfrak{X}})$ with a unit normal vector to $H_{\mathfrak{X}}$. Thus, $G^{\perp}_{n-1}(T_{\mathfrak{X}}\mathfrak{B})$ can be given smooth structure of $(n-1)$-sphere. Denote

$$G^{\perp}_{n-1}(T\mathfrak{B}) = \bigsqcup_{\mathfrak{X}\in\mathfrak{B}} G^{\perp}_{n-1}(T_{\mathfrak{X}}\mathfrak{B})\,.$$

One has the bundle $\pi\colon G^{\perp}_{n-1}(T\mathfrak{B}) \to \mathfrak{B}$ and the bundle morphism

$$p\colon G^{\perp}_{n-1}(T\mathfrak{B}) \to G_{n-1}(T\mathfrak{B})\,.$$

Let $\mathfrak{P} \in \mathrm{Part}(\mathfrak{B})$. Choose a point $\mathfrak{X} \in \partial\mathfrak{P}$. Consider the set $\mathfrak{Cu}_{\mathfrak{X};\mathfrak{P}}$ of all possible smooth curves $\chi\colon [0,\varepsilon[\to \mathfrak{B}$, $\varepsilon > 0$ that satisfy the conditions:

(i) $\chi(0) = \mathfrak{X}$;

(ii) $\chi(]0,\varepsilon[) \subset \mathfrak{B}\setminus\mathfrak{P}$.

In other words, all curves from the set $\mathfrak{Cu}_{\mathfrak{X};\mathfrak{P}}$ pass through $\mathfrak{X}$ and lie outside $\mathfrak{P}$. Each vector from[2] $T_{\mathfrak{X}}\mathfrak{B} \setminus T_{\mathfrak{X}}\partial\mathfrak{P}$ that is tangent to some curve $\chi$ from $\mathfrak{Cu}_{\mathfrak{X};\mathfrak{P}}$ at $\mathfrak{X}$, can be declared outward pointing (relatively to $\mathfrak{P}$). If $u$ is an outward pointing vector, then choose the inclined hyperplane $h_{\mathfrak{X}} := (T_{\mathfrak{X}}\partial\mathfrak{P}, [u])$. Repeating this procedure for all $\mathfrak{X} \in \partial\mathfrak{P}$ results in the desired collection $\{h_{\mathfrak{X}}\}_{\mathfrak{X}\in\partial\mathfrak{P}}$ of "material outwardly oriented area elements".

## 7.3 Piola Section on Inclined Hyperplanes

Again, we suppose that no Riemannian metric on the body $\mathfrak{B}$ is defined. Let $M = \{\varkappa_t\}_{t\in\mathbb{T}}$ be a smooth motion. The introduced notion of the inclined hyperplane allows one to give an exact meaning for the Cauchy postulate.

**Interaction density bundle.** Consider the vector bundle

$$\pi\colon \varkappa_t^* T^*\mathscr{P} \otimes \Lambda^{n-1}\left(G^*_{n-1}(T\mathfrak{B})\right) \to G_{n-1}(T\mathfrak{B})\,,$$

whose fiber is $T^*_{\varkappa_t(\mathfrak{X})}\mathscr{P} \otimes \Lambda^{n-1}(H^*_{\mathfrak{X}})$. The pullback of this bundle by means of the morphism $p\colon G^{\perp}_{n-1}(T\mathfrak{B}) \to G_{n-1}(T\mathfrak{B})$ gives

$$\tilde{\pi}\colon p^*\left(\varkappa_t^* T^*\mathscr{P} \otimes \Lambda^{n-1}\left(G^*_{n-1}(T\mathfrak{B})\right)\right) \to G^{\perp}_{n-1}(T\mathfrak{B})\,.$$

Following [124], we refer to this bundle as the *interaction density bundle*. All is ready for the mathematical formalization of the Cauchy postulate.

---

1 This inner product plays a technical role and does not require the metric on $\mathfrak{B}$.

2 Here we consider $T_{\mathfrak{X}}\partial\mathfrak{P}$ as an $(n-1)$-dimensional hyperplane of $T_{\mathfrak{X}}\mathfrak{B}$.

**The Cauchy postulate.** For each instant $t \in \mathbb{T}$, there exists a section

$$\tau_t^P : G_{n-1}^{\perp}(T\mathfrak{B}) \to p^* \left( \varkappa_t^* T^* \mathscr{P} \otimes \Lambda^{n-1} \left( G_{n-1}^*(T\mathfrak{B}) \right) \right) ,$$

called the *Piola section*[3], such that for each part $\mathfrak{P} \in \mathrm{Part}(\mathfrak{B})$ and the corresponding family $\{h_{\mathfrak{X}}\}_{\mathfrak{X} \in \partial\mathfrak{P}}$ of inclined hyperplanes for $\partial\mathfrak{P}$, one has

$$\forall \mathfrak{X} \in \partial\mathfrak{P} : \quad \tau_{\mathfrak{P};t}^{\mathrm{mat}}(\mathfrak{X}) = \tau_t^P(h_{\mathfrak{X}}) .$$

**The Piola stress form.** Our next purpose is to relate $\tau_{\mathfrak{P};t}^{\mathrm{mat}}$, $t \in \mathbb{T}$, with a covector-valued *Piola stress form*

$$P_t \in \mathrm{Sec}\left( \varkappa_t^* T^* \mathscr{P} \otimes \Lambda^{n-1}(T^*\mathfrak{B}) \right) , \tag{7.3}$$

defined on $\mathfrak{B}$. To this end, one can use the notion of *inclined restriction* [124]. For any $\mathfrak{X} \in \mathfrak{B}$, let $h_{\mathfrak{X}} = (H_{\mathfrak{X}}, [v])$ be some inclined hyperplane. The inclusion mapping $\iota_{H_{\mathfrak{X}}} : H_{\mathfrak{X}} \to T_{\mathfrak{X}}\mathfrak{B}$ generates the mapping

$$I^* : T^*_{\varkappa_t(\mathfrak{X})}\mathscr{P} \otimes \Lambda^{n-1}\left(T^*_{\mathfrak{X}}\mathfrak{B}\right) \to T^*_{\varkappa_t(\mathfrak{X})}\mathscr{P} \otimes \Lambda^{n-1}\left(H^*_{\mathfrak{X}}\right) ,$$

such that

$$I^*(\omega)(u_1, \ldots, u_{n-1}) := \omega(\iota_{H_{\mathfrak{X}}} u_1, \ldots, \iota_{H_{\mathfrak{X}}} u_{n-1}) .$$

The *inclined restriction* of $P_t(\mathfrak{X})$, the element

$$\iota^*_{h_{\mathfrak{X}}}(P_t(\mathfrak{X})) \in T^*_{\varkappa_t(\mathfrak{X})}\mathscr{P} \otimes \Lambda^{n-1}\left(H^*_{\mathfrak{X}}\right) ,$$

is defined as follows:

$$\iota^*_{h_{\mathfrak{X}}}(P_t(\mathfrak{X}))(u_1, \ldots, u_{n-1}) := I^*(P_t(\mathfrak{X}))(u_1, \ldots, u_{n-1}) ,$$

if $(v, u_1, \ldots, u_{n-1})$ are positively oriented, and

$$\iota^*_{h_{\mathfrak{X}}}(P_t(\mathfrak{X}))(u_1, \ldots, u_{n-1}) := -I^*(P_t(\mathfrak{X}))(u_1, \ldots, u_{n-1}) ,$$

if $(v, u_1, \ldots, u_{n-1})$ are negatively oriented. This definition implies the Newton's III law

$$\iota^*_{-h_{\mathfrak{X}}}(P_t(\mathfrak{X})) = -\iota^*_{h_{\mathfrak{X}}}(P_t(\mathfrak{X})) .$$

The layerwise mapping $h \mapsto \iota_h^*(P_t)$ plays the role of a Piola section. Let $\mathfrak{P} \in \mathrm{Part}(\mathfrak{B})$ and let $\{h_{\mathfrak{X}}\}_{\mathfrak{X} \in \partial\mathfrak{P}}$ be a collection of the corresponding inclined hyperplanes. One arrives at the desired relation [124]

$$\tau_{\mathfrak{P};t}^{\mathrm{mat}} = \iota_h^*(P_t) ,$$

or, pointwise,

$$\forall \mathfrak{X} \in \partial\mathfrak{P} : \quad \tau_{\mathfrak{P};t}^{\mathrm{mat}}(\mathfrak{X}) = \iota^*_{h_{\mathfrak{X}}}(P_t(\mathfrak{X}))$$

This is the generalization of the relation $\boldsymbol{\tau}_{\mathfrak{P};t}^{\mathrm{mat}} = \boldsymbol{P}_t \boldsymbol{N}$ in Euclidean mechanics.

3 In [124], the section $\tau_t$ is called the Cauchy section. Since we are dealing with a material description, the term "Piola section" is more preferable.

## 7.4 The Cauchy Section

One needs to consider shapes of the body $\mathfrak{B}$ and its parts with the aim of introducing the notion of the Cauchy stress form. In contrast to the case of the Piola stress form, one can use the additional data given by the Riemannian metric $g$ of the physical space $\mathscr{P}$. In this book, we use the approach of inclined hyperplanes for both the material and the spatial description. The reduction for the metric case can be found in [124].

**Inclined hyperplanes for shapes.** Let $M = \{\varkappa_t\}_{t\in\mathbb{T}}$ be a smooth motion. Fix $t \in \mathbb{T}$ and consider the shape $\mathcal{S}_{\varkappa_t}$ of the body $\mathfrak{B}$. If $x \in \mathcal{S}_{\varkappa_t}$, then the collection of all $(n-1)$-dimensional vector subspaces $H_x \subset T_x\mathcal{S}_{\varkappa_t}$ (hyperplanes at $x$) form the Grassmann manifold $G_{n-1}(T_x\mathcal{S}_{\varkappa_t})$. Thus, the Grassmann bundle of hyperplanes is $G_{n-1}(\mathcal{S}_{\varkappa_t}) = \bigsqcup_{x\in\mathcal{S}_{\varkappa_t}} G_{n-1}(T_x\mathcal{S}_{\varkappa_t})$.

Since the physical space $\mathscr{P}$ is equipped with metric $g$, one can pull back this metric on $\mathcal{S}_{\varkappa_t}$: $g^{(\mathcal{S}_{\varkappa_t})} = \iota^*_{\mathcal{S}_{\varkappa_t}} g$. Let $x \in \mathcal{S}_{\varkappa_t}$. For any hyperplane $H_x \subset T_x\mathcal{S}_{\varkappa_t}$ one can write the decomposition $T_x\mathcal{S}_{\varkappa_t} = H_x \oplus H_x^\perp$, where $H_x^\perp$ is the one-dimensional vector space, such that

$$H_x^\perp := \left\{u \in T_x\mathcal{S}_{\varkappa_t} \mid \forall v \in H_x:\quad g^{(\mathcal{S}_{\varkappa_t})}(v,u) = 0\right\} .$$

Choose some vector $n \in H_x^\perp$. By the definition, the covector $n^\flat := g_x^{(\mathcal{S}_{\varkappa_t})}(n,\cdot)$ annihilates $H_x$; $\ker n^\flat = H_x$. This implies $n^\flat \in \mathrm{Ann}(H_{\mathfrak{x}})$. Thus, the covector $n^\flat$ may be used to test vectors $u_1, u_2 \in T_x\mathcal{S}_{\varkappa_t} \setminus H_x$ for being on the same side of $H_x$ $(n^\flat(u_1)n^\flat(u_2) > 0)$ or being on the opposite sides of $H_x$ $(n^\flat(u_1)n^\flat(u_2) < 0)$. The equivalence relation $\sim_{H_x}$ on $T_x\mathcal{S}_{\varkappa_t} \setminus H_x$ can be introduced in a specific way:

$$\forall u_1, u_2 \in T_x\mathcal{S}_{\varkappa_t} \setminus H_x:\quad (u_1 \sim_{H_x} u_2) \Leftrightarrow \left(n^\flat(u_1)n^\flat(u_2) > 0\right) .$$

The quotient set consists of two classes – two *sides* of $H_x$:

$$(T_x\mathcal{S}_{\varkappa_t} \setminus H_x)/\sim_{H_x} = \{[u], [-u]\} .$$

An *inclined hyperplane* is any of the pairs $h_x = (H_x, [u])$ and $-h_x = (H_x, [-u])$. One arrives at the collection $G_{n-1}^\perp(T_x\mathcal{S}_{\varkappa_t})$ of all inclined hyperplanes at $x \in \mathcal{S}_{\varkappa_t}$ and defines the bundle

$$G_{n-1}^\perp(T\mathcal{S}_{\varkappa_t}) = \bigsqcup_{x\in\mathcal{S}_{\varkappa_t}} G_{n-1}^\perp(T_x\mathcal{S}_{\varkappa_t}) .$$

**The interaction density bundle and the boundary inclination.** We have the vector bundle

$$\pi\colon \iota^*_{\mathcal{S}_{\varkappa_t}} T^*\mathscr{P} \otimes \Lambda^{n-1}\left(G_{n-1}^*(T\mathcal{S}_{\varkappa_t})\right) \to G_{n-1}(T\mathcal{S}_{\varkappa_t}) ,$$

whose fiber is $T_x^*\mathscr{P} \otimes \Lambda^{n-1}(H_x^*)$. Introducing the morphism

$$p\colon G_{n-1}^\perp(T\mathcal{S}_{\varkappa_t}) \to G_{n-1}(T\mathcal{S}_{\varkappa_t}) ,$$

which layerwisely sends $(H_X, [u])$ to $H_X$, one arrives at the interaction density bundle

$$\tilde{\pi}\colon p^*\left(\iota^*_{\mathcal{S}_{\varkappa_t}} T^*\mathscr{P} \otimes \Lambda^{n-1}\left(G^*_{n-1}(T\mathcal{S}_{\varkappa_t}) \to G^{\perp}_{n-1}(T\mathcal{S}_{\varkappa_t}\right)\,.$$

Let $t \in \mathbb{T}$ and $\mathfrak{P} \in \mathrm{Part}(\mathfrak{B})$. Choose a point $x \in \partial\varkappa_t(\mathfrak{P})$. As in material case, consider the set $\mathfrak{Cu}_{x;\varkappa_t(\mathfrak{P})}$ of all possible smooth curves $\chi\colon [0,\varepsilon[\to \mathcal{S}_{\varkappa_t}$, $\varepsilon > 0$, which satisfy the conditions:
(i) $\chi(0) = x$;
(ii) $\chi(]0,\varepsilon[) \subset \mathcal{S}_{\varkappa_t} \setminus \varkappa_t(\mathfrak{P})$.

Each vector from $T_x\mathcal{S}_{\varkappa_t} \setminus T_x\partial\varkappa_t(\mathfrak{P})$ that is tangent to some curve $\chi$ from $\mathfrak{Cu}_{x;\varkappa_t(\mathfrak{P})}$ at $x$, can be declared outward pointing (relatively to $\varkappa_t(\mathfrak{P})$). If $u$ is an outward pointing vector, then choose the inclined hyperplane $h_{t,x} := (T_x\partial\varkappa_t(\mathfrak{P}), [u])$. Repeating this procedure for all $x \in \partial\varkappa_t(\mathfrak{P})$ results in the desired collection $\{h_{t,x}\}_{x\in\partial\varkappa_t(\mathfrak{P})}$ of "spatial outwardly-oriented area elements".

**The Cauchy postulate.** For each instant $t \in \mathbb{T}$, there exists a section

$$\tau^C_t\colon G^{\perp}_{n-1}(T\mathcal{S}_{\varkappa_t}) \to p^*\left(\iota^*_{\mathcal{S}_{\varkappa_t}} T^*\mathscr{P} \otimes \Lambda^{n-1}\left(G^*_{n-1}(T\mathcal{S}_{\varkappa_t})\right)\right)\,,$$

called the *Cauchy section*, such that for each part $\mathfrak{P} \in \mathrm{Part}(\mathfrak{B})$ and the corresponding family $\{h_{t,x}\}_{x\in\mathcal{S}_{\varkappa_t}}$ of inclined hyperplanes built for boundary $\partial\varkappa_t(\mathfrak{P})$, one has

$$\forall x \in \partial\varkappa_t(\mathfrak{P})\colon \quad \tau^{\mathrm{spat}}_{\mathfrak{P};t}(x) = \tau^C_t(h_{t,x})\,.$$

**The Cauchy stress form.** In a similar manner to the material description, one can relate $\tau^{\mathrm{spat}}_{\mathfrak{P};t}$, $t \in \mathbb{T}$, with a covector-valued *Cauchy stress form*

$$T_t \in \mathrm{Sec}\left(\iota^*_{\mathcal{S}_{\varkappa_t}} T^*\mathscr{P} \otimes \Lambda^{n-1}\left(T^*\mathcal{S}_{\varkappa_t}\right)\right)\,, \tag{7.4}$$

defined on $\mathcal{S}_{\varkappa_t}$. Using the notion of inclined restriction one arrives at

$$\tau^{\mathrm{spat}}_{\mathfrak{P};t} = \iota^*_{h_t}(T_t)\,,$$

or, pointwise,

$$\forall x \in \partial\varkappa_t(\mathfrak{P})\colon \tau^{\mathrm{spat}}_{\mathfrak{P};t}(x) = \iota^*_{h_{t,x}}(T_t(x))$$

This is the generalization of the relation $\tau^{\mathrm{spat}}_{\mathfrak{P};t} = \boldsymbol{T}_t\boldsymbol{n}$ in Euclidean mechanics.

## 7.5 Cauchy and Piola Stresses

Let $M = \{\varkappa_t\}_{t\in\mathbb{T}}$ be a smooth motion and

$$\{(\beta^{\mathrm{mat}}_t, \tau^{\mathrm{mat}}_{\mathfrak{P};t})\}_{\mathfrak{P}\in\mathrm{Part}(\mathfrak{B}),t\in\mathbb{T}}\,, \quad \{(\beta^{\mathrm{spat}}_t, \tau^{\mathrm{spat}}_{\mathfrak{P};t})\}_{\mathfrak{P}\in\mathrm{Part}(\mathfrak{B}),t\in\mathbb{T}}$$

be force systems. The Piola stress form (7.3) acts on the body $\mathfrak{B}$ and the Cauchy stress form (7.4) acts on the shapes $\mathcal{S}_{\varkappa_t}$, $t \in \mathbb{T}$.

If $\mathfrak{P} \in \operatorname{Part}(\mathfrak{B})$, then the powers (7.1) and (7.2) take the form

$$F_t^{\mathrm{mat}}(V) = \int\limits_{\mathfrak{P}} V_{M;t} \lrcorner \beta_t^{\mathrm{mat}} + \int\limits_{\partial\mathfrak{P}} V_{M;t} \lrcorner \iota_h^*(P_t)\,,$$

$$F_t^{\mathrm{spat}}(v) = \int\limits_{\varkappa_t(\mathfrak{P})} v_t \lrcorner \beta_t^{\mathrm{spat}} + \int\limits_{\partial\varkappa_t(\mathfrak{P})} v_t \lrcorner \iota_{h_t}^*(T_t)\,.$$

In the following, we omit inclined restrictions and simply write

$$\begin{aligned} F_t^{\mathrm{mat}}(V) &= \int\limits_{\mathfrak{P}} V_{M;t} \lrcorner \beta_t^{\mathrm{mat}} + \int\limits_{\partial\mathfrak{P}} V_{M;t} \lrcorner P_t\,,\\ F_t^{\mathrm{spat}}(v) &= \int\limits_{\varkappa_t(\mathfrak{P})} v_t \lrcorner \beta_t^{\mathrm{spat}} + \int\limits_{\partial\varkappa_t(\mathfrak{P})} v_t \lrcorner T_t\,. \end{aligned} \tag{7.5}$$

Stress forms can be considered as images of Hodge stars applied to fields of linear mappings. Choose a particular element $\varkappa \in M$. Sections

$$\mathcal{t} \in \operatorname{Sec}\left(\iota_{\mathcal{S}_\varkappa}^* T^*\mathcal{P} \otimes T^*\mathcal{S}_\varkappa\right) \quad \text{and} \quad \mathcal{p} \in \operatorname{Sec}(\varkappa^* T^*\mathcal{P} \otimes T^*\mathfrak{B})\,.$$

correspond to the Cauchy and first Piola–Kirchhoff stress forms as follows [126]:

$$T \in \operatorname{Sec}\left(\iota_{\mathcal{S}_\varkappa}^* T^*\mathcal{P} \otimes \Lambda^{n-1}\left(T^*\mathcal{S}_\varkappa\right)\right)\,, \quad T = *_2 \mathcal{t}\,, \tag{7.6}$$

$$P \in \operatorname{Sec}\left(\varkappa^* T^*\mathcal{P} \otimes \Lambda^{n-1}\left(T^*\mathfrak{B}\right)\right)\,, \quad P = *_2 \mathcal{p}\,, \tag{7.7}$$

where $*_2$ is the Hodge star operation applied to the "second leg". In (7.6), the Hodge star is associated with the induced metric on the shape $\mathcal{S}_\varkappa$, and in (7.7), the Hodge star is associated with the material metric on $\mathfrak{B}$.

Let $\mathcal{t} = t_{a\beta} e^a \otimes \tilde{e}^\beta$, $\mathcal{p} = p_{a\beta} e^a \otimes E^\beta$, where $(E_\alpha)_{\alpha=1}^n \subset \operatorname{Vec}(\mathfrak{B})$ and $(\tilde{e}_\alpha)_{\alpha=1}^n \subset \operatorname{Vec}(\mathcal{S}_\varkappa)$ are frames on the body and the shape, respectively, and $(e_i)_{i=1}^m \subset \operatorname{Vec}(\mathcal{P})$ is a frame on $\mathcal{P}$. Then in coordinates one has

$$T = t_{a\beta} e^a \otimes (*\tilde{e}^\beta)\,, \quad P = p_{a\beta} e^a \otimes (*E^\beta)\,,$$

Since [123]

$$*E^\beta = \sqrt{G} \sum_{\gamma=1}^{n} (-1)^{\gamma-1} G^{\beta\gamma} E^1 \wedge \cdots \wedge \widehat{E^\gamma} \wedge \cdots \wedge E^n\,,$$

$$*\tilde{e}^\beta = \sqrt{g} \sum_{\alpha=1}^{n} (-1)^{\alpha-1} g^{\beta\alpha} \tilde{e}^1 \wedge \cdots \wedge \widehat{\tilde{e}^\alpha} \wedge \cdots \wedge \tilde{e}^n\,,$$

we arrive at the following representations:

$$T = \sum_{\alpha=1}^{n} (-1)^{\alpha-1}\, g^{\beta\alpha}\, t_{a\beta}\, \sqrt{g}\, e^a \otimes \left(\tilde{e}^1 \wedge \cdots \wedge \widehat{\tilde{e}^{\alpha}} \wedge \cdots \wedge \tilde{e}^n\right) , \tag{7.8}$$

$$P = \sum_{\gamma=1}^{n} (-1)^{\gamma-1}\, G^{\beta\gamma}\, p_{a\beta}\, \sqrt{G}\, e^a \otimes \left(E^1 \wedge \cdots \wedge \widehat{E^{\gamma}} \wedge \cdots \wedge E^n\right) . \tag{7.9}$$

Note that the “exterior” part of the Cauchy stress form, the element of $\Omega^{n-1}(T^*\mathcal{S}_\varkappa)$, is a differential form on the shape $\mathcal{S}_\varkappa$. In other words, it is an element of internal geometry of the shape. If $n = \dim \mathfrak{B} < m = \dim \mathscr{P}$, then one does not identify spaces $T_x\mathcal{S}_\varkappa$ and $T_x\mathscr{P}$. The additional data is required to transform $\Omega^{n-1}(T^*\mathcal{S}_\varkappa)$ into $\Omega^{m-1}(\iota^*_{\mathcal{S}_\varkappa} T^*\mathscr{P})$. If $m = n$, then the spaces $\Omega^{n-1}(T^*\mathcal{S}_\varkappa)$ and $\Omega^{m-1}(\iota^*_{\mathcal{S}_\varkappa} T^*\mathscr{P})$ can be identified. In this situation, one is able to change the frame $(\tilde{e}_\alpha)_{\alpha=1}^n$ into $(e_i)_{i=1}^m$.

## 7.6 Transformation from Spatial Description to Material

**Piola transformation.** The Piola and Cauchy stress forms were introduced individually and appear in expressions for powers $p_t^{\mathrm{mat}}$ and $p_t^{\mathrm{spat}}$. These powers are realizations of the total force in material and spatial descriptions. In this regard, they represent one and the same quantity and can differ by change of variables only. Let $M = \{\varkappa_t\}_{t\in\mathbb{T}}$ be a smooth motion and the collections $\{\tau^{\mathrm{mat}}_{\mathfrak{P};t}\}_{\mathfrak{P}\in\mathrm{Part}(\mathfrak{B}),t\in\mathbb{T}}$, $\{\tau^{\mathrm{spat}}_{\mathfrak{P};t}\}_{\mathfrak{P}\in\mathrm{Part}(\mathfrak{B}),t\in\mathbb{T}}$, be force systems. The body forces are absent. Fix $t \in \mathbb{T}$. One can require that

$$\int_{\partial\mathfrak{P}} V_{M;t} \lrcorner P_t = \int_{\partial\varkappa_t(\mathfrak{P})} v_t \lrcorner T_t \,.$$

Recall that $v_t = V_{M;t} \circ \hat{\varkappa}_t^{-1}$. Assume that $\hat{\varkappa}_t$ is an orientation-preserving diffeomorphism[4] for each instant $t \in \mathbb{T}$. The change of variables theorem gives

$$\int_{\partial\mathfrak{P}} V_{M;t} \lrcorner P_t = \int_{\partial\mathfrak{P}} \hat{\varkappa}_t^*(v_t \lrcorner T_t) \,.$$

Let $(e_i)_{i=1}^m \subset \mathrm{Vec}(\mathscr{P})$ be a frame on $\mathscr{P}$. Then, one has decompositions:

$$v_t = v_t^i e_i \,, \quad T_t = T_{i\alpha} e^i \otimes \omega_t^\alpha \,,$$

where $\omega_t^\alpha \in \Omega^{n-1}(T^*\mathcal{S}_{\varkappa_t})$. One has $v_t \lrcorner T_t = v_t^i T_{i\alpha} \omega_t^\alpha$. Thus,

$$\hat{\varkappa}_t^*(v_t \lrcorner T_t) = \left(v_t^i \circ \hat{\varkappa}_t\right)(T_{i\alpha} \circ \hat{\varkappa}_t)\omega_t^\alpha = \left(v_t^i \circ \hat{\varkappa}_t\right)(e_i \circ \hat{\varkappa}_t) \lrcorner (T_{j\alpha} \circ \hat{\varkappa}_t)\left(e^j \circ \hat{\varkappa}_t\right) \otimes \hat{\varkappa}_t^*\omega_t^\alpha \,.$$

4 A diffeomorphism is called orientation-preserving if its tangent map takes oriented bases of the domain to oriented bases of the codomain [3].

Introduce the operation that performs a pullback of the "geometrical part" of the covector-valued form:

$$\begin{aligned} \widehat{\varkappa}_t^{*2} :\ & \mathrm{Sec}\left(\iota^*_{\mathcal{S}_{\varkappa_t}} T^*\mathscr{P} \otimes \Lambda^{n-1}\left(T^*\mathcal{S}_{\varkappa_t}\right)\right) \to \mathrm{Sec}\left(\varkappa_t^* T^*\mathscr{P} \otimes \Lambda^{n-1}(T^*\mathfrak{B})\right) , \\ & \widehat{\varkappa}_t^{*2}(\vartheta \otimes \omega) := (\vartheta \circ \widehat{\varkappa}_t) \otimes \widehat{\varkappa}_t^* \omega . \end{aligned} \tag{7.10}$$

We refer to $\widehat{\varkappa}_t^{*2}$ as the *partial pullback*. Here, $\vartheta \circ \widehat{\varkappa}_t$ is the composition of the mapping $\widehat{\varkappa}_t$ and the section $\vartheta$ of the cotangent bundle $\iota^*_{\mathcal{S}_{\varkappa_t}} T^*\mathscr{P}$. Using the operation (7.10) and the relation $V_{M;t} = (v_t^i \circ \widehat{\varkappa}_t)(e_i \circ \widehat{\varkappa}_t)$ gives

$$\widehat{\varkappa}_t^* (v_t \lrcorner T_t) = V_{M;t} \lrcorner \widehat{\varkappa}_t^{*2}(T_t) .$$

Finally,

$$\int\limits_{\partial\mathfrak{P}} V_{M;t} \lrcorner P_t = \int\limits_{\partial\mathfrak{P}} V_{M;t} \lrcorner \widehat{\varkappa}_t^{*2}(T_t) .$$

Since $\mathfrak{P}$ and $V_{M;t}$ can be chosen arbitrarily, one obtains

$$P_t = \widehat{\varkappa}_t^{*2} T_t . \tag{7.11}$$

This relation corresponds to the Piola transformation. The relations between stresses for the case when the body and physical space have Euclidean structure and $\dim \mathfrak{B} = \dim \mathscr{P} = 3$ are illustrated in the diagram

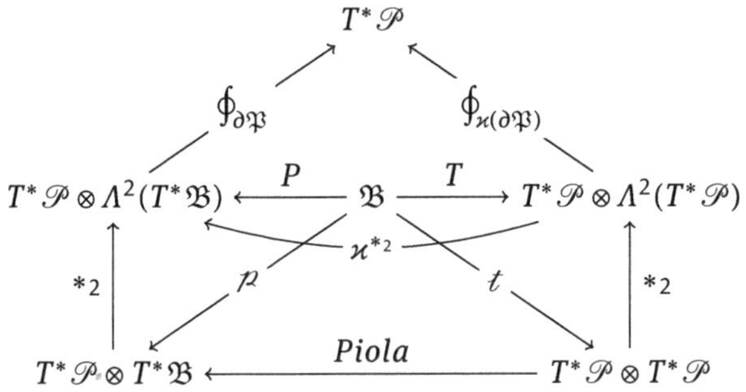


The diagram schematically illustrates the relations $P = *_2 \mathscr{p}$ and $T = *_2 \mathscr{t}$. Note that all mappings act from $\mathfrak{B}$. The geometrical sense for $*_2$ is shown. Such an operator transforms $T^*\mathscr{X} \otimes T^*\mathscr{Y}$ into $T^*\mathscr{X} \otimes \Lambda^2(T^*\mathscr{Y})$. For economy, one symbol $*_2$ is used here; but actually such mappings for $\mathfrak{B}$ and $\mathscr{P}$ are different. The mapping $\varkappa^{*2}$, as it is shown in the diagram, performs a pullback of the such part of the object that corresponds to the exterior 2-form in the physical space. This corresponds to the classical Piola transformation. If both the body and the physical space have Euclidean structure, then the integral of covector-valued form from $T^*\mathscr{X} \otimes \Lambda^2(T^*\mathscr{Y})$ is no more than the surface integral of covector density. The result of integration is a resulting contact force covector from $T^*\mathscr{X}$, which is shown in the diagram.

**Body force transformation.** Let $M = \{\varkappa_t\}_{t\in\mathbb{T}}$ be a smooth motion and the collections $\{\beta_t^{\mathrm{mat}}\}_{t\in\mathbb{T}}$, $\{\beta_t^{\mathrm{spat}}\}_{t\in\mathbb{T}}$, be force systems. The surface forces are absent. Fix $t \in \mathbb{T}$. One can require that

$$\int_{\mathfrak{P}} V_{M;t} \lrcorner \beta_t^{\mathrm{mat}} = \int_{\varkappa_t(\mathfrak{P})} v_t \lrcorner \beta_t^{\mathrm{spat}} .$$

Assuming again that $\widehat{\varkappa}_t$ is an orientation-preserving diffeomorphism for each instant $t \in \mathbb{T}$ and using the change of variables theorem, one arrives at

$$\int_{\mathfrak{P}} V_{M;t} \lrcorner \beta_t^{\mathrm{mat}} = \int_{\mathfrak{P}} \widehat{\varkappa}_t^* \left(v_t \lrcorner \beta_t^{\mathrm{spat}}\right) .$$

We will use reasonings similar to Piola transformation. Let $(e_i)_{i=1}^m \subset \mathrm{Vec}(\mathscr{P})$ be a frame on $\mathscr{P}$. Then one has decompositions:

$$v_t = v_t^i e_i , \quad \beta_t^{\mathrm{spat}} = e^i \otimes \eta_{t;i} ,$$

where $\eta_{t;i} \in \Omega^n(T^* \mathscr{S}_{\varkappa_t})$. Thus,

$$\widehat{\varkappa}_t^* \left(v_t \lrcorner \beta_t^{\mathrm{spat}}\right) = \left(v_t^i \circ \widehat{\varkappa}_t\right) (e_i \circ \widehat{\varkappa}_t) \lrcorner \left(e^j \circ \widehat{\varkappa}_t\right) \otimes \widehat{\varkappa}_t^* \eta_{t;j} .$$

Introduce the operation that performs a pullback of the "geometrical part":

$$\begin{gathered} \widehat{\varkappa}_t^{*2} : \mathrm{Sec}\left(\iota_{\mathscr{S}_{\varkappa_t}}^* T^* \mathscr{P} \otimes \Lambda^n \left(T^* \mathscr{S}_{\varkappa_t}\right)\right) \to \mathrm{Sec}\left(\varkappa_t^* T^* \mathscr{P} \otimes \Lambda^n (T^* \mathfrak{B})\right) , \\ \widehat{\varkappa}_t^{*2} (\vartheta \otimes \omega) := (\vartheta \circ \widehat{\varkappa}_t) \otimes \widehat{\varkappa}_t^* \omega . \end{gathered} \tag{7.12}$$

Using the operation (7.12) results in

$$\widehat{\varkappa}_t^* \left(v_t \lrcorner \beta_t^{\mathrm{spat}}\right) = V_{M;t} \lrcorner \widehat{\varkappa}_t^{*2} \left(\beta_t^{\mathrm{spat}}\right) .$$

Finally, we obtain

$$\int_{\mathfrak{P}} V_{M;t} \lrcorner \beta_t^{\mathrm{mat}} = \int_{\mathfrak{P}} V_{M;t} \lrcorner \widehat{\varkappa}_t^{*2} (\beta_t^{\mathrm{spat}}) .$$

Since $\mathfrak{P}$ and $V_{M;t}$ can be chosen arbitrarily, one obtains

$$\beta_t^{\mathrm{mat}} = \widehat{\varkappa}_t^{*2} \beta_t^{\mathrm{spat}} . \tag{7.13}$$

The relation (7.13) is an equality between covector-valued $n$-forms. In this regard, one can obtain a coordinate form using the following proposition. Suppose that $\mathfrak{M}$ and $\mathfrak{N}$ are two smooth $n$-dimensional manifolds, and $f : \mathfrak{M} \to \mathfrak{N}$ is a smooth map. If $(x^i)$ and $(y^i)$ are local coordinates on $\mathfrak{M}$ and $\mathfrak{N}$, respectively, then, [3]

$$f^*(h\, dy^1 \wedge \cdots \wedge dy^n) = (h \circ f) \det\left(\frac{\partial y^i}{\partial x^j}\right) dx^1 \wedge \cdots \wedge dx^n .$$

Choose the local coordinates $(\mathfrak{X}^\alpha)_{\alpha=1}^n$ on $\mathfrak{B}$ and $(X^\beta)_{\beta=1}^n$ on $\mathcal{S}_{\varkappa_t}$. Then $\eta_{t;j} = h_{t;j}\, dX^1 \wedge \cdots \wedge dX^n$. One has

$$\hat{\varkappa}_t^*(\eta_{t;j}) = (h_{t;j} \circ \hat{\varkappa}_t) \det\left(\frac{\partial X^\alpha}{\partial \mathfrak{X}^\beta}\right) d\mathfrak{X}^1 \wedge \cdots \wedge d\mathfrak{X}^n .$$

Finally, one obtains the following representation for $\beta_t^{\text{mat}}$:

$$\beta_t^{\text{mat}} = (h_{t;j} \circ \hat{\varkappa}_t) \det\left(\frac{\partial X^\alpha}{\partial \mathfrak{X}^\beta}\right) (e^j \circ \hat{\varkappa}_t) \otimes \left(d\mathfrak{X}^1 \wedge \cdots \wedge d\mathfrak{X}^n\right) .$$

## 7.7 Example: Neo-Hookean Solids

Let $\mathcal{t}$ be defined by the expression

$$\mathcal{t} = -p_0 \mathcal{I}^{\flat_p} + \mu\, \mathcal{B}^{\flat_p} ,$$

where $\mu$ is a constant, $\mathcal{I}$ is the identity tensor in the physical space, $\mathcal{B}$ is the left Cauchy–Green tensor, and $p_0$ is a scalar-valued mapping (hydrostatic loading). This relation is analogous to the classical constitutive equation for an incompressible neo-Hookean material. Since

$$\mathcal{I} = \partial_{X^j} \otimes dX^j , \quad \mathcal{B} = G^{im} g_{lj} F^l{}_m F^k{}_i \partial_{X^k} \otimes dX^j ,$$

where $(\partial_{X^k})$ and $(dX^k)$ are coordinate frame and coframe in the physical space $\mathcal{P}$, then

$$\mathcal{I}^{\flat_p} = g_{sj} dX^s \otimes dX^j , \quad \mathcal{B}^{\flat_p} = G^{im} g_{ks} g_{lj} F^l{}_m F^k{}_i dX^s \otimes dX^j .$$

Thus, the coordinate representation for $\mathcal{t}$ has the form

$$\mathcal{t} = t_{sj} dX^s \otimes dX^j = \left\{-p_0 g_{sj} + \mu G^{im} g_{ks} g_{lj} F^l{}_m F^k{}_i\right\} dX^s \otimes dX^j .$$

The corresponding covector-valued form is defined by the relation

$$T = *_2 \mathcal{t} = t_{sj} dX^s \otimes (* dX^j) .$$

One has that

$$* dX^j = \sqrt{g} \sum_{a=1}^{m} (-1)^{a-1} g^{ja} dX^1 \wedge \cdots \wedge \widehat{dX^a} \wedge \cdots \wedge dX^m ,$$

and thus, the following equality holds:

$$T = \sqrt{g} \sum_{a=1}^{m} (-1)^{a-1} g^{ja} \left\{-p_0 g_{sj} + \mu G^{im} g_{ks} g_{lj} F^l{}_m F^k{}_i\right\} dX^s \otimes (dX^1 \wedge \cdots \wedge \widehat{dX^a} \wedge \cdots \wedge dX^m) .$$

This relation shows that $T$ depends on $g$ and $\mathcal{G}$.

Since $P = \varkappa^{*2} T$, one has

$$P = \varkappa^{*2} T = \varkappa^{*2} \left\{ t_{sj} dX^s \otimes (*dX^j) \right\} = t_{sj} dX^s \otimes \varkappa^* (*dX^j) .$$

Finally, one arrives at the formula for $P$:

$$P = \sqrt{g} \sum_{a=1}^{m} (-1)^{a-1} g^{ja} \left\{ -p_0 g_{sj} + \mu G^{im} g_{ks} g_{lj} F^l{}_m F^k{}_i \right\} dX^s \\ \otimes \varkappa^* (dX^1 \wedge \cdots \wedge \widehat{dX^a} \wedge \cdots \wedge dX^m) ,$$

where the exterior $(n-1)$-form $\varkappa^* (dX^1 \wedge \cdots \wedge \widehat{dX^a} \wedge \cdots \wedge dX^m)$ is defined on the body $\mathfrak{B}$.

## 7.8 The Case dim $\mathfrak{B}$ = dim $\mathscr{P}$ = 3

Obtain the explicit formula that allows one to calculate the Piola–Kirchhoff exterior stress form components by using the known Cauchy stress form, material and spatial metrics, and the configuration gradient. The particular case $m = n = 3$ is considered. As we will show, this formula looks similar to the classical one. We begin with the relation (7.8), which in our considerations has the form:

$$T = t_{ab} \sqrt{g}\, dX^a \otimes (g^{b1} dX^2 \wedge dX^3 - g^{b2} dX^1 \wedge dX^3 + g^{b3} dX^1 \wedge dX^2) . \tag{7.14}$$

To use (7.11) we need the formula for $\varkappa^* (dX^i \wedge dX^j)$. From the definition of pullback, it follows that

$$\varkappa^* (dX^i \wedge dX^j) = \frac{\partial \varkappa^i}{\partial \mathfrak{X}^\alpha} \frac{\partial \varkappa^j}{\partial \mathfrak{X}^\beta} d\mathfrak{X}^\alpha \wedge d\mathfrak{X}^\beta .$$

Substituting formula (7.14) into (7.11), using the linearity of $\varkappa^*$ and the above formula for $\varkappa^* (dX^i \wedge dX^j)$, we arrive at the relation

$$P = J\, t_{ab} \sqrt{g}\, g^{bk}\, dX^a \otimes ((\mathscr{F}^{-1})^1{}_k\, d\mathfrak{X}^2 \wedge d\mathfrak{X}^3 - (\mathscr{F}^{-1})^2{}_k\, d\mathfrak{X}^1 \wedge d\mathfrak{X}^3 + (\mathscr{F}^{-1})^3{}_k\, d\mathfrak{X}^1 \wedge d\mathfrak{X}^2), \tag{7.15}$$

where $J = \det(\frac{\partial \varkappa^i}{\partial \mathfrak{X}^a})$, and $\mathscr{F}^{-1} = (\mathscr{F}^{-1})^\alpha{}_k \partial_{\mathfrak{X}^\alpha} \otimes dX^k$ is the inverse to $\mathscr{F} = T\varkappa$. Since components for $\mathscr{F}^{-1}$ and its transpose, $\mathscr{F}^{-T}$, are related by the expression $g^{ki} (\mathscr{F}^{-1})^\gamma{}_i = G^{\gamma\beta} (\mathscr{F}^{-T})^k{}_\beta$, the relation (7.15) may be rewritten in terms of $(\mathscr{F}^{-T})^k{}_\beta$. Now let us note that (7.9) in our particular case has the form:

$$P = p_{a\beta} \sqrt{G}\, e^a \otimes (G^{\beta 1} d\mathfrak{X}^2 \wedge \mathfrak{X}^3 - G^{\beta 2} \mathfrak{X}^1 \wedge \mathfrak{X}^3 + G^{\beta 3} \mathfrak{X}^1 \wedge \mathfrak{X}^2) .$$

Then we obtain

$$p_{a\beta} = \sqrt{\frac{g}{G}}\, J\, t_{ab} (\mathscr{F}^{-T})^b{}_\beta , \tag{7.16}$$

where $g = \det[g_{ij}]$, and $G = \det[G_{ij}]$.

## 7.9 The Eshelby Energy-Momentum Tensor on Manifolds

The Eshelby energy-momentum tensor can be generalized for the body $\mathfrak{B}$ as follows. The balance equation on the body has the form:

$$\operatorname{Div} \mathcal{p}^\sharp + \rho_0 b_0^\sharp = 0\,. \tag{7.17}$$

Here, $\mathcal{p}$ is the first Piola–Kirchhoff stress tensor, $\mathcal{p}^\sharp = p^{a\kappa}\partial_{\chi^a} \otimes \partial_{\mathfrak{X}^\kappa}$ and[5]

$$\operatorname{Div} \mathcal{p}^\sharp = \left[\partial_\kappa p^{a\kappa} + p^{a\kappa}\Gamma^\beta{}_{\kappa\beta} + p^{b\kappa}F^c{}_\kappa \gamma^a{}_{bc}\right]\partial_{\chi^a}\,,$$

where $\Gamma^q{}_{kl}$ and $\gamma^q{}_{kl}$ are Levi–Civita connection coefficients on $\mathfrak{B}$ and $\mathscr{P}$, respectively.

Applying contraction with $\mathscr{F}^{\mathrm{T}} = (\mathscr{F}^{\mathrm{T}})^a{}_j \partial_{\mathfrak{X}^a} \otimes dX^j$ to (7.17) gives:

$$\mathscr{F}^{\mathrm{T}} \llcorner \operatorname{Div} \mathcal{p} + \rho_0 \mathscr{F}^{\mathrm{T}} \llcorner b_0^\sharp = 0\,. \tag{7.18}$$

For future needs, we define the bilinear operation $\underset{\llcorner}{\llcorner}$, which satisfies the following property:

$$\partial_{\mathfrak{X}^a} \otimes dX^a \otimes d\mathfrak{X}^\beta \underset{\llcorner}{\llcorner} \partial_{X^b} \otimes \partial_{\mathfrak{X}^\kappa} = \left\langle dX^a, \partial_{X^b}\right\rangle \left\langle d\mathfrak{X}^\beta, \partial_{\mathfrak{X}^\kappa}\right\rangle \partial_{\mathfrak{X}^a}\,.$$

The following formula may be verified directly:

$$\operatorname{Div}(\mathscr{F}^{\mathrm{T}} \llcorner \mathcal{p}^\sharp) = \mathscr{F}^{\mathrm{T}} \llcorner \operatorname{Div} \mathcal{p}^\sharp + (\nabla\mathscr{F}^{\mathrm{T}}) \underset{\llcorner}{\llcorner} \mathcal{p}^\sharp\,. \tag{7.19}$$

By (7.19) for the equation (7.18), we have

$$\operatorname{Div}(\mathscr{F}^{\mathrm{T}} \llcorner \mathcal{p}^\sharp) - (\nabla\mathscr{F}^{\mathrm{T}}) \underset{\llcorner}{\llcorner} \mathcal{p}^\sharp + \rho_0 \mathscr{F}^{\mathrm{T}} \llcorner b_0^\sharp = 0\,.$$

Subtracting the field[6] $(\nabla_{\mathfrak{B}} W)^\sharp$ from the left and right-hand sides of the equation obtained, where $W\colon \mathfrak{B} \to \mathbb{R}$ is the elastic energy density defined at $\mathfrak{B}$, we obtain

$$\operatorname{Div}(\mathscr{F}^{\mathrm{T}} \lrcorner \mathcal{p}^\sharp) - (\nabla_{\mathfrak{B}} W)^\sharp - (\nabla\mathscr{F}^{\mathrm{T}}) \underset{\llcorner}{\llcorner} \mathcal{p}^\sharp + \rho_0 \mathscr{F}^{\mathrm{T}} \llcorner b_0^\sharp = -(\nabla_{\mathfrak{B}} W)^\sharp\,. \tag{7.20}$$

Let $\mathscr{I}$ denote the identity tensor field on $\mathfrak{B}$, $\mathscr{I} = \partial_{\mathfrak{X}^\kappa} \otimes d\mathfrak{X}^\kappa$. Since

$$\operatorname{Div}(W\mathscr{I}^\sharp) = \left\{G^{a\beta}\partial_{\mathfrak{X}^\beta} W\right\} \partial_{\mathfrak{X}^a} = (\nabla_{\mathfrak{B}} W)^\sharp\,,$$

5 According to Chapter 9, Section 9.2.5, the covariant derivative of a two-point tensor field $\mathcal{Q} = Q^{a\kappa}\partial_{\chi^a} \otimes \partial_{\mathfrak{X}^\kappa}$ has the representation $\nabla\mathcal{Q} = \nabla_\beta Q^{a\kappa}\partial_{\chi^a} \otimes \partial_{\mathfrak{X}^\kappa} \otimes d\mathfrak{X}^\beta$, in which

$$\nabla_\beta Q^{a\kappa} = \partial_\beta Q^{a\kappa} + Q^{a\lambda}\Gamma^\kappa{}_{\lambda\beta} + Q^{b\kappa}F^c{}_\beta \gamma^a{}_{bc}\,.$$

It follows that $\operatorname{Div}\mathcal{Q} = \nabla_\kappa Q^{a\kappa}$.

6 Note that $\nabla_{\mathfrak{B}} W$ is a covector field, and $(\nabla_{\mathfrak{B}} W)^\sharp$ is a vector field.

the expression (7.20) takes the final form:

$$\operatorname{Div} \mathscr{e} + (\nabla \mathscr{F}^{\mathrm{T}}) \llcorner_{\llcorner} \mathscr{p}^{\sharp} - \rho_0 \mathscr{F}^{\mathrm{T}} \llcorner b_0^{\sharp} = (\nabla_{\mathfrak{B}} W)^{\sharp} , \tag{7.21}$$

where $\mathscr{e} = W \mathscr{I}^{\sharp} - \mathscr{F}^{\mathrm{T}} \llcorner \mathscr{p}^{\sharp}$ is *Eshelby energy-momentum tensor.*

Let $\widetilde{W}$ be the elastic energy density represented as the function of the points and fields defined on $\mathfrak{B}$: $W(\mathfrak{X}) = \widetilde{W}(\mathfrak{X}; \mathscr{F}(\mathfrak{X}), \dots)$. Following [8], define $\nabla_{\mathfrak{B}} W|_{\mathrm{impl}}$ as[7]:

$$\nabla_{\mathfrak{B}} W|_{\mathrm{impl}} = \nabla_{\mathfrak{B}} W - \nabla_{\mathfrak{B}} W|_{\mathrm{expl}} ,$$

where

$$\nabla_{\mathfrak{B}} W|_{\mathrm{expl}} = \nabla_{\mathfrak{B}} \widetilde{W}(\mathfrak{X}; \mathscr{F}(\mathfrak{X}_0), \dots)\big|_{\mathfrak{X}_0 = \mathfrak{X}} .$$

In this case, the relation (7.21) may be rewritten in the form [8]:

$$-\operatorname{Div} \mathscr{e} = f_{\mathrm{int}}^{\sharp} + f_{\mathrm{ext}}^{\sharp} + f_{\mathrm{inh}}^{\sharp} ,$$

where $f_{\mathrm{int}}^{\sharp}$ is the internal force, $f_{\mathrm{ext}}^{\sharp}$ is the body force, and $f_{\mathrm{inh}}^{\sharp}$ is the *Eshelby force*. Such fields are represented by expressions

$$\begin{aligned} f_{\mathrm{int}}^{\sharp} &= (\nabla \mathscr{F}^{\mathrm{T}}) \llcorner_{\llcorner} \mathscr{p}^{\sharp} - \nabla_{\mathfrak{B}} W|_{\mathrm{impl}}^{\sharp} , \\ f_{\mathrm{ext}}^{\sharp} &= -\rho_0 \mathscr{F}^{\mathrm{T}} \llcorner b_0^{\sharp} , \\ f_{\mathrm{inh}}^{\sharp} &= -\nabla_{\mathfrak{B}} W|_{\mathrm{expl}}^{\sharp} . \end{aligned}$$

7 The expression $\nabla_{\mathfrak{B}} \widetilde{W}(\mathfrak{X}; \mathscr{F}(\mathfrak{X}_0), \dots)\big|_{\mathfrak{X}_0 = \mathfrak{X}}$ has the following sense: the covariant derivative is taken from the field $\mathfrak{X} \mapsto \widetilde{W}(\mathfrak{X}; \mathscr{F}(\mathfrak{X}_0), \dots)$, where $\mathscr{F}(\mathfrak{X}_0), \dots$ are field values at a fixed point $\mathfrak{X}_0$. After calculation of the covariant derivative, all entries of $\mathfrak{X}_0$ are replaced by $\mathfrak{X}$.

# 8 Material Uniformity and Inhomogeneity

## 8.1 Equivalence Relation Between Smooth Embeddings

Suppose that $\mathfrak{M}$ and $\mathfrak{N}$ are smooth manifolds, $s = \dim \mathfrak{M} \leq k = \dim \mathfrak{N}$. Let $p \in \mathfrak{M}$ be a point. Establish an equivalence relation $\sim_p$ between smooth embeddings from $C^\infty(\mathfrak{M}; \mathfrak{N})$ in such a way: for all smooth embeddings $\varkappa_1, \varkappa_2 \in C^\infty(\mathfrak{M}; \mathfrak{N})$,

$$(\varkappa_1 \sim_p \varkappa_2) \Leftrightarrow ((\varkappa_1(p) = \varkappa_2(p)) \wedge (T_p\varkappa_1 = T_p\varkappa_2)) \,. \tag{8.1}$$

Denote an equivalence class by $\mathrm{cl}_p$ or by $[\varkappa]_p$. In coordinates, the relation (8.1) can be represented as follows: $\varkappa_1 \sim_p \varkappa_2$ *iff*

$$\varkappa_1^i(p^1, \dots, p^s) = \varkappa_2^i(p^1, \dots, p^s)\,, \quad \text{and} \quad \partial_j \varkappa_1^i|_{(p^1,\dots,p^s)} = \partial_j \varkappa_2^i|_{(p^1,\dots,p^s)}\,,$$

for all $i = 1, \dots, k$, $j = 1, \dots, s$. Here, the mappings $\varkappa_1^i$, $\varkappa_2^i$ are coordinate representations of $\varkappa_1$ and $\varkappa_2$; $(p^i)$ are coordinates of $p$.

The equivalence relation between diffeomorphisms of $\mathfrak{M}$ to itself is defined similarly. Suppose that $\Phi\colon \mathfrak{M} \to \mathfrak{M}$ is a diffeomorphism. Then, we apply[1]

$$\mathrm{cl}_{\Phi(p)} \circ [\Phi]_p := [\varkappa \circ \Phi]_p \,,$$

for some $\varkappa \in \mathrm{cl}_{\Phi(p)}$. Such a definition is well defined and is aligned with the composition of linear mappings. Moreover, the following properties hold:

i) For any $p \in \mathfrak{M}$, one has

$$\mathrm{cl}_p \circ [\mathrm{Id}_{\mathfrak{M}}]_p = \mathrm{cl}_p \,;$$

ii) Suppose that $p \in \mathfrak{M}$ is any point and $\Phi$, $\Psi$ are diffeomorphisms of $\mathfrak{M}$ on itself. Then,

$$\mathrm{cl}_{\Phi\circ\Psi(p)} \circ [\Phi \circ \Psi]_p = (\mathrm{cl}_{\Phi\circ\Psi(p)} \circ [\Phi]_{\Psi(p)}) \circ [\Psi]_p \,.$$

This equality is true, since one has $[\varkappa \circ \Phi \circ \Psi]_p = [\varkappa \circ \Phi]_{\Psi(p)} \circ [\Psi]_p$ for $\varkappa \in \mathrm{cl}_{\Phi\circ\Psi(p)}$.

## 8.2 Local Configurations and Simple Bodies

Suppose that $\mathfrak{X} \in \mathfrak{B}$ is a material point. Local configuration at $\mathfrak{X}$ formalizes the idea of *infinitesimal localization* If the “observer” is placed into an infinitesimal neighborhood of the point $\mathfrak{X}$, then it can distinguish configurations only within this neighborhood. Assume that this localization has the first order: the “observer” cannot distinguish configurations that have the same values at $\mathfrak{X}$ and the same linear approaches. The

1 The mapping $\varkappa \circ \Phi$ is a smooth embedding and $[\varkappa \circ \Phi]_p$ is understood in the sense of (8.1).

https://doi.org/10.1515/9783110563214-008

mathematical formalization of this idea can be provided as follows. The infinitesimal neighborhood of $\mathfrak{X}$ is formalized as the tangent space $T_{\mathfrak{X}}\mathfrak{B}$. The first-order localization is formalized by means of the equivalence relation $\sim_{\mathfrak{X}}$ (8.1) on the set $\mathfrak{C}(\mathfrak{B};\mathscr{P})$. Following W. Noll [104], we refer to each equivalence class $[\varkappa]_{\mathfrak{X}}$ as a *local configuration at* $\mathfrak{X}$ and denote it by $\mathscr{K}_{\mathfrak{X}}$. Let $\mathfrak{C}_{\mathfrak{X}}(\mathfrak{B};\mathscr{P})$ be the quotient set (the set of all local configurations at $\mathfrak{X}$).

The material at any point $\mathfrak{X} \in \mathfrak{B}$ is characterized by its response. To obtain the response one uses measurers that give numerical data. Measured quantities at point $\mathfrak{X}$ are represented mathematically by tensors. Among them may be scalar mappings (elastic energy density), second-rank tensors (Cauchy stress tensor). The ordered set of all such objects is denoted by $R$; its elements are called *response descriptors*. Assume that the response of the material at $\mathfrak{X}$ depends on the first-order approximation of the configuration. That is, there is a mapping

$$\mathscr{R}_{\mathfrak{X}}: \mathfrak{C}_{\mathfrak{X}}(\mathfrak{B};\mathscr{P}) \to R\,, \quad \mathfrak{C}_{\mathfrak{X}}(\mathfrak{B};\mathscr{P}) \ni \mathscr{K}_{\mathfrak{X}} \mapsto \mathscr{R}_{\mathfrak{X}}(\mathscr{K}_{\mathfrak{X}}) = r \in R\,.$$

If this mapping is specified for all $\mathfrak{X} \in \mathfrak{B}$, then the body $\mathfrak{B}$, endowed with the collection $\{\mathscr{R}_{\mathfrak{X}}\}_{\mathfrak{X}}$ of mappings, is called a *simple body* [104].

## 8.3 Material Uniformity

Recall that in this book, we consider only simple bodies. Let $\mathfrak{X}, \mathfrak{Y} \in \mathfrak{B}$ be material points. The notion of material isomorphism gives the precise meaning to the statement "material at $\mathfrak{X}$ is the same as the material at $\mathfrak{Y}$." More precisely, the points $\mathfrak{X}$ and $\mathfrak{Y}$ are *materially isomorphic* [6, 104], if there exists a diffeomorphism $\Phi_{\mathfrak{X}\mathfrak{Y}}: \mathfrak{B} \to \mathfrak{B}$, such that $\mathfrak{X} = \Phi_{\mathfrak{X}\mathfrak{Y}}(\mathfrak{Y})$, and for all[2] $\mathscr{K}_{\mathfrak{X}} \in \mathfrak{C}_{\mathfrak{X}}(\mathfrak{B};\mathscr{P})$,

$$\mathscr{R}_{\mathfrak{X}}(\mathscr{K}_{\mathfrak{X}}) = \mathscr{R}_{\mathfrak{Y}}(\mathscr{K}_{\mathfrak{X}} \circ [\Phi_{\mathfrak{X}\mathfrak{Y}}]_{\mathfrak{Y}})\,.$$

The equivalence class $[\Phi_{\mathfrak{X}\mathfrak{Y}}]_{\mathfrak{Y}}$ is called a *material isomorphism from* $T_{\mathfrak{Y}}\mathfrak{B}$ *to* $T_{\mathfrak{X}}\mathfrak{B}$. We call the mapping $\Phi_{\mathfrak{X}\mathfrak{Y}}$ a *material automorphism from* $\mathfrak{Y}$ *to* $\mathfrak{X}$. One has the following properties:

a) $\Phi_{\mathfrak{X}\mathfrak{X}} = \mathrm{Id}_{\mathfrak{B}}$ is a material automorphism from $\mathfrak{X}$ to $\mathfrak{X}$.
b) If $\Phi_{\mathfrak{X}\mathfrak{Y}}$ is a material automorphism from $\mathfrak{Y}$ to $\mathfrak{X}$, $\Phi_{\mathfrak{Y}\mathfrak{Z}}$ is a material automorphism from $\mathfrak{Z}$ to $\mathfrak{Y}$, and then $\Phi_{\mathfrak{X}\mathfrak{Y}} \circ \Phi_{\mathfrak{Y}\mathfrak{Z}}$ is a material automorphism from $\mathfrak{Z}$ to $\mathfrak{X}$.
c) If $\Phi_{\mathfrak{X}\mathfrak{Y}}$ is a material automorphism from $\mathfrak{Y}$ to $\mathfrak{X}$, then $\Phi_{\mathfrak{X}\mathfrak{Y}}^{-1}$ is a material automorphism from $\mathfrak{X}$ to $\mathfrak{Y}$.

2 Note that the composition $\mathscr{K}_{\mathfrak{X}} \circ [\Phi_{\mathfrak{X}\mathfrak{Y}}]_{\mathfrak{Y}}$ is a local configuration at $\mathfrak{Y}$.

**Remark 8.1.** Here, we will give the sketch of the proof of properties a)–c).

a) For any $\mathscr{K}_{\mathfrak{X}} \in \mathfrak{C}_{\mathfrak{X}}(\mathfrak{B};\mathscr{P})$ according to property i), Section 8.1, one has $\mathscr{K}_{\mathfrak{X}} \circ [\mathrm{Id}_{\mathfrak{B}}]_{\mathfrak{X}} = \mathscr{K}_{\mathfrak{X}}$. Here, $\Phi_{\mathfrak{X}\mathfrak{X}} = \mathrm{Id}_{\mathfrak{B}}$.
b) For any $\mathscr{K}_{\mathfrak{X}} \in \mathfrak{C}_{\mathfrak{X}}(\mathfrak{B};\mathscr{P})$, using property ii) from Section 8.1, one obtains

$$\mathscr{K}_{\mathfrak{X}} \circ [\Phi_{\mathfrak{X}\mathfrak{Y}} \circ \Phi_{\mathfrak{Y}\mathfrak{Z}}]_{\mathfrak{Z}} = (\mathscr{K}_{\mathfrak{X}} \circ [\Phi_{\mathfrak{X}\mathfrak{Y}}]_{\mathfrak{Y}}) \circ [\Phi_{\mathfrak{Y}\mathfrak{Z}}]_{\mathfrak{Z}} \in \mathfrak{C}_{\mathfrak{Z}}(\mathfrak{B};\mathscr{P}),$$

and $\mathscr{R}_{\mathfrak{X}}(\mathscr{K}_{\mathfrak{X}}) = \mathscr{R}_{\mathfrak{Z}}((\mathscr{K}_{\mathfrak{X}} \circ [\Phi_{\mathfrak{X}\mathfrak{Y}}]_{\mathfrak{Y}}) \circ [\Phi_{\mathfrak{Y}\mathfrak{Z}}]_{\mathfrak{Z}})$.
c) For any $\mathscr{K}_{\mathfrak{Y}} \in \mathfrak{C}_{\mathfrak{Y}}(\mathfrak{B};\mathscr{P})$, applying properties a) and b), one obtains

$$\mathscr{R}_{\mathfrak{Y}}(\mathscr{K}_{\mathfrak{Y}}) = \mathscr{R}_{\mathfrak{Y}}\left(\left(\mathscr{K}_{\mathfrak{Y}} \circ \left[\Phi^{-1}_{\mathfrak{X}\mathfrak{Y}}\right]_{\mathfrak{X}}\right) \circ [\Phi_{\mathfrak{X}\mathfrak{Y}}]_{\mathfrak{Y}}\right) = \mathscr{R}_{\mathfrak{X}}\left(\mathscr{K}_{\mathfrak{Y}} \circ \left[\Phi^{-1}_{\mathfrak{X}\mathfrak{Y}}\right]_{\mathfrak{X}}\right),$$

and this implies that $\Phi^{-1}_{\mathfrak{X}\mathfrak{Y}}$ is a material automorphism from $\mathfrak{X}$ to $\mathfrak{Y}$.

**Remark 8.2.** Suppose that the physical space $\mathscr{P}$ is a Euclidean affine space $\mathscr{E}$. A local configuration $\mathscr{K}_{\mathfrak{X}}$ can be identified with a linear mapping $\boldsymbol{K}_{\mathfrak{X}}: T_{\mathfrak{X}}\mathfrak{B} \to \mathscr{V}$ as follows. Choose some configuration $\varkappa \in \mathscr{K}_{\mathfrak{X}}$ and define $\boldsymbol{K}_{\mathfrak{X}}$ by $\boldsymbol{K}_{\mathfrak{X}} := T_{\mathfrak{X}}\varkappa$. According to the definition of the relation $\sim_{\mathfrak{X}}$, the mapping $\boldsymbol{K}_{\mathfrak{X}}$ does not depend on the choice of a representative $\varkappa$ and is fully determined by the class $\mathscr{K}_{\mathfrak{X}}$. In the case of the physical space $\mathscr{E}$, we identify a local configuration $\mathscr{K}_{\mathfrak{X}}$ with the corresponding linear mapping $\boldsymbol{K}_{\mathfrak{X}}$. A material isomorphism can be defined as a vector space isomorphism $\Phi_{\mathfrak{X}\mathfrak{Y}}: T_{\mathfrak{Y}}\mathfrak{B} \to T_{\mathfrak{X}}\mathfrak{B}$. Such a definition corresponds to the conventional approach [104].

Suppose that all points of the body $\mathfrak{B}$ are materially isomorphic to each other. That is, the material (response) at any two points $\mathfrak{X}, \mathfrak{Y} \in \mathfrak{B}$ is the same. Such a body is called *materially uniform*. From now, we consider materially uniform simple bodies. Physically, a materially uniform simple body consists of infinitesimal representative volumes with the following property. Each of them, if separated from others and placed in a "mechanical test set up", gives the same response on the same deformation.

Now assume that one fixed a point $\mathfrak{X}_0 \in \mathfrak{B}$ and chosen some local configuration $\mathscr{K}^R_{\mathfrak{X}_0}$. Let $\{\Phi_{\mathfrak{X}_0\mathfrak{X}}\}_{\mathfrak{X}\in\mathfrak{B}}$ be a family of material automorphisms. For points $\mathfrak{X} \in \mathfrak{B}$, $\mathfrak{X} \neq \mathfrak{X}_0$, denote $\mathscr{K}^R_{\mathfrak{X}} := \mathscr{K}^R_{\mathfrak{X}_0} \circ [\Phi_{\mathfrak{X}_0\mathfrak{X}}]_{\mathfrak{X}}$. Thus, one has the family $\mathrm{Ref}_{\mathfrak{B}} = \{\mathscr{K}^R_{\mathfrak{X}}\}_{\mathfrak{X}\in\mathfrak{B}}$ of local configurations. Such local configurations give the same response:

$$\forall \mathfrak{X} \in \mathfrak{B}: \quad \mathscr{R}_{\mathfrak{X}}(\mathscr{K}^R_{\mathfrak{X}}) = \mathscr{R}_{\mathfrak{X}_0}(\mathscr{K}^R_{\mathfrak{X}_0}).$$

The family $\mathrm{Ref}_{\mathfrak{B}} = \{\mathscr{K}^R_{\mathfrak{X}}\}_{\mathfrak{X}\in\mathfrak{B}}$ depends on a family of material automorphisms and a point $\mathfrak{X}_0$. We refer to the family $\mathrm{Ref}_{\mathfrak{B}}$ as the *uniform reference*[3].

Let $\mathfrak{X}_{00} \in \mathfrak{B}$ be a point, such that $\mathfrak{X}_{00} \neq \mathfrak{X}_0$. The family $\mathrm{Ref}_{\mathfrak{B}} = \{\mathscr{K}^R_{\mathfrak{X}}\}_{\mathfrak{X}\in\mathfrak{B}}$ can be represented in terms of $\mathfrak{X}_{00}$: if $\mathfrak{X} \neq \mathfrak{X}_0$, then $\mathscr{K}^R_{\mathfrak{X}} = \mathscr{K}^R_{\mathfrak{X}_{00}} \circ [\Psi_{\mathfrak{X}_{00}\mathfrak{X}}]_{\mathfrak{X}}$, where $\Psi_{\mathfrak{X}_{00}\mathfrak{X}} = \Phi^{-1}_{\mathfrak{X}_0\mathfrak{X}_{00}} \circ \Phi_{\mathfrak{X}_0\mathfrak{X}}$. The above properties a)–c) were used for this relation.

3 This corresponds to the classical Nolls definition [104].

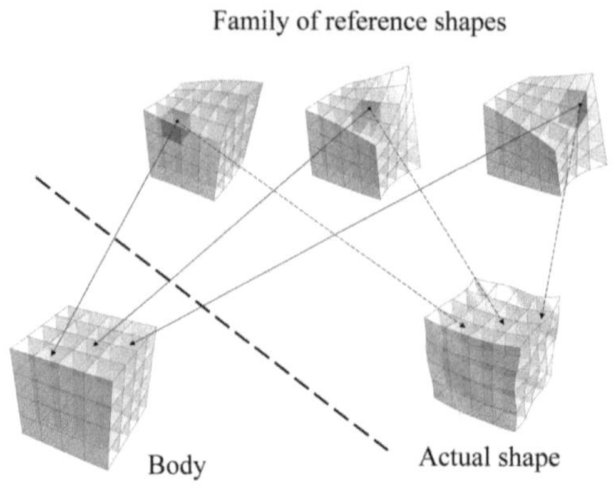


**Fig. 8.1:** A materially uniform body.

Suppose that $\mathrm{Ref}_{\mathfrak{B}} = \{\mathscr{K}^R_{\mathfrak{X}}\}_{\mathfrak{X}\in\mathfrak{B}}$. Choosing for any $\mathfrak{X} \in \mathfrak{B}$ a representative $\varkappa^R_{\mathfrak{X}} \in \mathscr{K}^R_{\mathfrak{X}}$, one arrives at the family $\overline{\mathrm{Ref}}_{\mathfrak{B}} = \{\varkappa^R_{\mathfrak{X}}\}_{\mathfrak{X}\in\mathfrak{B}}$. One observes the family $\{\mathcal{S}_{\varkappa^R_{\mathfrak{X}}}\}_{\mathfrak{X}\in\mathfrak{B}}$ of shapes and splits them into elementary volumes. Volumes that correspond to every point $\varkappa^R_{\mathfrak{X}}(\mathfrak{X})$, give the same response. However, if one fixes a point $\mathfrak{X}$ and considers elementary volumes at points $\varkappa^R_{\mathfrak{X}}(\mathfrak{Y})$, then their responses do not coincide with the response at $\mathfrak{X}$, in general. These reasonings are illustrated in Figure 8.1.

A uniform reference $\mathrm{Ref}_{\mathfrak{B}} = \{\mathscr{K}^R_{\mathfrak{X}}\}_{\mathfrak{X}\in\mathfrak{B}}$ generates a field of $m \times n$ matrices of rank $n$. Let $\mathfrak{X} \in \mathfrak{B}$. Choosing $\varkappa \in \mathscr{K}^R_{\mathfrak{X}}$, apply

$$K^i{}_\beta(\mathfrak{X}) := \left.\frac{\partial \varkappa^i}{\partial \mathfrak{X}^\beta}\right|_{\varphi(\mathfrak{X})},$$

where $\varkappa^i$ is a coordinate representation of $\varkappa$, $(\mathfrak{X}^\beta)$ are local coordinates on $\mathfrak{B}$, and $\varphi$ is a coordinate mapping on $\mathfrak{B}$. By virtue of the definition, the matrix $[K^i{}_\beta(\mathfrak{X})]$ does not depend on a representative $\varkappa$. Thus, the family $\mathrm{Ref}_{\mathfrak{B}}$ gives the family $\overline{\overline{\mathrm{Ref}}}_{\mathfrak{B}} = \{[K^i{}_\beta(\mathfrak{X})]\}_{\mathfrak{X}\in\mathfrak{B}}$ of matrices.

If there exists a configuration $\varkappa_R \in \mathfrak{C}(\mathfrak{B};\mathscr{P})$ such that $\varkappa_R \in \mathscr{K}^R_{\mathfrak{X}}$ for every $\mathfrak{X} \in \mathfrak{B}$, then each elementary volume of the shape $\mathcal{S}_{\varkappa_R}$ gives the same response. Such a body

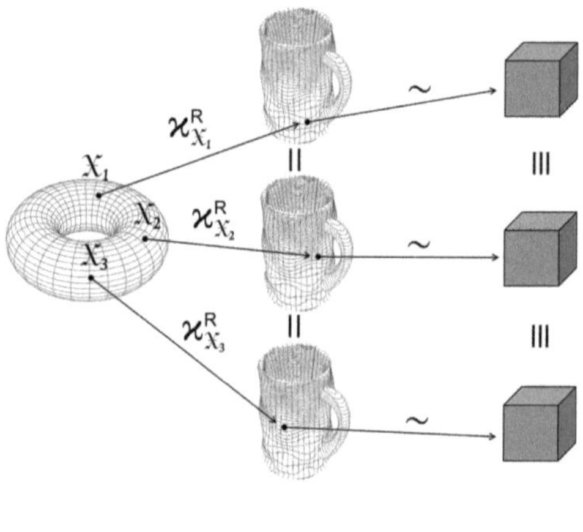

**Fig. 8.2:** A homogeneous body.

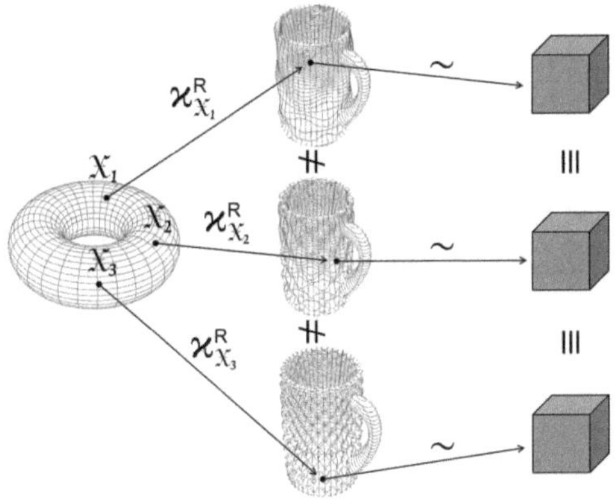


**Fig. 8.3:** An inhomogeneous body.

is called *homogeneous*. If such a configuration does not exist, then the body is called *inhomogeneous*. Figures 8.2 and 8.3 illustrate the difference between homogeneous and inhomogeneous bodies. Cubes represent uniform shapes.

## 8.4 A Material Metric

In reasonings provided in the previous chapters, a body as a manifold is endowed by some abstract metric $\mathscr{G}$. Despite of the spatial metric $g$, which is given *a priori*, the material metric is not presupposed to be defined independently of the physical process. For a materially uniform simple body, the material metric can be constructed in terms of a uniform reference.

Choose a uniform reference $\mathrm{Ref}_{\mathfrak{B}} = \{\mathscr{K}^R_{\mathfrak{X}}\}_{\mathfrak{X}\in\mathfrak{B}}$ and construct the family $\overline{\mathrm{Ref}}_{\mathfrak{B}} = \{\varkappa^R_{\mathfrak{X}}\}_{\mathfrak{X}\in\mathfrak{B}}$. Using elements of the family $\overline{\mathrm{Ref}}_{\mathfrak{B}}$ one can obtain the collection $\{(\mathfrak{B}, \mathscr{G}^{(\mathfrak{X})})\}_{\mathfrak{X}\in\mathfrak{B}}$ of Riemannian spaces with metrics $\mathscr{G}^{(\mathfrak{X})} := (\varkappa^R_{\mathfrak{X}})^* g$ that are pullbacks of the physical metric $g$, relative to the embeddings $\varkappa^R_{\mathfrak{X}}$. Thus,

$$\forall \mathfrak{Y} \in \mathfrak{B} \quad \forall u, v \in T_{\mathfrak{Y}}\mathfrak{B} : \quad \mathscr{G}^{(\mathfrak{X})}|_{\mathfrak{Y}}(u, v) = g\left(T_{\mathfrak{Y}}\varkappa^R_{\mathfrak{X}}(u), T_{\mathfrak{Y}}\varkappa^R_{\mathfrak{X}}(v)\right) .$$

The corresponding Levi–Civita connection coincides with the Levi–Civita connection on[4] $\mathscr{P}$. Now one can synthesize the Riemannian metric, which is non-Euclidean in general. Assume that the section $\mathscr{G} : \mathfrak{B} \to T^*\mathfrak{B} \otimes T^*\mathfrak{B}$, defined by the formula

$$\mathscr{G}_{\mathfrak{X}} := \mathscr{G}^{(\mathfrak{X})}|_{\mathfrak{X}} ,$$

is smooth. Each of the values $\mathscr{G}^{(\mathfrak{X})}|_{\mathfrak{X}}$ represents a symmetric positive definite bilinear form. Hence, $\mathscr{G}$ is a Riemannian metric. Since it is defined on the body $\mathfrak{B}$, we refer to $\mathscr{G}$

4 In the particular case of $\mathscr{P} = \mathscr{E}$, we have that the considering connection is Euclidean. In this case, the pullback transformation introduces some specific curvilinear coordinates on $\mathfrak{B}$.

as a *material metric*. By the definition, the material metric is combined from the family of Riemannian metrics $\{\mathscr{G}^{(\mathfrak{X})}\}_{\mathfrak{X}\in\mathfrak{B}}$, *i.e.*,

$$\forall \mathfrak{X} \in \mathfrak{B} \quad \forall u, v \in T_{\mathfrak{X}}\mathfrak{B}: \quad \mathscr{G}_{\mathfrak{X}}(u, v) = g\left(T_{\mathfrak{X}}\varkappa^R_{\mathfrak{X}}(u), T_{\mathfrak{X}}\varkappa^R_{\mathfrak{X}}(v)\right) . \tag{8.2}$$

The material metric (8.2) depends on the uniform reference $\mathrm{Ref}_{\mathfrak{B}}$, but does not depend on certain representatives from each equivalence class. This motivates one to rewrite (8.2) in terms of $\mathrm{Ref}_{\mathfrak{B}}$ as follows:

$$\forall \mathfrak{X} \in \mathfrak{B} \quad \forall u, v \in T_{\mathfrak{X}}\mathfrak{B}: \quad \mathscr{G}_{\mathfrak{X}}(u, v) = g\left(\mathscr{K}^R_{\mathfrak{X}}(u), \mathscr{K}^R_{\mathfrak{X}}(v)\right) . \tag{8.3}$$

Here, $\mathscr{K}^R_{\mathfrak{X}}(u) := T_{\mathfrak{X}}\varkappa(u)$, for some $\varkappa \in \mathscr{K}^R_{\mathfrak{X}}$.

**Remark 8.3.** If the physical space is affine Euclidean, $\mathscr{E}$, then (8.3) is represented in terms of $\boldsymbol{K}^R : \mathfrak{X} \mapsto \boldsymbol{K}^R_{\mathfrak{X}}$ in the following way:

$$\forall u, v \in T\mathfrak{B}: \quad \mathscr{G}(u, v) = [\boldsymbol{K}^R(u)] \cdot [\boldsymbol{K}^R(v)] .$$

This is the classical formula that was introduced by Noll [104].

Since the body $\mathfrak{B}$ is endowed with the material metric $\mathscr{G}$, one can introduce the Levi–Civita connection on it. The connection coefficients are defined via the formula

$$\Gamma^{\alpha}{}_{\beta\gamma} = \frac{G^{\alpha\tau}}{2}\left(\partial_\beta G_{\gamma\tau} + \partial_\gamma G_{\beta\tau} - \partial_\tau G_{\beta\gamma}\right) ,$$

where $G_{\alpha\beta} = g_{sp}K^s{}_\alpha K^p{}_\beta$ and $K^i{}_\beta = \frac{\partial \varkappa^i}{\partial \mathfrak{X}^\beta}$.

## 8.5 Bodies with Variable Material Composition

### 8.5.1 Formalization of Bodies with Variable Material Composition

Suppose that material manifold $\mathfrak{B}$ can vary in time. This means that material points can be added to or removed from $\mathfrak{B}$. This situation arises, for example, in the modeling of growing solids [25, 127–129], biological tissues [130] the processes of freezing, melting, and ablation, *etc.* Such varying in time bodies must be characterized not only by their geometrical structure, but also by the process of their variation in time. In the following, we will refer to this process as the *evolutionary process*.

Bodies with variable material composition can be represented by a family $\mathfrak{G} = \{\mathfrak{B}_\alpha\}_{\alpha\in\mathbb{I}}$ of $n$-dimensional bodies. Here, $\mathbb{I}$ is a linearly ordered set of indices. For each index $\alpha \in \mathbb{I}$, the body $\mathfrak{B}_\alpha$ characterizes an instantaneous material composition of the solid. Assume that $\mathbb{I}$ is a subset of $\mathbb{R}$. The number $\mathrm{Card}(\mathbb{I})$ characterizes the type of evolutionary process. When $\mathrm{Card}(\mathbb{I}) = N < \infty$, the process may be qualified as *discrete*, while $\mathrm{Card}(\mathbb{I}) = \aleph_1$ corresponds to a *continuous* process.

Assume that during the evolutionary process no material can be removed. That is, family $\mathfrak{G}$ is ordered by the subset relation, *i.e.*,

$$\forall \alpha, \beta \in \mathbb{I}: \quad (\alpha < \beta) \Rightarrow (\mathfrak{B}_\alpha \subsetneq \mathfrak{B}_\beta)\,.$$

Without loss of generality one may assume that $\mathbb{I} = \{1, \ldots, N\}$ in the case of the discrete process and $\mathbb{I} = [a_0, a_1]$ in the case of the continuous one. Thus, the sets

$$\mathfrak{B}_* = \bigcap_{\alpha \in \mathbb{I}} \mathfrak{B}_\alpha\,, \quad \mathfrak{B}^* = \bigcup_{\alpha \in \mathbb{I}} \mathfrak{B}_\alpha$$

are non-empty and represent the initial and final bodies, respectively. They coincide with some uniquely determined bodies from $\mathfrak{G}$. The body $\mathfrak{B}_*$ represents a substrate, while the body $\mathfrak{B}^*$ represents the total body composition. Following technological interpretation we refer to each element of $\mathfrak{G}$ as an *assembly*.

For each pair of indices $\alpha, \beta \in \mathbb{I}$, the corresponding bodies $\mathfrak{B}_\alpha, \mathfrak{B}_\beta$ have independent topological structures, in general. In particular, they have different topological invariants. Figure 8.4 shows the evolution of a body that is topologically equivalent to the $n$-solid torus, during which the Euler characteristic $\chi$ changes due to the attachment of extra material. In this book, we consider a particular case when $\mathfrak{B}_\alpha$ is a "*part*" of $\mathfrak{B}_\beta$, if $\alpha < \beta$. It can be formalized as follows: if $\alpha < \beta$, then $\mathfrak{B}_\alpha$ is an embedded submanifold of $\mathfrak{B}_\beta$. Thus, for each $\alpha \in \mathbb{I}$, the space $\mathfrak{B}_\alpha$ is an embedded submanifold of $\mathfrak{B}^*$. Since $\dim \mathfrak{B}_\alpha = \dim \mathfrak{B}^*$, this implies that $\mathfrak{B}_\alpha$ is an open subset of $\mathfrak{B}^*$.

In the case of the continuous process, we wish to represent each element of the family $\mathfrak{G}$ that corresponds to $\alpha \in \mathbb{I}$, as $\mathfrak{B}_\alpha = \mathfrak{B}_* \vee \widetilde{\mathfrak{B}}_\alpha$. Here, the symbol $\widetilde{\mathfrak{B}}_\alpha$ represents an extra body, formed by extra material that flowed in to the substrate $\mathfrak{B}_* = \mathfrak{B}_{a_0}$. In

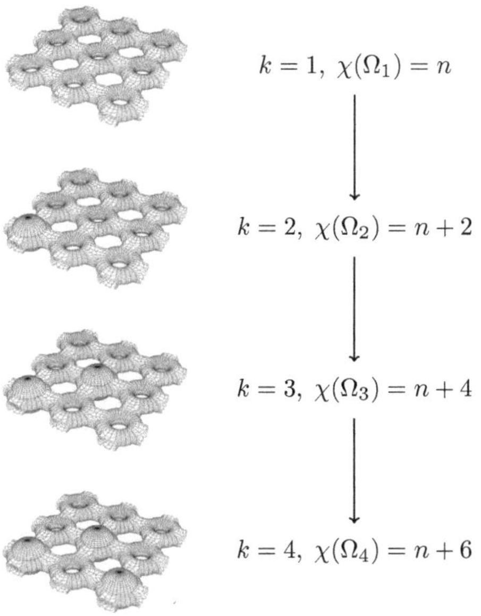


**Fig. 8.4:** Change in the topological structure of the body in the process of its evolution.

the case of discrete process, we want the following expression: $\mathfrak{B}_{s+1} = \mathfrak{B}_s \vee \mathfrak{L}_s$. Here, the symbol $\mathfrak{L}_s$ corresponds to the layer that is attached to the previous assembly, $\mathfrak{B}_s$. In both cases, the symbol $\vee$ denotes the operation of joining, which assigns to each pair of bodies a new body, which contains both of them.

Although the partition of the body into layers and its assembling seems to be natural operations on the set of material points, constituting the body, their formal implementations have technical issues. In particular, the set difference of the body and its part results in a set, which is not open and, therefore, cannot be a body. Moreover, after the union operation applied to two bodies with a partially coinciding boundary, the intersection of boundaries would not be a part of the result (recall that we treat a body as an open set).

Intuitive notions of an extra body (or layer, in the discrete case) and the joining operation may be formalized as follows. Denote the topology of $\mathfrak{B}^*$ by $\mathfrak{S}$. Since each $n$-dimensional body contained in $\mathfrak{B}^*$ is an open set, and vice versa, it is sufficient to define the operation of the detachment of an extra body (or layer) that acts on open set, represented a body, and results in open set, represented a part of this body. Let us consider the case when one body (or "substrate") is contained in another. Let $\mathfrak{B}_1$ and $\mathfrak{B}_2$ be $n$-dimensional smooth submanifolds of $\mathfrak{B}^*$ (that is, bodies), and $\mathfrak{B}_1 \subset \mathfrak{B}_2$. Then, the set $\mathfrak{B}_1$ is open in the topological space $\mathfrak{B}_2$. Therefore, the difference $\mathfrak{B}_2 \setminus \mathfrak{B}_1$ represents a closed set relative to the *induced topology* $\mathfrak{S}_{\mathfrak{B}_2}$ of $\mathfrak{B}_2$. Hence, one has that $\overline{\mathfrak{B}_2 \setminus \mathfrak{B}_1} = \mathfrak{B}_2 \setminus \mathfrak{B}_1$, and we arrive at the following representation:

$$\mathfrak{B}_2 \setminus \mathfrak{B}_1 = \mathrm{Int}_{\mathfrak{B}_2}(\mathfrak{B}_2 \setminus \mathfrak{B}_1) \cup \partial_{\mathfrak{B}_2}(\mathfrak{B}_2 \setminus \mathfrak{B}_1)\,,$$

where the topological operations of closure, interior ($\mathrm{Int}_{\mathfrak{B}_2}$), and boundary ($\partial_{\mathfrak{B}_2}$) are considered in the topology $\mathfrak{S}_{\mathfrak{B}_2}$. The set $\mathrm{Int}_{\mathfrak{B}_2}(\mathfrak{B}_2 \setminus \mathfrak{B}_1)$ is open in $\mathfrak{B}^*$. Since the interior operation is defined as the union of all open sets contained in the given set, we have that

$$\mathrm{Int}_{\mathfrak{B}_2}(\mathfrak{B}_2 \setminus \mathfrak{B}_1) = \mathrm{Int}(\mathfrak{B}_2 \setminus \mathfrak{B}_1)\,,$$

where Int is the interior operation relative to the topology of $\mathfrak{B}^*$. These reasonings allow one to introduce the operation $\mathfrak{J}$ of layer selection, which for $\mathfrak{B}_1 \subset \mathfrak{B}_2$ returns the set

$$\mathfrak{J}(\mathfrak{B}_2, \mathfrak{B}_1) := \mathrm{Int}(\mathfrak{B}_2 \setminus \mathfrak{B}_1)\,,$$

to which we refer as the *layer* or *extra body* (in the continuous case).

The operation $\mathfrak{J}$ allows one to partition any $n$-dimensional body $\mathfrak{B}_2$ into a body $\mathfrak{B}_1 \in \mathfrak{S}_{\mathfrak{B}_2}$ and the body $\mathfrak{J}(\mathfrak{B}_2, \mathfrak{B}_1)$. The symbol $\vee$ indicates this partition and is referred to as the *joining operation*:

$$\mathfrak{B}_2 = \mathfrak{B}_1 \vee \mathfrak{J}(\mathfrak{B}_2, \mathfrak{B}_1) := \mathfrak{B}_1 \cup \mathrm{Int}(\mathfrak{B}_2 \setminus \mathfrak{B}_1) \cup \partial_{\mathfrak{B}_2}(\mathfrak{B}_2 \setminus \mathfrak{B}_1)\,.$$

Finally,

- For a continuous process, one has $\mathfrak{B}_\alpha := \mathfrak{B}_* \vee \widetilde{\mathfrak{B}}_\alpha$, where $\widetilde{\mathfrak{B}}_\alpha := \mathfrak{J}(\mathfrak{B}_\alpha, \mathfrak{B}_*)$.
- For a discrete process, $\mathfrak{B}_{s+1} := \mathfrak{B}_s \vee \mathfrak{L}_s$, in which $\mathfrak{L}_s = \mathfrak{J}(\mathfrak{B}_{s+1}, \mathfrak{B}_s)$. In addition, since no material is removed, $\mathfrak{B}_* = \mathfrak{B}_1$. Thus, in this case,

$$\mathfrak{B}^* = \mathfrak{B}_N = (\ldots(\mathfrak{B}_* \vee \mathfrak{L}_1) \vee \cdots \vee \mathfrak{L}_{N-2}) \vee \mathfrak{L}_{N-1}\,, \qquad \mathfrak{L}_0 := \mathfrak{B}_*\,.$$

**Remark 8.4.** The operation $\vee$ is not equal to the conventional set union and has the following physical sense, which we will explain in the case of the discrete process. When we obtain a sequentially new assembly $\mathfrak{B}_{s+1}$ from the previous assembly $\mathfrak{B}_s$ and the layer $\mathfrak{L}_s$, we have to include all interior points of $\mathfrak{B}_s$ and $\mathfrak{L}_s$ and, in addition, such points of common boundary of $\overline{\mathfrak{B}_s}$ and $\overline{\mathfrak{L}_s}$ that represent a thin film between "gluing" solids.

**Remark 8.5.** Consider an example. Suppose that one has a discrete process and

$$\mathfrak{B}_s \,\tilde{\vee}\, \mathrm{Int}\,[\mathfrak{B}_{s+1} \setminus \mathfrak{B}_s] := \mathrm{Int}\,(\overline{\mathfrak{B}_s \cup \mathrm{Int}\,[\mathfrak{B}_{s+1} \setminus \mathfrak{B}_s]})\,,$$

where the closure and interior (Int) are taken in $\mathfrak{B}^* = \mathfrak{B}_N$. In the simple case one can use such a definition, but in general it may give an erroneous result. Let $\mathfrak{B}^* = \mathbb{R}^2$, the set $\mathfrak{B}_{s+1}$ be an open disk with unit radius, and the center punctured at zero, and the set $\mathfrak{B}_s$ be an open disk with radius $1/2$ and the center punctured at zero. In this case, $\mathfrak{B}_{s+1} \neq \mathfrak{B}_s \,\tilde{\vee}\, \mathrm{Int}\,[\mathfrak{B}_{s+1} \setminus \mathfrak{B}_s]$, but at the same time, $\mathfrak{B}_{s+1} = \mathfrak{B}_s \vee \mathrm{Int}\,[\mathfrak{B}_{s+1} \setminus \mathfrak{B}_s]$.

In the examples and the remainder of the chapter, we will consider the case when the physical space is Euclidean, $\mathscr{E}$.

### 8.5.2 The Discrete Process

Suppose that a discrete evolutionary structure is represented by a family $\mathfrak{G} = \{\mathfrak{B}_s\}_{s=1}^N$. Assume that each element of $\mathfrak{G}$ is a materially uniform inhomogeneous simple three-dimensional body. Then it can be represented by the following decomposition:

$$\mathfrak{B}_s = (\ldots(\mathfrak{B}_* \vee \mathfrak{L}_1) \vee \cdots \vee \mathfrak{L}_{s-2}) \vee \mathfrak{L}_{s-1}\,, \qquad s \in \{1, \ldots, N\}\,. \tag{8.4}$$

Each layer $\mathfrak{L}_k$ is assumed to be a materially uniform simple body that has a stress-free shape in $\mathscr{E}$. Moreover, assume that for each $s \in \{1, \ldots, N\}$ and any $k \in \{0, \ldots, s-1\}$, there exists a configuration $\varkappa^R_{s,k} \in \mathfrak{C}(\mathfrak{B}_s; \mathscr{E})$ of the body $\mathfrak{B}_s$ such that $\varkappa^R_{s,k}(\mathfrak{L}_k)$ is a uniform shape of $\mathfrak{L}_k$. For the body $\mathfrak{B}_4$ such an assumption is illustrated in the diagram:

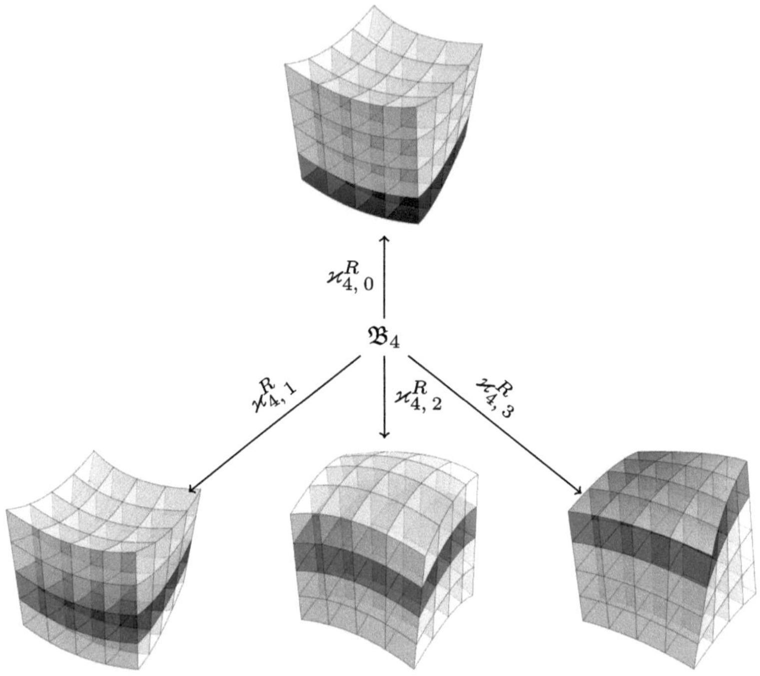


The areas in dark color correspond to a uniform shape.

### 8.5.3 The Continuous Process

**Evolving bodies.** Consider the case when material flows continuously into a substrate. This case might match the models for a variety of technological processes, like LbL processing, stereolithography, chemical vapor deposition (CVD), physical vapor deposition (PVD) and others, when the number of successive depositions of thin layers is so big that the whole process, despite of its step-by-step character, is similar to a continuous one. Such an evolving body can be represented by a continuous family $\mathfrak{G} = \{\mathfrak{B}_\alpha\}_{\alpha\in[a_0,a_1]}$ of three-dimensional materially uniform simple bodies. Each element of the family contains the substrate $\mathfrak{B}_* = \mathfrak{B}_{a_0}$ as a subbody, which will be referred to as the initial body. With the operation $\vee$ one can decompose an element $\mathfrak{B}_\alpha$ of the family $\mathfrak{G}$ into the initial body $\mathfrak{B}_*$ and an extra body $\widetilde{\mathfrak{B}}_\alpha$ as $\mathfrak{B}_\alpha = \mathfrak{B}_* \vee \widetilde{\mathfrak{B}}_\alpha$. The collection of all extra bodies constitutes the family $\widetilde{\mathfrak{G}} = \{\widetilde{\mathfrak{B}}_\alpha\}_{\alpha\in]a_0,a_1]}$.

Since $\mathfrak{B}_*$ represents a substrate, its shape at the beginning of the deposition process may be regarded as uniform (free from residual stresses). However, it cannot be said about any shape of extra bodies, because various physical and chemical processes that accompanied the deposition, cause local shrinkage. Denote[5] $\widetilde{\mathfrak{B}}^* = \mathfrak{J}(\mathfrak{B}^*, \mathfrak{B}_*) = \bigcup_{\alpha \in ]a_0, a_1]} \widetilde{\mathfrak{B}}_\alpha$. Such a body contains all extra material that was continuously added to the substrate $\mathfrak{B}_*$ during the whole evolving process. With a view to the step-by-step microstructure of the technological process one can accept the additional assumption: $\widetilde{\mathfrak{B}}^*$ as a three-dimensional manifold has the structure of a foliated manifold with 2D foils, treated as material surfaces [110]. Thus, $\widetilde{\mathfrak{B}}^*$ can be endowed with the structure of the principal bundle over the one-dimensional base. If, beyond that, each foil is diffeomorphic to some smooth 2D manifold, then $\widetilde{\mathfrak{B}}^*$ can be endowed with the structure of a fiber bundle with a typical fiber $\mathfrak{F}$. These considerations can be stated more strictly in the following way:

(i) There exists a smooth fiber bundle $(\widetilde{\mathfrak{B}}^*, \mathbb{J}, \pi, \mathfrak{F})$ with the total space $\widetilde{\mathfrak{B}}^*$, the one-dimensional base $\mathbb{J}$, and the projection $\pi\colon \widetilde{\mathfrak{B}}^* \to \mathbb{J}$, which is smooth and surjective. Each fiber $\mathfrak{S}_i := \pi^{-1}(\{i\})$, $i \in \mathbb{J}$, is diffeomorphic to the typical fiber $\mathfrak{F}$, which is a smooth two-dimensional manifold.

(ii) The fiber bundle $(\widetilde{\mathfrak{B}}^*, \mathbb{J}, \pi, \mathfrak{F})$ is related with the family $\widetilde{\mathfrak{G}}$ as follows. For each $\alpha \in ]a_0, a_1]$ there exists a submanifold $\mathbb{J}_\alpha \subset \mathbb{J}$, such that[6]

$$\widetilde{\mathfrak{B}}_\alpha = \bigcup_{i \in \mathbb{J}_\alpha} \mathfrak{S}_i \,.$$

(iii) There exists a family $\overline{\mathrm{Ref}}_{\widetilde{\mathfrak{B}}^*} = \{\varkappa^R_{\mathfrak{X}}\}_{\mathfrak{X} \in \widetilde{\mathfrak{B}}^*}$ of configurations $\varkappa^R_{\mathfrak{X}} \in \mathfrak{C}(\widetilde{\mathfrak{B}}^*; \mathscr{E})$ with the index set $\widetilde{\mathfrak{B}}^*$ and the following property. If $\mathfrak{X} \in \widetilde{\mathfrak{B}}^*$, then each point from a two-dimensional submanifold $\varkappa^R_{\mathfrak{X}}(\mathfrak{S}_{\pi(\mathfrak{X})})$ in $\mathscr{E}$ has some uniform spatial neighborhood. Moreover, if $\mathfrak{X}, \mathfrak{Y} \in \mathfrak{S}_i$ for some $i \in \mathbb{J}$, then $\varkappa^R_{\mathfrak{X}} = \varkappa^R_{\mathfrak{Y}}$.

Assumption (i) asserts that $\widetilde{\mathfrak{B}}^*$ is represented as a union of disjoint two-dimensional manifolds $\mathfrak{S}_i$:

$$\widetilde{\mathfrak{B}}^* = \bigcup_{i \in \mathbb{J}} \mathfrak{S}_i \,.$$

Assumption (ii) set the fiber bundle $(\widetilde{\mathfrak{B}}^*, \mathbb{J}, \pi, \mathfrak{F})$ in accordance with foliated structure of the elements of $\widetilde{\mathfrak{G}}$. Indeed, for each $\alpha \in ]a_0, a_1]$, there exists a smooth fiber bundle $(\widetilde{\mathfrak{B}}_\alpha, \mathbb{J}_\alpha, \pi_\alpha, \mathfrak{F})$ with the one-dimensional base $\mathbb{J}_\alpha$ and the projection $\pi_\alpha\colon \widetilde{\mathfrak{B}}_\alpha \to \mathbb{J}_\alpha$. Such a bundle is the restriction of $(\widetilde{\mathfrak{B}}^*, \mathbb{J}, \pi, \mathfrak{F})$. Assumption (iii) formalizes the suggestion related with the technological process of deposition. This claims that for every point $\mathfrak{X} \in \widetilde{\mathfrak{B}}^*$, there exists a configuration that brings the neighborhood of the point $\mathfrak{X}$, together with neighborhoods of all points from the fiber to which $\mathfrak{X}$ belongs, to the uniform state.

Let $\alpha \in ]a_0, a_1]$. For each material point $\mathfrak{X} \in \widetilde{\mathfrak{B}}_\alpha$ denote $\varkappa^R_{\alpha;\mathfrak{X}} := \varkappa^R_{\mathfrak{X}}|_{\widetilde{\mathfrak{B}}_\alpha}$.

5 For each $\alpha \in ]a_0, a_1]$ one has $\widetilde{\mathfrak{B}}_\alpha \subset \widetilde{\mathfrak{B}}^*$.

6 Here $\mathbb{J}_{a_1} := \mathbb{J}$.

**Intermediate configuration.** Fix $\alpha \in ]a_0, a_1]$ and let $\varkappa^\circ_\alpha \in \mathfrak{C}(\mathfrak{B}_\alpha; \mathscr{E})$ be a configuration. The shape $\varkappa^\circ_\alpha(\mathfrak{B}_\alpha)$ may serve as a "model" of the body $\mathfrak{B}_\alpha$ in the following sense. Any coordinate net of this shape can be transferred to the body $\mathfrak{B}_\alpha$. Figuratively speaking, the body is "dressed up" with a coordinate net, taken from the shape $\varkappa^\circ_\alpha(\mathfrak{B}_\alpha)$. Consider this in detail.

Choose local coordinates $(\xi_\alpha)^3_{i=1}$ on the shape $\varkappa^\circ_\alpha(\mathfrak{B}_\alpha)$. Let $H^\circ : U \to \mathbb{R}^3$ be a corresponding coordinate map on some open set $U \subset \varkappa^\circ_\alpha(\mathfrak{B}_\alpha)$. Then, the composition $H^\circ \circ \widehat{\varkappa}^\circ_\alpha|_{(\widehat{\varkappa}^\circ_\alpha)^{-1}(U)}$ is a coordinate map on the open set $(\widehat{\varkappa}^\circ_\alpha)^{-1}(U)$ of $\mathfrak{B}_\alpha$. The corresponding frame and coframe fields on the body are denoted by $(\partial_{\xi^i_\alpha})^3_{i=1}$, and, consequently, $(d\xi^i_\alpha)^3_{i=1}$. Despite the fact that local coordinates on the body and its intermediate shape have the same notations, the corresponding frames and coframes are different, see Tab. 8.1. We will refer to the configuration $\varkappa^\circ_\alpha$ as *intermediate configuration* and to the shape $\varkappa^\circ_\alpha(\mathfrak{B}_\alpha)$ as *intermediate shape.*

**Tab. 8.1:** Correspondence between frames and coframes on the body and the intermediate shape.

| | $\mathfrak{B}_\alpha$ | $\varkappa^\circ_\alpha(\mathfrak{B}_\alpha)$ |
|---|---|---|
| Coordinate map | $H^\circ \circ \widehat{\varkappa}^\circ_\alpha : \mathfrak{X} \mapsto (\xi^1_\alpha, \xi^2_\alpha, \xi^3_\alpha)$ | $H^\circ : x \mapsto (\xi^1_\alpha, \xi^2_\alpha, \xi^3_\alpha)$ |
| Coordinates | $(\xi^1_\alpha, \xi^2_\alpha, \xi^3_\alpha)$ | $(\xi^1_\alpha, \xi^2_\alpha, \xi^3_\alpha)$ |
| Frame | $\left(\partial_{\xi^i_\alpha}\right)^3_{i=1}$ | $\left(\boldsymbol{e}_{\xi^i_\alpha}\right)^3_{i=1}$ |
| Coframe | $(d\xi^i_\alpha)^3_{i=1}$ | $\left(\boldsymbol{e}^{\xi^i_\alpha}\right)^3_{i=1}$ |

**Parameterized intermediate configuration.** Let an intermediate configuration be such that the corresponding deformations from the intermediate shape to locally stress-free reference shapes would depend on scalar parameters, which represent the unknown parameters of the model. Such parameters (more precisely, smooth functions), should be obtained from the *evolutionary* problem. Thus, we assume that for each $\alpha \in ]a_0, a_1]$, there exists a configuration $\varkappa^\circ_\alpha \in \mathfrak{C}(\mathfrak{B}_\alpha; \mathscr{E})$, such that **1)** the shape $\varkappa^\circ_\alpha(\mathfrak{B}_*)$ is uniform; **2)** for each $\mathfrak{X} \in \widetilde{\mathfrak{B}}_\alpha$, the deformation

$$\widetilde{\gamma}_{\alpha;\mathfrak{X}} = \widehat{\varkappa}^R_{\alpha;\mathfrak{X}} \circ (\widehat{\varkappa}^\circ_\alpha|_{\widetilde{\mathfrak{B}}_\alpha})^{-1} : \varkappa^\circ_\alpha(\widetilde{\mathfrak{B}}_\alpha) \to \varkappa^R_{\alpha;\mathfrak{X}}(\widetilde{\mathfrak{B}}_\alpha),$$

from the shape $\varkappa^\circ_\alpha(\widetilde{\mathfrak{B}}_\alpha)$ into the shape $\varkappa^R_{\alpha;\mathfrak{X}}(\widetilde{\mathfrak{B}}_\alpha)$ is prescribed and depends on $k \geq 1$ free scalar parameters $\rho^1_{\alpha;\mathfrak{X}}, \ldots, \rho^k_{\alpha;\mathfrak{X}} \in \mathbb{R}$; **3)** the mappings

$$\widetilde{b}^j_\alpha : \widetilde{\mathfrak{B}}_\alpha \to \mathbb{R}, \quad \widetilde{b}^j_\alpha : \mathfrak{X} \mapsto \rho^j_{\alpha;\mathfrak{X}},$$

are smooth for all $j = 1, \ldots, k$ and $\alpha \in ]a_0, a_1]$.

Since the deformation $\widetilde{\gamma}_{\alpha;\mathfrak{X}}$ is prescribed up to parameters $\rho^j_{\alpha;\mathfrak{X}}$, the family $\{\widetilde{\boldsymbol{K}}^R_\alpha\}_{\alpha\in]a_0,a_1]}$ of uniform references can be obtained in a chart, induced by the intermediate configuration. Now, define local coordinates $(\Xi^i_{\alpha;\mathfrak{X}})_{i=1}^3$ on each shape $\varkappa^R_{\alpha;\mathfrak{X}}(\widetilde{\mathfrak{B}}_\alpha)$. Thus, the deformation gradient $T_y\widetilde{\gamma}_{\alpha;\mathfrak{X}}$ at point $y \in \varkappa^R_{\alpha;\mathfrak{X}}(\widetilde{\mathfrak{B}}_\alpha)$ is represented by

$$T_y\widetilde{\gamma}_{\alpha;\mathfrak{X}} = \left.\frac{\partial\widetilde{\gamma}^i_{\alpha;\mathfrak{X}}\left(\xi^1_\alpha, \xi^2_\alpha, \xi^3_\alpha; b^1_\alpha(\mathfrak{X}), \dots, b^k_\alpha(\mathfrak{X})\right)}{\partial\xi^j_\alpha}\right|_y \boldsymbol{e}_{\Xi^i_{\alpha;\mathfrak{X}}}|_y \otimes \boldsymbol{e}^{\xi^j_\alpha}|_{\widetilde{\gamma}_{\alpha;\mathfrak{X}}(y)} ,$$

where $\widetilde{\gamma}^i_{\alpha;\mathfrak{X}}$ is the coordinate representation of $\widetilde{\gamma}_{\alpha;\mathfrak{X}}$. For any point $\mathfrak{X} \in \mathfrak{B}_\alpha$, the configuration gradient $T_\mathfrak{X}\varkappa^\circ_\alpha : T_\mathfrak{X}\mathfrak{B}_\alpha \to \mathscr{V}$ has the dyadic decomposition

$$T_\mathfrak{X}\varkappa^\circ_\alpha = \boldsymbol{e}_{\xi^i_\alpha}|_x \otimes d\xi^i_\alpha|_\mathfrak{X} ,$$

where $x = \varkappa^\circ_\alpha(\mathfrak{X})$. Thus, the tangent map $T_\mathfrak{Y}\varkappa^R_{\alpha;\mathfrak{X}} : T_\mathfrak{Y}\widetilde{\mathfrak{B}}_\alpha \to \mathscr{V}$ to $\varkappa^R_{\alpha;\mathfrak{X}}$ at point $\mathfrak{Y} \in \widetilde{\mathfrak{B}}_\alpha$ can be derived as a composition $T_\mathfrak{Y}\varkappa^R_{\alpha;\mathfrak{X}} = T_y\widetilde{\gamma}_{\alpha;\mathfrak{X}} \circ T_\mathfrak{X}\widetilde{\varkappa}^\circ_\alpha|_{\widetilde{\mathfrak{B}}_\alpha}$:

$$T_\mathfrak{Y}\varkappa^R_{\alpha;\mathfrak{X}} = \left.\frac{\partial\widetilde{\gamma}^i_{\alpha;\mathfrak{X}}\left(\xi^1_\alpha, \xi^2_\alpha, \xi^3_\alpha; \widetilde{b}^1_\alpha(\mathfrak{X}), \dots, \widetilde{b}^k_\alpha(\mathfrak{X})\right)}{\partial\xi^j_\alpha}\right|_y \boldsymbol{e}_{\Xi^i_{\alpha;\mathfrak{X}}}|_y \otimes d\xi^j_\alpha|_\mathfrak{Y} ,$$

where $y = \varkappa^R_{\alpha;\mathfrak{X}}(\mathfrak{Y})$. Hence, the uniform reference $\widetilde{\boldsymbol{K}}^R_\alpha : \mathfrak{X} \mapsto T_\mathfrak{X}\varkappa^R_{\alpha;\mathfrak{X}}$ is defined at point $\mathfrak{X} \in \widetilde{\mathfrak{B}}_\alpha$ by the following expression:

$$\widetilde{\boldsymbol{K}}^R_\alpha|_\mathfrak{X} = \left.\frac{\partial\widetilde{\gamma}^i_{\alpha;\mathfrak{X}}\left(\xi^1_\alpha, \xi^2_\alpha, \xi^3_\alpha; \widetilde{b}^1_\alpha(\mathfrak{X}), \dots, \widetilde{b}^k_\alpha(\mathfrak{X})\right)}{\partial\xi^j_\alpha}\right|_x \boldsymbol{e}_{\Xi^i_{\alpha;\mathfrak{X}}}|_x \otimes d\xi^j_\alpha|_\mathfrak{X} , \tag{8.5}$$

where $x = \varkappa^R_{\alpha;\mathfrak{X}}(\mathfrak{X})$.

**The material metric.** So, uniform reference was determined up to mappings $\widetilde{b}^j_\alpha$, $j = 1, \dots, k$. It is convenient to extend them over the whole $\mathfrak{B}_\alpha$, *i.e.*, to construct the mapping $b^j_\alpha : \mathfrak{B}_\alpha \to \mathbb{R}$ defined by the relation

$$b^j_\alpha(\mathfrak{X}) = \begin{cases} 0, & \mathfrak{X} \in \mathfrak{B}_* \cup \partial_{\mathfrak{B}_\alpha}(\mathfrak{B}_\alpha \setminus \mathfrak{B}_*) , \\ \widetilde{b}^j_\alpha(\mathfrak{X}), & \mathfrak{X} \in \widetilde{\mathfrak{B}}_\alpha . \end{cases}$$

The set $\partial_{\mathfrak{B}_\alpha}(\mathfrak{B}_\alpha \setminus \mathfrak{B})$ represents the surface of discontinuity, which appears between the substrate and the extra body. It follows from the homogeneity of the body $\mathfrak{B}_*$ that its material Riemannian metric is Euclidean. Thus, in the general case, the material metric in the whole body $\mathfrak{B}_\alpha$ would have points of discontinuity in $\partial_{\mathfrak{B}_\alpha}(\mathfrak{B}_\alpha \setminus \mathfrak{B})$. Denote the metric on $\mathscr{E}$ by $\boldsymbol{g}$, then

$$\mathscr{G}_\alpha(\mathfrak{X}) = \begin{cases} (\varkappa^\circ_\alpha)^*\boldsymbol{g}(\mathfrak{X}), & \mathfrak{X} \in \mathfrak{B}_* \cup \partial_{\mathfrak{B}_\alpha}(\mathfrak{B}_\alpha \setminus \mathfrak{B}_*) , \\ \widetilde{\mathscr{G}}_\alpha(\mathfrak{X}), & \mathfrak{X} \in \widetilde{\mathfrak{B}}_\alpha , \end{cases} \tag{8.6}$$

represents the total material metric $\mathscr{G}_\alpha$ in $\mathfrak{B}_\alpha$. Here, $\widetilde{\mathscr{G}}_\alpha$ is defined by (8.3).

The functions $b^j_\alpha$, $j = 1, \dots, k$, on which (8.5) and (8.6) depend, define the distortion parameters. They can be defined by the modeling of physical and chemical processes during adhering of the layer. One of the most simple factors defining such processes is shrinkage, which occurs during a relatively short time interval (approximately instantly in the time scale of the whole adhering process). The idea of instantaneous shrinkage leads to the following formalization:

$$\forall \alpha < \beta \in ]a_0, a_1] : b^j_\beta|_{\mathfrak{B}_\alpha} = b^j_\alpha , \quad j = 1, \dots, k .$$

These equalities mean that parameters governing inhomogeneity in the solid cannot vary in internal parts of it.

# 9 Material Connections

**Introductory remarks.** Let $\mathfrak{M}$ be a smooth manifold. For a smooth vector field $u \in \mathrm{Vec}(\mathfrak{M})$, the partial derivatives $\partial_i u^j$ do not form the components of a second-order tensor, in general. Similarly, for a 1-form $v \in \mathrm{CVec}(\mathfrak{M})$, the partial derivatives $\partial_i v_j$ do not form any tensor, although the differences $\partial_i v_j - \partial_j v_i$ form the exterior derivative $dv$. The notion of derivation needs to be generalized, since we are dealing with a body $\mathfrak{B}$ and a physical space $\mathscr{P}$, which are not Euclidean in general, and intend to formulate differential balance equations. This generalization is provided by *covariant differentiation.*

Let us explain the difficulty of introducing the covariant derivative straightforwardly, as a limit of the relative increment. Let $(\mathfrak{E}, \mathfrak{M}, \pi)$ be some smooth vector bundle over a smooth manifold $\mathfrak{M}$ and $s\colon \mathfrak{M} \to \mathfrak{E}$ be its section. If $p, q \in \mathfrak{M}$ are different material points, then the sum $s_p + s_q$ has no meaning. Indeed, for $p \neq q$, the values $s_p$ and $s_q$ belong to the non-intersecting vector spaces $\mathfrak{E}_p$ and $\mathfrak{E}_q$. From this follows the difficulty of defining the derivative via a limit of the relative increment. One needs an extra intrinsic rule to connect $\mathfrak{E}_p$ and $\mathfrak{E}_q$. Such a rule is provided with a *connection* on the vector bundle.

In Chapter 14, we introduce the operation of covariant differentiation, drawing on a very general notion of the principal bundle. Since a vector bundle is a particular case of the principal bundle, the reasonings of aforementioned chapter can be directly applied to it. However, the operation of covariant differentiation on vector bundles can be constructed axiomatically. This is realized in this chapter.

In Chapter 8, the material metric $\mathscr{G}$ was introduced. This metric, in turn, generates the Levi–Civita connection on the body. The Riemann curvature tensor of such a connection gives a measure of inhomogeneity, the scalar curvature. In this chapter, we provide a second way of constructing the connection on the tangent bundle of the body. This way is intimately related with Cartan's idea of the moving frame method, and as a result, allows one to construct a uniform shape that would be a non-Euclidean space.

## 9.1 Connections on Vector Bundles

**Axiomatic definition.** The *connection* $\nabla$ on a smooth vector bundle $(\mathfrak{E}, \mathfrak{M}, \pi)$ is a mapping

$$\nabla\colon \mathrm{Vec}(\mathfrak{M}) \times \mathrm{Sec}(\mathfrak{E}) \to \mathrm{Sec}(\mathfrak{E})\,, \quad (u, s) \mapsto \nabla_u s\,,$$

which satisfies the following axioms [61, 131, 132]:

($\nabla_1$) $\forall u \in \mathrm{Vec}(\mathfrak{M})\ \forall s_1, s_2 \in \mathrm{Sec}(\mathfrak{E})\colon\ \nabla_u(s_1 + s_2) = \nabla_u s_1 + \nabla_u s_2;$

($\nabla_2$) $\forall u_1, u_2 \in \mathrm{Vec}(\mathfrak{M})\ \forall s \in \mathrm{Sec}(\mathfrak{E})\colon\ \nabla_{u_1+u_2} s = \nabla_{u_1} s + \nabla_{u_2} s;$

https://doi.org/10.1515/9783110563214-009

($\nabla_3$) $\forall u \in \text{Vec}(\mathfrak{M})\ \forall f \in C^\infty(\mathfrak{M})\ \forall s \in \text{Sec}(\mathfrak{E})$: $\nabla_u(fs) = u(f)s + f\nabla_u s$;
($\nabla_4$) $\forall u \in \text{Vec}(\mathfrak{M})\ \forall f \in C^\infty(\mathfrak{M})\ \forall s \in \text{Sec}(\mathfrak{E})$: $\nabla_{fu} s = f\nabla_u s$.

We refer to $\nabla_u$ as a *covariant derivative along u*. The Leibniz rule ($\nabla_3$) and the property ($\nabla_4$) imply that $\nabla_u$ satisfies the locality property. That is, let $U \subset \mathfrak{M}$ be any open set. Suppose that vector fields $u_1$, $u_2$, and sections $s_1$, $s_2$ of $\mathfrak{E}$ are equal in $U$: $u_1|_U = u_2|_U$ and $s_1|_U = s_2|_U$. Then,

$$\nabla_u s_1|_U = \nabla_u s_2|_U \,.$$

This property allows one to define connection[1] [131]

$$\nabla|_U \colon \text{Vec}(U) \times \text{Sec}(\mathfrak{E}|_U) \to \text{Sec}(\mathfrak{E}|_U)$$

on an open set $U \subset \mathfrak{M}$ by using the original connection $\nabla$. We refer to $\nabla|_U$ as a restriction of $\nabla$ and in the following, we omit the character $U$ after the vertical bar.

**Coordinate representation.** Let $s \in \text{Sec}(\mathfrak{E})$ and $u \in \text{Vec}(\mathfrak{M})$, while $U \subset \mathfrak{M}$ be a coordinate domain. Choose a smooth local frame $(e_a)$ for $\mathfrak{E}$ over $U$, and coordinate frames $(\partial_c)$, $(dx^c)$, respectively, for $T\mathfrak{M}$, $T^*\mathfrak{M}$ over $U$. Now obtain coordinate representations for the restriction of $\nabla_u s$ on $U$. Firstly, note that there exist smooth mappings $\omega^b_{\ ca} \in C^\infty(U)$, such that

$$\nabla_{\partial_a} e_b = \omega^c_{\ ab} e_c \,. \tag{9.1}$$

Then, applying the Leibniz rule ($\nabla_3$), we arrive at the expression

$$\nabla_u s = u^a \left\{ \partial_a s^c + s^b \omega^c_{\ ab} \right\} e_c \,,$$

where $u = u^a \partial_a$, $s = s^b e_b$.

**The covariant derivative of tensorial fields.** Given is that connection $\nabla$ on a smooth vector bundle can be extended to the other levels of the tensorial "tower" inductively. First, define the mapping

$$\nabla^{\text{sc}} \colon \text{Vec}(\mathfrak{M}) \times C^\infty(\mathfrak{M}) \to C^\infty(\mathfrak{M}) \,, \quad \nabla^{\text{sc}}_u f := u(f) \,.$$

Next, define the connection

$$\nabla^* \colon \text{Vec}(\mathfrak{M}) \times \text{Sec}(\mathfrak{E}^*) \to \text{Sec}(\mathfrak{E}^*)$$

on the dual vector bundle $(\mathfrak{E}^*, \mathfrak{M}, \pi)$ by the following characteristic property:

$$\forall s \in \text{Sec}(\mathfrak{E})\ \forall g \in \text{Sec}(\mathfrak{E}^*)\ \forall u \in \text{Vec}(\mathfrak{M}) : \quad \nabla^{\text{sc}}_u \langle g, s \rangle = \langle \nabla^*_u g, s \rangle + \langle g, \nabla_u s \rangle \,.$$

1 Here, $\mathfrak{E}|_U$ is the restriction of vector bundle $\mathfrak{E}$, *i.e.*, the smooth vector bundle with base $U$ and total space $\mathfrak{E}|_U = \bigsqcup_{p \in U} \mathfrak{E}_p$.

Let $U \subset \mathfrak{M}$ be a coordinate domain, $(e_a)$ be a smooth local frame for $\mathfrak{E}$ over $U$, and $(e^a)$ be its dual. Introduce the notations $u = \partial_a$, $g = e^b$, $s = e_c$. Then, in $U$ one has

$$\nabla^*_{\partial_a} e^b = v^b{}_{ad} e^d ,$$

where $v^b{}_{ad} \in C^\infty(U)$ are unknown functions. Since $\langle e^b, e_c\rangle = \delta^b_c$, with (9.1), one obtains

$$v^b{}_{ac} + \omega^b{}_{ac} = 0 ,$$

which implies

$$\nabla^*_{\partial_a} e^b = -\omega^b{}_{ad} e^d . \tag{9.2}$$

For a smooth section $g \in \mathrm{Sec}(\mathfrak{E}^*)$ and a vector field $u \in \mathrm{Vec}(\mathfrak{M})$, the representation

$$\nabla^*_u g = u^a \left\{\partial_a g_c - g_b \omega^b{}_{ac}\right\} e^c ,$$

is a direct consequence of (9.2) and the Leibniz rule ($\nabla_3$). Here, $g = g_a e^a$, $u = u^b \partial_b$.

Suppose now that the connections $\nabla^1$ and $\nabla^2$ are defined on smooth vector bundles $(\mathfrak{E}_1, \mathfrak{M}, \pi_1)$ and $(\mathfrak{E}_2, \mathfrak{M}, \pi_2)$ over the same base $\mathfrak{M}$. These connections induce new ones,

$$\nabla^{\mathfrak{E}_1 \oplus \mathfrak{E}_2} : \mathrm{Vec}(\mathfrak{M}) \times \mathrm{Sec}(\mathfrak{E}_1 \oplus \mathfrak{E}_2) \to \mathrm{Sec}(\mathfrak{E}_1 \oplus \mathfrak{E}_2) ,$$
$$\nabla^{\mathfrak{E}_1 \otimes \mathfrak{E}_2} : \mathrm{Vec}(\mathfrak{M}) \times \mathrm{Sec}(\mathfrak{E}_1 \otimes \mathfrak{E}_2) \to \mathrm{Sec}(\mathfrak{E}_1 \otimes \mathfrak{E}_2) ,$$

which are defined on the direct sum $(\mathfrak{E}_1 \oplus \mathfrak{E}_2, \mathfrak{M}, \pi^{\mathfrak{E}_1 \oplus \mathfrak{E}_2})$ and the tensor product $(\mathfrak{E}_1 \otimes \mathfrak{E}_2, \mathfrak{M}, \pi^{\mathfrak{E}_1 \otimes \mathfrak{E}_2})$ bundles as follows:

$$\begin{aligned} \nabla_u^{\mathfrak{E}_1 \oplus \mathfrak{E}_2}(s_1 \oplus s_2) &:= \nabla^1_u s_1 \oplus \nabla^2_u s_2 , \\ \nabla_u^{\mathfrak{E}_1 \otimes \mathfrak{E}_2}(s_1 \otimes s_2) &:= (\nabla^1_u s_1) \otimes s_2 + s_1 \otimes \nabla^2_u s_2 , \end{aligned} \tag{9.3}$$

where $u \in \mathrm{Vec}(\mathfrak{M})$ and $s_i \in \mathrm{Sec}(\mathfrak{E}_i)$, $i = 1, 2$.

**Notations for components.** Let $u \in \mathrm{Vec}(\mathfrak{M})$, $s \in \mathrm{Sec}(\mathfrak{E})$ and $g \in \mathrm{Sec}(\mathfrak{E}^*)$. Then,

$$\nabla_u s = u^a \nabla_a s^c e_c , \quad \nabla^*_u g = u^a \nabla^*_a g_c e^c ,$$

where

$$\nabla_a s^c := \partial_a s^c + s^b \omega^c{}_{ab} , \quad \nabla^*_a g_c := \partial_a g_c - g_b \omega^b{}_{ac} .$$

**Connection in operator-like form.** If $\nabla$ is a connection on $\mathfrak{E}$, then one can set

$$\widetilde{\nabla} s := \nabla_a s^c e_c \otimes dx^a ,$$

on every coordinate domain. Thus, the mapping [131, 133]

$$\widetilde{\nabla} : \mathrm{Sec}(\mathfrak{E}) \to \mathrm{Sec}(\mathfrak{E} \otimes T^*\mathfrak{M}) ,$$

is well defined. Such a mapping satisfies the following properties:

$\tilde{\nabla}_1$) *Additivity:* $\forall s_1, s_2 \in \mathrm{Sec}(\mathfrak{E})$: $\tilde{\nabla}(s_1 + s_2) = \tilde{\nabla}s_1 + \tilde{\nabla}s_2$;
$\tilde{\nabla}_2$) *Homogeneity:* $\forall s \in \mathrm{Sec}(\mathfrak{E})\ \forall \lambda \in \mathbb{R}$: $\tilde{\nabla}(\lambda s) = \lambda\tilde{\nabla}s$;
$\tilde{\nabla}_3$) *The Leibniz rule:* $\forall s \in \mathrm{Sec}(\mathfrak{E})\ \forall f \in C^\infty(\mathfrak{M})$: $\tilde{\nabla}(fs) = f\tilde{\nabla}s + s \otimes df$.

Conversely, given $\tilde{\nabla}$ generates the mapping $\nabla$ via formula

$$\nabla_u s := \tilde{\nabla}s \llcorner u\,.$$

In the following, we omit the tilde symbol in $\tilde{\nabla}$ and simply write $\nabla$.

**Parallel transport.** Let $\gamma: \mathbb{I} \to \mathfrak{M}$ be some smooth curve on $\mathfrak{M}$. A section $s \in \mathrm{Sec}(\mathfrak{E})$ is said to be *parallel along* $\gamma$, if

$$\forall t \in \mathbb{I}: \nabla_{\gamma'(t)} s = 0\,.$$

Choosing the local coordinate patch $(U, x^i)$, one can express the above equation componentwise:

$$\dot{s}^a + \dot{x}^c s^b \omega^a{}_{cb} = 0\,, \quad a = 1, \dots, \mathrm{rank}(\mathfrak{E})\,.$$

Here, $s = s^a e_a$, $\gamma' = \dot{x}^a \partial_a$ and $(\cdot)^\cdot$ denotes $d/dt$. Let $p = \gamma(t_0)$. Then one has the following initial condition: $s_p = s_0$. Thus, locally, there exists a unique solution for the Cauchy problem

$$\begin{aligned} \dot{s}^a + \dot{x}^c s^b \omega^a{}_{cb} &= 0\,, \quad a = 1, \dots, \mathrm{rank}(\mathfrak{E})\,, \\ s^a_p &= s^a_0\,, \quad a = 1, \dots, \mathrm{rank}(\mathfrak{E})\,. \end{aligned} \tag{9.4}$$

Therefore, one can define the *parallel transport map* $P^t_{t_0}: \mathfrak{E}_{\gamma(t_0)} \to \mathfrak{E}_{\gamma(t)}$ as follows: with each element $s_0 \in \mathfrak{E}_{\gamma(t_0)}$ it assigns the value $s(t)$ of the solution $s$ for the Cauchy problem (9.4); $P^t_{t_0}: s_0 \mapsto s(t)$.

Now, let $u = \gamma'(t_0)$ be the velocity vector of the curve $\gamma$ at the point $p$ and $s \in \mathrm{Sec}(\mathfrak{E})$ be a smooth section. Then, the value $\nabla_u s|_p$ can be defined via the limit

$$\nabla_u s|_p = \lim_{h \to 0} \frac{P^{t_0}_{t_0+h} s_{\gamma(t_0+h)} - s_{\gamma(t_0)}}{h} = \left.\frac{d}{dt} P^{t_0}_t s_{\gamma(t)}\right|_{t=t_0}\,.$$

Here, $P^{t_0}_{t_0+h} s_{\gamma(t_0+h)}$ belongs to $\mathfrak{E}_{\gamma(t_0)}$, and this means that for a sufficiently small $h$, the difference $P^{t_0}_{t_0+h} s_{\gamma(t_0+h)} - s_{\gamma(t_0)}$ is well defined and belongs to $\mathfrak{E}_{\gamma(t_0)}$. Thus, the element $\nabla_u s|_p \in \mathfrak{E}_{\gamma(t_0)}$ may be considered as a derivative of the section $s$ at the point $p$ along the vector $u$.

## 9.2 Affine Connection

Let $\mathfrak{M}$ be a smooth $n$-dimensional manifold. In the remainder of the chapter, we consider the case when the connection $\nabla$ is given on the tangent bundle $T\mathfrak{M}$. That is,

$$\nabla: \mathrm{Vec}(\mathfrak{M}) \times \mathrm{Vec}(\mathfrak{M}) \to \mathrm{Vec}(\mathfrak{M})\,.$$

In this case, we refer to $\nabla$ as an *affine connection*, and coefficients $\omega^a{}_{bc}$ are denoted by $\Gamma^a{}_{bc}$. Previous reasonings show that $\nabla$ can be induced on $T^*\mathfrak{M}$ and any other vector bundle that is constructed from copies of $T\mathfrak{M}$ and $T^*\mathfrak{M}$ via direct sum and tensor product operations. Such induced connections would be denoted similarly, by $\nabla$.

**Expressions for covariant derivatives.** If $(\partial_i)$, $(dx^i)$ are, respectively, the coordinate frame and the coframe, then from (9.1) and (9.2), it follows that

$$\nabla_{\partial_a}\partial_b = \Gamma^c{}_{ab}\partial_c\,, \quad \nabla_{\partial_a}dx^b = -\Gamma^b{}_{ac}dx^c\,.$$

If $\mathcal{T} \in \mathrm{Sec}(T^{(k,l)}(T\mathfrak{M}))$,

$$\mathcal{T} = T^{i_1\ldots i_k}{}_{j_1\ldots j_l}\partial_{i_1}\otimes\cdots\otimes\partial_{i_k}\otimes dx^{j_1}\otimes\cdots\otimes dx^{j_l}\,,$$

then using (9.3), one obtains

$$\begin{aligned}\nabla\mathcal{T} &= \nabla_i T^{i_1\ldots i_k}{}_{j_1\ldots j_l}\partial_{i_1}\otimes\cdots\otimes\partial_{i_k}\otimes dx^{j_1}\otimes\cdots\otimes dx^{j_l}\otimes dx^i\,,\\ \nabla_i T^{i_1\ldots i_k}{}_{j_1\ldots j_l} &= \partial_i T^{i_1\ldots i_k}{}_{j_1\ldots j_l} + T^{b\ldots i_k}{}_{j_1\ldots j_l}\Gamma^{i_1}{}_{ib} + \cdots + T^{i_1\ldots b}{}_{j_1\ldots j_l}\Gamma^{i_k}{}_{ib}\\ &\quad - T^{i_1\ldots i_k}{}_{b\ldots j_l}\Gamma^b{}_{ij_1} - \cdots - T^{i_1\ldots i_k}{}_{j_1\ldots b}\Gamma^b{}_{ij_l}\,.\end{aligned}$$

### 9.2.1 The Transformation Law

**Transformation of connection coefficients.** Let $(e_i)_{i=1}^n$ and $(\vartheta^i)_{i=1}^n$ be the local frame and coframe defined, respectively, on an open set $U \subset \mathfrak{M}$, which is a domain for some chart. Assume that $n^2$ smooth functions $\Omega^i{}_j \colon U \to \mathbb{R}$ that define the smooth field $[\Omega^i{}_j]\colon U \to \mathrm{GL}(n;\mathbb{R})$ of invertible $n\times n$ matrices are given. This field generates the local frame

$$U \ni p \mapsto \left(e'_i|_p\right)_{i=1}^n\,, \quad e'_i = \Omega^j{}_i e_j\,.$$

The corresponding dual coframe $U \ni p \mapsto (\vartheta'^i|_p)_{i=1}^n$ can be expressed through $(\vartheta^i)$ via the relations $\vartheta'^i = \mho^i{}_j\vartheta^j$, where $[\mho^i{}_j] = [\Omega^i{}_j]^{-1}$. Denote the corresponding connection coefficients by $\Gamma^k{}_{ij}$ and $\Gamma'^k{}_{ij}$, that is,

$$\nabla_{e_i}e_j = \Gamma^k{}_{ij}e_k\,, \quad \text{and} \quad \nabla_{e'_i}e'_j = \Gamma'^k{}_{ij}e'_k\,.$$

Obtain the transformation law from $\Gamma^k{}_{ij}$ to $\Gamma'^k{}_{ij}$. If $(\partial_{x^i})$ is a coordinate frame defined on $U$, then $e_i = A^j{}_i\partial_{x^j}$, where $[A^i{}_j]\colon U \to \mathrm{GL}(n;\mathbb{R})$ is a uniquely determined smooth field of matrices. One obtains

$$\nabla_{e'_j}e'_k = \nabla_{(\Omega^s{}_j e_s)}\left(\Omega^q{}_k e_q\right) = \Omega^s{}_j\nabla_{e_s}\left(\Omega^q{}_k e_q\right) = \Omega^s{}_j\left\{\Omega^q{}_k\nabla_{e_s}e_q + e_s\left(\Omega^q{}_k\right)e_q\right\}\,.$$

Since $\Gamma'^{i}{}_{jk} = \langle \vartheta'^{i}, \nabla_{e_j'} e_k' \rangle = \mho^{i}{}_{m} \langle \vartheta^{m}, \nabla_{e_j'} e_k' \rangle$ and $e_s(\Omega^{q}{}_{j}) = A^{l}{}_{s} \partial_{x^l} \Omega^{q}{}_{j}$, it follows that

$$\Gamma'^{i}{}_{jk} = \mho^{i}{}_{m} \Omega^{s}{}_{j} \left\{ \Omega^{q}{}_{k} \langle \vartheta^{m}, \nabla_{e_s} e_q \rangle + \langle \vartheta^{m}, e_q \rangle A^{l}{}_{s} \partial_{x^l} \Omega^{q}{}_{k} \right\} .$$

Taking relations $\Gamma^{m}{}_{sq} = \langle \vartheta^{m}, \nabla_{e_s} e_q \rangle$, and $\langle \vartheta^{m}, e_q \rangle = \delta^{m}_{q}$ into account leads us to the final expression

$$\Gamma'^{i}{}_{jk} = \Gamma^{m}{}_{sq} \mho^{i}{}_{m} \Omega^{s}{}_{j} \Omega^{q}{}_{k} + A^{l}{}_{s} \mho^{i}{}_{m} \Omega^{s}{}_{j} \partial_{x^l} \Omega^{m}{}_{k} . \tag{9.5}$$

**The case of holonomic frames.** The formula (9.5) determines the transformation law in the general case, when one changes the non-holonomic frame into a non-holonomic one. If the frame $(e_i)$ is holonomic, that is, $A^{l}{}_{s} = \delta^{l}_{s}$, then the relation (9.5) can be simplified:

$$\Gamma'^{i}{}_{jk} = \Gamma^{m}{}_{sq} \mho^{i}{}_{m} \Omega^{s}{}_{j} \Omega^{q}{}_{k} + \mho^{i}{}_{m} \Omega^{s}{}_{j} \partial_{x^s} \Omega^{m}{}_{k} . \tag{9.6}$$

If local frames $(e_i)$ and $(e_i')$ are both holonomic, then we are dealing with a change of coordinates. Thus, $\Omega^{i}{}_{j} = \dfrac{\partial x^{i}}{\partial x'^{j}}$, where $(x^{i})$ generate $(e_i)$ and $(x'^{i})$ generate $(e_i')$. The expression (9.6) takes the form

$$\Gamma'^{i}{}_{jk} = \Gamma^{m}{}_{sq} \frac{\partial x'^{i}}{\partial x^{m}} \frac{\partial x^{s}}{\partial x'^{j}} \frac{\partial x^{q}}{\partial x'^{k}} + \frac{\partial x'^{i}}{\partial x^{m}} \frac{\partial^2 x^{m}}{\partial x'^{j} \partial x'^{k}} . \tag{9.7}$$

Formula (9.7) defines the connection transformation induced by the change of coordinates. Note that due to the last item in the right hand side of (9.7), such a transformation does not satisfy the transformation rule of the third-rank tensor field.

### 9.2.2 Torsion, Curvature, and Non-Metricity

Let $\mathfrak{M}$ be a smooth manifold with the Riemannian metric $g$ and the connection $\nabla$. The fact that the values of $\Gamma^{i}{}_{jk}$ do not vanish is no evidence that the geometry on $\mathfrak{M}$ is non-Euclidean. Indeed, it is easy to check that in Euclidean space $\Gamma^{i}{}_{jk} = 0$ only in Cartesian coordinates, while curvilinear coordinates make $\Gamma^{i}{}_{jk}$ non-vanishing. The property of geometry, with which one associates the term “non-Euclidean”, is intimately related with tensor fields of torsion,

$$\mathfrak{T} \colon \mathrm{Vec}(\mathfrak{M}) \times \mathrm{Vec}(\mathfrak{M}) \to \mathrm{Vec}(\mathfrak{M}) ,$$

Riemann curvature,

$$\mathfrak{R} \colon \mathrm{Vec}(\mathfrak{M}) \times \mathrm{Vec}(\mathfrak{M}) \times \mathrm{Vec}(\mathfrak{M}) \to \mathrm{Vec}(\mathfrak{M}) ,$$

and non-metricity,

$$\mathfrak{Q} \colon \mathrm{Vec}(\mathfrak{M}) \times \mathrm{Vec}(\mathfrak{M}) \times \mathrm{Vec}(\mathfrak{M}) \to C^{\infty}(\mathfrak{M}) .$$

These fields are defined via the following relations [15]:

$$\mathfrak{T}(u, v) = \nabla_u v - \nabla_v u - [u, v] \ ; \tag{9.8}$$

$$\mathfrak{R}(u, v, w) = \nabla_u \nabla_v w - \nabla_v \nabla_u w - \nabla_{[u,v]} w \ ; \tag{9.9}$$

$$\mathfrak{Q} = -\nabla \mathcal{g} \ . \tag{9.10}$$

Here, $[\cdot, \cdot]$ is the Lie bracket (13.20).

**Remark 9.1.** The torsion tensor (9.8) is a measure of the connection symmetry. If $(\partial_i)$ is a coordinate frame on $\mathfrak{M}$, then

$$\begin{aligned}\mathfrak{T}(u, v) &= \nabla_u v - \nabla_v u - [u, v] \\ &= u^i \left(\partial_i v^k + v^j \Gamma^k{}_{ij}\right) \partial_k - v^j \left(\partial_j u^k + u^i \Gamma^k{}_{ji}\right) \partial_k - \left(u^i \partial_i v^k - v^j \partial_j u^k\right) \partial_k \\ &= u^i v^j \left(\Gamma^k{}_{ij} - \Gamma^k{}_{ji}\right) \partial_k \ ,\end{aligned}$$

from which follows the expression $\mathfrak{T} = \left(\Gamma^k{}_{ij} - \Gamma^k{}_{ji}\right) \partial_k \otimes dx^i \otimes dx^j$. This means that $\mathfrak{T} = 0$ *iff* $\Gamma^k{}_{ij} = \Gamma^k{}_{ji}$, for all $i, j, k$.

**Remark 9.2.** As follows from (9.10), the non-metricity tensor $\mathfrak{Q}$ determines an incoherence of the connection $\nabla$ with the Riemannian metric $\mathcal{g}$.

One can obtain the torsion $\mathfrak{T}$ and the non-metricity $\mathfrak{Q}$ via formulae (9.8) and (9.10) using a prescribed Riemannian metric (only non-metricity requires a metric) $\mathcal{g}$ and a connection $\nabla$. Consider the opposite situation, when one has to define the affine connection $\nabla$, corresponding to given Riemannian metric $\mathcal{g}$, torsion $\mathfrak{T}$, and non-metricity $\mathfrak{Q}$.

Denote by $\nabla$ the desired connection. Choose the coordinate frame $(\partial_i)$ and the coframe $(dx^i)$. Taking into account (9.3), and deriving the expressions for $\mathfrak{Q}$ and $\mathfrak{T}$ in components, one can obtain the following system of $2n^3$ equations:

$$\begin{aligned}\Gamma^k{}_{ij} - \Gamma^k{}_{ji} &= T^k{}_{ij} \ , \quad i, j, k = 1, \ldots, n \ , \\ \partial_i g_{jk} - \Gamma^m{}_{ij} g_{mk} - \Gamma^m{}_{ik} g_{mj} &= -Q_{ijk} \ , \quad i, j, k = 1, \ldots, n \ .\end{aligned}$$

Cyclic permutations give the expressions

$$\begin{aligned}-Q_{ijk} &= \partial_i g_{jk} - \Gamma^m{}_{ij} g_{mk} - \Gamma^m{}_{ik} g_{mj} \ , \\ -Q_{jki} &= \partial_j g_{ki} - \Gamma^m{}_{jk} g_{mi} - \Gamma^m{}_{ji} g_{mk} \ , \\ -Q_{kij} &= \partial_k g_{ij} - \Gamma^m{}_{ki} g_{mj} - \Gamma^m{}_{kj} g_{mi} \ .\end{aligned}$$

Summation of the second and third equations and subtracting the first one from the result leads to the equality

$$Q_{ijk} - Q_{jki} - Q_{kij} = \partial_j g_{ik} + \partial_k g_{ij} - \partial_i g_{jk} + T^m{}_{ij} g_{mk} + T^m{}_{ik} g_{mj} - \Gamma^m{}_{jk} g_{mi} - \Gamma^m{}_{kj} g_{mi} \ .$$

Thus, finally one has

$$2\Gamma^m{}_{jk}g_{mi} = \partial_j g_{ik} + \partial_k g_{ij} - \partial_i g_{jk} + T^m{}_{ij}g_{mk} + T^m{}_{ik}g_{mj} + T^m{}_{jk}g_{mi} + Q_{jki} + Q_{kij} - Q_{ijk}\,. \tag{9.11}$$

Since $[g_{ij}]$ is invertible, the following relations for $\Gamma^i{}_{jk}$ can be also derived:

$$\Gamma^i{}_{jk} = \frac{g^{im}}{2}(\partial_j g_{mk} + \partial_k g_{mj} - \partial_m g_{jk}) + \frac{g^{im}}{2}(T_{mjk} + T_{kmj} + T_{jmk}) + \frac{g^{im}}{2}(Q_{jkm} + Q_{kmj} - Q_{mjk})\,, \tag{9.12}$$

where $T_{kij} = g_{km}T^m{}_{ij}$. Formula (9.12) shows that it is sufficient to define the Riemannian metric and independent tensors $\mathfrak{Q}$ and $\mathfrak{T}$ to introduce an affine connection on the manifold $\mathfrak{M}$. Thus, $(\mathcal{g}, \mathfrak{T}, \mathfrak{Q})$ can be considered as a triple of independent variables.

### 9.2.3 A Particular Case: Euclidean Space

Let $\mathscr{E}$ be an affine Euclidean space with $\dim \mathscr{E} = m$. In the framework of conventional Euclidean geometry, a vector $\boldsymbol{u}$ is identified with the covector $\boldsymbol{u}^\flat = \mathbf{g}(\boldsymbol{u}, \cdot)$, and vice versa. Here, $\mathbf{g}$ is a Riemannian metric that defines the scalar product $(\cdot)$. Accordingly, if $(\boldsymbol{e}_k)_{k=1}^m$ is a frame in $\mathscr{E}$, then its corresponding dual is considered as a frame $(\boldsymbol{e}^k)_{k=1}^m$ that satisfies the relations $\boldsymbol{e}_k \cdot \boldsymbol{e}^s = \delta_k^s$, $s, k = 1, \dots, m$. This means that $\langle(\boldsymbol{e}^s)^\flat, \boldsymbol{e}_k\rangle = \mathbf{g}(\boldsymbol{e}^s, \boldsymbol{e}_k) = \delta_k^s$.

Introduce the curvilinear coordinates[2] $(q^i)$ in $\mathscr{E}$. They are defined by functions $q^k = q^k(x^1, \dots, x^m)$, which relate the Cartesian coordinates $(x^i)$ with $(q^i)$ and can be defined by $m$ smooth diffeomorphisms. Frame and coframe fields are determined by the expressions

$$\boldsymbol{e}_k = \boldsymbol{i}_s \frac{\partial x^s}{\partial q^k}\,, \quad \boldsymbol{e}^k = \boldsymbol{i}_s \frac{\partial q^k}{\partial x^s}\,.$$

Transition functions that match the curvilinear coordinates $(q^k)$ with Cartesian ones, in fact, determine all geometrical quantities that correspond to connection and related fields. Indeed, it is easy to show that

(i) The metric tensor $\mathbf{g} = g_{ij}\boldsymbol{e}^i \otimes \boldsymbol{e}^j$ and its corresponding dual, $\mathbf{g}^\sharp = g^{ij}\boldsymbol{e}_i \otimes \boldsymbol{e}_j$, which in conventional considerations is identified with $\mathbf{g}$, have the following compo-

2 Recall that curvilinear coordinates were considered in Chapter 2, Section 2.2.

nents:

$$g_{ij} = \frac{\partial x^k}{\partial q^i}\frac{\partial x^s}{\partial q^j}\delta_{ks}\,, \qquad g^{ij} = \frac{\partial q^i}{\partial x^k}\frac{\partial q^j}{\partial x^s}\delta^{ks}\,.$$

(ii) The covariant derivative $\nabla$ is defined by the relation

$$\nabla_{\boldsymbol{u}}\boldsymbol{v} = \boldsymbol{u}\cdot\nabla\boldsymbol{v}\,, \quad \text{where} \quad \nabla := \boldsymbol{e}^k\partial_k\,,$$

where $\partial_k \equiv \frac{\partial}{\partial q^k}$.

(iii) Christoffel symbols are represented by the formula

$$\gamma^i{}_{jk} = \boldsymbol{e}^i\cdot\nabla_{\boldsymbol{e}_j}\boldsymbol{e}_k = \boldsymbol{e}^i\cdot\frac{\partial\boldsymbol{e}_k}{\partial q^j} = \frac{\partial q^i}{\partial x^p}\frac{\partial^2 x^p}{\partial q^j\partial q^k}\,.$$

Despite the fact that specific quantities like the metric and Christoffel symbols differ from accustomed Cartesian values, they do not correspond to a space more general than the Euclidean one. In order to show this, it is enough to calculate curvature, torsion, and non-metricity. To calculate the curvature note that

$$\begin{aligned}\nabla_{\boldsymbol{u}}\nabla_{\boldsymbol{v}}\boldsymbol{w} &= u^k\left(\partial_k v^i\right)\left(\partial_i w^j\right)(\partial_j\boldsymbol{x}) + u^k v^i\left(\partial_k\partial_i w^j\right)(\partial_j\boldsymbol{x}) + u^k v^i\left(\partial_i w^j\right)(\partial_k\partial_j\boldsymbol{x})\\ &\quad + u^k\left(\partial_k v^i\right)w^j(\partial_i\partial_j\boldsymbol{x}) + u^k v^i\left(\partial_k w^j\right)(\partial_i\partial_j\boldsymbol{x}) + u^k v^i(\partial_k\partial_i\partial_j\boldsymbol{x})\,,\\ \nabla_{\boldsymbol{v}}\nabla_{\boldsymbol{u}}\boldsymbol{w} &= v^k\left(\partial_k u^i\right)\left(\partial_i w^j\right)(\partial_j\boldsymbol{x}) + v^k u^i\left(\partial_k\partial_i w^j\right)(\partial_j\boldsymbol{x}) + v^k u^i\left(\partial_i w^j\right)(\partial_k\partial_j\boldsymbol{x})\\ &\quad + v^k\left(\partial_k u^i\right)w^j(\partial_i\partial_j\boldsymbol{x}) + v^k u^i\left(\partial_k w^j\right)(\partial_i\partial_j\boldsymbol{x}) + v^k u^i(\partial_k\partial_i\partial_j\boldsymbol{x})\,,\\ \nabla_{[\boldsymbol{u},\boldsymbol{v}]}\boldsymbol{w} &= u^i\left(\partial_i v^j\right)\left(\partial_j w^k\right)(\partial_k\boldsymbol{x}) + u^i(\partial_i)w^k(\partial_j\partial_k\boldsymbol{x})\\ &\quad - v^i\left(\partial_i u^j\right)\left(\partial_j w^k\right)(\partial_k\boldsymbol{x}) - v^i\left(\partial_i u^j\right)w^k(\partial_j\partial_k\boldsymbol{x})\,.\end{aligned}$$

Taking the symmetry of the second-order partial derivatives, $\partial_i\partial_j\boldsymbol{x} = \partial_j\partial_i\boldsymbol{x}$, into account and eliminating similar terms, one obtains

$$\forall\boldsymbol{u},\boldsymbol{v},\boldsymbol{w}\in\mathrm{Vec}(\mathscr{E}):\quad \mathfrak{R}(\boldsymbol{u},\boldsymbol{v},\boldsymbol{w}) = \nabla_{\boldsymbol{u}}\nabla_{\boldsymbol{v}}\boldsymbol{w} - \nabla_{\boldsymbol{v}}\nabla_{\boldsymbol{u}}\boldsymbol{w} - \nabla_{[\boldsymbol{u},\boldsymbol{v}]}\boldsymbol{w} \equiv \boldsymbol{0}\,.$$

To calculate torsion one uses the following equalities:

$$\begin{aligned}\nabla_{\boldsymbol{u}}\boldsymbol{v} &= u^i\left(\partial_i v^j\right)(\partial_j\boldsymbol{x}) + u^i v^j(\partial_i\partial_j\boldsymbol{x})\,,\\ \nabla_{\boldsymbol{v}}\boldsymbol{u} &= v^i\left(\partial_i u^j\right)(\partial_j\boldsymbol{x}) + v^i u^j(\partial_i\partial_j\boldsymbol{x})\,,\\ [\boldsymbol{u},\boldsymbol{v}] &= u^i\left(\partial_i v^j\right)(\partial_j\boldsymbol{x}) - v^i\left(\partial_i u^j\right)(\partial_j\boldsymbol{x})\,.\end{aligned}$$

Thus, from the symmetry of the second derivatives, one obtains

$$\forall \boldsymbol{u}, \boldsymbol{v} \in \mathrm{Vec}(\mathscr{E}): \quad \mathfrak{T}(\boldsymbol{u}, \boldsymbol{v}) = \nabla_{\boldsymbol{u}} \boldsymbol{v} - \nabla_{\boldsymbol{v}} \boldsymbol{u} - [\boldsymbol{u}, \boldsymbol{v}] \equiv \boldsymbol{0} .$$

Finally, calculate the non-metricity. Taking the relations

$$\begin{aligned}
\boldsymbol{g}(\nabla_{\boldsymbol{u}} \boldsymbol{v}, \boldsymbol{w}) &= \underline{w_j u^i \left(\partial_i v^j\right)} + w^k u^i v^j (\partial_k \boldsymbol{x}) \cdot (\partial_i \partial_j \boldsymbol{x}) , \\
\boldsymbol{g}(\boldsymbol{v}, \nabla_{\boldsymbol{u}} \boldsymbol{w}) &= \underline{\underline{v_j u^i \left(\partial_i w^j\right)}} + v^k u^i w^j (\partial_k \boldsymbol{x}) \cdot (\partial_i \partial_j \boldsymbol{x}) , \\
\boldsymbol{u}[\boldsymbol{g}(\boldsymbol{v}, \boldsymbol{w})] &= \underline{u^i w_j \left(\partial_i v^j\right)} + \underline{\underline{u^i v_k \left(\partial_i w^k\right)}} + u^i v^j w^k (\partial_i \partial_j \boldsymbol{x}) \cdot (\partial_k \boldsymbol{x}) \\
&\quad + u^i v^j w^k (\partial_j \boldsymbol{x}) \cdot (\partial_i \partial_k \boldsymbol{x}) ,
\end{aligned}$$

into account and using the symmetry of the second derivatives again, one arrives at

$$\forall \boldsymbol{u}, \boldsymbol{v}, \boldsymbol{w} \in \mathrm{Vec}(\mathscr{E}): \quad \mathfrak{Q}(\boldsymbol{u}, \boldsymbol{v}, \boldsymbol{w}) = \mathscr{g}(\nabla_{\boldsymbol{u}} \boldsymbol{v}, \boldsymbol{w}) + \mathscr{g}(\boldsymbol{v}, \nabla_{\boldsymbol{u}} \boldsymbol{w}) - \boldsymbol{u}[\boldsymbol{g}(\boldsymbol{v}, \boldsymbol{w})] \equiv 0 .$$

### 9.2.4 A Particular Case: Riemannian Space

Riemannian space $\mathscr{P}$ is a Riemannian manifold together with a Levi–Civita connection on it. In Chapter 2, Section 2.4 such a connection was introduced from the requirement that autoparallel curves coincide with geodesics. It was shown in 2.4.4 that if $\mathscr{g}$ is a metric, then the Levi–Civita connection coefficients are given by

$$y^k{}_{sq} = \frac{g^{kl}}{2} \left(\partial_q g_{ls} + \partial_s g_{lq} - \partial_l g_{sq}\right) .$$

In this case, the curvature $\mathfrak{R}$ differs from zero generally, while torsion and non-metricity vanish. To show that torsion is identically zero calculate the following expressions:

$$\begin{aligned}
\nabla_u v &= \underline{\underline{u^i \left(\partial_i v^j\right) e_j}} + u^i v^j y^k{}_{ij} e_k , \\
\nabla_v u &= \underline{v^i \left(\partial_i u^j\right) e_j} + v^i u^j y^k{}_{ij} e_k , \\
[u, v] &= \underline{\underline{u^i \left(\partial_i v^j\right) e_j}} - \underline{v^i \left(\partial_i u^j\right) e_j} .
\end{aligned}$$

Substitute them into (9.8) and taking into account the symmetry of the connection coefficients, $y^k{}_{ij} = y^k{}_{ji}$, cancel out similar items. As a result, we obtain the desired conclusion:

$$\forall u, v \in \mathrm{Vec}(\mathscr{P}): \quad \mathfrak{T}(u, v) = \nabla_u v - \nabla_v u - [u, v] \equiv 0 .$$

In order to show that non-metricity also vanishes, we provide the following calculations:

$$\mathscr{g}(\nabla_u v, w) = \underline{w_k u^i \left(\partial_i v^k\right)} + \underset{\sim\sim\sim}{w_k u^i v^j \gamma^k{}_{ij}}\,,$$

$$\mathscr{g}(v, \nabla_u w) = \underline{\underline{v_k u^i \left(\partial_i w^k\right)}} + \underset{\sim\sim\sim}{v_k u^i w^j \gamma^k{}_{ij}}\,,$$

$$u[g(v, w)] = \underline{u^i w_j \left(\partial_i v^j\right)} + \underline{\underline{u^i v_k \left(\partial_i w^k\right)}} + \underset{\sim\sim\sim}{u^i v^j w^k (\partial_i g_{jk})}\,.$$

Since $\gamma_{ijk} + \gamma_{kji} - \partial_j g_{ij} = 0$, we arrive at

$$\forall u, v, w \in \mathrm{Vec}(\mathscr{P}): \quad \mathfrak{Q}(u, v, w) = \mathscr{g}(\nabla_u v, w) + \mathscr{g}(v, \nabla_u w) - u[g(v, w)] \equiv 0\,.$$

**Riemannian physical space as a model of the curvilinear substrate.** Suppose that the physical space is represented by a three-dimensional affine-Euclidean space $\mathscr{E}$. Let us consider shapes of a two-dimensional body $\mathfrak{B}$ (material surface, in the terminology of [110]), which are placed on some fixed two-dimensional embedded submanifold $\mathscr{S}$ of $\mathscr{E}$. Such a situation may hold when an LbL-process is considered: a two-dimensional assembly is constructed from layers that are deposited on a substrate. In this situation, it may be convenient to consider the Riemannian space $(\mathscr{S}, \mathscr{g}^{(s)})$ as a physical space itself. Here, $\mathscr{g}^{(s)} = \iota_{\mathscr{S}}^* \boldsymbol{g}$ is the metric induced from $\mathscr{E}$. Suppose that $(x^1, x^2, x^3)$ are Cartesian coordinates on $\mathscr{E}$, and $(u^1, u^2)$ are local coordinates on $\mathscr{S}$, which where obtained, for example, by applying the local $k$-slice condition (see Chapter 4, Section 4.5.2). Thus, the metric $\mathscr{g}^{(s)}$ has the following dyadic decomposition (see (4.3)):

$$\mathscr{g}^{(s)} = g_{\alpha\beta} du^\alpha \otimes du^\beta\,, \quad \text{where} \quad g_{\alpha\beta} = \delta_{ij} \frac{\partial x^i}{\partial u^\alpha} \frac{\partial x^j}{\partial u^\beta}\,.$$

The mapping $x^i\colon (u^1, u^2) \mapsto x^i(u^1, u^2)$, $i = 1, 2, 3$, is the coordinate representation of the inclusion map $\iota_{\mathscr{S}}$.

**3D Euclidean description for $\mathscr{S}$ as for a surface.** Since $\mathscr{S}$ is embedded into $\mathscr{E}$, each point of $\mathscr{S}$ is described by the position vector field $\boldsymbol{p}\colon \mathscr{S} \to \mathscr{V}$; $\boldsymbol{p}(x) = x^k(u^1, u^2)\boldsymbol{i}_k$. Define the vector fields $\boldsymbol{e}_\alpha\colon \mathscr{S} \to \mathscr{V}$, $\alpha = 1, 2$, as follows:

$$\boldsymbol{e}_\alpha = \frac{\partial \boldsymbol{p}(u^1, u^2)}{\partial u^\alpha}\,, \qquad \alpha = 1, 2\,.$$

For a point $x \in \mathscr{S}$, the pair $(\boldsymbol{e}_1|_x, \boldsymbol{e}_2|_x)$ constitutes a basis of the vector space $\mathrm{In}_{\mathscr{S};x}(T_x\mathscr{S}) \subset \mathscr{V}$. The space $\mathscr{T}_x\mathscr{S} = \mathrm{In}_{\mathscr{S};x}(T_x\mathscr{S})$ is the image of the tangent space $T_x\mathscr{S}$ under the linear inclusion map (4.4) into the translation space $\mathscr{V}$. Recall that in classical surface theory, the set $x + \mathscr{T}_x\mathscr{S}$ (the subspace of the affine space $\mathscr{E}$) is referred to as the *tangent plane at* $x$. Thus, $(\boldsymbol{e}_\alpha)_{\alpha=1}^2$ is the local frame on the surface $\mathscr{S}$. In the framework of the classical surface theory, the role of the Riemannian metric is played by the first fundamental form $I$:

$$\forall \boldsymbol{u}, \boldsymbol{v} \in \mathcal{T}_x\mathcal{S}: \quad I(\boldsymbol{u}, \boldsymbol{v}) := \boldsymbol{u} \cdot \boldsymbol{v} .$$

Its components are equal to $I_{\alpha\beta} = \boldsymbol{e}_\alpha \cdot \boldsymbol{e}_\beta$ and, thus, $I_{\alpha\beta} = g_{\alpha\beta}$. The first fundamental form $I$ coincides with the Riemannian metric $g^{(s)}$ up to inclusion.

At each point $x \in \mathcal{S}$, the translation vector space $\mathcal{V}$ splits into the direct sum $\mathcal{T}_x\mathcal{S} \oplus \mathcal{N}_x\mathcal{S} = \mathcal{V}$. In this expression, the item $\mathcal{N}_x\mathcal{S}$ denotes the orthogonal complement of $\mathcal{T}_x\mathcal{S}$. It is a one-dimensional vector subspace of $\mathcal{V}$ with the basis $\boldsymbol{n}|_x = \|\boldsymbol{e}_1|_x \times \boldsymbol{e}_2|_x\|^{-1}(\boldsymbol{e}_1|_x \times \boldsymbol{e}_2|_x)$. By the definition, the line $x + \mathcal{N}_x\mathcal{S}$ is normal to the tangent plane $x + \mathcal{T}_x\mathcal{S}$. Thus, we obtain a unit normal vector field $\boldsymbol{n}: x \mapsto \boldsymbol{n}|_x$ on $\mathcal{S}$.

For every point $x \in \mathcal{S}$, the principal part of the deviation of the surface from the tangent plane $x + \mathcal{T}_x\mathcal{S}$ is given by the second fundamental form $II_x$, which in classical notation is written as

$$II|_x = L_x(du^1)^2 + 2M_x du^1 du^2 + N_x(du^2)^2 ,$$
$$L = \boldsymbol{n} \cdot \partial_{u^1}\boldsymbol{e}_1 , \quad M = \boldsymbol{n} \cdot \partial_{u^2}\boldsymbol{e}_1 , \quad N = \boldsymbol{n} \cdot \partial_{u^2}\boldsymbol{e}_2 .$$

The Riemannian metric $g^{(s)}$ generates the Levi–Civita connection on $\mathcal{S}$ with the coefficients

$$\Gamma^\alpha{}_{\beta\gamma} = \frac{g^{\alpha\lambda}}{2}\left(\partial_\gamma g_{\lambda\beta} + \partial_\beta g_{\lambda\gamma} - \partial_\lambda g_{\beta\gamma}\right) , \quad \alpha, \beta, \gamma, \lambda = 1, 2 .$$

In this case, the curvature $\mathfrak{R}$, the Ricci tensor, and the scalar curvature are determined by (components are written in the coordinate frame)

$$R_{\alpha\beta\gamma\delta} = K(g_{\alpha\gamma}g_{\delta\beta} - g_{\alpha\delta}g_{\gamma\beta}) ,$$
$$R_{\alpha\beta} = K g_{\alpha\beta} ,$$
$$S = 2K ,$$

where $K = \det II / \det I$ is the Gaussian curvature of the surface $\mathcal{S}$ [134].

### 9.2.5 Connection on the Pullback Bundle

**Definition.** Let $\mathfrak{M}$ and $\mathfrak{N}$ be smooth manifolds and $\varkappa: \mathfrak{M} \to \mathfrak{N}$ be a $C^\infty$-diffeomorphism. Suppose that an affine connection $\nabla$ is given on $T\mathfrak{N}$. Such a connection defines the connection

$$\nabla^{(\varkappa)}: \mathrm{Vec}(\mathfrak{M}) \times \mathrm{Sec}(\varkappa^* T\mathfrak{N}) \to \mathrm{Sec}(\varkappa^* T\mathfrak{N}) ,$$

on the pullback bundle $\varkappa^* T\mathfrak{N}$ as follows:

$$\nabla^{(\varkappa)}_u V := \left\{\nabla_{\varkappa_* u}(V \circ \varkappa^{-1})\right\} \circ \varkappa . \tag{9.13}$$

**Remark 9.3.** Pushforward $\varkappa_*$ is a $\mathbb{R}$-linear mapping, and this implies ($\nabla_1$) and ($\nabla_2$). Next, if $f \in C^\infty(\mathfrak{M})$, then $\varkappa_* u(f \circ \varkappa^{-1}) = u(f)$ and $\varkappa_*(fu) = f\varkappa_*(u)$. This implies ($\nabla_3$) and ($\nabla_4$).

**Coordinate representations.** Let $(e_i)$ be smooth local frame for $T\mathfrak{N}$, then $(E_i)$, where $E_i = e_i \circ \varkappa$ is a local frame for $\varkappa^* T\mathfrak{N}$. Let $(x^i)$ be local coordinates on $\mathfrak{M}$ and $(y^i)$ be local coordinates on $\mathfrak{N}$. Denote the affine connection coefficients for $\nabla$ by $\gamma^i{}_{jk}$. Then, one has

$$\nabla^{(\varkappa)}_{\partial_{x^a}} E_b = \{\nabla_{\varkappa_* \partial_{x^a}} e_b\} \circ \varkappa \,.$$

Denote $T\varkappa = F^i{}_j \partial_{y^i} \otimes dx^j$. Since $\varkappa_* \partial_{x^a} = (F^i{}_a \circ \varkappa^{-1}) \partial_{y^i}$, then

$$\nabla^{(\varkappa)}_{\partial_{x^a}} E_b = F^i{}_a (\nabla_{\partial_{y^i}} e_b) \circ \varkappa \,.$$

With the relation $\nabla_{\partial_{y^i}} e_b = \gamma^c{}_{ib} e_c$ one finally obtains

$$\nabla^{(\varkappa)}_{\partial_{x^a}} E_b = F^i{}_a \gamma^c{}_{ib} E_c \,.$$

Thus, $\Gamma^c{}_{ab} = F^i{}_a \gamma^c{}_{ib}$ are connection coefficients for $\nabla^{(\varkappa)}$. If $u$ is a smooth section for $T\mathfrak{M}$ and $V$ is a smooth section for $\varkappa^* T\mathfrak{N}$, then

$$\nabla^{(\varkappa)}_u V = u^a \left\{ \partial_a V^c + V^b F^i{}_a \gamma^c{}_{ib} \right\} E_c \,.$$

If $(e^i)$ is a frame dual to $(e^i)$, then the ordered tuple $(E^i)$ of sections $E^i = e^i \circ \varkappa$ is the corresponding dual frame for $(E_i)$. Using $\Gamma^c{}_{ab} = F^i{}_a \gamma^c{}_{ib}$ and formula (9.2), one obtains

$$\nabla^{(\varkappa)}_{\partial_{x^a}} E^b = -F^i{}_a \gamma^b{}_{id} E^d \,.$$

**The case of two-point tensors.** We restrict ourselves to the case of the vector bundle $\varkappa^* T\mathfrak{N} \otimes T\mathfrak{M}$. Define the connection on it using (9.3). Let $\nabla^{(\mathfrak{M})}$ be an affine connection on $\mathfrak{M}$. Then

$$\nabla_u (s_1 \otimes s_2) := \left( \nabla^{(\varkappa)}_u s_1 \right) \otimes s_2 + s_1 \otimes \nabla^{(\mathfrak{M})}_u s_2 \,.$$

Let $(\widetilde{E}_\alpha)$ be a frame on $\mathfrak{M}$ and let $\mathcal{T} : p \mapsto T^{i\alpha}(p) E_i|_p \otimes \widetilde{E}_\alpha|_p$ be a two-point tensor. Then, $\nabla \mathcal{T} = \nabla_\beta T^{a\kappa} E_a \otimes \widetilde{E}_\kappa \otimes dx^\beta$, in which

$$\nabla_\beta T^{a\kappa} = \partial_\beta T^{a\kappa} + T^{a\lambda} \Gamma^\kappa{}_{\lambda\beta} + T^{b\kappa} F^c{}_\beta \gamma^a{}_{bc} \,,$$

and $\Gamma^\kappa{}_{\lambda\beta}$ are connection coefficients of $\nabla^{(\mathfrak{M})}$.

**Application in continuum mechanics.** Let $\mathfrak{B}$ be a body, $\mathscr{P}$ be a physical space with the metric $g$ and the Levi–Civita connection $\nabla$, and $\varkappa$ be a configuration. As in Section 9.2.4, the Levi–Civita connection can be induced to the shape $\mathcal{S}_\varkappa$. Since $\hat{\varkappa}$ is a diffeomorphism, the above reasonings can be applied to the body $\mathfrak{B}$ and the shape $\mathcal{S}_\varkappa$. Thus, the pullback bundle $\hat{\varkappa}^* T\mathcal{S}_\varkappa$ can be endowed with the connection $\nabla^{(\hat{\varkappa})}$. Note that the elements of $\hat{\varkappa}^* T\mathcal{S}_\varkappa$ can be regarded as tangent to the $\mathcal{S}_\varkappa$ parts of the material velocities $V$.

Consider the particular case when $\dim \mathfrak{B} = \dim \mathscr{P}$. The shape $\mathcal{S}_\varkappa$ is an open subset of $\mathscr{P}$. This means that the Levi–Civita connection on $\mathcal{S}_\varkappa$ is $\nabla|_{\mathcal{S}_\varkappa}$. As we agreed previously, we omit the vertical bar and simply write $\nabla$. Since $T\mathcal{S}_\varkappa$ can be considered as just a restriction of $T\mathscr{P}$, *i.e.*, $T\mathscr{S}_\varkappa = \bigsqcup_{X\in\mathscr{S}_\varkappa} T_X\mathscr{P}$, one can identify $\hat{\varkappa}^* T\mathscr{S}_\varkappa$ with $\varkappa^* T\mathscr{P}$. Then, $\nabla^{(\hat{\varkappa})}$ is a connection on $\varkappa^* T\mathscr{P}$ with coefficients $\Gamma^c{}_{ab} = F^i{}_a \gamma^c{}_{ib}$, where $F^i{}_j$ are components of the configuration gradient $\boldsymbol{F} = F^i{}_j \partial_i \otimes d\mathfrak{X}^j$.

### 9.2.6 The Moving Frame Method

Let $\mathfrak{M}$ be a smooth *parallelizable* manifold with $\dim \mathfrak{M} = n$. From the parallelizability, it follows that there exists a *global* frame $(z_i)_{i=1}^n$ for $T\mathfrak{M}$. This means that $z_i \in \mathrm{Vec}(\mathfrak{M})$ for all $i = 1, \ldots, n$, and for each $p \in \mathfrak{M}$, the $n$-tuple $(z_i|_p)_{i=1}^n$ of the values is linearly independent. The *moving frame* method was proposed by E. Cartan in the case of $n$-dimensional affine-Euclidean space $\mathscr{E}$ and allows one to define an affine connection by a predefined field of linear transformations that act on the orthonormal frame $(\boldsymbol{i}_k)_{k=1}^n$ [135]. In our case, we have a smooth manifold with a non-trivial atlas, in general. To apply the idea of Cartans' moving frame one needs: **1)** to choose some admissible smooth atlas; **2)** to construct an affine connection $\nabla|_U$ on each charts' domain $U$; **3)** to "glue" these connections together. Thus, it is sufficient to restrict ourselves by considering a chart domain.

Let $U$ be some chart domain and $(\partial_i)_{i=1}^n$ be the corresponding coordinate frame. This is a holonomic frame, *i.e.*, $[\partial_i, \partial_j] = 0$ for all $i, j$. There exist such $n^2$ smooth functions $\Omega^i{}_j : U \to \mathbb{R}$ that form a smooth field $[\Omega^i{}_j] : U \to \mathrm{GL}(n; \mathbb{R})$, and

$$\forall i \in \{1, \ldots, n\}: \quad z_i = \Omega^j{}_i \partial_j \,.$$

The frame $(z_i)_{i=1}^n$ is non-holonomic in general, that is,

$$[z_i, z_j] = -c_{ij}{}^k z_k \,, \tag{9.14}$$

where $c_{ij}{}^k : U \to \mathbb{R}$ are the *objects of anholonomity*. Using the relation (13.21)

$$[fu, gv]_p = f(p)g(p)[u, v]_p + f(p)u_p(g)v_p - g(p)v_p(f)u_p \,,$$

in which $f, g$ are smooth functions defined on a neighborhood of $p$, one obtains the following system with unknowns $c_{ij}{}^m$ in fixed $i, j$:

$$-c_{ij}{}^m \Omega^k{}_m = \Omega^q{}_i(\partial_q \Omega^k{}_j) - \Omega^q{}_j(\partial_q \Omega^k{}_i)\,, \quad k = 1, \ldots, n \,.$$

The solution for this system has the form

$$c_{ij}{}^k = -\mho^k{}_m \left\{\Omega^q{}_i \left(\partial_q \Omega^m{}_j\right) - \Omega^q{}_j \left(\partial_q \Omega^m{}_i\right)\right\} \,, \tag{9.15}$$

in which $[\mho^i{}_j] = [\Omega^i{}_j]^{-1}$. Taking the relation

$$\partial_i \mho^k{}_j = -\mho^m{}_j \left(\partial_i \Omega^q{}_m\right) \mho^k{}_q$$

into account, one obtains a more convenient representation for the (9.15):

$$c_{ij}{}^{k} = \Omega^{m}{}_{i}\,\Omega^{q}{}_{j}\left(\partial_m \mho^{k}{}_{q} - \partial_q \mho^{k}{}_{m}\right) . \tag{9.16}$$

To the frame $(z_i)_{i=1}^{n}$ there corresponds a coframe $(\vartheta^i)_{i=1}^{n}$, which is defined by the relations $\langle \vartheta^i, z_j\rangle = \delta^i_j$, for all $i, j$. Herewith, $\vartheta^i = \mho^i{}_j dx^j$. Let us obtain the coframe analog of (9.14). One has

$$d\vartheta^k = d(\mho^k{}_j dx^j) = \sum_{i<j}\left(\partial_i \mho^k{}_j - \partial_j \mho^k{}_i\right) dx^i \wedge dx^j = \frac{1}{2}\left(\partial_i \mho^k{}_j - \partial_j \mho^k{}_i\right) dx^i \wedge dx^j ,$$

where in the last equality the summation is on all $i$ and $j$. After some calculations, one obtains

$$\begin{aligned} d\vartheta^k &= \frac{1}{2}\left(\partial_i \mho^k{}_j - \partial_j \mho^k{}_i\right) \delta^i_p \delta^j_q dx^p \wedge dx^q \\ &= \frac{1}{2}\left(\partial_i \mho^k{}_j - \partial_j \mho^k{}_i\right) \Omega^i{}_l \mho^l{}_p \Omega^j{}_s \mho^s{}_q dx^p \wedge dx^q \\ &= \frac{1}{2} c_{pq}{}^{k} \vartheta^p \wedge \vartheta^q = \sum_{p<q} c_{pq}{}^{k} \vartheta^p \wedge \vartheta^q . \end{aligned} \tag{9.17}$$

Now define the affine connection on $U$ by formulae

$$\nabla_{z_a} z_b = \omega^c{}_{ab} z_c , \quad a, b = 1, \ldots, n ,$$

where $n^3$ functions $\omega^c{}_{ab} \in C^\infty(U)$ are prescribed. If $\Gamma^k{}_{ij}$ are the connection coefficients in the coordinate frame $(\partial_i)_{i=1}^{n}$, then according to (9.6) one has

$$\omega^i{}_{jk} = \Gamma^m{}_{sq} \mho^i{}_m \Omega^s{}_j \Omega^q{}_k + \mho^i{}_m \Omega^s{}_j \partial_s \Omega^m{}_k .$$

By multiplication of both sides of this expression on $\Omega^o{}_i \mho^j{}_l \mho^k{}_r$ and summation by $i, j, k$, the result is

$$\Gamma^i{}_{jk} = \omega^a{}_{bc} \Omega^i{}_a \mho^b{}_j \mho^c{}_k - \mho^m{}_k \partial_j \Omega^i{}_m .$$

The relation obtained relation may be transformed as follows. Since

$$\mho^m{}_k \partial_j \Omega^i{}_m = -\mho^m{}_k \delta^i_r \partial_j \Omega^r{}_m = \mho^m{}_k \mho^c{}_r \Omega^i{}_c \partial_j \Omega^r{}_m ,$$

and with identity $\partial_i \mho^k{}_j = -\mho^m{}_j (\partial_i \Omega^q{}_m) \mho^k{}_q$, one has

$$\Gamma^i{}_{jk} = \omega^a{}_{bc} \Omega^i{}_a \mho^b{}_j \mho^c{}_k + \Omega^i{}_c \partial_j \mho^c{}_k . \tag{9.18}$$

Let us calculate the torsion, curvature, and non-metricity of the connection $\nabla$. According to the relation (9.8), one has

$$\mathfrak{T} = T^k{}_{ij} z_k \otimes \vartheta^i \otimes \vartheta^j , \quad \text{where} \quad T^k{}_{ij} = \omega^k{}_{ij} - \omega^k{}_{ji} + c_{ij}{}^{k} .$$

Here, $c_{ij}{}^{k}$ are the objects of anholonomity (9.16). Next, the curvature tensor is defined by the relation (9.9), so

$$\mathfrak{R} = R^{m}{}_{ijk}\, z_m \otimes \vartheta^i \otimes \vartheta^j \otimes \vartheta^k\,,$$

where the components $R^{m}{}_{ijk}$ are represented by the expressions

$$R^{m}{}_{ijk} = \Omega^{s}{}_{i}\partial_s \omega^{m}{}_{jk} - \Omega^{s}{}_{j}\partial_s \omega^{m}{}_{ik} + \omega^{s}{}_{jk}\omega^{m}{}_{is} - \omega^{s}{}_{ik}\omega^{m}{}_{js} + c_{ij}{}^{s}\omega^{m}{}_{sk}\,.$$

Finally, for the non-metricity equation (9.10) one has

$$\mathfrak{Q} = Q_{ijk}\vartheta^i \otimes \vartheta^j \otimes \vartheta^k\,, \quad \text{where} \quad -Q_{ijk} = \Omega^{s}{}_{i}\partial_s g_{jk} - g_{mk}\omega^{m}{}_{ij} - g_{jm}\omega^{m}{}_{ik}\,.$$

Here, $\mathscr{g} = g_{ij}\vartheta^i \otimes \vartheta^j$ is some prescribed Riemannian metric on $\mathfrak{M}$.

Consider some particular cases:

**1***. Assume that $(z_i)_{i=1}^n$ is an orthonormal frame, *i.e.*, $\mathscr{g}(z_i, z_j) = \delta_{ij}$ for all $i, j = 1, \ldots, n$. Then,

$$Q_{ijk} = \omega^{k}{}_{ij} + \omega^{j}{}_{ik}\,.$$

If we require the connection $\nabla$ to be metric compatible (that is, $\mathfrak{Q} = 0$), then we have the following compatibility constraints:

$$\omega^{k}{}_{ij} + \omega^{j}{}_{ik} = 0, \quad i, j, k = 1, \ldots, n\,.$$

**2***. Let $\omega^{a}{}_{bc} = 0$, for all $a, b, c = 1, \ldots, n$. Then, we have

$$\nabla_{z_a} z_b = 0\,, \quad a, b = 1, \ldots, n\,.$$

This means that the frame field $(z_i)_{i=1}^n$ is parallel everywhere. Expression (9.18) takes the form

$$\Gamma^{i}{}_{jk} = \Omega^{i}{}_{c}\partial_j \mho^{c}{}_{k}\,. \tag{9.19}$$

For torsion, curvature, and non-metricity, we have

$$T^{k}{}_{ij} = c_{ij}{}^{k}\,, \quad R^{m}{}_{ijk} = 0\,, \quad Q_{ijk} = -\Omega^{s}{}_{i}\partial_s g_{jk}\,. \tag{9.20}$$

The first equality shows that the objects of anholonomity $c_{ij}{}^{k}$ represent the components of the torsion of the so-obtained connection relative to the non-holonomic moving frame.

**3***. Let $(z_i)_{i=1}^n$ be an orthonormal frame and $\omega^{a}{}_{bc} = 0$, for all $a, b, c = 1, \ldots, n$. Then, the connection coefficients $\Gamma^{i}{}_{jk}$ may be calculated by the formula (9.19). In (9.20) only torsion tensor components are non-zero. That is, $\nabla$ has zero curvature and is compatible with the metric. Such a connection $\nabla$ is called the *Weitzenböck connection* [136, 137]. The space $(\mathfrak{M}, \mathscr{g}, \nabla)$ is called a *space with absolute parallelism* or the *Weitzenböck space*.

### 9.2.7 The Weitzenböck Connection as a Material Connection

Consider the case when the physical space is represented by a three-dimensional affine-Euclidean space $\mathscr{E}$. Let $\mathfrak{B}$ be a three-dimensional materially uniform inhomogeneous body and $\mathrm{Ref}_{\mathfrak{B}} = \{\boldsymbol{K}^R_{\mathfrak{X}}\}_{\mathfrak{X}\in\mathfrak{B}}$ be its uniform reference. Each $\boldsymbol{K}^R_{\mathfrak{X}}$ can be considered as a non-degenerate linear transformation $\boldsymbol{K}^R_{\mathfrak{X}}: T_{\mathfrak{X}}\mathfrak{B} \to \mathscr{V}$.

Fix a material point $\mathfrak{X}_0 \in \mathfrak{B}$. Any tangent space $T_{\mathfrak{X}}\mathfrak{B}$ can be obtained from $T_{\mathfrak{X}_0}\mathfrak{B}$ by means of a material isomorphism $\Phi_{\mathfrak{X}\mathfrak{X}_0}: T_{\mathfrak{X}_0}\mathfrak{B} \to T_{\mathfrak{X}}\mathfrak{B}$, consistent with $\mathrm{Ref}_{\mathfrak{B}}$. Thus, $T_{\mathfrak{X}_0}\mathfrak{B}$ can serve as a reference object, an *archetype* (in M Epstein's terminology [6]). The image $\boldsymbol{K}^R_{\mathfrak{X}_0}(T_{\mathfrak{X}_0}\mathfrak{B})$ is the spatial representation of the archetype. Since the body and the physical space have the same dimensions, this spatial representation coincides with the translation space $\mathscr{V}$. This holds for any other tangent space $T_{\mathfrak{X}_{00}}\mathfrak{B}$, accepted as an archetype. Thus, $\mathscr{V}$ is the common spatial representation for all archetypes of the body $\mathfrak{B}$. Any infinitesimal neighborhood $T_{\mathfrak{X}}\mathfrak{B}$ can be obtained from the spatial representation of an archetype, that is, from $\mathscr{V}$, by means of the linear mapping

$$\boldsymbol{P}_{\mathfrak{X}} = \left[\boldsymbol{K}^R_{\mathfrak{X}}\right]^{-1} : \mathscr{V} \to T_{\mathfrak{X}}\mathfrak{B}\,.$$

The field $\boldsymbol{P}: \mathfrak{X} \mapsto \boldsymbol{P}_{\mathfrak{X}}$ is referred to as an *implant* (in G. Maugin's terminology [8]).

Choose some configuration $\tilde{\varkappa} \in \mathfrak{C}(\mathfrak{B};\mathscr{E})$ and identify the body with its shape $\mathscr{S}_{\tilde{\varkappa}}$. In addition, identify all infinitesimal material neighborhoods with the corresponding from $T\mathscr{S}_{\tilde{\varkappa}}$ by means of $T\tilde{\varkappa}$. Previous reasonings give the following physical representation for $\mathscr{V}$. It is a uniform "brick" that is deformed by the implant $\boldsymbol{P}$ before it can be put into the assembly of deformed "bricks" constituting $\mathscr{S}_{\tilde{\varkappa}}$ without voids. The orthonormal frame $(\boldsymbol{i}_k)_{k=1}^3$ is deformed with this "brick" and results in the field of frames $(\boldsymbol{z}_i)_{i=1}^3$ defined as $\boldsymbol{z}_i = P^k{}_i \boldsymbol{i}_k$. Due to the inhomogeneity of the body, the field $(\boldsymbol{z}_i)_{i=1}^3$ is non-holonomic (Figure 9.1). Our aim is to turn $\mathscr{S}_{\tilde{\varkappa}}$ into a uniform shape. This is the state when frames $(\boldsymbol{z}_i)_{i=1}^3$ are mutually parallel at all points. To this end, one needs to introduce such an affine connection on $\mathscr{S}_{\tilde{\varkappa}}$ to achieve parallelizability. Clearly, this is the Weitzenböck connection.

Parallelize the frame $(\boldsymbol{z}_i)_{i=1}^3$ by means of the Weitzenböck connection. According to (9.19) one obtains

$$\Gamma^k{}_{ij} = P^k{}_m \partial_i [(\boldsymbol{P})^{-1}]^m{}_j\,,$$

or, in terms of $\boldsymbol{K}^R: \mathfrak{X} \mapsto \boldsymbol{K}^R_{\mathfrak{X}}$,

$$\Gamma^k{}_{ij} = [(\boldsymbol{K}^R)^{-1}]^k{}_m \partial_i [\boldsymbol{K}^R]^m{}_j\,.$$

The Weitzenböck space $(\mathscr{S}_{\tilde{\varkappa}}, \boldsymbol{g}, \nabla)$ represents the desired global uniform shape.

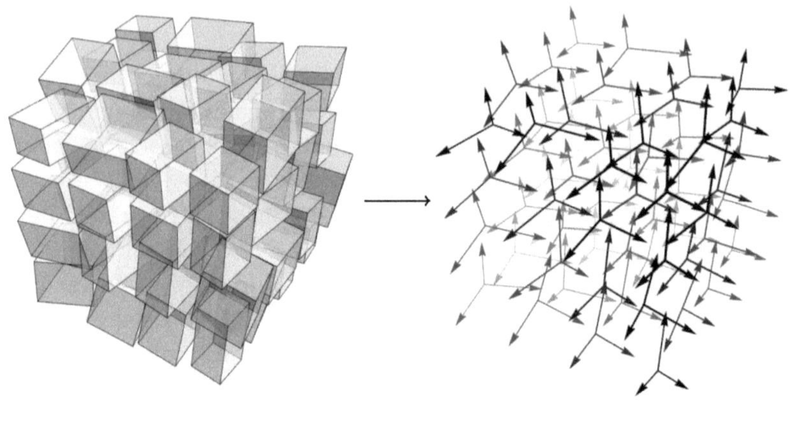

Natural local shapes    Continuous field of frames

**Fig. 9.1:** Correspondence between local shapes and space with absolute parallelism.

### 9.2.8 The Weitzenböck Connection: Example

Suppose that $(\mathscr{E}, \mathscr{V})$ is a three-dimensional affine-Euclidean space, $(\boldsymbol{i}_s)_{s=1}^3$ is an orthonormal basis, and $(x^1, x^2, x^3)$ are Cartesian coordinates. The orthonormal basis defines a frame (field of bases) which values are obtained by the translation of the original basis. The coframe is denoted by $(i^s)_{s=1}^3$. The position of a point from $\mathscr{E}$ is defined by the position vector field $\boldsymbol{p}$.

**1***. Consider the spherical coordinates $r = q^1$, $\theta = q^2$, $\varphi = q^3$ on $\mathscr{E}$ related with the Cartesian ones as

$$x^1 = r\cos\varphi\sin\theta\,, \quad x^2 = r\sin\varphi\sin\theta\,, \quad x^3 = r\cos\theta\,. \tag{9.21}$$

The position of a point $x$ is defined by the translation vector $\boldsymbol{p}(x) = \boldsymbol{i}_k x^k(q^1, q^2, q^3)$. The local frame $(e_s)_{s=1}^3$ in $\mathscr{E}$ is defined as $\boldsymbol{e}_s = \frac{\partial \boldsymbol{p}(x)}{\partial q^s}$, that is,

$$\begin{aligned}
\boldsymbol{e}_1 &= \boldsymbol{e}_r = \boldsymbol{i}_1\cos\varphi\sin\theta + \boldsymbol{i}_2\sin\varphi\sin\theta + \boldsymbol{i}_3\cos\theta\,,\\
\boldsymbol{e}_2 &= \boldsymbol{e}_\theta = \boldsymbol{i}_1 r\cos\varphi\cos\theta + \boldsymbol{i}_2 r\sin\varphi\cos\theta - \boldsymbol{i}_3 r\sin\theta\,,\\
\boldsymbol{e}_3 &= \boldsymbol{e}_\varphi = -\boldsymbol{i}_1 r\sin\varphi\sin\theta + \boldsymbol{i}_2 r\cos\varphi\sin\theta\,.
\end{aligned}$$

The transformation $\boldsymbol{i}_k \mapsto \boldsymbol{e}_s$ is defined by the linear mapping $\Omega$, and its inverse, by $\mho$. In the frames $(\boldsymbol{i}_s)_{s=1}^3$, $(\boldsymbol{e}_i)_{i=1}^3$, these mappings have the following dyadic decompositions:

$$\Omega = \boldsymbol{e}_s \otimes i^s = \Omega^i{}_j \boldsymbol{i}_i \otimes i^j\,, \quad \mho = \boldsymbol{i}_s \otimes e^s = \mho^i{}_j \boldsymbol{e}_i \otimes e^j\,.$$

The matrices of components of these mappings can be written as

$$[\Omega^i{}_j] = \begin{pmatrix} \cos\varphi\sin\theta & r\cos\varphi\cos\theta & -r\sin\varphi\sin\theta \\ \sin\varphi\sin\theta & r\sin\varphi\cos\theta & r\cos\varphi\sin\theta \\ \cos\theta & -r\sin\theta & 0 \end{pmatrix},$$
$$[\mho^i{}_j] = \begin{pmatrix} \cos\varphi\sin\theta & \sin\varphi\sin\theta & \cos\theta \\ \frac{\cos\varphi\cos\theta}{r} & \frac{\sin\varphi\cos\theta}{r} & -\frac{\sin\theta}{r} \\ -\frac{\sin\varphi}{r\sin\theta} & \frac{\cos\varphi}{r\sin\theta} & 0 \end{pmatrix}. \tag{9.22}$$

The matrices (9.22) result from calculations based on the relations (9.21), which determine the transition from Cartesian coordinates to spherical ones. However, now one can adopt a different point of view. Let the elements of the matrices (9.22) be given smooth functions. They determine the smooth field of linear transformations of the Cartesian frame $(\boldsymbol{i}_s)_{s=1}^3$ to its local counterpart. Then, one can benefit from Cartan's moving frame method and define the corresponding Weitzenböck connection.

Calculation of the Weitzenböck connection coefficients in a spherical coordinate system by the formula (9.19), in which the symbols $\Omega$ and $\mho$ must be interchanged, and the derivative is calculated by $r$, $\theta$, $\varphi$, gives (the non-zero expressions are given)

$$\Gamma^1{}_{22} = -r\,, \qquad \Gamma^1{}_{33} = -r\sin^2\theta\,, \qquad \Gamma^2{}_{12} = \Gamma^2{}_{21} = \frac{1}{r}\,,$$
$$\Gamma^2{}_{33} = -\sin\theta\cos\theta\,, \qquad \Gamma^3{}_{23} = \Gamma^3{}_{32} = \cot\theta\,, \qquad \Gamma^3{}_{13} = \Gamma^3{}_{31} = \frac{1}{r}\,.$$

As expected, one obtains the well-known expressions for the Euclidean connection coefficients in the spherical coordinates. Thus, the Weitzenböck connection obtained, which establishes absolute parallelism for the frame $(\boldsymbol{i}_s)_{s=1}^3$, coincides with the Euclidean one. Actually, it is easy to show by simple calculations that the related torsion $\mathfrak{T}$, curvature $\mathfrak{R}$, and non-metricity $\mathfrak{Q}$ equal zero.

However, the situation will improve, in principle, if one replace the local bases, induced from the coordinate relations, by their normalized counterparts. The normalized spherical frame is determined by the relations

$$\boldsymbol{e}_{\langle 1\rangle} = \boldsymbol{e}_{\langle r\rangle} = \boldsymbol{i}_1\cos\varphi\sin\theta + \boldsymbol{i}_2\sin\varphi\sin\theta + \boldsymbol{i}_3\cos\theta = \boldsymbol{e}_r\,,$$
$$\boldsymbol{e}_{\langle 2\rangle} = \boldsymbol{e}_{\langle \theta\rangle} = \boldsymbol{i}_1\cos\varphi\cos\theta + \boldsymbol{i}_2\sin\varphi\cos\theta - \boldsymbol{i}_3\sin\theta = \frac{1}{r}\boldsymbol{e}_\theta\,,$$
$$\boldsymbol{e}_{\langle 3\rangle} = \boldsymbol{e}_{\langle \varphi\rangle} = -\boldsymbol{i}_1\sin\varphi + \boldsymbol{i}_2\cos\varphi = \frac{1}{r\sin\theta}\boldsymbol{e}_\varphi\,.$$

The transformation $\boldsymbol{e}_k \mapsto \boldsymbol{e}_{\langle s\rangle}$ is defined by the linear mapping $\widehat{\Omega}$ and its inverse $\widehat{\mho}$. Such mappings have the following representation:

$$\widehat{\Omega} = \boldsymbol{e}_{\langle s\rangle}\otimes\boldsymbol{e}^s = \widehat{\Omega}^i{}_j\boldsymbol{e}_i\otimes\boldsymbol{e}^j\,, \qquad \widehat{\mho} = \boldsymbol{e}_s\otimes\boldsymbol{e}^{\langle s\rangle} = \widehat{\mho}^i{}_j\boldsymbol{e}_{\langle i\rangle}\otimes\boldsymbol{e}^{\langle j\rangle}\,.$$
$$[\widehat{\Omega}^i{}_j] = \operatorname{diag}\left\{1, r^{-1}, (r\sin\theta)^{-1}\right\}\,, \qquad [\widehat{\mho}^i{}_j] = \operatorname{diag}\{1, r, r\sin\theta\}\,.$$

Since the functions that determine the elements of $\widehat{\Omega}$ are smooth, they determine the smooth field of transformations. Apply these transformations to the field of frames obtained above with spherical coordinates. This results in the field of normalized frames that are non-holonomic. Corresponding to them connection coefficients are calculated by the formula (9.19) and are the following (non-zero expressions are shown only):

$$\Gamma^2{}_{12} = \frac{1}{r}\,, \quad \Gamma^3{}_{13} = \frac{1}{r}\,, \quad \Gamma^3{}_{23} = \cot\theta\,.$$

Such an obtained connection is not Euclidean yet. Indeed, the components listed below of the torsion do not vanish:

$$T^2{}_{12} = -T^2{}_{21} = \frac{1}{r}\,, \quad T^3{}_{13} = -T^3{}_{31} = \frac{1}{r}\,, \quad T^3{}_{23} = -T^3{}_{32} = \cot\theta\,.$$

At the same time, components of the curvature $\mathfrak{R}$ and the non-metricity $\mathfrak{Q}$ are zero. Thus, the non-holonomic field of frames, appearing due to the very often used procedure of normalizing, generates the example of non-Euclidean space with the Weitzenböck connection.

**2***. Introduce the cylindrical coordinates $r = q^1$, $\varphi = q^2$, $\xi = q^3$ in $\mathscr{E}$ by the following transformations:

$$x^1 = r\cos\varphi\,, \quad x^2 = r\sin\varphi\,, \quad x^3 = \xi\,,$$

The local frame can be defined by the relations

$$\begin{aligned} \boldsymbol{e}_1 &= \boldsymbol{e}_r = \boldsymbol{i}_1\cos\varphi + \boldsymbol{i}_2\sin\varphi\,, \\ \boldsymbol{e}_2 &= \boldsymbol{e}_\varphi = -\boldsymbol{i}_1 r\sin\varphi + \boldsymbol{i}_2 r\cos\varphi\,, \\ \boldsymbol{e}_3 &= \boldsymbol{e}_\xi = \boldsymbol{i}_3\,. \end{aligned}$$

The transformation $\Omega\colon \boldsymbol{i}_k \mapsto \boldsymbol{e}_s$ and its inverse, $\mho$, are defined by the expressions

$$\Omega = \boldsymbol{e}_s \otimes \boldsymbol{i}^s = \Omega^i{}_j \boldsymbol{i}_i \otimes \boldsymbol{i}^j\,, \qquad \mho = \boldsymbol{i}_s \otimes \boldsymbol{e}^s = \mho^i{}_j \boldsymbol{e}_i \otimes \boldsymbol{e}^j\,,$$

$$\left[\Omega^i{}_j\right] = \begin{pmatrix} \cos\varphi & -r\sin\varphi & 0 \\ \sin\varphi & r\cos\varphi & 0 \\ 0 & 0 & 1 \end{pmatrix}, \qquad \left[\mho^i{}_j\right] = \begin{pmatrix} \cos\varphi & \sin\varphi & 0 \\ -\frac{\sin\varphi}{r} & \frac{\cos\varphi}{r} & 0 \\ 0 & 0 & 1 \end{pmatrix}.$$

Repeating the reasoning provided above, one can obtain the field of transformations generated by the local basis of the cylindrical coordinates, and, upon them, calculate the Weitzenböck connection. This results in the following non-zero connection coefficients:

$$\Gamma^1{}_{22} = -r\,, \quad \Gamma^2{}_{12} = \Gamma^2{}_{21} = \frac{1}{r}\,.$$

As one would expect, the Weitzenböck connection thus obtained coincides with the Euclidean one, and its components coincide with the corresponding components of the Euclidean connection in cylindrical coordinates.

Meanwhile, the procedure of normalizing the local frames results in the non-holonomic frame field

$$\boldsymbol{e}_{\langle 1\rangle} = \boldsymbol{e}_{\langle r\rangle} = \boldsymbol{i}_1 \cos\varphi + \boldsymbol{i}_2 \sin\varphi = \boldsymbol{e}_r \,,$$
$$\boldsymbol{e}_{\langle 2\rangle} = \boldsymbol{e}_{\langle \varphi\rangle} = -\boldsymbol{i}_1 \sin\varphi + \boldsymbol{i}_2 \cos\varphi = \frac{1}{r}\boldsymbol{e}_\varphi \,,$$
$$\boldsymbol{e}_{\langle 3\rangle} = \boldsymbol{e}_{\langle \xi\rangle} = \boldsymbol{i}_3 = \boldsymbol{e}_\xi \,,$$

and, consequently, the Weitzenböck space with non-zero torsion. Indeed, the transformation $\widehat{\Omega}\colon \boldsymbol{e}_k \mapsto \boldsymbol{e}_{\langle s\rangle}$ and its inverse, $\widehat{\mho}$, are represented by the relations

$$\widehat{\Omega} = \boldsymbol{e}_{\langle s\rangle} \otimes \boldsymbol{e}^s = \widehat{\Omega}^i{}_j \boldsymbol{e}_i \otimes \boldsymbol{e}^j \,, \qquad \widehat{\mho} = \boldsymbol{e}_s \otimes \boldsymbol{e}^{\langle s\rangle} = \widehat{\mho}^i{}_j \boldsymbol{e}_{\langle i\rangle} \otimes \boldsymbol{e}^{\langle j\rangle} \,.$$
$$\left[\widehat{\Omega}^i{}_j\right] = \operatorname{diag}\left\{1, r^{-1}, 1\right\} \,, \qquad \left[\widehat{\mho}^i{}_j\right] = \operatorname{diag}\left\{1, r, 1\right\} \,.$$

The connection coefficients calculated by (9.19) are equal to zero except one:

$$\Gamma^2{}_{12} = \frac{1}{r} \,.$$

Consequently, torsion does not vanish. It has two non-zero components:

$$T^2{}_{12} = -T^2{}_{21} = \frac{1}{r} \,.$$

In turn, the components of the curvature $\mathfrak{R}$ and the non-metricity $\mathfrak{Q}$ are equal to zero. Thus, $\mathscr{E}$ together with the obtained connection results in a Weitzenböck space with non-trivial torsion.

**3**$^*$. The formal mathematical constructions provided above can be interpreted in terms of continuum mechanics. Actually, these constructions determine two different spaces, $(\mathscr{E}_0, \nabla)$ and $(\mathscr{E}_0, \widehat{\nabla})$, over one and the same underlying manifold $\mathscr{E}_0$. The definition for $\mathscr{E}_0$ can be extracted from the definition of the Euclidean affine space (Chapter 2, Section 2.1), taking only its topological properties and atlas, which contains the chart, matching the cylindrical (or spherical) coordinates. The difference between these spaces lies in the fact that the connection on the first is Euclidean (flat), while the connection on the second one is not. One can associate the flat space $(\mathscr{E}_0, \nabla)$ with the actual shape of the body, which is observable in the physical space $\mathscr{P}$ but is not uniform (is not free from residual stresses). In turn, the space $(\mathscr{E}_0, \widehat{\nabla})$ can be associated with the non-Euclidean reference shape, which is locally uniform but cannot be embedded into the Euclidean physical space as a whole. Metaphorically speaking, the shape will crumble into small parts that do not fit together.

Indeed, while attempting to make that happen, one would invariably get the discontinuous subset of $\mathscr{P}$. The result can be imagined only in some approach when one treats infinitesimal parts as sufficiently small but still finite. Such an artistic image corresponding to cylindrical coordinate transformations is shown in Figure 9.2. On the left-hand side one can see the image of the actual shape, while on the right, the sketch (according to the idea that small parts represent infinitesimal pieces) of the set resulting from the attempt to place a uniform shape into the Euclidean physical space. The corresponding image for spherical coordinates is shown in Figure 9.3.

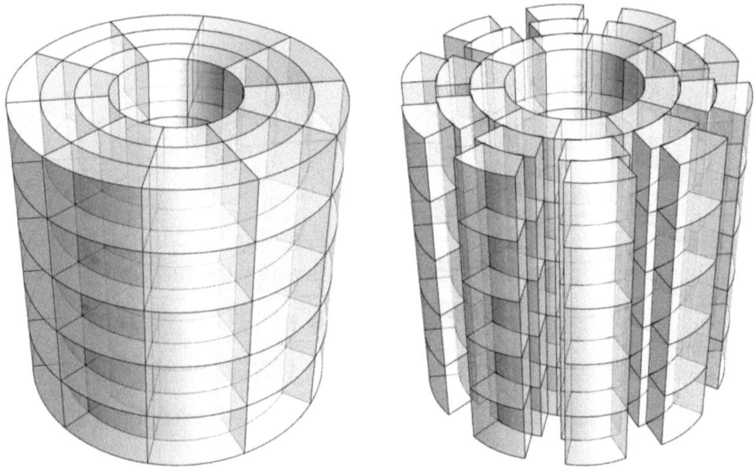


**Fig. 9.2:** The case of cylindrical coordinate transformations.

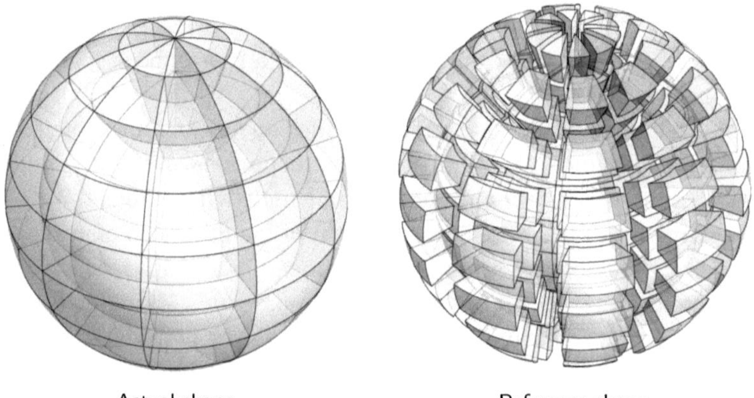


**Fig. 9.3:** The case of spherical coordinate transformations.

# 10 Balance Equations

Recall that in conventional elasticity, the balance equations for mass and linear momentum can be written in material description as follows:

$$\rho \det \boldsymbol{F} = \rho_0 ,$$
$$\operatorname{div}_R \boldsymbol{P} + \rho_0 \boldsymbol{b}_0 = \rho_0 \frac{\partial \boldsymbol{V}}{\partial t} .$$

Their spatial counterparts are given by equations

$$\frac{\partial \rho}{\partial t} + \operatorname{div}(\rho \boldsymbol{v}) = 0 ,$$
$$\operatorname{div} \boldsymbol{T} + \rho \boldsymbol{b} = \rho \frac{\partial \boldsymbol{v}}{\partial t} .$$

The divergence operators $\operatorname{div}_R$ and div are defined in Euclidean space and relate with each other via transformation induced by the "change of variables", *i.e.*, by transition functions that relate Cartesian coordinates with curvilinear ones (together with the chain rule). In present book, the body is assumed to be a smooth manifold endowed with a non-Euclidean connection that turns it into Riemannian or Weitzenböck spaces, and the physical space is considered as Riemannian, in general. Therefore, the mathematical formalization for stresses must be generalized as was shown in Chapter 7. This also requires a generalization of the notion of the divergence operator and the corresponding formalizations for balance laws. These generalizations are the subject of present chapter.

## 10.1 Divergence

### 10.1.1 The Case of a Vector Field

Let $\mathfrak{M}$ be a smooth $s$-dimensional manifold that is endowed with some orientation $[\mu]$, the Riemannian metric $g$, and the affine connection $\nabla$. The divergence of a vector field $u \in \mathrm{Vec}(\mathfrak{M})$ can be defined in two ways:

($\operatorname{div}_1$) The definition via the Lie derivative of the volume form $\mu$:

$$\operatorname{div}_\mu u \in C^\infty(\mathfrak{M}) : \quad \mathscr{L}_u \mu = (\operatorname{div}_\mu u)\mu .$$

This definition is closely related with the Gauss theorem [123]: if $\partial\mathfrak{M} \neq \emptyset$ then

$$\int_{\mathfrak{M}} (\operatorname{div}_\mu u)\mu = \int_{\partial\mathfrak{M}} i_u \mu .$$

https://doi.org/10.1515/9783110563214-010

(div$_2$) A contraction of the vector field $u$ with covariant derivative:

$$\mathrm{div}_\nabla u \in C^\infty(\mathfrak{M}) : \quad \mathrm{div}_\nabla u = \nabla_i u^i .$$

Here,

$$\nabla_i u^k = \left(\partial_i u^k + u^j \Gamma^k{}_{ij}\right) .$$

Let $\nabla$ be the Levi–Civita connection; the orientation is generated by the Riemannian volume form $dV = \sqrt{g}\, dx^1 \wedge \cdots \wedge dx^s$, in which $g = \det[g_{ij}]$. Calculate the divergence of $u \in \mathrm{Vec}(\mathfrak{M})$ by using the definitions (div$_1$) and (div$_2$).

**1***. By *Cartan's magic formula* (6.5) one has

$$\mathscr{L}_u dV = i_u d(dV) + d(i_u dV) = d(i_u dV) ,$$

since $d(dV) = 0$. By induction, one can prove that

$$i_u dV = \sqrt{g} \sum_{k=1}^{s} (-1)^{k-1} u^k\, dx^1 \wedge \cdots \wedge \widehat{dx^k} \wedge \cdots \wedge dx^s ,$$

where $u^i = \langle dx^i, u\rangle$. Thus,

$$\begin{aligned}\mathscr{L}_u dV &= d\left( \sqrt{g} \sum_{k=1}^{s} (-1)^{k-1} u^k\, dx^1 \wedge \cdots \wedge \widehat{dx^k} \wedge \cdots \wedge dx^s \right) \\ &= \sum_{k=1}^{s} (-1)^{k-1}\, d(u^k \sqrt{g}) \wedge dx^1 \wedge \cdots \wedge \widehat{dx^k} \wedge \cdots \wedge dx^s .\end{aligned}$$

The last expression transforms into

$$\sum_{k=1}^{s} (-1)^{k-1} \partial_k (u^k \sqrt{g})\, dx^k \wedge dx^1 \wedge \cdots \wedge \widehat{dx^k} \wedge \cdots \wedge dx^s ,$$

by means of the equalities $d(u^k \sqrt{g}) = \partial_j (u^k \sqrt{g}) dx^j$ and $dx^i \wedge dx^i = 0$. Thus, using permutations one obtains

$$\mathscr{L}_u dV = \partial_k (u^k \sqrt{g})\, dx^1 \wedge \cdots \wedge dx^s .$$

Applying the definition (div$_1$) and taking into account that $dx^1 \wedge \cdots \wedge dx^s$ is the local frame in $\Lambda^s(T^*\mathfrak{M})$, one arrives at

$$\mathrm{div}_{dV}\, u = \frac{1}{\sqrt{g}} \partial_k \left(u^k \sqrt{g}\right) .$$

**2***. Using the Voss–Weyl formula (2.22) $\partial_k \sqrt{g} = \sqrt{g}\, \Gamma^l{}_{lk}$ allows one to represent $\nabla_k u^k$ as

$$\nabla_k u^k = \left(\partial_k u^k + u^k \Gamma^l{}_{lk}\right) = \frac{1}{\sqrt{g}} \partial_k \left(u^k \sqrt{g}\right) .$$

Thus,

$$\mathrm{div}_\nabla u = \frac{1}{\sqrt{g}} \partial_k \left(u^k \sqrt{g}\right) . \tag{10.1}$$

**Remark 10.1.** Let us comment on the Voss–Weyl formula (2.22). It was previously derived for the case of curvilinear coordinates. However, the proof can be still applied to the present considerations. Indeed, from (9.12), in which all $T_{ijk}$ and $Q_{ijk}$ vanish, one has

$$\Gamma^i{}_{jk} = \frac{g^{ir}}{2}(\partial_j g_{rk} + \partial_k g_{rj} - \partial_r g_{jk}) .$$

A similar formula links Christoffel symbols and the metric. One needs to use only this formula and the equality $g^{lr}\partial_k g_{lr} = \partial_k \ln g$ to derive the desired relation.

It follows from **1*** and **2*** that

$$\mathrm{div}_{dV}\, u = \mathrm{div}_\nabla\, u .$$

This means that both definitions of the divergence, $(\mathrm{div}_1)$ and $(\mathrm{div}_2)$, in the case of the Levi–Civita connection give the same smooth scalar field.

Now, consider the more general situation when $\nabla$ is not a Levi–Civita connection (*i.e.*, autoparallel curves do not yet coincide with geodesics). If $\mathfrak{M}$ represents the body $\mathfrak{B}$, then it corresponds to the case when the material space is endowed with an absolute parallel connection (the Weitzenböck connection). According to (9.12), one has

$$\Gamma^j{}_{jk} = \frac{1}{\sqrt{g}}\partial_k \sqrt{g} + \frac{1}{2}\left(T^j{}_{jk} + T_k{}^j{}_j + T_j{}^j{}_k\right) + \frac{1}{2}\left(Q_{jk}{}^j + Q_k{}^j{}_j - Q^j{}_{jk}\right) .$$

Thus, the relation (10.1) transforms into

$$\mathrm{div}_\nabla\, u = \frac{1}{\sqrt{g}}\partial_k\left(u^k\sqrt{g}\right) + \frac{u^k}{2}\left(T^j{}_{jk} + T_k{}^j{}_j + T_j{}^j{}_k\right) + \frac{u^k}{2}\left(Q_{jk}{}^j + Q_k{}^j{}_j - Q^j{}_{jk}\right) . \tag{10.2}$$

The expression (10.2) coincides with $\mathrm{div}_{dV}\, u$ only when $\mathfrak{T} = 0$ and $\mathfrak{Q} = 0$. Thus, one has two different definitions for the divergence fields. Since the affine connection $\nabla$ and the orientation $[dV]$ are fixed, a possible way to obtain equal scalar fields of divergences is to find such a volume form $\mu \in [dV]$ that

$$\forall u \in \mathrm{Vec}(\mathfrak{M}) : \; \mathrm{div}_\mu\, u = \mathrm{div}_\nabla\, u .$$

We will refer to this exterior $n$-form $\mu$ as *the compatible with the affine connection* $\nabla$ [138].

The requirement $\mu \in [dV]$ is equivalent to the statement that there exists such $f \in C^\infty(\mathfrak{M})$ that $f > 0$ in $\mathfrak{M}$ and $\mu = f\, dV$. Then, for any $u \in \mathrm{Vec}(\mathfrak{M})$, one has[1]

$$\mathrm{div}_{f\,dV}\, u = \frac{1}{\sqrt{g}}\partial_k\left(u^k\sqrt{g}\right) + \frac{u^k\partial_k f}{f} .$$

1 Here the relation $\mathrm{div}_{f\,dV}\, u = \mathrm{div}_{dV}\, u + \dfrac{u(f)}{f}$ was used.

The requirement $\mathrm{div}_{f\,dV}\, u = \mathrm{div}_\nabla u$ implies that for any $k \in \{1, \ldots, s\}$, the equality $\mathrm{div}_{f\,dV}\, \partial_k = \mathrm{div}_\nabla \partial_k$ holds. Then, using (10.2) one obtains

$$k \in \{1, \ldots, s\} : \partial_k \ln f = S_k + V_k \,,$$

where

$$S_k = \frac{1}{2}\left(T^{j}{}_{jk} + T_{k}{}^{j}{}_{j} + T_{j}{}^{j}{}_{k}\right)\,, \quad V_k = \frac{1}{2}\left(Q_{jk}{}^{j} + Q_{k}{}^{j}{}_{j} - Q^{j}{}_{jk}\right)\,.$$

Denoting $\theta = \ln f$, one arrives at the following system of restrictions on $\theta$:

$$\begin{aligned} \partial_1\theta &= S_1 + V_1\,,\\ \partial_2\theta &= S_2 + V_2\,,\\ &\cdots\cdots\cdots\cdots\\ \partial_s\theta &= S_s + V_s\,. \end{aligned} \tag{10.3}$$

This means that $d\theta = (S_k + V_k)dx^k$. *That is, $\theta$ exists if and only if the 1-form $(S_k + V_k)dx^k$ is exact*[2]. If such a $\theta$ exists, then the volume form in question is defined as

$$\mu = e^{\theta}\, \sqrt{g}\, dx^1 \wedge \cdots \wedge dx^s\,.$$

One obtains the following necessary condition: if there exists $\theta \in C^\infty(\mathfrak{M})$, such that $d\theta = (S_k + V_k)dx^k$, then $(S_k + V_k)dx^k$ is a closed 1-form. Next, one can see that if $\theta_0$ satisfies the equation $d\theta = (S_k + V_k)dx^k$, then so do $\theta_0 + \gamma$, where $\gamma$ is a constant. Thus, each element of the set $\{\theta_0 + \gamma \mid \gamma \in \mathbb{R}\}$ satisfies the equation $d\theta = (S_k + V_k)dx^k$.

The sufficient conditions of existence of the function $\theta$ are provided in the framework of de Rham cohomology theory[3]. For the equation $d\theta = (S_k + V_k)dx^k$, one needs to consider the first de Rham group $H^1_{dR}(\mathfrak{M})$. In particular, if $\mathfrak{M}$ is contractible, it is sufficient to check that $(S_k + V_k)dx^k$ is closed to establish the existence of $\theta$.

### 10.1.2 The Case of a Covector-Valued Form

In this paragraph, the generalization of the divergence operator for covector-valued exterior forms is given. Suppose again that $\mathfrak{M}$ is a smooth manifold. In the case of covector-valued forms introduce the analog for divergence.

**Inner-exterior product.** The definition of inner-exterior operation depends on the type of tensorial objects to which this operation applies. Consider the following cases:

---

2 Recall that the smooth differential form $\omega$ is closed, if $d\omega = 0$, and is exact, if $\omega = d\eta$ for some smooth differential form $\eta$.

3 For brief information see Chapter 3, Section 3.6.1.

($\dot{\wedge}_1$) The inner-exterior product $\dot{\wedge}$ is the bilinear mapping of a vector-valued 1-form and a covector-valued $(k-1)$-form, which results in a $k$-form, *i.e.*,

$$\dot{\wedge}\colon (\mathrm{Vec}(\mathfrak{M}) \otimes \mathrm{CVec}(\mathfrak{M})) \times (\mathrm{CVec}(\mathfrak{M}) \otimes \Omega^{k-1}(\mathfrak{M})) \to \Omega^k(\mathfrak{M}) ,$$

$(\mathscr{T}, S) \mapsto \mathscr{T}\dot{\wedge}S$, which satisfies the property

$$(u \otimes \beta)\dot{\wedge}(\vartheta \otimes \omega) = \langle u, \vartheta\rangle \beta \wedge \omega ,$$

for each $u \in \mathrm{Vec}(\mathfrak{M})$, $\vartheta, \beta \in \mathrm{CVec}(\mathfrak{M})$, $\omega \in \Omega^{k-1}(\mathfrak{M})$.

($\dot{\wedge}_2$) Inner-exterior product $\dot{\wedge}$ is the bilinear mapping of a covector-valued 1-form and a vector-valued $(k-1)$-form, which results in a $k$-form, *i.e.*,

$$\dot{\wedge}\colon (\mathrm{CVec}(\mathfrak{M}) \otimes \mathrm{CVec}(\mathfrak{M})) \times (\mathrm{Vec}(\mathfrak{M}) \otimes \Omega^{k-1}(\mathfrak{M})) \to \Omega^k(\mathfrak{M}) ,$$

$(\mathscr{T}, S) \mapsto \mathscr{T}\dot{\wedge}S$, which satisfies

$$(\vartheta \otimes \beta)\dot{\wedge}(u \otimes \omega) = \langle u, \vartheta\rangle \beta \wedge \omega ,$$

for each $u \in \mathrm{Vec}(\mathfrak{M})$, $\vartheta, \beta \in \mathrm{CVec}(\mathfrak{M})$, $\omega \in \Omega^{k-1}(\mathfrak{M})$.

($\dot{\wedge}_3$) Finally, if the Riemannian metric $\mathscr{g}$ is chosen on $\mathfrak{M}$, then the inner-exterior product $\dot{\wedge}$ is the bilinear mapping, which can be defined as follows:

$$\dot{\wedge}\colon (\mathrm{Vec}(\mathfrak{M}) \otimes \mathrm{CVec}(\mathfrak{M})) \times (\mathrm{Vec}(\mathfrak{M}) \otimes \Omega^{k-1}(\mathfrak{M})) \to \Omega^k(\mathfrak{M}) ,$$

$(\mathscr{T}, S) \mapsto \mathscr{T}\dot{\wedge}S$, which satisfies the property

$$(u \otimes \beta)\dot{\wedge}(v \otimes \omega) = \mathscr{g}(u, v)\beta \wedge \omega ,$$

for each $u, v \in \mathrm{Vec}(\mathfrak{M})$, $\beta \in \mathrm{CVec}(\mathfrak{M})$, and $\omega \in \Omega^{k-1}(\mathfrak{M})$.

**Cartan's exterior derivative: definition.** Let $\Gamma^s$ denote $\mathrm{Vec}(\mathfrak{M})$ or $\mathrm{CVec}(\mathfrak{M})$. Define the differential operator, namely, *Cartan's exterior derivative*:

$$\mathfrak{d}\colon \Gamma^s \otimes \Omega^{k-1}(\mathfrak{M}) \to \Gamma^s \otimes \Omega^k(\mathfrak{M}) , \quad T \mapsto \mathfrak{d}T , \tag{10.4}$$

where $1 \le k$, by the relation

$$d\,\langle u, T\rangle = \langle u, \mathfrak{d}T\rangle + \nabla u \dot{\wedge} T . \tag{10.5}$$

Note that if $k = 1$ the operator, (10.4) coincides with the covariant derivative.

Since affine connections on the body $\mathfrak{B}$ and the physical space $\mathscr{P}$ are not the same, the corresponding operators (10.4) also differ. We use the symbol $\mathfrak{D}$ for Cartan's exterior derivative on $\mathfrak{B}$ and use the symbol $\mathfrak{d}$ for its counterpart on $\mathscr{P}$. These operators are related by the equality: $\forall u \in (\Gamma^s)^*\ \forall v_1, \ldots, v_k \in T\mathfrak{B}$,

$$\mathfrak{d}(\vartheta \otimes \omega)(u, \varkappa_* v_1, \ldots, \varkappa_* v_k) = \mathfrak{D}(\vartheta \otimes \varkappa^* \omega)(u, v_1, \ldots, v_k) , \tag{10.6}$$

which holds for all fixed $\vartheta \in \Gamma^s$ and $\omega \in \Omega^{k-1}(\mathfrak{B})$. The corresponding diagram (composed for $\dim \mathscr{P} = 3$)

$$\begin{array}{ccc} \mathrm{Sec}(T^*\mathscr{P} \otimes \Lambda^2 T^*\mathfrak{B}) & \xleftarrow{\varkappa^{*2}} & \mathrm{Sec}(T^*\mathscr{P} \otimes \Lambda^2 T^*\mathscr{P}) \\ \downarrow \mathfrak{D} & & \downarrow \mathfrak{d} \\ \mathrm{Sec}(T^*\mathscr{P} \otimes \Lambda^3 T^*\mathfrak{B}) & \xleftarrow{\varkappa^{*2}} & \mathrm{Sec}(T^*\mathscr{P} \otimes \Lambda^3 T^*\mathscr{P}) \end{array}$$

is commutative. The mapping $\varkappa^{*2}$ represents the partial pullback (see Chapter 7, Formula (7.10)), which acts on the right-hand part of the tensorial object, *i.e.*, on the exterior form.

Assume that both the body and the physical space are three-dimensional. Obtain coordinate representations for Cartan's exterior derivative (10.4) and (10.5) of the covector-valued Cauchy stress form.

Let $(e_a)_{a=1}^3$ be a coordinate frame on the physical space $\mathscr{P}$ and $(e^a)_{a=1}^3$ be the corresponding coordinate coframe. Cauchy stresses are defined by the covector-valued exterior 2-form $T = e^a \otimes (*t_a)$, where $t_a = t_{ab}e^b$. For a smooth vector field $v = v^q e_q$, which conforms to the spatial velocity field, and its contraction with $T$ results in an exterior 2-form

$$\langle v, T\rangle = \left\langle v^q e_q, e^a \otimes \left(*t_{ab}e^b\right)\right\rangle = t_{ab}\delta^a_q v^q(*e^b) = t_{ab}v^a(*e^b)\,.$$

In order to derive the divergence-like expressions of familiar appearance calculate the exterior differential of $\langle v, T\rangle$. One has

$$d\,\langle v, T\rangle = [d\,(t_{ab}v^a)] \wedge (*e^b) + t_{ab}v^a d(*e^b)\,.$$

For the first term, one obtains

$$d\,(t_{ab}v^a) = \partial_k\,(t_{ab}v^a)\,e^k\,.$$

The Hodge star acts on $e^b$ as follows:

$$*e^b = \sqrt{g}(g^{b1}e^2 \wedge e^3 - g^{b2}e^1 \wedge e^3 + g^{b3}e^1 \wedge e^2)\,,$$

and one has

$$d(*e^b) = \partial_k(g^{bk}\sqrt{g})\omega\,,$$

where $\omega = e^1 \wedge e^2 \wedge e^3$. Hence,

$$[d\,(t_{ab}v^a)] \wedge (*e^b) = g^{bk}\sqrt{g}\,\partial_k\,(t_{ab}v^a)\,\omega\,, \quad t_{ab}v^a d(*e^b) = t_{ab}v^a\partial_k(g^{bk}\sqrt{g})\omega\,.$$

Taking into account the expressions obtained, one arrives at the following formula for $d\,\langle v, T\rangle$:

$$d\,\langle v, T\rangle = \left[g^{bk}\sqrt{g}\,\partial_k\,(t_{ab}v^a) + t_{ab}v^a\partial_k(g^{bk}\sqrt{g})\right]\omega = \left[\partial_k\left(t_{ab}v^a g^{bk}\sqrt{g}\right)\right]\omega\,.$$

Since $t_{ab}g^{bk} = t_a{}^k$, we obtain the final relation:

$$d\langle v, T\rangle = \left[\partial_k\left(t_a{}^k v^a \sqrt{g}\right)\right]\omega\,.$$

Now, calculate $\nabla v\dot{\wedge}T$. Using the equality

$$\nabla v = \nabla_k v^a e_a \otimes e^k = \left(\partial_k v^a + \gamma^a{}_{bk}v^b\right)e_a \otimes e^k\,,$$

one obtains

$$\nabla v\dot{\wedge}T = \left(\nabla_k v^a e_a \otimes e^k\right)\dot{\wedge}\left[t_{sl}e^s(*e^l)\right] = t_{sl}\nabla_k v^a\,\langle e_a, e^s\rangle\,[e^k \wedge (*e^l)]\,.$$

Since $\langle e_a, e^s\rangle = \delta^s_a$, and $e^k \wedge (*e^l) = g^{kl}\sqrt{g}\,\omega$, one obtains

$$\nabla v\dot{\wedge}T = t_{al}\nabla_k v^a g^{kl}\sqrt{g}\,\omega = \left(\partial_k v^b + \gamma^b{}_{ak}v^a\right)t_b{}^k\sqrt{g}\,\omega\,.$$

Using relations obtained for $d\langle v, T\rangle$ and $\nabla v\dot{\wedge}T$, according to the formula (10.5) one obtains

$$\langle v, \eth T\rangle = d\langle v, T\rangle - \nabla v\dot{\wedge}T = \left[\partial_k\left(t_a{}^k\sqrt{g}\right) - \gamma^b{}_{ak}t_b{}^k\sqrt{g}\right]v^a\omega\,.$$

Taking into account the Voss–Weyl formula,

$$\partial_k\sqrt{g} = \sqrt{g}\,\gamma^q{}_{qk}\,,$$

and the equality $\mu = \sqrt{g}\,\omega$, one finally obtains

$$\langle v, \eth T\rangle = \left[\partial_k t_a{}^k + t_a{}^k\gamma^q{}_{qk} - \gamma^b{}_{ak}t_b{}^k\right]v^a\mu\,.$$

Note that the expression in square brackets corresponds to the divergence of the stress field in the curvilinear coordinates of the Euclidean space.

The case of the first Piola–Kirchhoff covector-valued form $P$ needs to be considered individually. In such a case, the operator $\nabla$, which appears in formula (10.5), is defined by (9.13) in Chapter 9. The expression $\nabla V$ has the form

$$\nabla V = \left(\partial_\alpha V^k + \gamma^k{}_{ij}V^i F^j{}_\alpha\right)e_k \otimes E^\alpha\,.$$

Here, $(E_\alpha)$ and $(e_a)$ are frames in $\mathfrak{B}$ and $\mathscr{P}$, respectively, and $\gamma^k{}_{ij}$ are coefficients of the connection defined on $\mathscr{P}$.

Obtain the coordinate representation for $\langle V, \mathfrak{D}P\rangle$, where $P$ is the first Piola–Kirchhoff exterior form, and $V$ is a material velocity. Let $(e_a)_{a=1}^3$ be a coordinate frame in $\mathscr{P}$ and $(e^a)_{a=1}^3$ be its corresponding coframe. In addition, let $(E_a)_{a=1}^3$ be a coordinate frame on $\mathfrak{B}$ and $(E^a)_{a=1}^3$ be the coframe. The first Piola–Kirchhoff stresses are defined by the covector-valued 2-form $P = e^a \otimes (*\not{p}_a)$, where $p_a = p_{a\beta}E^\beta$. For a vector field $V = V^q e_q$, which conforms to the velocity field, one obtains the exterior 2-form

$$\langle V, P\rangle = \left\langle V^q e_q, e^a \otimes (*p_{a\beta}E^\beta)\right\rangle = p_{a\beta}V^a(*E^\beta)\,.$$

Calculate the exterior differential from $\langle V, P\rangle$. One has

$$d\langle V, P\rangle = \left[d\left(p_{a\beta}V^a\right)\right]\wedge(*E^\beta) + p_{a\beta}V^a d(*E^\beta)\,.$$

For the first term, the following equality holds:

$$d\left(p_{a\beta}V^a\right) = \partial_\kappa\left(p_{a\beta}V^a\right)E^\kappa\,.$$

From the properties of the Hodge star, it follows that

$$*E^\beta = \sqrt{G}(G^{\beta 1}E^2\wedge E^3 - G^{\beta 2}E^1\wedge E^3 + G^{\beta 3}E^1\wedge E^2)\,,$$

and the cofactor in the second term can be represented as follows:

$$d(*E^\beta) = \partial_\kappa(G^{\beta\kappa}\sqrt{G})\omega\,,$$

where $\omega = E^1\wedge E^2\wedge E^3$. Hence,

$$[d(p_{a\beta}V^a)]\wedge(*E^\beta) = G^{\beta\kappa}\sqrt{G}\,\partial_\kappa\left(p_{a\beta}V^a\right)\omega\,,$$
$$p_{a\beta}V^a d(*E^\beta) = p_{a\beta}V^a\partial_\kappa(G^{\beta\kappa}\sqrt{G})\omega\,.$$

Taking into account the relations obtained, one arrives at the following expressions for $d\langle V, \mathscr{P}\rangle$:

$$d\langle V, P\rangle = \left[G^{\beta\kappa}\sqrt{G}\,\partial_\kappa\left(p_{a\beta}V^a\right) + p_{a\beta}V^a\partial_\kappa(G^{\beta\kappa}\sqrt{G})\right]\omega = \left[\partial_\kappa\left(p_{a\beta}V^a G^{\beta\kappa}\sqrt{G}\right)\right]\omega\,.$$

Since $p_{a\beta}G^{\beta\kappa} = p_a{}^\kappa$, one obtains

$$d\langle V, P\rangle = \left[\partial_\kappa\left(p_a{}^\kappa V^a\sqrt{G}\right)\right]\omega\,.$$

Note that, unlike in the previous case, the metric *of the material manifold* appears in the resulting formula.

Now calculate $\nabla V\dot\wedge P$. Since $\nabla V = \left(\partial_\alpha V^k + \gamma^k{}_{ij}V^iF^j{}_\alpha\right)e_k\otimes E^\alpha$, then

$$\nabla V\dot\wedge P = \left(\left(\partial_\alpha V^k + \gamma^k{}_{ij}V^iF^j{}_\alpha\right)e_k\otimes E^\alpha\right)\dot\wedge\left[p_{s\lambda}e^s(*E^\lambda)\right]$$
$$= p_{s\lambda}\left(\partial_\alpha V^k + \gamma^k{}_{ij}V^iF^j{}_\alpha\right)\langle e_k, e^s\rangle\,[E^\alpha\wedge(*E^\lambda)]\,.$$

Taking into account the expressions $\langle e_k, e^s\rangle = \delta^s_k$ and $E^\alpha\wedge(*E^\lambda) = G^{\alpha\lambda}\sqrt{G}\,\omega$, one obtains

$$\nabla V\dot\wedge P = p_{k\lambda}\left(\partial_\alpha V^k + \gamma^k{}_{ij}V^iF^j{}_\alpha\right)G^{\alpha\lambda}\sqrt{G}\,\omega = \left(\partial_\kappa V^a + \gamma^a{}_{ij}V^iF^j{}_\kappa\right)p_a{}^\kappa\sqrt{G}\,\omega\,.$$

It follows from (10.5) and the expressions obtained for $d\langle V, P\rangle$ and $\nabla V\dot\wedge P$ that

$$\langle V, \mathfrak{D}P\rangle = d\langle V, P\rangle - \nabla V\dot\wedge P = \left[\partial_\kappa\left(p_a{}^\kappa\sqrt{G}\right) - \gamma^b{}_{aj}F^j{}_\kappa p_b{}^\kappa\sqrt{G}\right]V^a\omega\,.$$

Applying the Voss–Weyl formula

$$\partial_K \sqrt{G} = \sqrt{G}\,\Gamma^{T}{}_{TK}\,,$$

and the equality $\mu_0 = \sqrt{G}\,\omega$, one finally arrives at

$$\langle V, \mathfrak{D}P\rangle = \left[\partial_K p_a{}^K + p_a{}^K \Gamma^{T}{}_{TK} - \gamma^b{}_{aj} F^j{}_K p_b{}^K\right] V^a \mu_0\,.$$

Note that the expression in square brackets depends on both the affine connection coefficients related to the physical manifold and their material counterparts.

The above derivations show that $\mathfrak{d}$ and $\mathfrak{D}$ play the role of spatial and material divergence operators div, $\mathrm{div}_R$, which arise in the balance equations of classical continuum mechanics. However, in contrast to them, the operators $\mathfrak{d}$ and $\mathfrak{D}$ do not presuppose that the affine connection on $\mathfrak{B}$ and $\mathscr{P}$ is Euclidean.

## 10.2 The Reynolds Transport Theorem

In continuum mechanics, the Reynolds transport theorem plays a key role, particularly as a tool for deriving a variety of forms for the balance equations. This theorem was stated by Reynolds and proven by Spielrein (see the footnote in [59, p.347]). In the framework of conventional mechanics, this theorem is used in the following form. Let $\mathfrak{B}_0$ be a subset of the reference shape, $\lambda_t$ a family of deformations, *i.e.*, mappings from the reference shape to the actual one, and $f$ a smooth scalar-valued function that depends on points from $\mathfrak{B}_0$ and on time. Then, the following equality holds:

$$\frac{d}{dt}\int\limits_{\lambda_t(\mathfrak{B}_0)} f(X,t)\,dV(X) = \int\limits_{\lambda_t(\mathfrak{B}_0)} \left\{\frac{\partial f(X,t)}{\partial t} + \mathrm{div}(fv)\right\} dV(X)\,.$$

Commonly, this equality is derived in one of two ways. In the first one, by using the change of variables, the integral reduces to an integral over a fixed volume. In the second method, the moving region is considered, and the increment of its volume is calculated. Below a sketch of the generalization of Reynolds' transport theorem is given.

Let $\mathfrak{B}$ be a body and $\mathscr{P}$ be a physical space, which are represented by oriented manifolds with volume forms $\mu_0$ and $\mu$. Assume that $\dim \mathfrak{B} = \dim \mathscr{P}$. Let $M = \{\varkappa_t\}_{t\in\mathbb{T}}$ be a smooth motion. Beyond that, suppose that each diffeomorphism $\hat{\varkappa}_t$ is orientation preserving. Let $\{\psi_{t,s}\}_{t,s\in\mathbb{T}}$ be the corresponding time-dependent flow. In this case, we can use the following theorem of the change of variables [123]. Fix $s \in \mathbb{T}$. Let $\omega$ be a differential $n$-form on $\varkappa_s(\mathfrak{B})$ with a compact support. Then, $\psi^*_{t,s}\omega$ has a compact support, and the following equality holds:

$$\int\limits_{\psi_{t,s}(\varkappa_s(\mathfrak{B}))} \omega = \int\limits_{\varkappa_s(\mathfrak{B})} \psi^*_{t,s}\omega\,. \tag{10.7}$$

Taking into account (10.7) and (6.7), using the linearity[4] of $\psi^*_{t,s}$, the Lie derivatives' properties,[5] and using the definition of the divergence, Reynolds' transport theorem can be proven. Let $f\colon \mathscr{P} \times \mathbb{T} \to \mathbb{R}$ be a smooth mapping. Then, the following equality holds [123] for $\mathfrak{P} \in \mathrm{Part}(\mathfrak{B})$:

$$\frac{d}{dt} \int\limits_{\psi_{t,s}(\varkappa_s(\mathfrak{P}))} f(X,t)\,\mu = \int\limits_{\psi_{t,s}(\varkappa_s(\mathfrak{P}))} \left\{\frac{\partial f(X,t)}{\partial t} + \mathrm{div}_\mu(fv)\right\} \mu\,,$$

where $v$ is the spatial velocity. Such a equality can be rewritten in the form

$$\frac{d}{dt} \int\limits_{\varkappa_t(\mathfrak{P})} f(X,t)\,\mu = \int\limits_{\varkappa_t(\mathfrak{P})} \left\{\frac{\partial f(X,t)}{\partial t} + \mathrm{div}_\mu(fv)\right\} \mu\,.$$

The latter equality can be represented in a more simple form by using the operation $\mathbb{L}$, which is defined by (6.8). Indeed, assuming that $\mu$ does not depend on $t$, one obtains

$$\frac{d}{dt} \int\limits_{\varkappa_t(\mathfrak{P})} f(X,t)\,\mu = \int\limits_{\varkappa_t(\mathfrak{P})} \mathbb{L}_v(f\mu)\,. \tag{10.8}$$

## 10.3 Balance Equations in Integral Form

### 10.3.1 Power Balance

Let $\mathfrak{B}$ be a body and $\mathscr{P}$ be a physical space. Suppose that $\mathscr{G}$ is a Riemannian metric on $\mathfrak{B}$ and $g$ is a Riemannian metric on $\mathscr{P}$. Consider a smooth motion $\{\varkappa_t\}_{t\in\mathbb{T}}$ on $\mathfrak{B}$ and the force systems

$$\left\{\left(\beta_t^{\mathrm{mat}}, \tau_{\mathfrak{P};t}^{\mathrm{mat}}\right)\right\}_{\mathfrak{P}\in\mathrm{Part}(\mathfrak{B}),t\in\mathbb{T}}\,, \quad \left\{\left(\beta_t^{\mathrm{spat}}, \tau_{\mathfrak{P};t}^{\mathrm{spat}}\right)\right\}_{\mathfrak{P}\in\mathrm{Part}(\mathfrak{B}),t\in\mathbb{T}}\,,$$

on $\mathfrak{B}$ in material and spatial descriptions, where $\beta_t^{\mathrm{mat}}$ and $\beta_t^{\mathrm{spat}}$ are body force densities and $\tau_{\mathfrak{P};t}^{\mathrm{mat}}$ and $\tau_{\mathfrak{P};t}^{\mathrm{spat}}$ are surface force densities (Chapter 7, Section 7.1). The powers (total forces) in material and spatial descriptions are represented by the expressions (7.5). Let $\mu_0$ and $\mu_t$ be volume forms on $\mathfrak{B}$ and $\mathscr{S}_{\varkappa_t}$.

**Kinetic energy.** The kinetic energy density on $\mathfrak{P} \in \mathrm{Part}(\mathfrak{B})$ in the material description is given by $\frac{1}{2}\mathscr{G}(V,V)\rho_0\mu_0$, and therefore, the total kinetic energy, can be defined

---

4 Pullback of smooth mapping is a linear operation between spaces of exterior forms; it preserves the rank of any form [123, p. 354].

5 In particular for the smooth mapping $f\colon \mathscr{P} \to \mathbb{R}$ and volume form $\mu$ one has $\mathscr{L}_u(f\mu) = (\mathscr{L}_u f)\mu + f\mathscr{L}_u\mu$.

as

$$\mathfrak{K}^{\text{mat}} := \frac{1}{2} \int_{\mathfrak{P}} \mathscr{G}(V, V) \rho_0 \mu_0 \,,$$

where $\rho_0$ is the mass density in $\mathfrak{B}$ (material description). Similarly, the kinetic energy density on $\mathfrak{P} \in \text{Part}(\mathfrak{B})$ in the spatial description is given by $\frac{1}{2}g(v, v)\rho_t\mu_t$, and the total kinetic energy can be defined as

$$\mathfrak{K}^{\text{spat}} := \frac{1}{2} \int_{\varkappa_t(\mathfrak{P})} g(v, v) \rho_t \mu_t \,,$$

where $\rho_t$ is the mass density in $\mathfrak{B}$ (spatial description).

**Internal energy.** Similarly to the kinetic energy, we define the internal energy density as $e_0\rho_0\mu_0$ in the material description and $e_t\rho_t\mu_t$ in the spatial description. Here, $e_0$ and $e_t$ are scalar mappings;

$$e_0 \colon (\mathfrak{X}, t, \mathscr{G}) \mapsto e_0(\mathfrak{X}, t, \mathscr{G}) \,, \quad e \colon (X, t, g) \mapsto e(X, t, g) \,.$$

Thus,

$$\mathfrak{E}^{\text{mat}} := \int_{\mathfrak{P}} e_0 \rho_0 \mu_0 \,,$$

is the total internal energy in the material description, and

$$\mathfrak{E}^{\text{spat}} := \int_{\varkappa_t(\mathfrak{P})} e_t \rho_t \mu_t \,,$$

is the total energy in the spatial description.

**Power balance.** For any part $\mathfrak{P} \in \text{Part}(\mathfrak{B})$ the power balance in material and spatial descriptions is defined as follows:

$$F_t^{\text{mat}}(V) = \frac{d}{dt}\left(\mathfrak{K}^{\text{mat}} + \mathfrak{E}^{\text{mat}}\right) \,,$$
$$F_t^{\text{spat}}(v) = \frac{d}{dt}\left(\mathfrak{K}^{\text{spat}} + \mathfrak{E}^{\text{spat}}\right) \,.$$

It is convenient to introduce the time-dependent covectors $b_0 \colon \mathfrak{B} \times \mathbb{T} \to T^*\mathscr{P}$ and $b \colon \mathscr{S}_{\varkappa_t} \times \mathbb{T} \to T^*\mathscr{P}$, such that $\beta_t^{\text{mat}} = b_0 \otimes \mu_0$ and $\beta_t^{\text{spat}} = b \otimes \mu_t$. Similarly, the mass densities, stress forms, and volume forms are represented as time-dependent fields. Next, applying Stokes' theorem to the second summands of (7.5), we obtain

$$\int_{\partial\mathfrak{P}} V \lrcorner P = \int_{\mathfrak{P}} d(V \lrcorner P) \,, \qquad \int_{\partial\varkappa_t(\mathfrak{P})} v \lrcorner T = \int_{\varkappa_t(\mathfrak{P})} d(v \lrcorner T) \,.$$

Thus, for a part $\mathfrak{P} \in \mathrm{Part}(\mathfrak{B})$ and the motion $\{\varkappa_t\}_{t\in\mathbb{T}}$, one has a power balance in the material and spatial descriptions:

$$\int_{\mathfrak{P}} (\langle V, b_0\rangle \rho_0\mu_0 + d(V \lrcorner P)) = \frac{d}{dt}\left( \int_{\mathfrak{P}} \left(\frac{1}{2}\mathscr{G}(V, V) + e_0\right) \rho_0\mu_0 \right) . \tag{10.9}$$

$$\int_{\varkappa_t(\mathfrak{P})} (\langle v, b\rangle \rho\mu + d(v \lrcorner T)) = \frac{d}{dt}\left( \int_{\varkappa_t(\mathfrak{P})} \left(\frac{1}{2}g(v, v) + e\right) \rho\mu \right) . \tag{10.10}$$

### 10.3.2 Mass Conservation

Let $\mathfrak{P} \in \mathrm{Part}(\mathfrak{B})$ be a part. The integrals

$$\mathfrak{m}_0 = \int_{\mathfrak{P}} \rho_0\mu_0 , \quad \mathfrak{m}(t) = \int_{\varkappa_t(\mathfrak{P})} \rho_t\mu_t$$

represent the total mass of the part $\mathfrak{P}$ in the material and spatial descriptions. The mass conservation law requires that

$$\forall t \in \mathbb{T}: \mathfrak{m}(t) = \mathfrak{m}_0 ,$$

for arbitrary $\mathfrak{P} \in \mathrm{Part}(\mathfrak{B})$. Using the change of variables theorem, one obtains

$$\int_{\mathfrak{P}} (\rho_0\mu_0 - \hat{\varkappa}_t^*(\rho_t\mu_t)) = 0 .$$

Since $\mathfrak{P}$ is arbitrary, one arrives at the relation

$$\rho_0\mu_0 = \hat{\varkappa}_t^*(\rho_t\mu_t) . \tag{10.11}$$

Another integral formulation of the mass conservation law is the following:

$$\forall t_0 \in \mathbb{T}: \left.\frac{d}{dt}\mathfrak{m}(t)\right|_{t_0} = 0 .$$

Assuming that
(i) $\dim \mathfrak{B} = \dim \mathscr{P}$;
(ii) volume form $\mu$ and metric $g$ doesn't depend on $t$,

and using (10.8), one obtains

$$\int_{\varkappa_t(\mathfrak{P})} \mathbb{L}_v(\rho\mu) = 0 ,$$

that holds for arbitrary $\mathfrak{P} \in \mathrm{Part}(\mathfrak{B})$. Thus,

$$\mathbb{L}_v(\rho\mu) = 0\,. \tag{10.12}$$

Finally, we have obtained two differential representations (10.11) and (10.12) of the integral mass conservation laws. The first one is the analog of the classical relation $\rho_0 = \rho J$, while the second one in the classical case gives $\partial_t\rho + \mathrm{div}(\rho v) = 0$.

### 10.3.3 Transformation of the Spatial Power Balance Equation

Our aim is to interchange the operator $d/dt$ with the integral operator in the expression (10.10). It is assumed that:

(i) $\dim \mathfrak{B} = \dim \mathscr{P}$;

(ii) volume form $\mu$ and metric $g$ do not depend on $t$.

Recall that all shapes $\mathscr{S}_{\varkappa_t}$ are open sets. Denote $f = (\frac{1}{2}g(v,v)+e)\rho\mu$. Then the following sequence of equalities hold (here $s \in \mathbb{T}$ is fixed):

$$\frac{d}{dt}\int\limits_{\varkappa_t(\mathfrak{B}_0)} f(X,t) = \frac{d}{dt}\int\limits_{\psi_{t,s}(\varkappa_s(\mathfrak{B}_0))} f(X,t) = \frac{d}{dt}\int\limits_{\varkappa_s(\mathfrak{B}_0)} \psi_{t,s}^* f(X,t)\,.$$

Note that in the last equality, the change of variables Theorem (10.7) was applied. One has

$$\frac{d}{dt}\int\limits_{\varkappa_t(\mathfrak{B}_0)} f(X,t) = \int\limits_{\varkappa_s(\mathfrak{B}_0)} \frac{d}{dt}\{\psi_{t,s}^* f(X,t)\}\,.$$

According to the definition (6.8) of derivation $\mathbb{L}_v$ and equalities (6.7) and (6.9), one has the following representation for $\frac{d}{dt}\psi_{t,s}^* f$:

$$\frac{d}{dt}\psi_{t,s}^* f = \psi_{t,s}^*\left(\mathscr{L}_v f + \frac{\partial f}{\partial t}\right) = \psi_{t,s}^*(\mathbb{L}_v f)\,,$$

where $v$ is the spatial velocity. Since $\mathbb{L}_v$ is a derivation, the expression $\mathbb{L}_v f$ can be transformed as follows:

$$\mathbb{L}_v f = \left(\frac{1}{2}g(v,v) + e\right)\mathbb{L}_v(\rho\mu) + (\mathbb{L}_v e)\rho\mu + \left(\mathbb{L}_v\left[\frac{1}{2}g(v,v)\right]\right)\rho\mu\,.$$

Next, we will be interested in the expression $\mathbb{L}_v[\frac{1}{2}g(v,v)]$. By the definition of $\mathbb{L}$, it has the form

$$\mathbb{L}_v\left[\frac{1}{2}g(v,v)\right] = \frac{1}{2}\frac{\partial g(v,v)}{\partial t} + \mathscr{L}_v\left[\frac{1}{2}g(v,v)\right]\,.$$

Here, $\frac{\partial g(v,v)}{\partial t} = 2g_{ij}v^i\frac{\partial v^j}{\partial t} = 2\langle v^\flat, \frac{\partial v}{\partial t}\rangle$. Since $\frac{1}{2}g(v,v)$ is a scalar, and then

$$\mathscr{L}_v\left[\frac{1}{2}g(v,v)\right] = \frac{1}{2}v[g(v,v)]\,.$$

We have not said anything about connection up to this point. However, at this moment, the connection begins to play a role. The last equation can be rewritten using the non-metricity of the physical space (9.10):

$$\mathscr{L}_v\left[\frac{1}{2}\mathscr{g}(v, v)\right] = \mathscr{g}(v, \nabla_v v) - \frac{1}{2}\mathfrak{Q}(v, v, v)\,.$$

Thus, $\mathbb{L}_v[\frac{1}{2}\mathscr{g}(v, v)]$ has the form

$$\mathbb{L}_v\left[\frac{1}{2}\mathscr{g}(v, v)\right] = \left\langle v^\flat, \frac{\partial v}{\partial t} + \nabla_v v\right\rangle - \frac{1}{2}\mathfrak{Q}(v, v, v)\,.$$

Since the physical space is Riemannian, $\mathfrak{Q}(v, v, v) = 0$. We arrive at the vector field

$$a = \frac{\partial v}{\partial t} + \nabla_v v$$

which is the spatial acceleration (see Chapter 6, Section 6.3). Thus,

$$\mathbb{L}_v\left[\frac{1}{2}\mathscr{g}(v, v)\right] = \left\langle v^\flat, a\right\rangle\,.$$

Finally, by virtue of

$$\frac{d}{dt}\int\limits_{\varkappa_t(\mathfrak{P})} f(X, t) = \int\limits_{\varkappa_s(\mathfrak{P})} \psi_{t,s}^*(\mathbb{L}_v f) = \int\limits_{\varkappa_t(\mathfrak{P})} \mathbb{L}_v f\,,$$

one obtains the relation

$$\frac{d}{dt}\left(\int\limits_{\varkappa_t(\mathfrak{P})}\left(\frac{1}{2}\mathscr{g}(v, v) + e\right)\rho\mu\right) = \int\limits_{\varkappa_t(\mathfrak{P})}\left\{\left(\frac{1}{2}\mathscr{g}(v, v) + e\right)\mathbb{L}_v(\rho\mu) + (\mathbb{L}_v e)\rho\mu + \langle v^\flat, a\rangle\rho\mu\right\}.$$

Using the relation obtained and the definition of $\eth$, the formula (10.10) can be rewritten in the form (the equality $\langle v^\flat, a\rangle = \langle v, a^\flat\rangle$ is used):

$$\int\limits_{\varkappa_t(\mathfrak{P})}\langle v, \eth T + (b - a^\flat)\otimes\rho\mu\rangle = \int\limits_{\varkappa_t(\mathfrak{P})}\left\{\left(\frac{1}{2}\mathscr{g}(v, v) + e\right)\mathbb{L}_v(\rho\mu) + (\mathbb{L}_v e)\rho\mu\right\} - \int\limits_{\varkappa_t(\mathfrak{P})}\nabla v \dot{\wedge} T\,. \tag{10.13}$$

## 10.4 Derivation of Differential Balance Equations Using the Covariance Principle

### 10.4.1 The Principle of Covariance

Assume that there is a theory that is described via the functionally independent tensor fields $u_1, \ldots, u_r$, defined in space (or space-time) $\mathscr{M}$, and the equations of such a theory have the form

$$F_k(u_1, \ldots, u_r) = 0\,, \quad k = 1, \ldots, r\,. \tag{10.14}$$

These can be partial differential equations, integral equations, *etc.* The equations (10.14) and the corresponding theory are called *covariant* [60, p. 156], if for any diffeomorphism $\Phi\colon \mathscr{M} \to \mathscr{M}$ the following relations hold:

$$(\Phi^* F_k)(u_1, \ldots, u_r) = F_k(\Phi^* u_1, \ldots, \Phi^* u_r)\,, \quad k = 1, \ldots, r\,.$$

In other words, for a smooth change of coordinates, the form of the functions $F_k$ does not change (for a more comprehensive treatment and historical review, see [139]).

### 10.4.2 Change of Frame and Objective Transformations

Assume that $\dim \mathfrak{B} = \dim \mathscr{P} = n$. In the following, we present the results obtained by Marsden and Hughes [60] and Yavari et al. [126, 140].

**Generalization of the change of frame.** In Chapter 2, Section 2.5, we considered the Galilean change of frame. With the purpose of deriving differential conservation laws, we provide its generalization as follows. Let $\mathscr{M}$ be Newtonian space-time. Consider the two frames

$$\varphi_1\colon \mathscr{M} \to \mathscr{P} \times \mathbb{T}\,, \quad e \mapsto (x, t)\,,$$
$$\varphi_2\colon \mathscr{M} \to \mathscr{P} \times \mathbb{T}\,, \quad e \mapsto (x', t')\,.$$

The mapping $\varphi_{12} := \varphi_2 \circ \varphi_1^{-1}\colon \mathscr{P} \times \mathbb{T} \to \mathscr{P} \times \mathbb{T}$ represents a change of frame. Let us postulate that it has the form

$$(x', t') = (\xi(x, t), t)\,,$$

where $\xi\colon \mathscr{P} \times \mathbb{T} \to \mathscr{P}$ is a $C^\infty$-mapping with the property: for all $t \in \mathbb{T}$, the mapping $\xi_t := (\cdot, t)\colon \mathscr{P} \to \mathscr{P}$ is a diffeomorphism. In addition, for some fixed $t_0 \in \mathbb{T}$, we apply

$$\xi_t\big|_{t=t_0} = \mathrm{Id}_{\mathscr{P}}\,, \quad T\xi_t\big|_{t=t_0} = \mathrm{Id}_{T\mathscr{P}}\,.$$

**Charts on $\mathscr{P}$ and $\xi_t(\mathscr{P})$.** Let $A = \{(U_\alpha, \varphi_\alpha)\}_{\alpha\in I}$ be a smooth atlas on $\mathscr{P}$. This atlas is induced on $\xi_t(\mathscr{P}) = \mathscr{P}$ by means of the diffeomorphism $\xi_t$, *i.e.*, the set $A_t = \{(\xi_t(U_\alpha), \psi_\alpha)\}_{\alpha\in I}$ is an atlas on $\xi_t(\mathscr{P})$. Here, $\psi_\alpha = \varphi_\alpha \circ \xi_t^{-1}$. The mapping $\xi_t$ has a trivial coordinate representation with respect to the atlases $A$ and $A_t$:

$$\tilde{\xi}_t = \psi_\alpha \circ \xi_t \circ \varphi_\alpha^{-1} = \mathrm{Id}_{\varphi_\alpha(U_\alpha)}\colon (x^1, \ldots, x^n) \mapsto ((x')^1, \ldots, (x')^n)\,,$$

and the tangent map $T\xi_t$ is represented by the dyadic decomposition

$$T\xi_t = \partial_{(x')^i} \otimes dx^i\,.$$

Thus,

$$T\xi_t \llcorner \partial_{x^i} = \partial_{(x')^i}\,,$$

and in $t = t_0$,

$$(\partial_{(x')^i})|_{t=t_0} = \partial_{x^i} .$$

The relation between $A$ and $A_t$ is illustrated on the following diagram ($V_\alpha = \xi_t(U_\alpha)$):

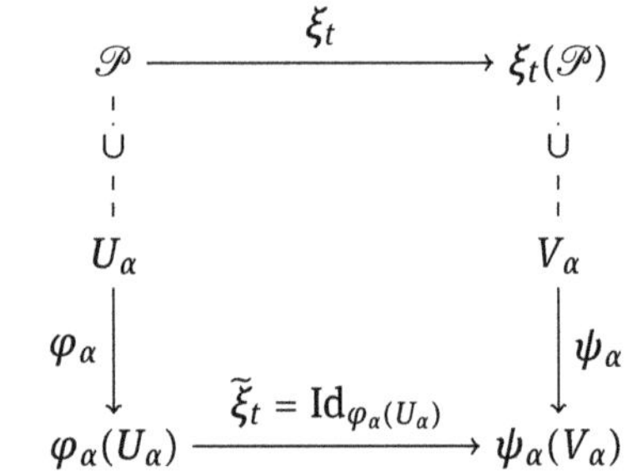


**Motion and velocity in the change of frame.** Let $M = \{\varkappa_t\}_{t\in\mathbb{T}}$ be a smooth motion with respect to $\varphi_1$ and $M' = \{\varkappa'_t\}_{t\in\mathbb{T}}$ be a smooth motion with respect to $\varphi_2$. Then, the relation

$$\forall \mathfrak{X} \in \mathfrak{B}\ \forall t \in \mathbb{T}\colon \varkappa'_t(\mathfrak{X}) = \xi(\varkappa_t(\mathfrak{X}), t)$$

holds. Differentiating this equality with respect to $t$ one obtains (note that the prime indicates that the quantity is considered in $\varphi_2$)

$$v' = (\xi_t)_* v + w_t , \quad \text{where} \quad w_t = \frac{\partial \xi}{\partial t} ,$$

where the left and right-hand sides are defined on $\xi_t(\mathscr{M})$.

**Objective transformations.** Assume that transformations of $\rho\mu$, $b - a^\flat$, $T$ and $e$ are *objective*. That is, the relations between primed and unprimed quantities have the form:

$$\begin{aligned} \rho'\mu' &= (\xi_t)_*(\rho\mu) , & b' - (a')^\flat &= (\xi_t)_*(b - a^\flat) , \\ T' &= (\xi_t)_* T , & e'(x', t, g') &= e(x, t, (\xi_t)_* g) . \end{aligned} \tag{10.15}$$

Note that in the latter relation, the field $g' = (\xi_t)_* g$ is presented. This is the metric used with respect to the frame $\varphi_2$.

We use the following result, which is proven in [60] (Theorem 6.19). If $\mathscr{T}$ is a time-dependent tensor field and $\mathscr{T}' = (\xi_t)_*\mathscr{T}$, then

$$\mathbb{L}_{v'}\mathscr{T}' = (\xi_t)_*\mathbb{L}_v\mathscr{T} .$$

### 10.4.3 Derivation of Spatial Conservation Laws

The starting point will be the relation (10.13), rewritten in the following form:

$$\int_{\varkappa_t(\mathfrak{P})} \langle v, (b - a^\flat) \otimes \rho\mu \rangle + \int_{\partial\varkappa_t(\mathfrak{P})} v \lrcorner T = \int_{\varkappa_t(\mathfrak{P})} \left\{ \left( \frac{1}{2} g(v, v) + e \right) \mathbb{L}_v(\rho\mu) + (\mathbb{L}_v e)\rho\mu \right\} . \tag{10.16}$$

Applying $\xi_t$, one has

$$\int\limits_{\varkappa'_t(\mathfrak{P})} \langle v', (b' - (a')^\flat) \otimes \rho'\mu' \rangle + \int\limits_{\partial\varkappa'_t(\mathfrak{P})} v' \lrcorner T' = \int\limits_{\varkappa'_t(\mathfrak{P})} \left\{ \left( \frac{1}{2} g'(v', v') + e' \right) \mathbb{L}_{v'}(\rho'\mu') + (\mathbb{L}_{v'} e')\rho'\mu' \right\} . \quad (10.17)$$

Taking into account expressions (10.15) and bearing in mind the relation

$$(\mathbb{L}_{v'} e')\Big|_{t=t_0} = \mathbb{L}_v e + \frac{\partial e}{\partial g} : \mathcal{L}_w g ,$$

with $t = t_0$, one arrives at

$$\int\limits_{\varkappa_t(\mathfrak{P})} \langle v + w, (b - a^\flat) \otimes \rho\mu \rangle + \int\limits_{\partial\varkappa_t(\mathfrak{P})} (v + w) \lrcorner T = \int\limits_{\varkappa_t(\mathfrak{P})} \left\{ \left( \frac{1}{2} g(v + w, v + w) + e \right) \mathbb{L}_v(\rho\mu) + \left( \mathbb{L}_v e + \frac{\partial e}{\partial g} : \mathcal{L}_w g \right) \rho\mu \right\} , \quad (10.18)$$

where $w = w_{t_0}$. Subtracting (10.16) from (10.18) one obtains

$$\int\limits_{\varkappa_t(\mathfrak{P})} \langle w, (b - a^\flat) \otimes \rho\mu \rangle + \int\limits_{\partial\varkappa_t(\mathfrak{P})} w \lrcorner T = \int\limits_{\varkappa_t(\mathfrak{P})} \left\{ \left( \frac{1}{2} g(w, w) + g(v, w) \right) \mathbb{L}_v(\rho\mu) + \left( \frac{\partial e}{\partial g} : \mathcal{L}_w g \right) \rho\mu \right\} .$$

Applying Stokes' theorem and using the definition of $\mathfrak{d}$ one obtains

$$\int\limits_{\varkappa_t(\mathfrak{P})} \left\{ \langle w, \mathfrak{d}T + (b - a^\flat) \otimes \rho\mu \rangle + \nabla w \dot{\wedge} T - \left( \frac{1}{2} g(w, w) + g(v, w) \right) \mathbb{L}_v(\rho\mu) - \left( \frac{\partial e}{\partial g} : \mathcal{L}_w g \right) \rho\mu \right\} = 0 . \quad (10.19)$$

The part $\mathfrak{P} \in \operatorname{Part}(\mathfrak{B})$ is arbitrary. In this regard, (10.19) reduces to

$$\langle w, \mathfrak{d}T + (b - a^\flat) \otimes \rho\mu \rangle + \nabla w \dot{\wedge} T - \left( \frac{1}{2} g(w, w) + g(v, w) \right) \mathbb{L}_v(\rho\mu) - \left( \frac{\partial e}{\partial g} : \mathcal{L}_w g \right) \rho\mu = 0 . \quad (10.20)$$

The gradient $(\nabla w)^\flat$ can be decomposed into symmetrical and antisymmetrical parts as follows:

$$(\nabla w)^\flat = \frac{1}{2} \mathcal{L}_w g + d(w^\flat) . \quad (10.21)$$

**Remark 10.2.** In the classical Euclidean case, the relation (10.21) represents the decomposition of the displacement gradient into the infinitesimal strain tensor and the vortex tensor.

Since quantities $w$, $\langle w^\flat, w\rangle$, $dw^\flat$, and $\mathscr{L}_w g$ can take arbitrary values (from the physical point of view, the first one corresponds to a value of an arbitrary vector field, the second one corresponds to its magnitude, the third one to its curl, and the fourth one to its gradient), the equality (10.20) is equivalent to the system of four independent equations [126]:

$$\mathbb{L}_v(\rho\mu) = 0\,, \tag{10.22}$$

$$\eth\mathscr{T} + b \otimes \rho\mu = \dot{v}^\flat \otimes \rho\mu\,, \tag{10.23}$$

$$(u \otimes v)\dot{\wedge} T = (v \otimes u)\dot{\wedge} T\,, \tag{10.24}$$

$$*_2\left(2\rho\frac{\partial e}{\partial g}\right) = T\,. \tag{10.25}$$

The equation (10.22) represents the mass conservation law, (10.23) is the balance of linear momentum, (10.24) is the balance of angular momentum (here $u, v \in T\mathscr{P}$ are arbitrary), and finally, (10.25) is the Doyle–Ericksen relation analog [37].

If the differential conservation law (10.23) holds, then from the equality (10.13), the relation for the power of stress is as follows:

$$\frac{d}{dt}\int\limits_{\varkappa_t(\mathfrak{P})} e\rho\mu = \int\limits_{\varkappa_t(\mathfrak{P})} \nabla v \dot{\wedge} T\,.$$

# 11 The Evolutionary Problem – Examples

## 11.1 Example: The Cylindrical Problem

### 11.1.1 Hollow Cylinders with Discrete Inhomogeneity

Assume that the solid is an inhomogeneous hollow cylinder, which inhomogeneity appears during the process of its assembly from a number of thin unstressed hollow cylindrical parts [141]. Its incompatibility may be clarified from the technological point of view as follows. Hollow cylindrical layers join consequently, and each layer undergoes shrinkage at once after joining. Such a scenario of accretion is implemented, for example, in a stereolithographic process. We assume that[1]

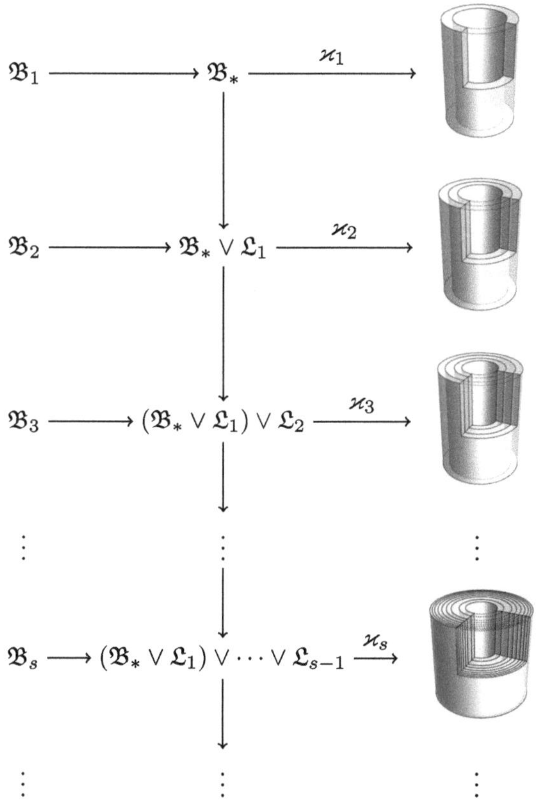


**Fig. 11.1:** Cylindrical assemblies.

1 These assumptions contain considerations of Chapter 8, Section 8.5.2. Moreover, they allow us to use the semi-inverse method for determining the stress–strain state of the cylinder.

https://doi.org/10.1515/9783110563214-011

(DG1) For each $s \in \{1, \dots, N\}$, and for any $k \in \{0, \dots, s-1\}$ there exists a configuration

$$\varkappa^R_{s,k} : \mathfrak{B}_s \to \mathscr{E}$$

such that the shape $\varkappa^R_{s,k}(\mathfrak{L}_k)$ is stress free.

(DG2) Images of the layers $\mathfrak{L}_r$ under configurations $\varkappa^R_{s,k}$, $\varkappa^R_{s,k}(\mathfrak{L}_r) \subset \mathscr{E}$ are represented by hollow cylinders with inner radii $R^r_i$ and exterior radii $R^r_e$, and common to all height $h$.

(DG3) All bodies (*i.e.*, assemblies) $\mathfrak{B}_s$, $s = 2, \dots, N$ result from a sequential joining of layers $\mathfrak{L}_k$ to $\mathfrak{B}_1$. The image of any body $\mathfrak{B}_s$ under the actual configuration $\varkappa_s$ is a hollow cylinder. Cylinders $\varkappa_s(\mathfrak{L}_k)$ are coaxial.

(DG4) Before joining the assembly, each layer undergoes axisymmetric plane deformation.

(DG5) The inner radius of the body $\mathfrak{L}_0 = \mathfrak{B}_1$ in reference configuration, $R^0_i = \rho$, is known, and reference thicknesses $\Delta^k = R^k_e - R^k_i$, $k = 0, \dots, N-1$, of the layers are prescribed. Other reference radii are unknown, but they can be determined from the following recurrence relation: the inner reference radius of layer $\mathfrak{L}_s$ is equal to $R^s_i = S^s\, r^{s-1}_{e,s}$, where $0 < S^s < 1$ is a shrinkage coefficient, and $r^{s-1}_{e,s}$ is the exterior actual radius of the layer $\mathfrak{L}_{s-1}$ in assembly $\mathfrak{B}_s$.

Denote by $r^k_{i,s}$ the actual inner radius of the $k$-th layer and by $r^k_{e,s}$ the actual exterior radius of the $k$-th layer in assembly $\mathfrak{B}_s$. Figure 11.1 illustrates the assembly process.

For the assembly with number 3, the assumptions (DG1)–(DG2) are illustrated by the diagram:

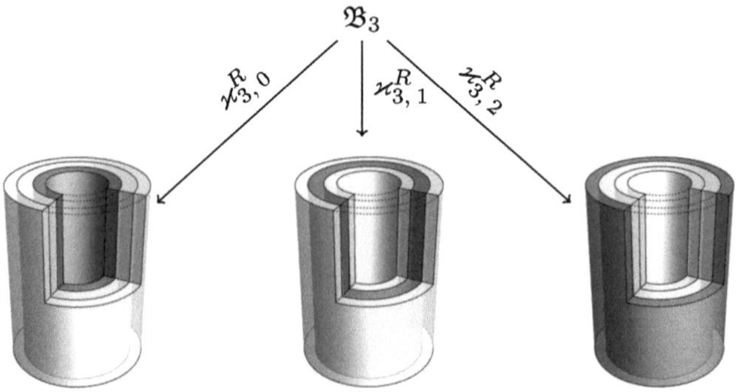


The area in dark color corresponds to a stress-free shape.

**Universal deformation.** Consider the assembly with the number $s$ ($s \in \{1, \dots, N\}$). Since $\varkappa_s(\mathfrak{B}_s)$ is a hollow cylinder, it is convenient to use cylindrical coordinates $(r, \theta, z)$, which are related with Cartesian coordinates via formulae:

$$x^1 = r\cos\theta\,, \quad x^2 = r\sin\theta\,, \quad x^3 = z\,.$$

The elements of the local frame $(\boldsymbol{e}_r, \boldsymbol{e}_\theta, \boldsymbol{e}_z)$, which corresponds to the cylindrical coordinates, can be derived as $\boldsymbol{e}_c = (\partial x^k)/(\partial c)\boldsymbol{i}_k$. Index $c$ designates one of the symbols $r, \theta$, or $z$, and $x = x^k(r, \theta, z)\boldsymbol{i}_k$ is the position vector field, represented by the Cartesian frame. By a $(e^r, e^\theta, e^z)$, we denote the coframe field $e^c\colon \varkappa_s(\mathfrak{B}_s) \to \mathscr{V}^*$, $e^c(\boldsymbol{e}_j) = \delta^c_j$, $c, j \in \{r, \theta, z\}$.

Let $k$ be a number of the layer in the assembly $\mathfrak{B}_s$; $k \in \{1, \ldots, s\}$. According to the assumptions stated above, the deformation of the reference shape $\varkappa^R_{s,k}(\mathfrak{L}_k)$ into the actual one, $\varkappa_s(\mathfrak{L}_k)$, is flat axisymmetric. To this end, we introduce the cylindrical coordinates $(R^{s,k}, \Theta^{s,k}, Z^{s,k})$ on the set $\varkappa^R_{s,k}(\mathfrak{L}_k)$. The deformation

$$d_{s,k}\colon \varkappa^R_{s,k}(\mathfrak{L}_k) \to \varkappa_s(\mathfrak{L}_k)$$

belongs to the class of universal deformation [37] and can be represented in coordinate-free notation as follows:

$$\boldsymbol{x}(\boldsymbol{X}_{s,k}) = \boldsymbol{e}_{R^{s,k}}\sqrt{(e^{R^{s,k}} \llcorner \boldsymbol{X}_{s,k})^2 + a^k_s} + \boldsymbol{e}_{Z^{s,k}} \otimes e^{Z^{s,k}} \llcorner \boldsymbol{X}_{s,k}\,,$$

where $a^k_s$ is the deformation parameter. In coordinates $(R^{s,k}, \Theta^{s,k}, Z^{s,k})$ and $(r, \theta, z)$, the deformation $d_{s,k}$ can be written as:

$$\widetilde{d_{s,k}}\colon (R^{s,k}, \Theta^{s,k}, Z^{s,k}) \mapsto (r, \theta, z) = \left(\sqrt{(R^{s,k})^2 + a^k_s}, \Theta^{s,k}, Z^{s,k}\right).$$

**Strain measures.** The deformation gradient $\boldsymbol{F}_{s,k}$, its transpose, $\boldsymbol{F}^{\mathrm{T}}_{s,k}$, and the left Cauchy–Green tensor $\boldsymbol{B}_{s,k} = \boldsymbol{F}_{s,k}\boldsymbol{F}^{\mathrm{T}}_{s,k}$ in coordinates $(R^k, \Theta^k, Z^k)$ and $(r, \theta, z)$ may be written in the forms:

$$\boldsymbol{F}_{s,k} = \frac{R^{s,k}}{\sqrt{(R^{s,k})^2 + a^k_s}}\boldsymbol{e}_r \otimes e^{R^{s,k}} + \boldsymbol{e}_\theta \otimes e^{\Theta^{s,k}} + \boldsymbol{e}_z \otimes e^{Z^{s,k}}\,, \tag{11.1}$$

$$\boldsymbol{F}^{\mathrm{T}}_{s,k} = \frac{R^{s,k}}{\sqrt{(R^{s,k})^2 + a^k_s}}\boldsymbol{e}_{R_{s,k}} \otimes e^r + \frac{(R^{s,k})^2 + a^k_s}{(R^{s,k})^2}\boldsymbol{e}_{\Theta_{s,k}} \otimes e^\theta + \boldsymbol{e}_{Z_{s,k}} \otimes e^z\,,$$

$$\boldsymbol{B}_{s,k} = \frac{r^2 - a^k_s}{r^2}\boldsymbol{e}_r \otimes e^r + \frac{r^2}{r^2 - a^k_s}\boldsymbol{e}_\theta \otimes e^\theta + \boldsymbol{e}_z \otimes e^z\,. \tag{11.2}$$

In the derivation of (11.1)–(11.2) we take into account that metric tensors $\boldsymbol{G}_{s,k}$ and $\boldsymbol{g}$ on $\varkappa^R_{s,k}(\mathfrak{L}_k)$ and $\varkappa_s(\mathfrak{L}_k)$ have the following representations:

$$\boldsymbol{G}_{s,k} = e^{R^{s,k}} \otimes e^{R^{s,k}} + (R^{s,k})^2 e^{\Theta^{s,k}} \otimes e^{\Theta^{s,k}} + e^{Z^{s,k}} \otimes e^{Z^{s,k}}\,,$$
$$\boldsymbol{g} = e^r \otimes e^r + r^2 e^\theta \otimes e^\theta + e^z \otimes e^z\,.$$

**Cauchy stresses.** Assume that layers are made of hyperelastic material. The constitutive relation has the form: $\boldsymbol{T} = -p\boldsymbol{I} + \beth_1\boldsymbol{B} + \beth_{-1}\boldsymbol{B}^{-1}$, where $\boldsymbol{I}$ is the identity tensor, $p$ is a hydrostatic component, $\daleth_1 = (1+\beta)\mu/2$, $\daleth_{-1} = (\beta-1)\mu/2$, and $\beta$ and $\mu$ are material constants. Taking into account the diagonality of $\boldsymbol{B}$, one can derive the expression for the Cauchy stress tensor[2] in the layer with number $k$:

$$\boldsymbol{T}^{\sharp}_{s,k} = T^{rr}_{s,k}\boldsymbol{e}_r \otimes \boldsymbol{e}_r + T^{\theta\theta}_{s,k}\boldsymbol{e}_\theta \otimes \boldsymbol{e}_\theta + T^{zz}_{s,k}\boldsymbol{e}_z \otimes \boldsymbol{e}_z\,,$$

$$T^{rr}_{s,k} = -p_{s,k} + \daleth_1 \frac{r^2 - a^k_s}{r^2} + \daleth_{-1}\frac{r^2}{r^2 - a^k_s}\,,$$

$$T^{\theta\theta}_{s,k} = -\frac{p_{s,k}}{r^2} + \daleth_1 \frac{1}{r^2 - a^k_s} + \daleth_{-1}\frac{r^2 - a^k_s}{r^4}\,, \qquad T^{zz}_{s,k} = -p_{s,k} + \daleth_1 + \daleth_{-1}\,.$$

From the balance of linear momentum, $\nabla \cdot \boldsymbol{T}^{\sharp} = \boldsymbol{0}$, it follows that

$$\frac{\partial T^{rr}_{s,k}}{\partial r} + \frac{T^{rr}_{s,k}}{r} - r\,T^{\theta\theta}_{s,k} = 0\,, \qquad \frac{\partial T^{\theta\theta}_{s,k}}{\partial \theta} = 0\,, \qquad \frac{\partial T^{zz}_{s,k}}{\partial z} = 0\,. \tag{11.3}$$

Using expressions (11.3) one can obtain components of $\boldsymbol{T}^{\sharp}_k$:

$$T^{rr}_{s,k} = \frac{\mu}{2}\left(\ln \frac{r^2 - a^k_s}{r^2} - \frac{a^k_s}{r^2}\right) + p^k_{0,s}\,,$$

$$T^{\theta\theta}_{s,k} = \frac{T^{rr}_{s,k}}{r^2} + \frac{\mu}{r^2}\left(\frac{r^2}{r^2 - a^k_s} - \frac{r^2 - a^k_s}{r^2}\right),$$

$$T^{zz}_{s,k} = T^{rr}_{s,k} + \mu\, a^k_s \frac{r^2 - (1+\beta)a^k_s/2}{r^2\left(r^2 - a^k_s\right)}\,.$$

Here, $p^k_{0,s}$ are the constants of integration.

**Boundary conditions.** The Cauchy stress tensor $\boldsymbol{T}_k$ is defined on each layer up to parameters $a^k_s$ and $p^k_{0,s}$. Choosing them one can satisfy some boundary conditions. We assume that on the inner, $r = r^0_{i,s}$, and on the exterior, $r = r^{s-1}_{e,s}$, boundaries of assembly $\mathfrak{B}_s$, uniform hydrostatic loading $p_{i,s}$ and $p_{e,s}$ are given:

$$\boldsymbol{T}^{\sharp}_{s,0} \lrcorner\, e^r\Big|_{r=r^0_{i,s}} = p_{i,s}\boldsymbol{e}_r\,, \qquad \boldsymbol{T}^{\sharp}_{s,s-1} \lrcorner\, e^r\Big|_{r=r^{s-1}_{e,s}} = p_{e,s}\boldsymbol{e}_r\,, \tag{11.4}$$

and also that layers in the actual shape are in ideal contact:

$$\begin{aligned} \boldsymbol{T}^{\sharp}_{s,k} \lrcorner\, e^r\Big|_{r=r^k_{e,s}} &= \boldsymbol{T}^{\sharp}_{s,k+1} \lrcorner\, e^r\Big|_{r=r^{k+1}_{i,s}}\,, \\ r^k_{e,s} &= r^{k+1}_{i,s}\,, \qquad k = 0,\dots,s-2\,. \end{aligned} \tag{11.5}$$

2 We consider the Cauchy stress tensor as a symmetric field $\boldsymbol{T}\colon \varkappa(\mathfrak{B}) \to \mathscr{V} \otimes \mathscr{V}^*$. For a normal vector field $\boldsymbol{n}\colon \partial\varkappa(\mathfrak{B}) \to \mathscr{V}$ we have the following representation for the stress vector: $\boldsymbol{t_n} = \boldsymbol{T} \llcorner \boldsymbol{n}$.

Conditions (11.4) and (11.5) may be satisfied if $a_s^k$ and $p_{0,s}^k$, $k = 0, \dots, s-1$ are the solutions of the following system of non-linear equations:

$$\begin{aligned} \mathscr{T}\left[R_i^1; a_s^0\right] + p_{0,s}^0 &= p_{i,s}\,, \\ \mathscr{T}\left[R_e^s; a_s^{s-1}\right] + p_{0,s}^{s-1} &= p_{e,s}\,, \\ \mathscr{T}\left[R_e^k; a_s^k\right] + p_{0,s}^k &= \mathscr{T}\left[R_i^{k+1}; a_s^{k+1}\right] + p_{0,s}^{k+1}\,, \\ \left(R_e^k\right)^2 + a_s^k &= \left(R_i^{k+1}\right)^2 + a_s^{k+1}\,, \quad k = 0, \dots, s-2\,, \end{aligned} \tag{11.6}$$

where $\mathscr{T}[x; a] = \frac{\mu}{2}[\ln \frac{x^2}{x^2+a} - \frac{a}{x^2+a}]$.

By solving the system (11.6), one can consistently exclude the unknowns $p_{0,s-1}^k$ and arrive at the system of $s$ equations relative to the parameters $a_{s-1}^k$:

$$\begin{aligned} \prod_{k=0}^{s-1} \gamma_k \frac{\alpha_k + \nu_k x_s}{\beta_k + \nu_k x_s} &= W_s \exp\left[\sum_{k=0}^{s-1} (1-\gamma_k) \frac{\alpha_k + \nu_k x_s - 1}{(\alpha_k + \nu_k x_s)(\beta_k + \nu_k x_s)}\right], \\ a_s^k &= a_s^0 + A_k\,, \quad k = 1, \dots, s-1\,, \end{aligned} \tag{11.7}$$

where $A_0 = 0$, $A_k = \sum_{p=0}^{k-1}((R_e^p)^2 - (R_i^{p+1})^2)$, $k = 1, \dots, s-1$, $\gamma_k = (R_i^k/R_e^k)^2$, $\alpha_k = 1 + A_k/(R_e^k)^2$, $\beta_k = \gamma_k + A_k/(R_e^k)^2$, $\nu_k = (R_e^0/R_e^k)^2$, $x_s = a_s^0/(R_e^0)^2$, and $W_s = \exp[2\frac{p_{i,s}-p_{e,s}}{\mu}]$.

Taking into account the known reference radii and pressures, one can uniquely determine parameters $a_s^k$ from the system (11.7). Hereafter, parameters $p_{0,s}^k$ can be determined from the equations (11.6).

**Recurrent sequence of systems.** The conditions of (DG5) with equation (11.7) allows one to determine the parameters $a_s^k$ and $p_{0,s}^k$. Values for $\alpha_s$, $\beta_s$, $\gamma_s$, and $\nu_s$ can be obtained from the recurrence expressions

$$\begin{aligned} \alpha_{s+1} &= 1 + \frac{\alpha_s - (S^{s+1})^2(\alpha_s + \nu_s x_s)}{\zeta_s}\,, & \gamma_{s+1} &= \frac{(S^{s+1})^2(\alpha_s + \nu_s x_s)}{\zeta_s}\,, \\ \beta_{s+1} &= \gamma_{s+1} + \frac{\alpha_s - (S^{s+1})^2(\alpha_s + \nu_s x_s)}{\zeta_s}\,, & \nu_{s+1} &= \frac{\nu_s}{\zeta_s}\,, \\ \zeta_s &= \left(S^{s+1}\sqrt{\alpha_s + \nu_s x_s} + \sqrt{\nu_s \xi_s}\right)^2, & \xi_s &= \left(\frac{\Delta^{s+1}}{\rho + \Delta^0}\right)^2, \end{aligned}$$

where $s \geq 0$. The initial values for recurrent parameters are the following: $\alpha_0 = \nu_0 = 1$, $\beta_0 = \gamma_0 = (\rho/(\rho + \Delta^0))^2$.

**Piola–Kirchhoff stresses.** Since the Cauchy stress tensor $\boldsymbol{T}_{s,k}$ in each layer $\mathfrak{L}_k$ is determined, one can calculate the first Piola–Kirchhoff stress tensor $\boldsymbol{P}_{s,k}$ by formula $\boldsymbol{P}_{s,k} = J_{s,k}\boldsymbol{T}_{s,k}\boldsymbol{F}_{s,k}^{-T}$, where $J_{s,k} = \det \boldsymbol{F}_{s,k}$. For incompressible material $J_{s,k} = 1$. Thus, $\boldsymbol{P}_{s,k}$ has the following dyadic decomposition (indexes corresponding to layer numbers are suppressed for the sake of brevity):

$$
\begin{aligned}
\boldsymbol{P}^{\sharp}_{s,k} &= P^{rR}\boldsymbol{e}_r \otimes \boldsymbol{e}_R + P^{\theta\Theta}\boldsymbol{e}_\theta \otimes \boldsymbol{e}_\Theta + P^{zZ}\boldsymbol{e}_z \otimes \boldsymbol{e}_Z\,,\\
P^{rR} &= \frac{\sqrt{R^2+a}}{R}T^{rr} = p_0\frac{\sqrt{R^2+a}}{R} + \frac{\mu}{2}\frac{\sqrt{R^2+a}}{R}\left(\ln\frac{R^2}{R^2+a} - \frac{a}{R^2+a}\right),\\
P^{\theta\Theta} &= T^{\theta\theta} = \frac{p_0}{R^2+a} + \frac{\mu}{2(R^2+a)}\left(\ln\frac{R^2}{R^2+a} - \frac{a}{R^2+a}\right)\\
&\quad + \frac{\mu}{R^2+a}\left(\frac{R^2+a}{R^2} - \frac{R^2}{R^2+a}\right),\\
P^{zZ} &= T^{zz} = p_0 + \frac{\mu}{2}\left(\ln\frac{R^2}{R^2+a} - \frac{a}{R^2+a}\right) + \mu\, a\frac{R^2+(1-\beta)a/2}{R^2(R^2+a)}\,.
\end{aligned}
$$

**Eshelby energy–momentum tensor.** We now proceed to construct a representation for the Eshelby energy-momentum tensor. To this end, recall that the density of elastic energy stored in each layer $\mathfrak{L}_k$ is defined by the expression

$$
W_{s,k} = \frac{\beth_1}{2}(I_1(\boldsymbol{B}_{s,k})-3) - \frac{\beth_{-1}}{2}(I_2(\boldsymbol{B}_{s,k})-3)\,,
$$

where $I_1(\boldsymbol{B}_{s,k})$, $I_2(\boldsymbol{B}_{s,k})$ are invariants of $\boldsymbol{B}_{s,k}$:

$$
I_1(\boldsymbol{B}_{s,k}) = I_2(\boldsymbol{B}_{s,k}) = \frac{r^2-a_s^k}{r^2} + \frac{r^2}{r^2-a_s^k} + 1\,.
$$

Therefore,

$$
W_{s,k} = \frac{\mu(a_s^k)^2}{2r^2\left(r^2-a_s^k\right)} = \frac{\mu(a_s^k)^2}{2(R^k)^2\left((R^k)^2+a_s^k\right)}\,.
$$

Taking into account the expressions for the first Piola–Kirchhoff stress tensor and elastic energy, one can obtain Eshelby energy-momentum tensor by the formula

$$
\boldsymbol{E} = W\boldsymbol{I} - \boldsymbol{F}^{\mathrm{T}}\boldsymbol{P}\,. \tag{11.8}
$$

Thus, one arrives at the following decomposition:

$$
\begin{aligned}
\boldsymbol{E}^{\sharp} &= E^{RR}\boldsymbol{e}_R \otimes \boldsymbol{e}_R + E^{\Theta\Theta}\boldsymbol{e}_\Theta \otimes \boldsymbol{e}_\Theta + E^{ZZ}\boldsymbol{e}_Z \otimes \boldsymbol{e}_Z\,,\\
E^{RR} &= \frac{\mu a^2}{2R^2(R^2+a)} - \frac{R}{\sqrt{R^2+a}}P^{rR}\,,\\
E^{\Theta\Theta} &= \frac{\mu a^2}{2R^4(R^2+a)} - \frac{R^2+a}{R^2}P^{\theta\Theta}\,, \quad E^{ZZ} = \frac{\mu a^2}{2R^2(R^2+a)} - P^{zZ}\,.
\end{aligned}
$$

Let us make intermediate conclusions for the case of discrete structural inhomogeneity. Relations for strain and stresses were obtained in the framework of conventional Euclidean continuum mechanics. Indeed, measures of strain and stresses are defined for each layer, and, in totality, form piecewise continuous mappings defined on the actual shape. To satisfy boundary conditions one needs to solve the system (11.6). Nevertheless, for a large amount of layers one may need a large amount of time to solve the system. This may represent considerable computational problems. Thereby, it may be preferable to model the appropriate continuous process. Methods for modeling such continuous processes are discussed in Chapter 8.

### 11.1.2 Hollow Cylinders with Continuous Inhomogeneity

Now consider the continuous assembling process and the corresponding continuous inhomogeneity. According to Chapter 8, Section 8.5.1, the family $\mathfrak{G} = \{\mathfrak{B}_\alpha\}_{\alpha\in\mathbb{I}}$ is continual in this case: $\mathbb{I} \subset \mathbb{R}$, $\operatorname{Card}\mathbb{I} = \aleph_1$. The following assumptions are used for each $\alpha \in \mathbb{I}$:

(CG1) For each point $\mathfrak{X} \in \mathfrak{B}_\alpha$, a configuration that brings a neighborhood of $\mathfrak{X}$ into the stress-free state exists.

(CG2) If $\varkappa^R_{\alpha;\mathfrak{X}}$ is a configuration that brings a neighborhood of $\mathfrak{X} \in \mathfrak{B}_\alpha$ into the stress-free state, it also brings such a set of points into the stress-free state, which in totality maps by the configuration $\varkappa^R_{\alpha;\mathfrak{X}}$ onto the cylindrical surface.

(CG3) Cylindrical surfaces corresponding to different $\mathfrak{X}$ either coincide or not intersect.

(CG4) The family $\{\varkappa_\alpha\}_{\alpha\in\mathbb{I}}$ of actual configurations is such that their images are hollow cylinders with a height common to all.

Note that the notion of smooth bundle (Chapter 14, Section 14.2.1) may serve as a convenient mathematical formalization of assumptions (CG1)–(CG3). One can postulate these assumptions in alternative form:

(CGB1) For each $\alpha \in \mathbb{I}$, there is a smooth bundle $(\mathfrak{B}_\alpha, \mathbb{J}_\alpha, \pi_\alpha)$ with base $\mathbb{J}_\alpha$, diffeomorphic to an one-dimensional submanifold of $\mathbb{R}$, total space $\mathfrak{B}_\alpha$, and projection $\pi_\alpha\colon \mathfrak{B}_\alpha \to \mathbb{J}_\alpha$. For any $i \in \mathbb{J}_\alpha$, the fiber $\mathfrak{S}_{\alpha,i} := \pi_\alpha^{-1}(\{i\})$ is a smooth two-dimensional manifold[3].

(CGB2) For each $\alpha \in \mathbb{I}$, $i \in \mathbb{J}_\alpha$, there exists a configuration $\varkappa^R_{\alpha,i} \in \mathfrak{C}^r(\mathfrak{B}_\alpha; \mathscr{E})$, such that $\varkappa^R_{\alpha,i}(\mathfrak{B}_\alpha)$ is a hollow cylinder with height $h$, and for any point $X$ of the cylinder $\varkappa^R_{\alpha,i}(\mathfrak{S}_{\alpha,i})$, there exists a stress-free neighborhood (in $\mathscr{E}$).

Let $\varkappa^\circ_\alpha \in \mathfrak{C}^r(\mathfrak{B}_\alpha; \mathscr{E})$ be any configuration, such that $\varkappa^\circ_\alpha(\mathfrak{B}_\alpha)$ is a hollow cylinder. Like in Chapter 8, Section 8.5.3, we will refer to $\varkappa^\circ_\alpha$ as to the *intermediate configuration*. One can endow the set $\varkappa^\circ_\alpha(\mathfrak{B}_\alpha)$ with cylindrical coordinates $(R^\circ_\alpha, \Theta^\circ_\alpha, Z^\circ_\alpha)$. The corresponding coordinate map is denoted by $H^\circ : X \mapsto (R^\circ_\alpha, \Theta^\circ_\alpha, Z^\circ_\alpha)$. Cylindrical coordinates are

3 Hence, $\mathfrak{B}_\alpha = \bigcup_{i\in\mathbb{J}_\alpha} \mathfrak{S}_{\alpha,i}$.

related with the Cartesian as follows:

$$X^1 = R^\circ_\alpha \cos \Theta^\circ_\alpha\,, \quad X^2 = R^\circ_\alpha \sin \Theta^\circ_\alpha\,, \quad X^3 = Z^\circ_\alpha\,. \tag{11.9}$$

Since $\varkappa^\circ_\alpha(\mathfrak{B}_\alpha)$ is open in $\mathscr{E}$, the ordered pair $(\varkappa^\circ_\alpha(\mathfrak{B}_\alpha), (H^\circ)^{-1})$ is the chart on $\mathscr{E}$.

Using the map $\varkappa^\circ_\alpha$, one can transfer the cylindrical coordinates $(R^\circ_\alpha, \Theta^\circ_\alpha, Z^\circ_\alpha)$ to the body $\mathfrak{B}_\alpha$. The corresponding coordinate map is $H^\circ \circ \widetilde{\varkappa}^\circ_\alpha : \mathfrak{X} \mapsto (R^\circ_\alpha, \Theta^\circ_\alpha, Z^\circ_\alpha)$. The frame and coframe on the body are consequently denoted by $(\partial_{R^\circ_\alpha}, \partial_{\Theta^\circ_\alpha}, \partial_{Z^\circ_\alpha})$, and $(dR^\circ_\alpha, d\Theta^\circ_\alpha, dZ^\circ_\alpha)$. The analog of the Tab. 8.1 for this consideration is provided by Tab. 11.1.

**Tab. 11.1:** Correspondence between frames and coframes on the body and the intermediate shape.

| | $\mathfrak{B}_\alpha$ | $\varkappa^\circ_\alpha(\mathfrak{B}_\alpha)$ |
|---|---|---|
| Coordinate map | $H^\circ \circ \widetilde{\varkappa}^\circ_\alpha : \mathfrak{X} \mapsto (R^\circ_\alpha, \Theta^\circ_\alpha, Z^\circ_\alpha)$ | $H^\circ : X \mapsto (R^\circ_\alpha, \Theta^\circ_\alpha, Z^\circ_\alpha)$ |
| Coordinates | $(R^\circ_\alpha, \Theta^\circ_\alpha, Z^\circ_\alpha)$ | $(R^\circ_\alpha, \Theta^\circ_\alpha, Z^\circ_\alpha)$ |
| Frame | $(\partial_{R^\circ_\alpha}, \partial_{\Theta^\circ_\alpha}, \partial_{Z^\circ_\alpha})$ | $(\boldsymbol{e}_{R^\circ_\alpha}, \boldsymbol{e}_{\Theta^\circ_\alpha}, \boldsymbol{e}_{Z^\circ_\alpha})$ |
| Coframe | $(dR^\circ_\alpha, d\Theta^\circ_\alpha, dZ^\circ_\alpha)$ | $(\boldsymbol{e}^{R^\circ_\alpha}, \boldsymbol{e}^{\Theta^\circ_\alpha}, \boldsymbol{e}^{Z^\circ_\alpha})$ |

**Remark 11.1.** Recall, that transferring local coordinates from the intermediate shape to the body by means of the mapping $\widetilde{\varkappa}^\circ_\alpha$ is equivalent to the consideration of $\varkappa^\circ_\alpha(\mathfrak{B}_\alpha)$ as a "model" of $\mathfrak{B}_\alpha$. The body "dresses up" the coordinate net taken from the set $\varkappa^\circ_\alpha(\mathfrak{B}_\alpha)$.

In the pair of charts $(\mathfrak{B}_\alpha, (H^\circ \circ \widetilde{\varkappa}^\circ_\alpha)^{-1})$, $(\varkappa^\circ_\alpha(\mathfrak{B}_\alpha), (H^\circ)^{-1})$, the coordinate representation of $\varkappa^\circ_\alpha$ has the form: $\widetilde{\varkappa}^\circ_\alpha = H^\circ \circ j_{\varkappa^\circ_\alpha(\mathfrak{B}_\alpha)} \circ (H^\circ)^{-1} = \mathrm{Id}$, i.e., this is the identity map. Hence, for any point $\mathfrak{X} \in \mathfrak{B}_\alpha$, the tangent map (configuration gradient) $T_\mathfrak{X}\varkappa^\circ_\alpha : T_\mathfrak{X}\mathfrak{B}_\alpha \to \mathscr{V}$ has the following dyadic decomposition:

$$T_\mathfrak{X}\varkappa^\circ_\alpha = \boldsymbol{e}_{R^\circ_\alpha}|_X \otimes dR^\circ_\alpha|_\mathfrak{X} + \boldsymbol{e}_{\Theta^\circ_\alpha}|_X \otimes d\Theta^\circ_\alpha|_\mathfrak{X} + \boldsymbol{e}_{Z^\circ_\alpha}|_X \otimes dZ^\circ_\alpha|_\mathfrak{X}\,, \tag{11.10}$$

where $X = \varkappa^\circ_\alpha(\mathfrak{X})$.

We will use the intermediate configurations to obtain the families $\overline{\mathrm{Ref}}_{\mathfrak{B}_\alpha}$. Let us fix $\alpha \in \mathbb{I}$. The assumptions (CGB1) and (CGB2) allow one to define the family $\overline{\mathrm{Ref}}_{\mathfrak{B}_\alpha} = \{\varkappa^R_{\alpha;\mathfrak{X}}\}_{\mathfrak{X}\in\mathfrak{B}_\alpha}$ as follows:

$$\varkappa^R_{\alpha;\mathfrak{X}} = \varkappa^R_{\alpha,\pi_\alpha(\mathfrak{X})}\,.$$

It is clear that if $\mathfrak{X}, \mathfrak{Y} \in \mathfrak{S}_{\alpha,i}$, then $\varkappa^R_{\alpha;\mathfrak{X}} = \varkappa^R_{\alpha;\mathfrak{Y}}$.

**Local configuration.** According to the assumption (CGB2) each of the cylinders $\varkappa^R_{\alpha;\mathfrak{X}}(\mathfrak{B}_\alpha)$ may be endowed with an ordered triple of cylindrical coordinates $(R_{\alpha,\mathfrak{X}}, \Theta_{\alpha,\mathfrak{X}}, Z_{\alpha,\mathfrak{X}})$. In the pair of coordinates $(R^\circ_\alpha, \Theta^\circ_\alpha, Z^\circ_\alpha)$, and $(R_{\alpha,\mathfrak{X}}, \Theta_{\alpha,\mathfrak{X}}, Z_{\alpha,\mathfrak{X}})$ defined on the sets $\varkappa^\circ_\alpha(\mathfrak{B}_\alpha)$, and $\varkappa^R_{\alpha;\mathfrak{X}}(\mathfrak{B}_\alpha)$, respectively, the deformation $d^R_{\alpha;\mathfrak{X}} = \varkappa^R_{\alpha;\mathfrak{X}} \circ (\widetilde{\varkappa}^\circ_\alpha)^{-1}$ has a representation

$$\tilde{d}^R_{\alpha;\mathfrak{X}}:\ (R^\circ_\alpha, \Theta^\circ_\alpha, Z^\circ_\alpha) \mapsto (R_{\alpha,\mathfrak{X}}, \Theta_{\alpha,\mathfrak{X}}, Z_{\alpha,\mathfrak{X}}),$$
$$R_{\alpha,\mathfrak{X}} = \sqrt{(R^\circ_\alpha)^2 - \rho_{\alpha,\mathfrak{X}}}, \quad \Theta_{\alpha,\mathfrak{X}} = \Theta^\circ_\alpha, \quad Z_{\alpha,\mathfrak{X}} = Z^\circ_\alpha, \tag{11.11}$$

where $\rho_{\alpha,\mathfrak{X}} \in \mathbb{R}$. For $\mathfrak{X} \in \mathfrak{S}_{\alpha,i}$, the tangent map $T_{\mathfrak{Y}}\varkappa^R_{\alpha;\mathfrak{X}}: T_{\mathfrak{Y}}\mathfrak{B}_\alpha \to \mathscr{V}$ to $\varkappa^R_{\alpha;\mathfrak{X}}$ at point $\mathfrak{Y} \in \mathfrak{B}_\alpha$ may be calculated as a composition $T_{\mathfrak{Y}}\varkappa^R_{\alpha;\mathfrak{X}} = \boldsymbol{F}_{\alpha;\mathfrak{X}} \circ T_{\mathfrak{Y}}\widehat{\varkappa}^\circ_\alpha$, where[4]

$$\boldsymbol{F}_{\alpha;\mathfrak{X}} = \frac{R^\circ_\alpha}{\sqrt{(R^\circ_\alpha)^2 - \rho_{\alpha,\mathfrak{X}}}}\boldsymbol{e}_{R_{\alpha,\mathfrak{X}}} \otimes \boldsymbol{e}^{R^\circ_\alpha} + \boldsymbol{e}_{\Theta_{\alpha,\mathfrak{X}}} \otimes \boldsymbol{e}^{\Theta^\circ_\alpha} + \boldsymbol{e}_{Z_{\alpha,\mathfrak{X}}} \otimes \boldsymbol{e}^{Z^\circ_\alpha}$$

is the deformation gradient of $d^R_{\alpha;\mathfrak{X}}$. Finally, $T_{\mathfrak{Y}}\varkappa^R_{\alpha;\mathfrak{X}}$ has the following representation:

$$T_{\mathfrak{Y}}\varkappa^R_{\alpha;\mathfrak{X}} = \left.\frac{R^\circ_\alpha}{\sqrt{(R^\circ_\alpha)^2 - \rho_{\alpha,\mathfrak{X}}}}\right|_{\mathfrak{Y}} \boldsymbol{e}_{R_{\alpha,\mathfrak{X}}}|_Y \otimes dR^\circ_\alpha|_{\mathfrak{Y}} + \boldsymbol{e}_{\Theta_{\alpha,\mathfrak{X}}}|_Y \otimes d\Theta^\circ_\alpha|_{\mathfrak{Y}} + \boldsymbol{e}_{Z_{\alpha,\mathfrak{X}}}|_Y \otimes dZ^\circ_\alpha|_{\mathfrak{Y}}, \tag{11.12}$$

in which $Y = \varkappa^R_{\alpha;\mathfrak{X}}(\mathfrak{Y})$, and $(\boldsymbol{e}_{R_{\alpha,\mathfrak{X}}}, \boldsymbol{e}_{\Theta_{\alpha,\mathfrak{X}}}, \boldsymbol{e}_{Z_{\alpha,\mathfrak{X}}})$ is a coordinate frame corresponding to the coordinates $(R_{\alpha,\mathfrak{X}}, \Theta_{\alpha,\mathfrak{X}}, Z_{\alpha,\mathfrak{X}})$.

The local configuration $\mathscr{K}_\alpha: \mathfrak{X} \mapsto T_{\mathfrak{X}}\varkappa^R_{\alpha;\mathfrak{X}}$ is defined at point $\mathfrak{X} \in \mathfrak{B}_\alpha$ by the following expression:

$$\mathscr{K}_\alpha|_{\mathfrak{X}} = \left.\frac{R^\circ_\alpha}{\sqrt{(R^\circ_\alpha)^2 - \rho_{\alpha,\mathfrak{X}}}}\right|_{\mathfrak{X}} \boldsymbol{e}_{R_{\alpha,\mathfrak{X}}}|_X \otimes dR^\circ_\alpha|_{\mathfrak{X}} + \boldsymbol{e}_{\Theta_{\alpha,\mathfrak{X}}}|_X \otimes d\Theta^\circ_\alpha|_{\mathfrak{X}} + \boldsymbol{e}_{Z_{\alpha,\mathfrak{X}}}|_X \otimes dZ^\circ_\alpha|_{\mathfrak{X}}, \tag{11.13}$$

where $X = \varkappa^R_{\alpha;\mathfrak{X}}(\mathfrak{X})$.

**Distortion function.** Let $C := H^\circ \circ \widehat{\varkappa}^\circ_\alpha: \mathfrak{X} \mapsto (R^\circ_\alpha, \Theta^\circ_\alpha, Z^\circ_\alpha)$ be the coordinate map, which assigns to each point $\mathfrak{X} \in \mathfrak{B}_\alpha$ a triple of cylindrical coordinates $(R^\circ_\alpha, \Theta^\circ_\alpha, Z^\circ_\alpha)$. Then, one can calculate radial parameter by the relation $\rho_{\alpha,\mathfrak{X}} = \rho_{\alpha,C^{-1}(R^\circ_\alpha,\Theta^\circ_\alpha,Z^\circ_\alpha)}$ and define the mapping

$$b_\alpha:\ (R^\circ_\alpha, \Theta^\circ_\alpha, Z^\circ_\alpha) \mapsto \rho_{\alpha,C^{-1}(R^\circ_\alpha,\Theta^\circ_\alpha,Z^\circ_\alpha)}.$$

In the following, we use

**Assumption.** *The mapping $b_\alpha$ depends on $R^\circ_\alpha$ only.*

For each index $\alpha$ from $\mathbb{I}$, we apply such an assumption and obtain the family $\{b_\alpha\}_{\alpha\in\mathbb{I}}$. A mapping $b_\alpha: R^\circ_\alpha \mapsto b_\alpha(R^\circ_\alpha)$ is defined on the finite interval $\mathbb{I}_\alpha \subset \mathbb{R}$. We choose intermediate configurations with the following property:

$$\forall \alpha_1, \alpha_2 \in \mathbb{I}:\ (\alpha_1 < \alpha_2) \Rightarrow (b_{\alpha_2}|_{\mathbb{I}_{\alpha_1}} = b_{\alpha_1}). \tag{11.14}$$

4 Since we are dealing with shapes that are open sets, we identify $T_{\varkappa^\circ_\alpha(\mathfrak{Y})}\varkappa^\circ_\alpha(\mathfrak{B}_\alpha)$ with $\mathscr{V}$ (see Chapter 4, Section 4.5.4).

**Remark 11.2.** If $\alpha_1 < \alpha_2$ then $\varkappa^\circ_{\alpha_1}(\mathfrak{B}_{\alpha_1}) \subset \varkappa^\circ_{\alpha_2}(\mathfrak{B}_{\alpha_2})$ and the stress–strain state at points in $\varkappa^\circ_{\alpha_2}(\mathfrak{B}_{\alpha_2})$, which are in $\varkappa^\circ_{\alpha_1}(\mathfrak{B}_{\alpha_1})$, is fixed.

**Material metric.** Now, one can obtain the family $\{\mathscr{G}_\alpha\}_{\alpha\in\mathbb{I}}$ of material metrics:

$$\mathscr{G}_\alpha : \mathfrak{B}_\alpha \to T^*\mathfrak{B}_\alpha \otimes T^*\mathfrak{B}_\alpha\,, \quad \mathscr{G}_\alpha(u,v) = \mathbf{g}(\mathscr{K}_\alpha u, \mathscr{K}_\alpha v)\,,$$

where $\mathbf{g}$ is the spatial metric. In explicit form:

$$\mathscr{G}_\alpha = \frac{(R^\circ_\alpha)^2}{(R^\circ_\alpha)^2 - b_\alpha(R^\circ_\alpha)} dR^\circ_\alpha \otimes dR^\circ_\alpha + \left((R^\circ_\alpha)^2 - b_\alpha(R^\circ_\alpha)\right) d\Theta^\circ_\alpha \otimes d\Theta^\circ_\alpha + dZ^\circ_\alpha \otimes dZ^\circ_\alpha\,. \tag{11.15}$$

**Connection.** Thus, one has the family $\{(\mathfrak{B}_\alpha, \mathscr{G}_\alpha)\}_{\alpha\in\mathbb{I}}$ of Riemann spaces. The non-zero Levi–Civita connection $\nabla_\alpha$ coefficients are represented by expressions (for the sake of brevity, the indices $\alpha$ and $\circ$ are omitted):

$$\Gamma^1{}_{11} = \frac{-2b(r) + rb'(r)}{2(r^3 - rb(r))}\,, \quad \Gamma^1{}_{22} = \frac{(r^2 - b(r))(-2r + b'(r))}{2r^2}\,,$$
$$\Gamma^2{}_{12} = \Gamma^2{}_{21} = \frac{2r - b'(r)}{2r^2 - 2b(r)}\,.$$

The curvature of the so-obtained connection is non-zero in general. Its scalar curvature has the form (here $'$ denotes $d/dR^\circ_\alpha$):

$$\mathrm{Ric}_\alpha = \frac{R^\circ_\alpha b''_\alpha(R^\circ_\alpha) - b'_\alpha(R^\circ_\alpha)}{(R^\circ_\alpha)^3}\,, \tag{11.16}$$

and reaches zero only when $b_\alpha(R^\circ_\alpha) = C_1(R^\circ_\alpha)^2 + C_2$. This mathematical fact reflects the characteristic features of such bodies: the incompatibility of deformations and the presence of intrinsic stresses.

**Strain measures.** The intermediate configuration $\varkappa^\circ_\alpha$ was used to define the material metric tensor. It generates the configuration gradient and the Cauchy–Green tensors.

Since $(\partial(\widetilde{\varkappa}^\circ_\alpha)^i)/(\partial \mathfrak{X}^j) = \delta^i_j$, and according to the expressions (5.13) and (5.14), one has the following relations for the right and left Cauchy–Green tensors:

$$\begin{aligned} \mathscr{C}^\circ_\alpha &= \frac{(R^\circ_\alpha)^2 - b_\alpha(R^\circ_\alpha)}{(R^\circ_\alpha)^2} \partial_{R^\circ_\alpha} \otimes dR^\circ_\alpha + \frac{(R^\circ_\alpha)^2}{(R^\circ_\alpha)^2 - b_\alpha(R^\circ_\alpha)} \partial_{\Theta^\circ_\alpha} \otimes d\Theta^\circ_\alpha + \partial_{Z^\circ_\alpha} \otimes dZ^\circ_\alpha\,, \\ \mathbf{B}^\circ_\alpha &= \frac{(R^\circ_\alpha)^2 - b_\alpha(R^\circ_\alpha)}{(R^\circ_\alpha)^2} \mathbf{e}_{R^\circ_\alpha} \otimes \mathbf{e}^{R^\circ_\alpha} + \frac{(R^\circ_\alpha)^2}{(R^\circ_\alpha)^2 - b_\alpha(R^\circ_\alpha)} \mathbf{e}_{\Theta^\circ_\alpha} \otimes \mathbf{e}^{\Theta^\circ_\alpha} + \mathbf{e}_{Z^\circ_\alpha} \otimes \mathbf{e}^{Z^\circ_\alpha}\,. \end{aligned} \tag{11.17}$$

By virtue of the assumption (CG4) one can endow the set $\varkappa_\alpha(\mathfrak{B}_\alpha)$ with the cylindrical coordinates $(r_\alpha, \theta_\alpha, z_\alpha)$. In the coordinates, the deformation $d_\alpha = \varkappa_\alpha \circ (\widetilde{\varkappa}^\circ_\alpha)^{-1}$ is assumed to be in the form

$$\tilde{d}_\alpha : (R^\circ_\alpha, \Theta^\circ_\alpha, Z^\circ_\alpha) \mapsto (r_\alpha, \theta_\alpha, z_\alpha)\,, \quad r_\alpha = \sqrt{(R^\circ_\alpha)^2 + A_\alpha}\,, \quad \theta_\alpha = \Theta^\circ_\alpha\,, \quad z_\alpha = Z^\circ_\alpha\,,$$

where $A_\alpha \in \mathbb{R}$. Hence, the deformation gradient is defined as:

$$\boldsymbol{F}_\alpha = \frac{R^\circ_\alpha}{\sqrt{(R^\circ_\alpha)^2 + A_\alpha}} \boldsymbol{e}_{r_\alpha} \otimes e^{R^\circ_\alpha} + \boldsymbol{e}_{\theta_\alpha} \otimes e^{\Theta^\circ_\alpha} + \boldsymbol{e}_{z_\alpha} \otimes e^{Z^\circ_\alpha} . \tag{11.18}$$

The configuration gradient $T_{\mathfrak{X}}\varkappa_\alpha$ may be written as a multiplicative decomposition

$$T_{\mathfrak{X}}\varkappa_\alpha = \boldsymbol{F}_\alpha \circ T_{\mathfrak{X}}\widehat{\varkappa}^\circ_\alpha ,$$

and according to (11.10) and (11.18) it has the form:

$$T_{\mathfrak{X}}\varkappa_\alpha = \frac{R^\circ_\alpha}{\sqrt{(R^\circ_\alpha)^2 + A_\alpha}} \boldsymbol{e}_{r_\alpha}|_X \otimes dR^\circ_\alpha|_{\mathfrak{X}} + \boldsymbol{e}_{\theta_\alpha}|_X \otimes d\Theta^\circ_\alpha|_{\mathfrak{X}} + \boldsymbol{e}_{z_\alpha}|_X \otimes dZ^\circ_\alpha|_{\mathfrak{X}} , \tag{11.19}$$

where $X = \varkappa_\alpha(\mathfrak{X})$.

According to the relations (5.13) and (5.14), where $\mathcal{G}$ is defined by the expression (11.15), $\boldsymbol{g} = e^{r_\alpha} \otimes e^{r_\alpha} + r^2_\alpha e^{\theta_\alpha} \otimes e^{\theta_\alpha} + e^{z_\alpha} \otimes e^{z_\alpha}$, and $T_{\mathfrak{X}}\varkappa_\alpha$ was obtained above (formula (11.19)), one has the formulae

$$\mathscr{C}_\alpha = \frac{(R^\circ_\alpha)^2 - b_\alpha(R^\circ_\alpha)}{(R^\circ_\alpha)^2 + A_\alpha} \partial_{R^\circ_\alpha} \otimes dR^\circ_\alpha + \frac{(R^\circ_\alpha)^2 + A_\alpha}{(R^\circ_\alpha)^2 - b_\alpha(R^\circ_\alpha)} \partial_{\Theta^\circ_\alpha} \otimes d\Theta^\circ_\alpha + \partial_{Z^\circ_\alpha} \otimes dZ^\circ_\alpha ,$$

$$\begin{aligned}\boldsymbol{B}_\alpha &= \frac{r^2_\alpha - A_\alpha - b_\alpha\left(\sqrt{r^2_\alpha - A_\alpha}\right)}{r^2_\alpha} \boldsymbol{e}_{r_\alpha} \otimes e^{r_\alpha} \\ &\quad + \frac{r^2_\alpha}{r^2_\alpha - A_\alpha - b_\alpha\left(\sqrt{r^2_\alpha - A_\alpha}\right)} \boldsymbol{e}_{\theta_\alpha} \otimes e^{\theta_\alpha} + \boldsymbol{e}_{z_\alpha} \otimes e^{z_\alpha} .\end{aligned} \tag{11.20}$$

**Stress forms.** One has the following expression for the Cauchy stress tensor $\boldsymbol{T}^\circ_\alpha$ in the intermediate configuration:

$$(\boldsymbol{T}^\circ_\alpha)^\sharp = T^{R^\circ_\alpha R^\circ_\alpha}_\alpha \boldsymbol{e}_{R^\circ_\alpha} \otimes \boldsymbol{e}_{R^\circ_\alpha} + T^{\Theta^\circ_\alpha \Theta^\circ_\alpha}_\alpha \boldsymbol{e}_{\Theta^\circ_\alpha} \otimes \boldsymbol{e}_{\Theta^\circ_\alpha} + T^{Z^\circ_\alpha Z^\circ_\alpha}_\alpha \boldsymbol{e}_{Z^\circ_\alpha} \otimes \boldsymbol{e}_{Z^\circ_\alpha} , \tag{11.21}$$

$$T^{R^\circ_\alpha R^\circ_\alpha}_\alpha = -p^\circ_\alpha + \beth_1 \frac{(R^\circ_\alpha)^2 - b_\alpha(R^\circ_\alpha)}{(R^\circ_\alpha)^2} + \beth_{-1} \frac{(R^\circ_\alpha)^2}{(R^\circ_\alpha)^2 - b_\alpha r(R^\circ_\alpha)} ,$$

$$T^{\Theta^\circ_\alpha \Theta^\circ_\alpha}_\alpha = -\frac{p^\circ_\alpha}{(R^\circ_\alpha)^2} + \beth_1 \frac{1}{(R^\circ_\alpha)^2 - b_\alpha(R^\circ_\alpha)} + \beth_{-1} \frac{(R^\circ_\alpha)^2 - b_\alpha(R^\circ_\alpha)}{(R^\circ_\alpha)^4} ,$$

$$T^{Z^\circ_\alpha Z^\circ_\alpha}_\alpha = -p^\circ_\alpha + \beth_1 + \beth_{-1} .$$

With the equilibrium equation $\operatorname{div}(\boldsymbol{T}^\circ_\alpha)^\sharp = \mathbf{0}$ and the dyadic representation (11.21), we obtain

$$\frac{\partial T_\alpha^{R^\circ_\alpha R^\circ_\alpha}}{\partial R^\circ_\alpha} + \frac{T_\alpha^{R^\circ_\alpha R^\circ_\alpha}}{R^\circ_\alpha} - R^\circ_\alpha T_\alpha^{\Theta^\circ_\alpha \Theta^\circ_\alpha} = 0\,, \qquad \frac{\partial T_\alpha^{\Theta^\circ_\alpha \Theta^\circ_\alpha}}{\partial \Theta^\circ_\alpha} = 0\,, \qquad \frac{\partial T_\alpha^{Z^\circ_\alpha Z^\circ_\alpha}}{\partial Z^\circ_\alpha} = 0\,. \tag{11.22}$$

From (11.22), it follows that

$$T_\alpha^{R^\circ_\alpha R^\circ_\alpha}(R) = B^\circ_\alpha + \mu \int\limits_{R^\circ_{\alpha;i}}^{R} \left[ \frac{\xi}{\xi^2 - b_\alpha(\xi)} - \frac{\xi^2 - b_\alpha(\xi)}{\xi^3} \right] d\xi\,. \tag{11.23}$$

Here, $R^\circ_{\alpha;i}$ is the inner radius of the cylinder $\varkappa^\circ_\alpha(\mathfrak{B}_\alpha)$, and $B^\circ_\alpha$ is a free parameter. Then, for the other components one has:

$$\begin{aligned} T_\alpha^{\Theta^\circ_\alpha \Theta^\circ_\alpha} &= \frac{T_\alpha^{R^\circ_\alpha R^\circ_\alpha}}{(R^\circ_\alpha)^2} + \frac{\mu}{(R^\circ_\alpha)^2} \left( \frac{(R^\circ_\alpha)^2}{(R^\circ_\alpha)^2 - b_\alpha\,(R^\circ_\alpha)} - \frac{(R^\circ_\alpha)^2 - b_\alpha\,(R^\circ_\alpha)}{(R^\circ_\alpha)^2} \right), \\ T_\alpha^{Z^\circ_\alpha Z^\circ_\alpha} &= T_\alpha^{R^\circ_\alpha R^\circ_\alpha} + \mu\, b_\alpha\,(R^\circ_\alpha)\, \frac{(R^\circ_\alpha)^2 - (1+\beta) b_\alpha\,(R^\circ_\alpha)/2}{(R^\circ_\alpha)^2\,((R^\circ_\alpha)^2 - b_\alpha\,(R^\circ_\alpha))}\,. \end{aligned}$$

The Cauchy stress tensor obtained depends on only one parameter $B^\circ_\alpha$. For example, one may take $T_\alpha^{R^\circ_\alpha R^\circ_\alpha}(R^\circ_{\alpha;i}) = 0$, from which follows $B^\circ_\alpha = 0$. Then on the exterior boundary of the cylinder $\varkappa^\circ_\alpha(\mathfrak{B}_\alpha)$, with $R^\circ_\alpha = R^\circ_{\alpha;e}$, one has

$$T_\alpha^{R^\circ_\alpha R^\circ_\alpha}\left(R^\circ_{\alpha;e}\right) = \mu \int\limits_{R^\circ_{\alpha;i}}^{R^\circ_{\alpha;e}} \left[ \frac{\xi}{\xi^2 - b_\alpha(\xi)} - \frac{\xi^2 - b_\alpha(\xi)}{\xi^3} \right] d\xi\,.$$

The corresponding stress vector $\boldsymbol{T}^\circ_\alpha(R^\circ_{\alpha;e}) \llcorner \boldsymbol{e}_{R^\circ_\alpha}$ represents a force per unit area, which one needs to apply on the exterior cylindrical boundary of $\varkappa^\circ_\alpha(\mathfrak{B}_\alpha)$ in equilibrium.

Since $T_{R^\circ_\alpha R^\circ_\alpha} = T_\alpha^{R^\circ_\alpha R^\circ_\alpha}$, $T_{\Theta^\circ_\alpha \Theta^\circ_\alpha} = (R^\circ_\alpha)^4 T_\alpha^{\Theta^\circ_\alpha \Theta^\circ_\alpha}$, $T_{Z^\circ_\alpha Z^\circ_\alpha} = T_\alpha^{Z^\circ_\alpha Z^\circ_\alpha}$, and due to the formula (7.14), one can obtain:

$$\begin{aligned} T^\circ_\alpha = {} & R^\circ_\alpha T_\alpha^{R^\circ_\alpha R^\circ_\alpha} e^{R^\circ_\alpha} \otimes \left(e^{\Theta^\circ_\alpha} \wedge e^{Z^\circ_\alpha}\right) - (R^\circ_\alpha)^3\, T_\alpha^{\Theta^\circ_\alpha \Theta^\circ_\alpha} e^{\Theta^\circ_\alpha} \otimes \left(e^{R^\circ_\alpha} \wedge e^{Z^\circ_\alpha}\right) \\ & + R^\circ_\alpha T_\alpha^{Z^\circ_\alpha Z^\circ_\alpha} e^{Z^\circ_\alpha} \otimes \left(e^{R^\circ_\alpha} \wedge e^{\Theta^\circ_\alpha}\right)\,. \end{aligned}$$

This covector valued exterior 2-form represents the Cauchy stress in the intermediate configuration. If $\boldsymbol{v} = v^i \boldsymbol{e}_i$ is a spatial velocity, then the power density is represented by an exterior 2-form:

$$\begin{aligned} \boldsymbol{v} \lrcorner T^\circ_\alpha = {} & R^\circ_\alpha T_\alpha^{R^\circ_\alpha R^\circ_\alpha} v^{R^\circ_\alpha} \left(e^{\Theta^\circ_\alpha} \wedge e^{Z^\circ_\alpha}\right) - (R^\circ_\alpha)^3\, T_\alpha^{\Theta^\circ_\alpha \Theta^\circ_\alpha} v^{\Theta^\circ_\alpha} \left(e^{R^\circ_\alpha} \wedge e^{Z^\circ_\alpha}\right) \\ & + R^\circ_\alpha T_\alpha^{Z^\circ_\alpha Z^\circ_\alpha} v^{Z^\circ_\alpha} \left(e^{R^\circ_\alpha} \wedge e^{\Theta^\circ_\alpha}\right)\,. \end{aligned}$$

Finally, calculate the stresses in the actual configuration $\varkappa_\alpha$. Using (11.20) we obtain

$$(\boldsymbol{T}_\alpha)^\sharp = T_\alpha^{r_\alpha r_\alpha}\boldsymbol{e}_{r_\alpha}\otimes\boldsymbol{e}_{r_\alpha} + T_\alpha^{\theta_\alpha\theta_\alpha}\boldsymbol{e}_{\theta_\alpha}\otimes\boldsymbol{e}_{\theta_\alpha} + T_\alpha^{z_\alpha z_\alpha}\boldsymbol{e}_{z_\alpha}\otimes\boldsymbol{e}_{z_\alpha}\,, \tag{11.24}$$

$$T_\alpha^{r_\alpha r_\alpha}(r) = B_\alpha + \mu\int\limits_{r_{\alpha;i}}^{r}\left[\frac{\tau}{\tau^2 - A_\alpha - b_\alpha\left(\sqrt{\tau^2 - A_\alpha}\right)} - \frac{\tau^2 - A_\alpha - b_\alpha\left(\sqrt{\tau^2 - A_\alpha}\right)}{\tau^3}\right]d\tau\,, \tag{11.25}$$

$$T_\alpha^{\theta_\alpha\theta_\alpha} = \frac{T_\alpha^{r_\alpha r_\alpha}}{r_\alpha^2} + \frac{\mu}{r_\alpha^2}\left(\frac{r_\alpha^2}{r_\alpha^2 - A_\alpha - b_\alpha\left(\sqrt{r_\alpha^2 - A_\alpha}\right)} - \frac{r_\alpha^2 - A_\alpha - b_\alpha\left(\sqrt{r_\alpha^2 - A_\alpha}\right)}{r_\alpha^2}\right),$$

$$T_\alpha^{z_\alpha z_\alpha} = T_\alpha^{r_\alpha r_\alpha} + \mu\left\{A_\alpha + b_\alpha\left(\sqrt{r_\alpha^2 - A_\alpha}\right)\right\}\frac{r_\alpha^2 - (1+\beta)\left\{A_\alpha + b_\alpha\left(\sqrt{r_\alpha^2 - A_\alpha}\right)\right\}/2}{r_\alpha^2\left(r_\alpha^2 - A_\alpha - b_\alpha\left(\sqrt{r_\alpha^2 - A_\alpha}\right)\right)}\,.$$

The corresponding Cauchy exterior 2-form is represented as follows:

$$T_\alpha = r_\alpha\,T^{r_\alpha r_\alpha}\,e^{r_\alpha}\otimes\left(e^{\theta_\alpha}\wedge e^{z_\alpha}\right) - r_\alpha^3\,T^{\theta_\alpha\theta_\alpha}\,e^{\theta_\alpha}\otimes\left(e^{r_\alpha}\wedge e^{z_\alpha}\right) + r_\alpha\,T^{z_\alpha z_\alpha}\,e^{z_\alpha}\otimes\left(e^{r_\alpha}\wedge e^{\theta_\alpha}\right)\,.$$

Since the Cauchy tensor is known, we can obtain the Piola–Kirchhoff tensor by the (7.16). Here, $\mathscr{F} = T\varkappa_\alpha$, $\boldsymbol{g} = e^{r_\alpha}\otimes e^{r_\alpha} + r_\alpha^2 e^{\theta_\alpha}\otimes e^{\theta_\alpha} + e^{z_\alpha}\otimes e^{z_\alpha}$, and $\mathscr{G}_\alpha$ is represented by (11.15). Hence,

$$J_\alpha = \frac{R_\alpha^\circ}{\sqrt{(R_\alpha^\circ)^2 + A_\alpha}}\,,\qquad g_\alpha = r_\alpha^2 = (R_\alpha^\circ)^2 + A_\alpha\,,\qquad G_\alpha = (R_\alpha^\circ)^2\,.$$

According to (11.19),

$$\mathscr{F}^{-1} = \frac{\sqrt{(R_\alpha^\circ)^2 + A_\alpha}}{R_\alpha^\circ}\partial_{R_\alpha^\circ}\otimes e^{r_\alpha} + \partial_{\Theta_\alpha^\circ}\otimes e^{\theta_\alpha} + \partial_{Z_\alpha^\circ}\otimes e^{z_\alpha}\,.$$

Then, applying the transpose, we arrive at the formula

$$\mathscr{F}_\alpha^{-\mathrm{T}} = \frac{R_\alpha^\circ\sqrt{(R_\alpha^\circ)^2 + A_\alpha}}{(R_\alpha^\circ)^2 - b_\alpha(R_\alpha^\circ)}\boldsymbol{e}_{r_\alpha}\otimes dR_\alpha^\circ + \frac{(R_\alpha^\circ)^2 - b_\alpha(R_\alpha^\circ)}{(R_\alpha^\circ)^2 + A_\alpha}\boldsymbol{e}_{\theta_\alpha}\otimes d\Theta_\alpha^\circ + \boldsymbol{e}_{z_\alpha}\otimes dZ_\alpha^\circ\,.$$

Thus, the first Piola–Kirchhoff tensor has the form:

$$\begin{aligned}\mathscr{P} &= \frac{R_\alpha^\circ\sqrt{(R_\alpha^\circ)^2 + A_\alpha}}{(R_\alpha^\circ)^2 - b_\alpha(R_\alpha^\circ)}\,T^{r_\alpha r_\alpha}\,e^{r_\alpha}\otimes dR_\alpha^\circ\\ &\quad + \left((R_\alpha^\circ)^2 + A_\alpha\right)\left((R_\alpha^\circ)^2 - b_\alpha(R_\alpha^\circ)\right)\,T^{\theta_\alpha\theta_\alpha}\,e^{\theta_\alpha}\otimes d\Theta_\alpha^\circ + T^{z_\alpha z_\alpha}\,e^{z_\alpha}\otimes dZ_\alpha^\circ\,.\end{aligned}$$

Now we have everything to obtain Eshelby energy momentum tensor. Since

$$\mathcal{P}^\sharp = \frac{\sqrt{(R^\circ_\alpha)^2 + A_\alpha}}{R^\circ_\alpha} T^{r_\alpha r_\alpha} \boldsymbol{e}_{r_\alpha} \otimes dR^\circ_\alpha + T^{\theta_\alpha \theta_\alpha} \boldsymbol{e}_{\theta_\alpha} \otimes d\Theta^\circ_\alpha + T^{z_\alpha z_\alpha} \boldsymbol{e}_{z_\alpha} \otimes dZ^\circ_\alpha ,$$

$$\mathcal{F}^{\mathrm{T}} = \frac{(R^\circ_\alpha)^2 - b_\alpha (R^\circ_\alpha)}{R^\circ_\alpha \sqrt{(R^\circ_\alpha)^2 + A_\alpha}} \partial_{R^\circ_\alpha} \otimes \boldsymbol{e}^{r_\alpha} + \frac{(R^\circ_\alpha)^2 + A_\alpha}{(R^\circ_\alpha)^2 - b_\alpha (R^\circ_\alpha)} \partial_{\Theta^\circ_\alpha} \otimes \boldsymbol{e}^{\theta_\alpha} + \partial_{Z^\circ_\alpha} \otimes \boldsymbol{e}^{z_\alpha} ,$$

we obtain

$$\begin{aligned} \mathcal{E} = & \frac{(R^\circ_\alpha)^2 - b_\alpha (R^\circ_\alpha)}{(R^\circ_\alpha)^2} \{W - T^{r_\alpha r_\alpha}\} \partial_{R^\circ_\alpha} \otimes \partial_{R^\circ_\alpha} \\ & + \frac{1}{(R^\circ_\alpha)^2 - b_\alpha (R^\circ_\alpha)} \{W - ((R^\circ_\alpha)^2 + A_\alpha) T^{\theta_\alpha \theta_\alpha}\} \partial_{\Theta^\circ_\alpha} \otimes \partial_{\Theta^\circ_\alpha} \\ & + \{W - T^{z_\alpha z_\alpha}\} \partial_{Z^\circ_\alpha} \otimes \partial_{Z^\circ_\alpha} . \end{aligned}$$

**Evolutionary problem.** All the calculations carried out above assumed that the family of local reference shapes $\overline{\mathrm{Ref}_{\mathfrak{B}_\alpha}}$ is given. For example, with a hollow cylinder, this means that the family of functions $\{b_\alpha\}_{\alpha\in\mathbb{I}}$ is known. However, as a rule, they need to be found from the problem for the assembly process. Let us analyze this in detail.

Let the constant internal radius $R^\circ_{\mathrm{int}}$ of the intermdiate shape be known. Suppose that the following smooth functions are given: shrinkage $S\colon \mathbb{I} \to ]0, 1[$, external $R^\circ_{\mathrm{ext}} : \mathbb{I} \to \mathbb{R}_+$ radius of the intermediate shape, and internal $p_i : \mathbb{I} \to \mathbb{R}$ and external $p_e : \mathbb{I} \to \mathbb{R}$ hydrostatic loadings in actual configuration. The families $\{b_\alpha\}_{\alpha\in\mathbb{I}}$ and $\{A_\alpha\}_{\alpha\in\mathbb{I}}$ of the deformation parameters are unknown. On the interior and exterior boundaries of $\varkappa_\alpha(\mathfrak{B}_\alpha)$, consider the following conditions:

$$T^{r_\alpha r_\alpha}_\alpha (r_{\alpha;i}) = p_i(\alpha) , \quad T^{r_\alpha r_\alpha}_\alpha (r_{\alpha;e}) = p_e(\alpha) ,$$

where $T^{r_\alpha r_\alpha}_\alpha(r)$ is defined by the (11.25). These equalities lead to the relation

$$\int\limits_{R^\circ_{\mathrm{int}}}^{R^\circ_{\mathrm{ext}}(\alpha)} \left[ \frac{1}{\xi^2 - b_\alpha(\xi)} - \frac{\xi^2 - b_\alpha(\xi)}{(\xi^2 + A_\alpha)^2} \right] \xi \, d\xi = \frac{p_e(\alpha) - p_i(\alpha)}{\mu} . \tag{11.26}$$

Another condition relates $b_\alpha$ with shrinkage $S$:

$$b_\alpha (R^\circ_{\mathrm{ext}}(\alpha)) + A_\alpha = S(\alpha) . \tag{11.27}$$

Both (11.26) and (11.27) allow one to determine $A_\alpha$ and $b_\alpha$. Assumption (11.14) is used additionally.

For any $\mathfrak{X} \in \mathfrak{B}_\alpha$, the function $b_\alpha$ is the parameter of deformation of point's $\varkappa^\circ_\alpha(\mathfrak{X})$ neighborhood in the intermediate configuration to the corresponding local stress free one. The corresponding deformation gradient is represented by $\mathcal{K}_\alpha|_{\mathfrak{X}} \circ (T_{\mathfrak{X}}\varkappa^\circ_\alpha)^{-1}$, where $\mathscr{K}_\alpha|_\mathfrak{x}$ is local configuration (11.13):

$$\mathscr{K}_\alpha|_\mathfrak{x} = \left.\frac{R^\circ_\alpha}{\sqrt{(R^\circ_\alpha)^2 - b_\alpha(R^\circ_\alpha)}}\right|_\mathfrak{x} \boldsymbol{e}_{R_{\alpha,\mathfrak{x}}}|_X \otimes dR^\circ_\alpha|_\mathfrak{x} + \boldsymbol{e}_{\Theta_{\alpha,\mathfrak{x}}}|_X \otimes d\Theta^\circ_\alpha|_\mathfrak{x} + \boldsymbol{e}_{Z_{\alpha,\mathfrak{x}}}|_X \otimes dZ^\circ_\alpha|_\mathfrak{x} .$$

The existence of the initial body is the particular case with applications in the modeling of additive technology processes. This case may be considered in the framework of the above reasonings as follows. Let $\mathfrak{B}$ be a smooth manifold that represents initial body. For each $\alpha \in \mathbb{I}$, we take $\mathfrak{B}_\alpha := \mathfrak{B} \vee \mathfrak{L}_\alpha$ for elements of the family $\mathfrak{G}$ (see Chapter 8, Section 8.5.1). Here, $\mathfrak{L}_\alpha$ represents an extra body, which consists of infinitesimally thin layers that are attached to the initial body $\mathfrak{B}$. Assume that $\varkappa^\circ_\alpha(\mathfrak{B})$ is a stress-free hollow cylinder for each $\alpha \in \mathbb{I}$. Thus, the mapping $b_\alpha$ is represented as follows:

$$b_\alpha(R^\circ_\alpha) = \begin{cases} 0, & R^\circ_i < R^\circ_\alpha < R^\circ_{\alpha;e} , \\ \tilde{b}_\alpha(R^\circ_\alpha) , & R^\circ_\alpha > R^\circ_{\alpha;e} . \end{cases}$$

In this formula, $R^\circ_i = R^\circ_{\text{int}}$ and $R^\circ_{\alpha;e}$ are the prescribed inner and outer radii of $\varkappa^\circ_\alpha(\mathfrak{B})$, respectively, and $\tilde{b}_\alpha$ is a smooth function that represents the local parameter of deformation of $\mathfrak{L}_\alpha$. Thus, (11.26) takes the form:

$$\int\limits_{R^\circ_{\text{int}}}^{R^\circ_{\alpha;e}} \left[\frac{1}{\xi^2} - \frac{\xi^2}{(\xi^2 + A_\alpha)^2}\right] \xi\, d\xi + \int\limits_{R^\circ_{\alpha;e}}^{R^\circ_{\text{ext}}(\alpha)} \left[\frac{1}{\xi^2 - \tilde{b}_\alpha(\xi)} - \frac{\xi^2 - \tilde{b}_\alpha(\xi)}{(\xi^2 + A_\alpha)^2}\right] \xi\, d\xi = \frac{p_e(\alpha) - p_i(\alpha)}{\mu} ,$$

or, since the first integral may be calculated directly,

$$\ln \frac{R^\circ_{\alpha;e}\sqrt{(R^\circ_{\text{int}})^2 + A_\alpha}}{R^\circ_{\text{int}}\sqrt{(R^\circ_{\alpha;e})^2 + A_\alpha}} - \frac{A_\alpha}{2}\left(\frac{1}{(R^\circ_{\alpha;e})^2 + A_\alpha} - \frac{1}{(R^\circ_{\text{int}})^2 + A_\alpha}\right) + \int\limits_{R^\circ_{\alpha;e}}^{R^\circ_{\text{ext}}(\alpha)} \left[\frac{1}{\xi^2 - \tilde{b}_\alpha(\xi)} - \frac{\xi^2 - \tilde{b}_\alpha(\xi)}{(\xi^2 + A_\alpha)^2}\right] \xi\, d\xi = \frac{p_e(\alpha) - p_i(\alpha)}{\mu} . \quad (11.28)$$

Equation (11.27) should be rewritten in terms of $\tilde{b}_\alpha$:

$$\tilde{b}_\alpha(R^\circ_{\text{ext}}(\alpha)) + A_\alpha = S(\alpha) . \quad (11.29)$$

Hence, for this case one, should solve the evolutionary problem (11.28) and (11.29) to determine $\tilde{b}_\alpha$ and $A_\alpha$.

### 11.1.3 Results and Discussion

To compare strain and stress distributions in discrete structurally inhomogeneous solids with similar distributions in solids with continuously distributed inhomogeneity consider following examples:

a) A hollow cylinder assembled from a number of thin hollow cylindrical parts. We suppose that each part possesses a natural configuration. A sequence of assemblies with a common to all total volume $V$ and an increasing number of parts (10, 30, 200, and 1000) is studied.
b) A hollow cylinder with volume $V$, the same as in a), and piecewise continuous distribution for deformation parameter $b_\alpha$, which is taken to be the continuous approximation of the sequence of deformation parameters $a_s$ obtained for assembly from a) with a maximal numbers of parts.
c) A hollow cylinder with volume $V$, the same as in a) and b), and piecewise continuous distribution for $b_\alpha$, which is obtained as the numerical solution of the evolutionary problem (11.28) and (11.29).

From the general considerations discussed above, it is expected that the stress and strain distributions in solids with discrete inhomogeneity tend to the corresponding stress and strain distributions in a solid with some piecewise continuous $b_\alpha$, which can be obtained as a solution of the evolutionary problem (11.28) and (11.29) or can be represented as an approximation of the deformation parameters $a_s$ of sufficiently dense assembly. The following calculations illustrate this fact numerically.

For calculations, we took the following values of parameters. For each of the assemblies, the inner radius $\rho$ of the first cylindrical layer in reference configuration and its thickness $\Delta^0$ are taken equal to $\rho = \Delta^0 = 2$ mkm. The reference thicknesses $\Delta^k$, $k \geq 1$, for each of the assemblies are taken equal to[5] $\Delta^k = 22/(N-1)$ mkm, where $N$ is the number of layers. The shrinkage coefficients are taken equal to $S^k = 0.7$. Suppose that inner and outer hydrostatic loadings, $p_{i,s}$ and $p_{e,s}$, vanish. In computations, we put $\beta = 0.24$.

For assemblies, the recurrent equation systems (11.6) are solved. Strain and stress distributions are calculated by the equations obtained in Section 11.1.1. In Figures 11.2 and 11.3 the numerical results are shown, where the dotted line corresponds to ten layers, the dot-dashed line corresponds to 30 layers, and the dashed line corresponds to 200 layers. The case of 1000 layers is not shown on the graphs because its difference from continuous distributions discussed below is graphically indistinguishable. Angle brackets in $Y_{<ij>}$ denote that the corresponding component of $\boldsymbol{Y}$ is considered in a physical (with unit norm vectors) frame. The symbols $Y_{<rr>}$, $Y_{<\theta\theta>}$, and $Y_{<zz>}$ correspondingly denote radial, circumferential, and axial physical components of a tensor $\boldsymbol{Y}$. So do $Y_{<rR>}$, $Y_{<\theta\Theta>}$, $Y_{<zZ>}$, but relatively to the pair of cylindrical coordinate systems.

Results for the continuously structural inhomogeneous solid that corresponds to the discrete cases considered above can be obtained in the following way. Let the mapping $r \mapsto b_\alpha(r)$ given by $b_\alpha(r) = -3.165$, if $r < 3.583$ and $b_\alpha(r) = -4.568 + 1.335\, r +$

5 These values were chosen according to [26].

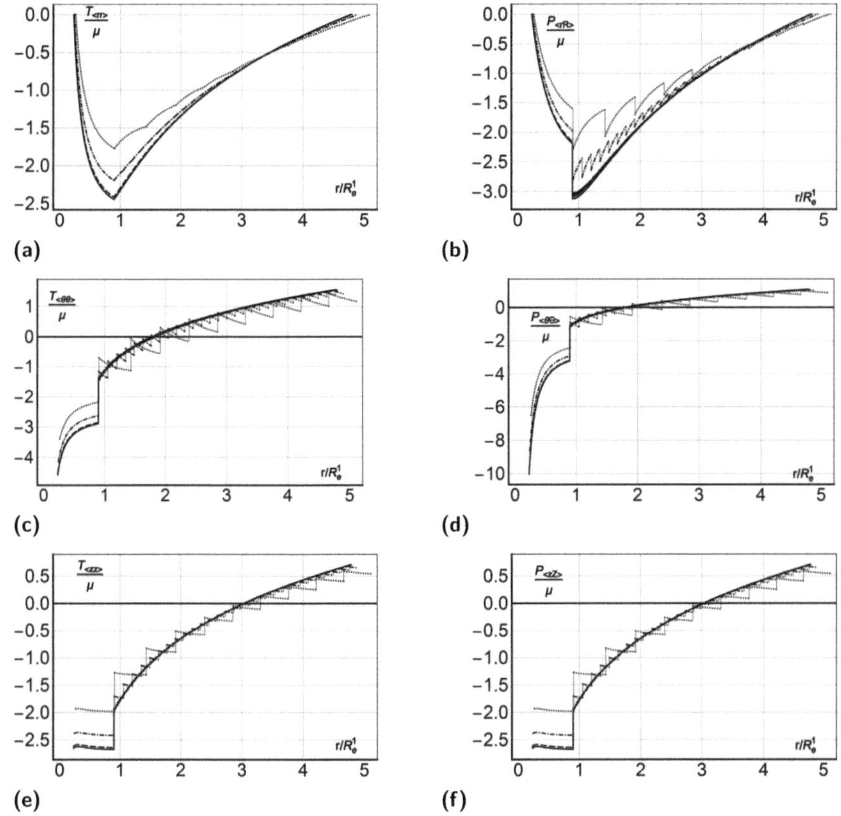


**Fig. 11.2:** Relative physical components $T_{<ij>}/\mu$ and $P_{<ij>}/\mu$ of Cauchy and first Piola–Kirchhoff stress tensors: (a) radial Cauchy stress component, (b) radial first Piola–Kirchhoff stress component, (c) circumferential Cauchy stress component, (d) circumferential first Piola–Kirchhoff stress component, (e) axial Cauchy stress component, (f) axial first Piola–Kirchhoff stress component.

$3.526\cdot 10^{-1}r^2 + 8.298\cdot 10^{-3}r^3 - 1.620\cdot 10^{-4}r^4$, if $r > 3.583$, be the numerical approximation of the values for deformation parameters in the most dense fragmentation (1000 layers). The material metric in this case is Riemannian with non-trivial curvature. This fact is illustrated by the distribution of the Ricci invariant (11.16) shown in Figure 11.3 (f). Note that the Ricci invariant is equal to zero on the part of the material manifold corresponding to the initial body. Adhered material forms the part of the body with non-Euclidean material metrics.

The corresponding Cauchy stress tensor $\boldsymbol{T}^{\mathrm{approx}}$ was obtained in Section 11.1.2, where for $T_\alpha^{\mathring{R}_\alpha \mathring{R}_\alpha}$, the expression (11.23) is used with $B^\circ_\alpha = 0$. The corresponding first Piola–Kirchhoff stress tensor and the Eshelby energy-momentum tensor were also obtained in Section 11.1.2, where $A_\alpha$ is to be equal to zero. These distributions are shown in Figures 11.2, 11.3 a, c, and e by solid lines.

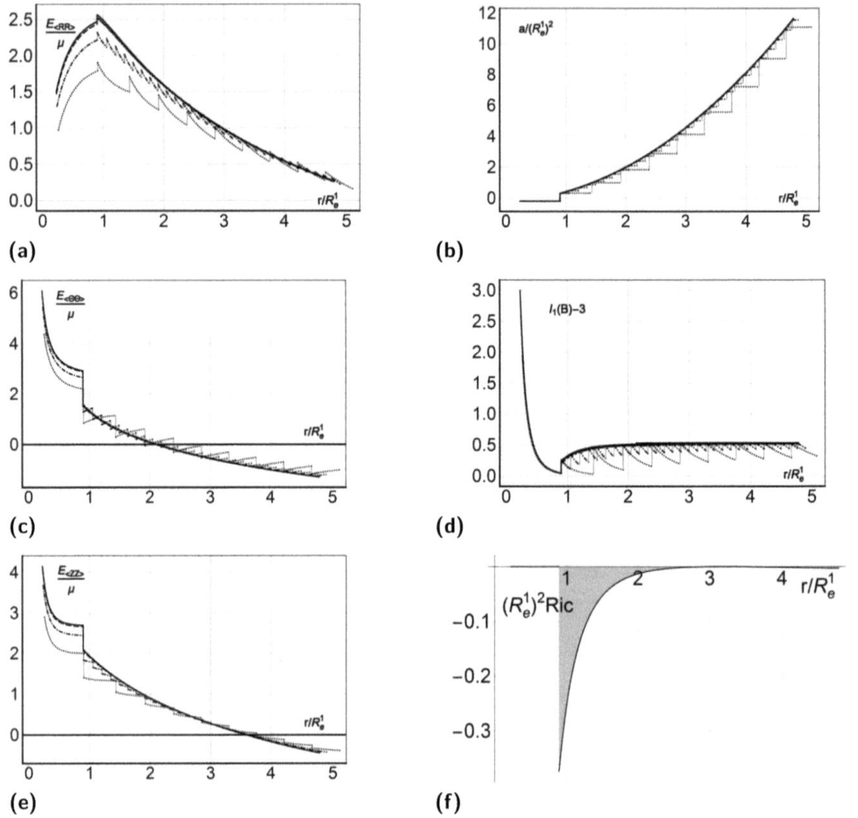


**Fig. 11.3:** Relative physical components $E_{\langle ij\rangle}/\mu$ of the Eshelby energy-momentum tensor; distributions of the deformation parameter, and the first and Ricci invariants: (a) radial Eshelby energy momentum tensor component, (b) deformation parameter, (c) circumferential Eshelby energy momentum tensor component, (d) the first invariant of left Cauchy-Green tensor B, (e) axial Eshelby energy momentum tensor component, (f) Ricci invariant (scalar curvature).

The function

$$x \mapsto \delta(x) = \frac{T_{rr}^{\text{approx}}(x) - T_{rr}^{\text{discr}}(x)}{\max T_{rr}^{\text{discr}}},$$

estimates the difference between stress states in solids with discrete and continuous inhomogeneities. A plot of this function is shown in Figure 11.4; label 1 designates 10 layers, label 2 designates 30 layers, label 3 designates 200 layers, and label 4 designates 1000 layers. In the right upper corner of Figure 11.4, the logarithms of absolute values for $\delta$, i.e., $\ln|\delta|$, are shown.

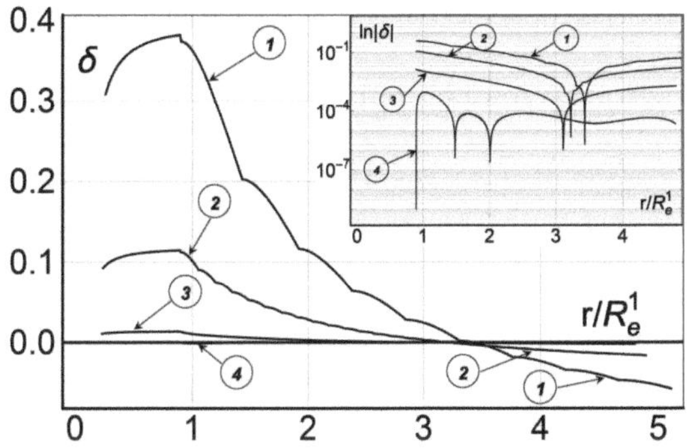


**Fig. 11.4:** Estimation of the difference between stress states in solids with discrete and continuous inhomogeneities.

## 11.2 Uniform Inflation of a Spherical Multilayered Structure

### 11.2.1 Layers and Assemblies

Consider assemblies that consist of a finite number of spherical layers. We assume that for each layer, there exists a configuration which image, namely, the shape of the layer, is free of stresses. In the general case, these shapes are geometrically incompatible, *i.e.*, they cannot be combined without gaps or overlaps. The totality of the reference shapes of the layers plays the same role as the natural shape of the body in the classical theory of elasticity; relative to these shapes, the strain measures, which are arguments of the function of the stored elastic energy, are calculated.

Consider a family $\mathfrak{G} = \{\mathfrak{B}_\alpha\}_{\alpha\in\mathbb{I}}$ of three-dimensional bodies, which are ordered with respect to the subset relation:

$$\forall\alpha,\beta \in \mathbb{I}\colon (\alpha < \beta) \Rightarrow (\mathfrak{B}_\alpha \subset \mathfrak{B}_\beta)\,.$$

In addition, assume that $\mathfrak{B}_\alpha$ is a submanifold of $\mathfrak{B}_\beta$. Speaking figuratively, this means that all bodies from $\mathfrak{G}$ are topologically "uniform" and can be obtained from some "model" body. In our case, such a body is, for example, a spherical shell with unit exterior radius and with interior radius equal to $1/2$. Consider the discrete case: $\mathbb{I} = \{0, 1, \ldots, N\}$, $N \le 1$.

Let the physical space be the Euclidean affine space $\mathscr{E}$ (Chapter 2, Section 2.1). Assume that the following conditions are satisfied:

a) Actual shapes of bodies $\mathfrak{B}_\alpha \in \mathfrak{G}$, as well as actual shapes of layers $\mathfrak{L}_\alpha$, are spherical shells in $\mathscr{E}$, *i.e.*, they can be defined as the sets

$$B_\alpha = \left\{x \in \mathscr{E} \mid r_\alpha^i \le \|x - o\| \le r_\alpha^e\right\}\,, \quad L_{\alpha\beta} = \left\{x \in \mathscr{E} \mid r_{\alpha,\beta}^i \le \|x - o\| \le r_{\alpha,\beta}^e\right\}\,,$$

respectively.

𝔟) Natural (stress-free) shapes of layers are represented by spherical shells in $\mathscr{E}$, *i.e.*,

$$L_\alpha = \left\{x \in \mathscr{E} \mid R^i_\alpha \leq \|x - o\| \leq R^e_\alpha\right\} .$$

𝔠) Internal, $R^i_{\alpha+1}$, and external, $R^e_{\alpha+1}$, radii of stress-free shapes of $\mathfrak{L}_{\alpha+1}$ are related with the external radius of the actual shape of the assembly $\mathfrak{B}_\alpha$ by relations

$$R^i_{\alpha+1} = S_\alpha r^e_\alpha , \quad R^e_{\alpha+1} = R^i_{\alpha+1} + \Delta_{\alpha+1} ,$$

where $\Delta_\alpha$, $1, 2, \ldots, N$ are prescribed thicknesses of stress-free shapes of layers, and $S_\alpha$ are prescribed quantities that represent gaps (if $S_\alpha > 1$) or overlaps (if $S_\alpha < 1$) of stress-free layers and characterize technological shrinkage due to the layer solidification.

𝔡) Internal, $R^i_0$, and external, $R^e_0$, reference radii for the initial body (the layer with number 0) are given.

The following notation is used: a single lower index denotes the number of the corresponding layer or assembly, and the double lower index, whose elements are separated by a comma, are the assembly number and layer number in this assembly. Taking into account the accepted designations, the sequence of shapes of assemblies can be represented by the diagram 11.5.

The assumptions 𝔞)–𝔡) imply that for each assembly $\mathfrak{B}_\alpha$, there exist configurations $\varkappa_{\alpha,\beta}$, $\beta = 0, \ldots, N$, whose images are spherical shells in $\mathscr{E}$. Parts of these shapes corresponding to layers with a number $\beta$, are free of stresses. In particular, for the assembly with number 3, there exist three such configurations; they are shown in the diagram 11.6. The area in dark color corresponds to a stress-free shape.

### 11.2.2 Coordinates and Vector Bases

Since the physical space is Euclidean, one can choose the Cartesian coordinates $(o, \boldsymbol{i}_1, \boldsymbol{i}_2, \boldsymbol{i}_3)$ in it. Moreover, because all shapes of assemblies $\mathfrak{B}_\alpha$ and layers $\mathfrak{L}_\alpha$ are centrally symmetric, in addition to Cartesian coordinates $(x^1, x^2, x^3)$, it is convenient to use spherical coordinates $(r, \theta, \varphi)$ that are related with Cartesian via the formulae:

$$x^1 = r \sin\theta \cos\varphi , \quad x^2 = r \sin\theta \sin\varphi , \quad x^3 = r \cos\theta ,$$

where $r > 0$, $\theta \in \,]0, \pi[$, $\varphi \in [0, 2\pi[$ together with "deformed" spherical coordinates $(r_{\alpha,\beta}, \theta_{\alpha,\beta}, \varphi_{\alpha,\beta})$ that are related with $(r, \theta, \varphi)$ by the expressions

$$r = \widetilde{f}_{\alpha,\beta}(r_{\alpha,\beta}) , \quad \theta = \theta_{\alpha,\beta} , \quad \varphi = \varphi_{\alpha,\beta} . \tag{11.30}$$

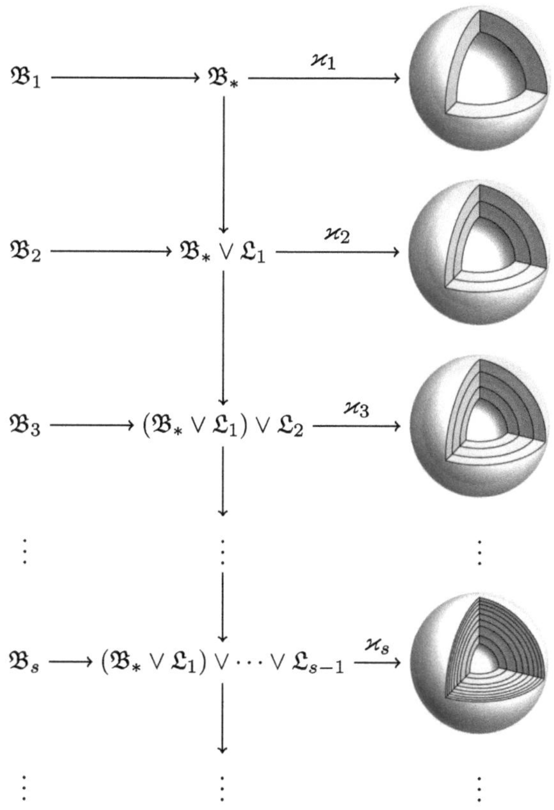


**Fig. 11.5:** Hollow ball assemblies.

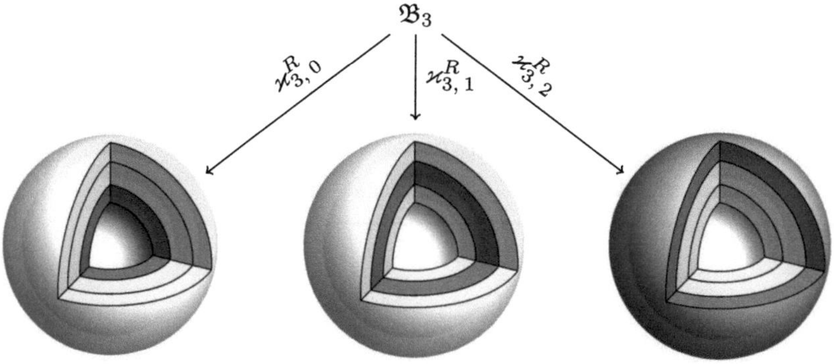


**Fig. 11.6:** Family of shapes with stress-free layers.

Here, $\widetilde{f}_{\alpha,\beta}\colon \mathbb{R}_+ \to \mathbb{R}_+$ is a homeomorphism onto its image that can be defined as the extension of a function $f_{\alpha,\beta}\colon [R^i_\beta, R^e_\beta] \to [r^i_{\alpha,\beta}, r^e_{\alpha,\beta}]$. The function[6] $f_{\alpha,\beta}$ specifies the distance from the origin $o$ to the actual positions of material points of the shape $L_{\alpha,\beta}$ of the layer $\mathfrak{L}_\beta$ in the assembly $\mathfrak{B}_\alpha$ depending on their reference positions in the reference shape $L_\beta$. Each of these coordinate systems is associated with the shape of the corresponding layer and plays the role of "frozen" (convected) coordinates. Thus, an extended atlas of physical space is constructed. In addition to the Cartesian chart, such an atlas contains charts that are defined by various variants of spherical coordinates. Such an extension makes it possible to simplify further calculations using different coordinates as needed.

Each coordinate system generates a field of local bases, *i.e.*, triples of vectors $(\boldsymbol{e}_r, \boldsymbol{e}_\theta, \boldsymbol{e}_\varphi)$ and $(\overset{\alpha,\beta}{\boldsymbol{e}}_r, \overset{\alpha,\beta}{\boldsymbol{e}}_\theta, \overset{\alpha,\beta}{\boldsymbol{e}}_\varphi)$, which are tangent to the corresponding coordinate lines and are defined by formulae

$$\boldsymbol{e}_c = \frac{\partial x^s}{\partial c}\boldsymbol{i}_s\,, \quad c \in \{r, \theta, \varphi\}\,; \quad \overset{\alpha,\beta}{\boldsymbol{e}}_c = \frac{\partial x^s}{\partial c_{\alpha,\beta}}\boldsymbol{i}_s,\, c_{\alpha,\beta} \in \{r_{\alpha,\beta}, \theta_{\alpha,\beta}, \varphi_{\alpha,\beta}\}\,.$$

Dual bases $(\boldsymbol{e}^r, \boldsymbol{e}^\theta, \boldsymbol{e}^\varphi)$ and $(\underset{\alpha,\beta}{\boldsymbol{e}}^r, \underset{\alpha,\beta}{\boldsymbol{e}}^\theta, \underset{\alpha,\beta}{\boldsymbol{e}}^\varphi)$ are defined by the biorthogonality relations

$$\boldsymbol{e}^s \cdot \boldsymbol{e}_c = \delta^s_c,\, s, c \in \{r, \theta, \varphi\}\,; \quad \underset{\alpha,\beta}{\boldsymbol{e}}^s \cdot \overset{\alpha,\beta}{\boldsymbol{e}}_c = \delta^s_c,\, s, c \in \{r_{\alpha,\beta}, \theta_{\alpha,\beta}, \varphi_{\alpha,\beta}\}\,.$$

The exact expressions for the bases are given in Tab. 11.2.

**Tab. 11.2:** Bases for different coordinates.

| | $\boldsymbol{i}_1$ | $\boldsymbol{i}_2$ | $\boldsymbol{i}_3$ |
|---|---|---|---|
| $\boldsymbol{e}_r = \boldsymbol{e}^r = \overset{\alpha,\beta}{\boldsymbol{e}}_r = \underset{\alpha,\beta}{\boldsymbol{e}}^r$ | $\sin\theta\cos\varphi$ | $\sin\theta\sin\varphi$ | $\cos\theta$ |
| $\boldsymbol{e}_\theta$ | $r\cos\theta\cos\varphi$ | $r\cos\theta\sin\varphi$ | $-r\sin\theta$ |
| $\boldsymbol{e}_\varphi$ | $-r\sin\theta\sin\varphi$ | $r\sin\theta\cos\varphi$ | $0$ |
| $\overset{\alpha,\beta}{\boldsymbol{e}}_\theta$ | $r_{\alpha,\beta}\cos\theta\cos\varphi$ | $r_{\alpha,\beta}\cos\theta\sin\varphi$ | $-r_{\alpha,\beta}\sin\theta$ |
| $\overset{\alpha,\beta}{\boldsymbol{e}}_\varphi$ | $-r_{\alpha,\beta}\sin\theta\sin\varphi$ | $r_{\alpha,\beta}\sin\theta\cos\varphi$ | $0$ |
| $\boldsymbol{e}^\theta$ | $r^{-1}\cos\theta\cos\varphi$ | $r^{-1}\cos\theta\sin\varphi$ | $-r^{-1}\sin\theta$ |
| $\boldsymbol{e}^\varphi$ | $-(r\sin\theta)^{-1}\sin\varphi$ | $(r\sin\theta)^{-1}\cos\varphi$ | $0$ |
| $\underset{\alpha,\beta}{\boldsymbol{e}}^\theta$ | $r^{-1}_{\alpha,\beta}\cos\theta\cos\varphi$ | $r^{-1}_{\alpha,\beta}\cos\theta\sin\varphi$ | $-r^{-1}_{\alpha,\beta}\sin\theta$ |
| $\underset{\alpha,\beta}{\boldsymbol{e}}^\varphi$ | $-(r_{\alpha,\beta}\sin\theta)^{-1}\sin\varphi$ | $(r_{\alpha,\beta}\sin\theta)^{-1}\cos\varphi$ | $0$ |

6 In the following, it is necessary to express the reference radius through the actual one and formulate the differential relations. In this regard, the function $f_{\alpha,\beta}$ must be bijective and infinitely many times continuously differentiable together with its inverse.

### 11.2.3 Strain Measures

Since the deformation of each layer $\mathfrak{L}_\beta$ in the body $\mathfrak{B}_\alpha$ in the natural shape is centrally symmetric (11.30) (radial displacements are given by the function $f_{\alpha,\beta}$), the deformation gradient, its transpose, and the Cauchy–Green strain tensors can be represented by diagonal dyadic expansions

$$
\begin{aligned}
\boldsymbol{F}_{\alpha,\beta} &= f'_{\alpha,\beta}\boldsymbol{e}_r \otimes \underset{\alpha,\beta}{\boldsymbol{e}^r} + \boldsymbol{e}_\theta \otimes \underset{\alpha,\beta}{\boldsymbol{e}^\theta} + \boldsymbol{e}_\varphi \otimes \underset{\alpha,\beta}{\boldsymbol{e}^\varphi}\,, \\
\boldsymbol{F}^{\mathrm{T}}_{\alpha,\beta} &= f'_{\alpha,\beta}\overset{\alpha,\beta}{\boldsymbol{e}_r} \otimes \boldsymbol{e}^r + \left(\frac{f_{\alpha,\beta}}{r_{\alpha,\beta}}\right)^2 \overset{\alpha,\beta}{\boldsymbol{e}_\theta} \otimes \boldsymbol{e}^\theta + \left(\frac{f_{\alpha,\beta}}{r_{\alpha,\beta}}\right)^2 \overset{\alpha,\beta}{\boldsymbol{e}_\varphi} \otimes \boldsymbol{e}^\varphi\,, \\
\boldsymbol{B}_{\alpha,\beta} &= \mathscr{F}_{\alpha,\beta}\mathscr{F}^{\mathrm{T}}_{\alpha,\beta} = (f'_{\alpha,\beta})^2\boldsymbol{e}_r \otimes \boldsymbol{e}^r + \left(\frac{f_{\alpha,\beta}}{r_{\alpha,\beta}}\right)^2 \boldsymbol{e}_\theta \otimes \boldsymbol{e}^\theta + \left(\frac{f_{\alpha,\beta}}{r_{\alpha,\beta}}\right)^2 \boldsymbol{e}_\varphi \otimes \boldsymbol{e}^\varphi\,, \qquad (11.31) \\
\boldsymbol{B}^{-1}_{\alpha,\beta} &= \left(\frac{1}{f'_{\alpha,\beta}}\right)^2 \boldsymbol{e}_r \otimes \boldsymbol{e}^r + \left(\frac{r_{\alpha,\beta}}{f_{\alpha,\beta}}\right)^2 \boldsymbol{e}_\theta \otimes \boldsymbol{e}^\theta + \left(\frac{r_{\alpha,\beta}}{f_{\alpha,\beta}}\right)^2 \boldsymbol{e}_\varphi \otimes \boldsymbol{e}^\varphi\,. \qquad (11.32)
\end{aligned}
$$

The principal invariants of tensors $\mathscr{B}_{\alpha,\beta}$ are defined by the relations

$$
\begin{aligned}
I_1(\boldsymbol{B}_{\alpha,\beta}) &= \operatorname{tr}\boldsymbol{B}_{\alpha,\beta} = (f'_{\alpha,\beta})^2 + 2\left(\frac{f_{\alpha,\beta}}{r_{\alpha,\beta}}\right)^2, \\
I_2(\boldsymbol{B}_{\alpha,\beta}) &= \frac{1}{2}\left(I_1^2(\boldsymbol{B}_{\alpha,\beta}) - I_1(\boldsymbol{B}^2_{\alpha,\beta})\right) = 2(f'_{\alpha,\beta})^2\left(\frac{f_{\alpha,\beta}}{r_{\alpha,\beta}}\right)^2 + \left(\frac{f_{\alpha,\beta}}{r_{\alpha,\beta}}\right)^4, \\
I_3(\boldsymbol{B}_{\alpha,\beta}) &= \det\boldsymbol{B}_{\alpha,\beta} = (f'_{\alpha,\beta})^2\left(\frac{f_{\alpha,\beta}}{r_{\alpha,\beta}}\right)^4.
\end{aligned}
$$

### 11.2.4 Incompressible Material

We assume that the material of the layers is incompressible, homogeneous, and isotropic, and the density of the elastic energy stored by it during deformation is determined by the two-constant Mooney–Rivlin potential,

$$W_{\alpha,\beta} = C_1\left(I_1(\boldsymbol{B}_{\alpha,\beta}) - 3\right) + C_2\left(I_2(\boldsymbol{B}_{\alpha,\beta}) - 3\right)\,. \qquad (11.33)$$

The constants (elastic modules) $C_1$ and $C_2$ can be expressed through another pair of modules $b$ and $\mu$ as follows: $C_1 = (1 + b)\mu/4$, $C_2 = (1 - b)\mu/4$. The use of the modules $C_1$ and $C_2$ later allows the expressions obtained to be written in a more concise form, while the $\mu$ and $b$ modules indicate the relationship between the non-linear hyperelastic material model and the linear elastic model; for small deformations from the unstressed state, $\mu$ coincides with the shear modulus of a linearly elastic incompressible material.

For the potential (11.33), the Rivlin formula gives the Cauchy stresses:

$$\boldsymbol{T}_{\alpha,\beta} = -p_{\alpha,\beta}\boldsymbol{I} + 2C_1\boldsymbol{B}_{\alpha,\beta} - 2C_2\boldsymbol{B}^{-1}_{\alpha,\beta}\,,$$

where $\boldsymbol{I}$ is the identity tensor, $p_{\alpha,\beta}$ is the hydrostatic loading, which is determined from the equilibrium equation $\nabla \cdot \boldsymbol{T} = \boldsymbol{0}$ and the prescribed boundary conditions. For a centrally symmetric field, they can be represented in the form

$$\forall r \in \left]r^i_{\alpha,\beta}, r^e_{\alpha,\beta}\right[ \ : \quad \frac{\partial \overset{\alpha,\beta}{T}{}^r{}_r}{\partial r} + \frac{2}{r}\left(\overset{\alpha,\beta}{T}{}^r{}_r - \overset{\alpha,\beta}{T}{}^\theta{}_\theta\right) = 0\,, \tag{11.34}$$

$$\left.\overset{\alpha,0}{T}{}^r{}_r\right|_{r=r^i_{\alpha,0}} = q^i_\alpha\,, \quad \left.\overset{\alpha,\alpha}{T}{}^r{}_r\right|_{r=r^e_{\alpha,\alpha}} = q^e_\alpha\,, \tag{11.35}$$

$$\boldsymbol{T}_{\alpha,\beta} \cdot \boldsymbol{e}^r\big|_{r=r^e_{\alpha,\beta}} = \boldsymbol{T}_{\alpha,\beta+1} \cdot \boldsymbol{e}^r\big|_{r=r^i_{\alpha,\beta+1}}\,, \quad r^e_{\alpha,\beta} = r^i_{\alpha,\beta+1}\,. \tag{11.36}$$

Boundary conditions (11.35) characterize the uniform hydrostatic loading of the assembly of spherical boundaries, while equalities (11.36) state the conditions of ideal contact between the layers.

Solutions of the equation (11.34) can be obtained in the form

$$\begin{aligned}
\overset{\alpha,\beta}{T}{}^r{}_r &= \overset{\alpha,\beta}{p}{}_0 + \frac{C_1\left(5r^3 - a_{\alpha,\beta}\right)\left(r^3 - a_{\alpha,\beta}\right)^{2/3} + 2C_2 r^2\left(r^3 + a_{\alpha,\beta}\right)}{r^4\left(r^3 - a_{\alpha,\beta}\right)^{1/3}}\,,\\
\overset{\alpha,\beta}{T}{}^\theta{}_\theta &= \overset{\alpha,\beta}{T}{}^r{}_r + 2\frac{a_{\alpha,\beta}\left(2r^3 - a_{\alpha,\beta}\right)}{r^4\left(r^3 - a_{\alpha,\beta}\right)^{4/3}}\left(C_1\left(r^3 - a_{\alpha,\beta}\right)^{2/3} - C_2 r^2\right)\,,\\
\overset{\alpha,\beta}{T}{}^\varphi{}_\varphi &= \overset{\alpha,\beta}{T}{}^\theta{}_\theta\,,
\end{aligned} \tag{11.37}$$

where $\overset{\alpha,\beta}{p}{}_0$, $a_{\alpha,\beta}$ are constants, which must be chosen so that the boundary conditions (11.35) and (11.36) are satisfied. Substitution of (11.37) into them results in a system of non-linear algebraic equations. For the initial body ($\alpha = 0$), the system consists of two equations with two unknowns, $\overset{0,0}{p}{}_0$, $a_{0,0}$:

$$\overset{0,0}{p}{}_0 + \mathscr{T}\left[R^i_0; a_{0,0}\right] = q^i_0\,, \quad \overset{0,0}{p}{}_0 + \mathscr{T}\left[R^e_0; a_{0,0}\right] = q^e_0\,, \tag{11.38}$$

where

$$\mathscr{T}[R; a] = \left(R(R^3 + a)^{4/3}\right)^{-1}\left\{C_1(5R^3 + 4a)R^2 + 2C_2(R^3 + a)^{2/3}(R^3 + 2a)\right\}\,.$$

Systems for other assemblies $\mathfrak{B}_\alpha$, $0 < \alpha \le N$ (the relations $R^e_\beta = R^i_\beta + \Delta_\beta$, an $0 < \beta < \alpha$ are taken into account) consist of $2\alpha$ equations relative to $3\alpha - 1$ unknowns $\overset{\alpha,0}{p}{}_0, \overset{\alpha,1}{p}{}_0, \dots, \overset{\alpha,\alpha}{p}{}_0, a_{\alpha,0}, a_{\alpha,1}, \dots, a_{\alpha,\alpha}, R^i_1, R^i_2, \dots, R^i_\alpha$:

$$\begin{gathered}
\overset{\alpha,0}{p}{}_0 + \mathscr{T}\left[R^i_0; a_{\alpha,0}\right] = q^i_\alpha\,, \quad \overset{\alpha,\alpha}{p}{}_0 + \mathscr{T}\left[R^e_\alpha; a_{\alpha,\alpha}\right] = q^e_\alpha\,,\\
\overset{\alpha,1}{p}{}_0 + \mathscr{T}\left[R^e_1; a_{\alpha,1}\right] = \overset{\alpha,2}{p}{}_0 + \mathscr{T}\left[R^i_2; a_{\alpha,2}\right]\,,\\
\left(R^e_1\right)^3 + a_{\alpha,1} = \left(R^i_2\right)^3 + a_{\alpha,2}\,,\\
\dots\\
\overset{\alpha,\alpha-1}{p}{}_0 + \mathscr{T}\left[R^e_{\alpha-1}; a_{\alpha,\alpha-1}\right] = \overset{\alpha,\alpha}{p}{}_0 + \mathscr{T}\left[R^i_\alpha; a_{\alpha,\alpha}\right]\,,\\
\left(R^e_{\alpha-1}\right)^3 + a_{\alpha,\alpha-1} = \left(R^i_\alpha\right)^3 + a_{\alpha,\alpha}\,.
\end{gathered} \tag{11.39}$$

The condition c) gives $N-1$ equations that complement the system (11.39):

$$R^i_\alpha = S_\alpha \sqrt[3]{(R^e_{\alpha-1})^3 + a_{\alpha-1,\alpha-1}}\,, \qquad 0 < \alpha \le N\,. \tag{11.40}$$

They transform the underdetermined equations (11.39) into a recurrent sequence of simpler subsystems of equations. The system (11.38) can be used as the base of the recursion. Solving this system, one obtains $\overset{0,0}{p}_0$ and $a_{0,0}$. With the known value $a_{0,0}$, one can determine $R^i_1$ by (11.40). The first recursion step involves the solution of the system (11.39) with $\alpha = 1$. This is the system of four equations with four unknowns $\overset{1,0}{p}_0$, $\overset{1,1}{p}_0$, $a_{1,0}$, $a_{1,1}$. With the known $a_{1,1}$, one can determine $R^i_2$ from (11.40). The second recursion step involves the solution of the system (11.39) with $\alpha = 2$. This is the system of six equations relative to six variables, namely, $\overset{2,0}{p}_0$, $\overset{2,1}{p}_0$, $\overset{2,2}{p}_0$, $a_{2,0}$, $a_{2,1}$, $a_{2,2}$. Further steps are performed similarly.

### 11.2.5 Compressible Material

Now consider a compressible material. Suppose that the elastic energy density of the layer $\mathfrak{L}_\beta$ is given by the three-constant Mooney–Rivlin potential

$$W_{\alpha,\beta} = C_1\left(J^{-2/3}_{\alpha,\beta} I_1 - 3\right) + C_2\left(J^{-4/3}_{\alpha,\beta} I_2 - 3\right) + \frac{d}{2}(J_{\alpha,\beta} - 1)^2\,,$$

where $J_{\alpha,\beta} = \det \boldsymbol{F}_{\alpha,\beta} = f'_{\alpha,\beta}\left(f_{\alpha,\beta}/r_{\alpha,\beta}\right)^2 \neq 1$ and the constant $d$ corresponds to the bulk expansion module. In this case, the expression for the Cauchy stress tensor $\boldsymbol{T}_{\alpha,\beta}$ takes the form

$$\boldsymbol{T}_{\alpha,\beta} = \beta_0 \boldsymbol{I} + \beta_1 \boldsymbol{B}_{\alpha,\beta} + \beta_{-1}\boldsymbol{B}^{-1}_{\alpha,\beta}\,, \tag{11.41}$$

where $\beta_0$, $\beta_1$, $\beta_{-1}$ are scalar reaction functions,

$$\beta_0 = \frac{3d(I_3^{1/2} - 1)I_3^{7/6} - 2C_1 I_1 I_3^{1/3} - 2C_2 I_2}{3I_3^{7/6}}\,, \qquad \beta_1 = \frac{2C_1}{I_3^{5/6}}\,, \qquad \beta_{-1} = \frac{2C_2}{I_3^{1/6}}\,.$$

Elementary calculations lead to the following expressions for Cauchy stresses:

$$\boldsymbol{T}_{\alpha,\beta} = \overset{\alpha,\beta}{T}{}^r{}_r\, \boldsymbol{e}_r \otimes \boldsymbol{e}^r + \overset{\alpha,\beta}{T}{}^\theta{}_\theta\, \boldsymbol{e}_\theta \otimes \boldsymbol{e}^\theta + \overset{\alpha,\beta}{T}{}^\varphi{}_\varphi\, \boldsymbol{e}_\varphi \otimes \boldsymbol{e}^\varphi\,, \tag{11.42}$$

$$\begin{aligned}
\overset{\alpha,\beta}{T}{}^r{}_r &= \frac{4}{3} r^{-4}_{\alpha,\beta} J^{-7/3}_{\alpha,\beta} \Big\{ f^4_{\alpha,\beta}\left(C_2 + 3/4\, d\, J^{1/3}_{\alpha,\beta}\,(J_{\alpha,\beta} - 1)\,(f'_{\alpha,\beta})^2\right) \\
&\quad + C_1\, r^8_{\alpha,\beta} J^{8/3}_{\alpha,\beta} f^{-4}_{\alpha,\beta} - r^2_{\alpha,\beta} f^2_{\alpha,\beta}\left(C_1 J^{2/3}_{\alpha,\beta} + C_2 (f'_{\alpha,\beta})^2\right)\Big\}\,, \\
\overset{\alpha,\beta}{T}{}^\theta{}_\theta = \overset{\alpha,\beta}{T}{}^\varphi{}_\varphi &= \frac{2}{3} r^{-4}_{\alpha,\beta} J^{-7/3}_{\alpha,\beta} \Big\{ f^4_{\alpha,\beta}\left(3/2\, d\, J^{1/3}_{\alpha,\beta}\,(J_{\alpha,\beta} - 1)\,(f'_{\alpha,\beta})^2 - C_2\right) \\
&\quad - C_1\, r^8_{\alpha,\beta} J^{8/3}_{\alpha,\beta} f^{-4}_{\alpha,\beta} + r^2_{\alpha,\beta} f^2_{\alpha,\beta}\left(C_1 J^{2/3}_{\alpha,\beta} + C_2 (f'_{\alpha,\beta})^2\right)\Big\}\,.
\end{aligned}$$

In the case of central symmetry, the equilibrium equation $\nabla \cdot \boldsymbol{T} = \boldsymbol{0}$ reduces to one scalar equation

$$\frac{\partial \overset{\alpha,\beta}{T^r}_r}{\partial r_{\alpha,\beta}} + 2\frac{f'_{\alpha,\beta}}{f_{\alpha,\beta}}\left(\overset{\alpha,\beta}{T^r}_r - \overset{\alpha,\beta}{T^\theta}_\theta\right) = 0\,.$$

Substituting the expression (11.42) into the latter equation results in a second-order ODE with the unknown function $f$:

$$\left(\beta_0 + \beta_1 (f'_{\alpha,\beta})^2 + \beta_{-1}\frac{1}{(f'_{\alpha,\beta})^2}\right)' \\ + 2\frac{f'_{\alpha,\beta}}{f_{\alpha,\beta}}\left\{\beta_1\left((f'_{\alpha,\beta})^2 - \frac{f^2_{\alpha,\beta}}{r^2_{\alpha,\beta}}\right) + \beta_{-1}\left(\frac{1}{(f'_{\alpha,\beta})^2} - \frac{r^2_{\alpha,\beta}}{f^2_{\alpha,\beta}}\right)\right\} = 0\,.$$

Boundary conditions are represented by expressions (11.35) and (11.36). Thus, for a fixed $\alpha$, the boundary value problem relative to functions $f_{\alpha,\beta}$ can be stated as follows:

$$\begin{aligned}
\forall R \in \left]R^i_\beta, R^e_\beta\right[: \quad & \Phi\left(f_{\alpha,\beta}, f'_{\alpha,\beta}\right) + f''_{\alpha,\beta} = 0\,, \\
\Psi\left(f_{\alpha,0}, f'_{\alpha,0}\right)\Big|_{R=R^i_0} = q^i_\alpha\,, \quad & \Psi\left(f_{\alpha,\alpha}, f'_{\alpha,\alpha}\right)\Big|_{R=R^e_\alpha} = q^e_\alpha\,, \\
\forall \beta \in \{0, 2, \ldots, \alpha - 1\}: \quad & f_{\alpha,\beta}\Big|_{R=R^e_\beta} = f_{\alpha,\beta+1}\Big|_{R=R^i_{\beta+1}}\,, \\
\Psi\left(f_{\alpha,\beta}, f'_{\alpha,\beta}\right)\Big|_{R=R^e_\beta} = \; & \Psi\left(f_{\alpha,\beta+1}, f'_{\alpha,\beta+1}\right)\Big|_{R=R^i_{\beta+1}}\,,
\end{aligned} \tag{11.43}$$

$$\begin{aligned}
\Phi(f,f') = {} & \left(Rf\left\{f^8\left(9/4\, d f'^2 J^{4/3} - 7C_2\right) + C_1 R^8 J^{8/3} + R^2 f^6\left(5C_1 J^{2/3} + C_2 f'^2\right)\right\}\right)^{-1} \\
& \times \left\{f'\left(f - Rf'\right)\left(2f^8\left(C_2 - 9/4\, d f'^2 J^{4/3}\right) + C_1 R^8 J^{8/3} - 9C_1 R^3 f^5 f' J^{2/3}\right.\right. \\
& \left.\left. - \left\{R^2 f^6\left(4C_1 J^{2/3} - C_2 f'^2\right) + 9C_2 R f^7 f'\right\}\right)\right\}\,, \\
\Psi(f,f') = {} & 2(3f^8 f'^4)^{-1} J^{5/3}\left(R^4 f^4\left(-3C_2 f'^6 - C_2 + 3/2\, d J^{1/3}(J-1) f'^2\right)\right. \\
& \left. + R^8 f'^2\left(2C_1 J^{2/3} + 3C_2 f'^2\right) - 2R^6 f^2\left(C_1 J^{2/3} + C_2 f'^2\right) + 3C_2 f^8 f'^2\right)\,.
\end{aligned}$$

This boundary value problem can be solved numerically, for example, by the shooting method. The initial approximation can be taken from the analytical solution of the problem for incompressible material, obtained above. Figures 11.7 and 11.8 illustrate the differences in the stress states obtained for a single hollow ball using linear, incompressible, and compressible models. The constant hydrostatic load was assumed to be equal to $0.2\,\mu$ and $0.7\,\mu$.

Recall that the components of the stress tensor corresponding to the linear model have the form

$$T^{\langle rr\rangle} = q\frac{R_2^3}{R_1^3 - R_2^3}\left(\frac{R_1^3}{R^3} - 1\right), \quad T^{\langle\theta\theta\rangle} = -q\frac{R_2^3}{R_1^3 - R_2^3}\left(1 + \frac{R_1^3}{2R^3}\right).$$

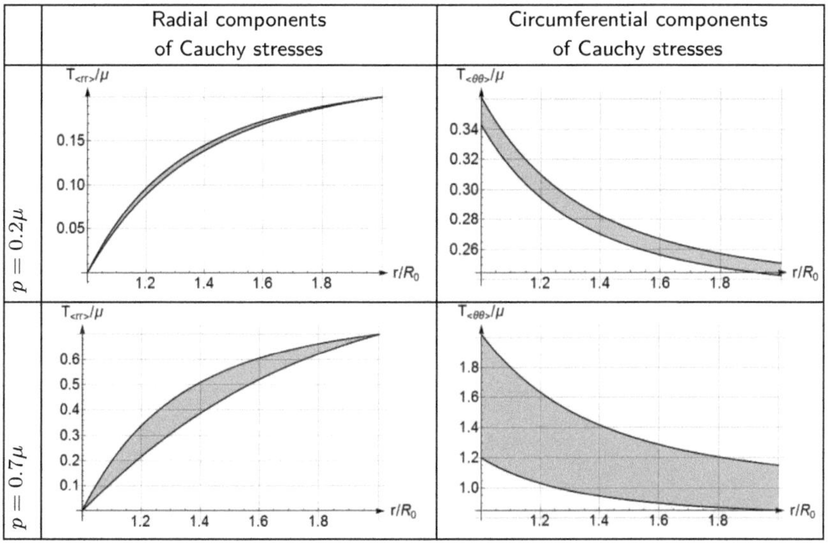


**Fig. 11.7:** Comparative graphs of stresses for linear and compressible models. The *blue line* corresponds to the linear model and the *orange line* corresponds to the compressible one.

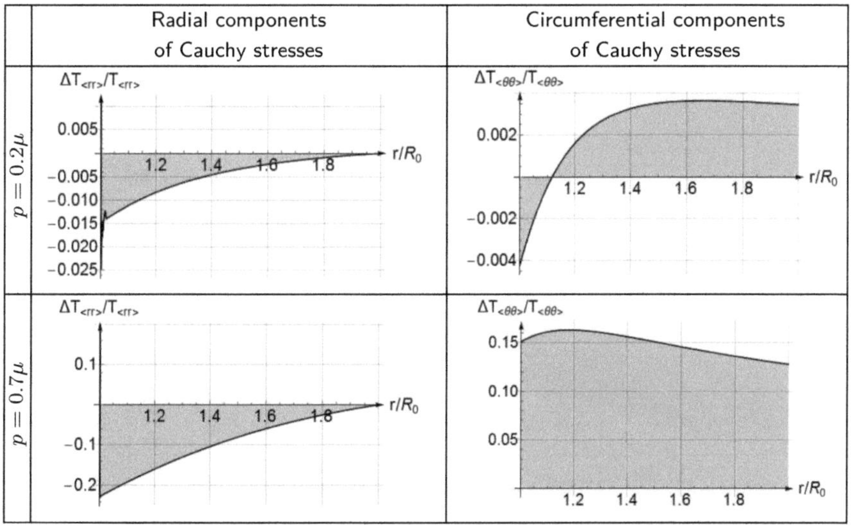


**Fig. 11.8:** Comparative graphs of stresses for incompressible and compressible models.

Here, $q$ is a hydrostatic loading and $R_1 < R_2$ is a reference radii of a spherical shell. The graphics in Figure 11.7 show the distributions of dimensionless stresses, corresponding to the linear and compressible models.

Figure 11.8 illustrates the distributions of differences between stresses corresponding for compressible and incompressible models. This data shows that the

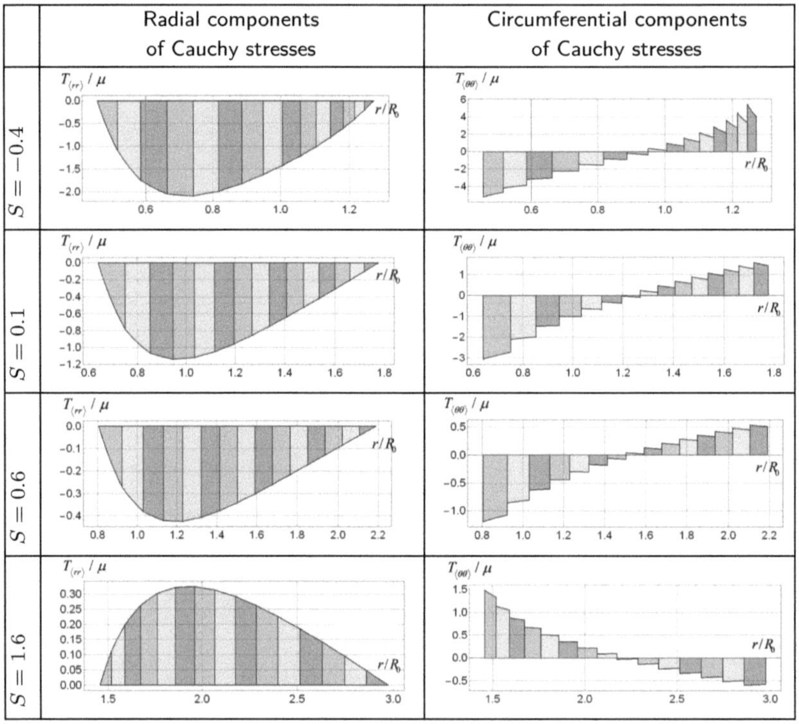


**Fig. 11.9:** Solution for the boundary value problem for a system of hollow balls: dimensionless stress distributions.

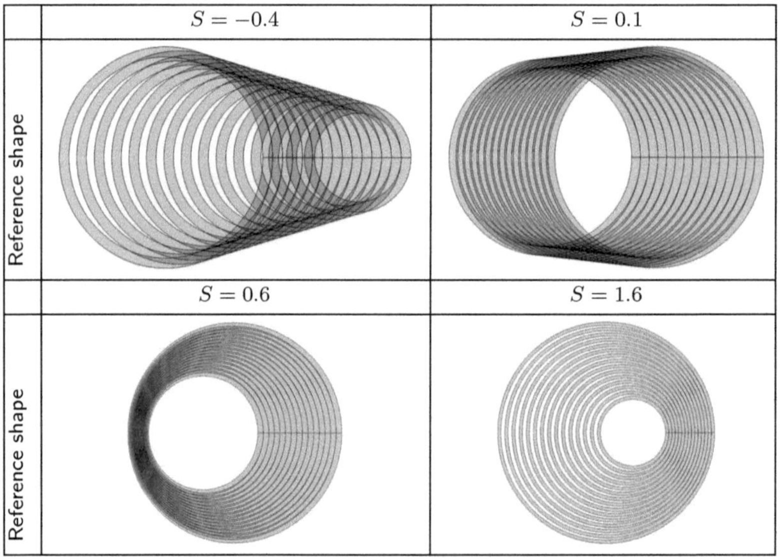


**Fig. 11.10:** Solution for the boundary value problem for a system of balls: reference shapes.

stress state obtained in the framework of the compressible model can differ substantially from the corresponding state calculated within the incompressible model.

Figures 11.9 and 11.10 illustrate the stress distributions for 30 layers and various gaps.

### 11.2.6 Continuous Non-Euclidean Structures

In order to determine the stress–strain state for multilayered structures, one needs to solve the boundary value problem that is determined by the system of non-linear equations (11.43). For a large number of layers, this system enlarges sufficiently, and the numerical solution of it becomes a hard computational problem. At the same time, "shredding" of layers, *i.e.*, increasing their amount and decreasing their thicknesses with a fixed total thickness of the layered structure, with the assumption that parameters $S_n$, $q_n^i$, and $q_n^e$ vary smoothly over the whole structure, lead to distributions for stresses and strains that are similar to the continuous ones. In this regard, it seems convenient to continualize multilayered structures. Such a continualization allows one to replace the discontinuous distributions for Cauchy–Green strain measures by their continuous approximations, for example, by polynomials with least square fitting of discrete distributions.

With the application to the spherical symmetry problem discussed above, one has to fit piecewise functions that determine the diagonal components of the discontinuous field $\boldsymbol{F}_N\colon [r_N^i, r_N^e] \to \mathrm{Lin}(\mathscr{V};\mathscr{V})$:

$$\boldsymbol{F}_N = A_N \boldsymbol{e}_r \otimes \boldsymbol{e}^r + a_N \boldsymbol{e}_\theta \otimes \boldsymbol{e}^\theta + a_N \boldsymbol{e}_\varphi \otimes \boldsymbol{e}^\varphi \,, \tag{11.44}$$

$$A_N(r) = \begin{cases} \dfrac{1}{\{f_{N,0}^{-1}\}'(r)}\,, & r \in [r_{N,0}^i, r_{N,0}^e]\,; \\ \dfrac{1}{\{f_{N,1}^{-1}\}'(r)}\,, & r \in [r_{N,1}^i, r_{N,1}^e]\,; \\ \dots & \dots \\ \dfrac{1}{\{f_{N,N}^{-1}\}'(r)}\,, & r \in [r_{N,N}^i, r_{N,N}^e]\,, \end{cases}$$

$$a_N(r) = \begin{cases} \dfrac{r}{f_{N,0}^{-1}(r)}\,, & r \in [r_{N,0}^i, r_{N,0}^e]\,; \\ \dfrac{r}{f_{N,1}^{-1}(r)}\,, & r \in [r_{N,1}^i, r_{N,1}^e]\,; \\ \dots & \dots \\ \dfrac{r}{f_{N,N}^{-1}(r)}\,, & r \in [r_{N,N}^i, r_{N,N}^e]\,. \end{cases} \tag{11.45}$$

Denote the continuous fittings for $A_N$ and $a_N$ by $A$ and $a$, respectively. These functions determine the continuous field $\boldsymbol{F}$,

$$\boldsymbol{F} = A\boldsymbol{e}_r \otimes \boldsymbol{e}^r + a\boldsymbol{e}_\theta \otimes \boldsymbol{e}^\theta + a\boldsymbol{e}_\varphi \otimes \boldsymbol{e}^\varphi \,. \tag{11.46}$$

**Remark 11.3.** The function (11.44) may be considered as a result of "gluing" deformation gradients of each layer defined in the actual shape. Such "gluing" cannot be defined on the Euclidean reference shape because of its non-connectivity resulting from gaps or overlaps. In contrast to this, the conversion of the approximated field $\boldsymbol{F}$ to the non-Euclidean reference shape is well defined. It is provided by pullback mapping.

For numerical computations, we consider assemblies consisting of 5, 10, 20, 100, and 200 layers. Their total thicknesses in the undistorted shape are equal to $0.3R_0$. Elastic modules were taken equal to $\lambda = 2\mu$, $d = 4\mu$, and $b = \mu/2$, and the stress-free shapes of the layers coincide with the stress-free shape of the substrate, $q^i_\alpha = q^e_\alpha = 0$. The approximating polynomials are given in Tabs. 11.3 and 11.4.

**Tab. 11.3:** Approximating polynomials for $A_N$.

| n | $A$ |
|---|---|
| 5 | $-15.093\,r^4 + 74.889\,r^3 - 134.463\,r^2 + 103.332\,r - 27.638$ |
| 10 | $-9.617\,r^4 + 42.167\,r^3 - 67.697\,r^2 + 46.103\,r - 9.954$ |
| 30 | $-1.267\,r^4 + 4.524\,r^3 - 4.654\,r^2 - 0.411\,r + 2.796$ |
| 50 | $-0.241\,r^4 + 0.058\,r^3 + 2.594\,r^2 - 5.605\,r + 4.180$ |
| 100 | $0.236\,r^4 - 2.008\,r^3 + 5.927\,r^2 - 7.979\,r + 4.808$ |
| 200 | $0.364\,r^4 - 2.561\,r^3 + 6.816\,r^2 - 8.610\,r + 4.973$ |

**Tab. 11.4:** Approximating polynomials for $a_N$.

| n | $a$ |
|---|---|
| 5 | $-1.768\,r^4 - 4.024\,r^3 + 24.203\,r^2 - 28.263\,r + 10.816$ |
| 10 | $4.450\,r^4 - 21.559\,r^3 + 38.210\,r^2 - 28.536\,r + 8.418$ |
| 30 | $0.951\,r^4 - 4.265\,r^3 + 7.101\,r^2 - 4.211\,r + 1.419$ |
| 50 | $0.378\,r^4 - 1.680\,r^3 + 2.774\,r^2 - 1.022\,r + 0.547$ |
| 100 | $0.101\,r^4 - 0.447\,r^3 + 0.735\,r^2 + 0.466\,r + 0.143$ |
| 200 | $0.026\,r^4 - 0.115\,r^3 + 0.189\,r^2 + 0.863\,r + 0.037$ |

Comparisons of discontinuous functions and their approximations for 5, 10, and 30 layers are shown in the Figure 11.11. Note that for a larger number of layers, the graphs of these functions are typographically indistinguishable.

The replacement of piecewise continuous distributions with continuous ones turns the reference shape of the whole layered structure from a non-connected collection of spherical parts into a connected set. The final assembly $\mathfrak{B}_N$ becomes a non-

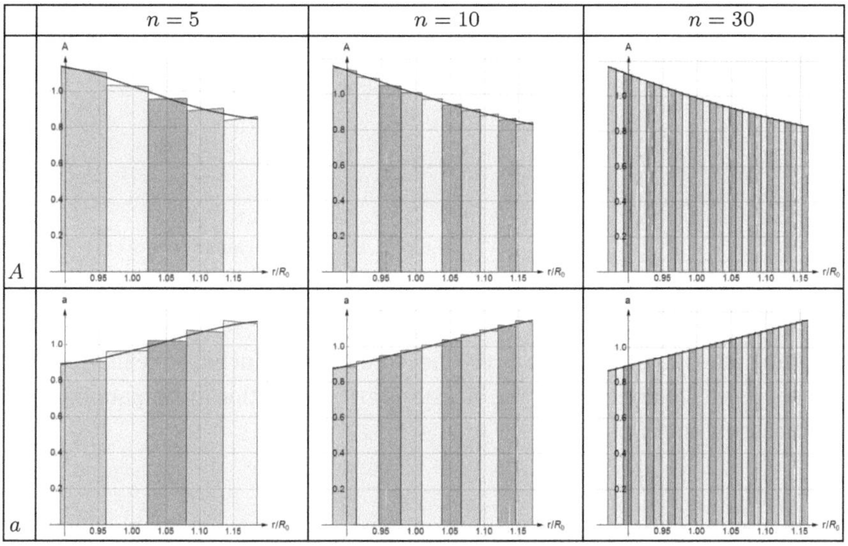


**Fig. 11.11:** Comparison of discrete and continuous distributions.

Euclidean one, $\widetilde{\mathfrak{B}}_N$, with the material connection that can be defined in a number of ways. One of them is the synthesis of the material metric $\mathcal{G}$ (Chapter 8, Section 8.4).

With the assumption of spherical symmetry, the metric $\mathcal{G}$ can be decomposed as follows:

$$\mathcal{G} = A^{-2} dr \otimes dr + r^2 a^{-2} d\theta \otimes d\theta + r^2 \sin^2\theta a^{-2} d\varphi \otimes d\varphi\,, \tag{11.47}$$

On the body $\widetilde{\mathfrak{B}}_N$, this metric defines the structure of Riemannian space with a Levi–Civita connection:

$$\Gamma^k{}_{ij} = \frac{G^{kl}}{2}\left(\partial_i G_{jl} + \partial_j G_{il} - \partial_l G_{ij}\right), \quad \text{where} \quad G_{ij} = g_{sp} K^s{}_i K^p{}_j\,, \quad \text{and} \quad \boldsymbol{K} = \boldsymbol{F}^{-1}\,,$$

which non-zero coefficients are defined by the relations

$$\Gamma^1{}_{11} = -\frac{A'(r)}{A(r)}\,, \quad \Gamma^1{}_{22} = \frac{rA(r)^2(ra'(r) - a(r))}{a(r)^3}\,, \quad \Gamma^1{}_{33} = \frac{rA(r)^2\sin^2\theta(ra'(r) - a(r))}{a(r)^3}\,,$$

$$\Gamma^2{}_{12} = \Gamma^2{}_{21} = \Gamma^3{}_{13} = \Gamma^3{}_{31} = \frac{1}{r} - \frac{a'(r)}{a(r)}\,,$$

$$\Gamma^2{}_{33} = -\sin\theta\cos\theta\,, \quad \Gamma^3{}_{23} = \Gamma^3{}_{32} = \cot\theta\,.$$

The specifics of the so-obtained material connection can be illustrated by the scalar curvature

$$\begin{aligned}\mathrm{Ric}(r) = 2(&-5r^2A(r)^2a'(r)^2 + 2ra(r)A(r)(rA(r)a''(r) + a'(r)(rA'(r) + 3A(r)))\\ &- a(r)^2A(r)(2rA'(r) + A(r)) + a(r)^4)/(r^2a(r)^2)\,.\end{aligned}$$

The distribution of $\mathrm{Ric}(r)$ with $n = 200$ is shown in Figure 11.12.

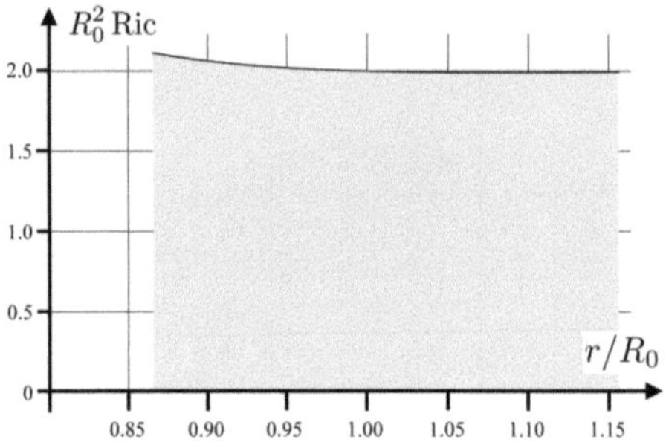


**Fig. 11.12:** Scalar curvature for the case $n = 200$.

Non-zero values can be treated as the measure for non-Euclidean features of the reference shape and, consequently, the measures for deformation incompatibility and residual stresses sources.

### 11.2.7 Small Perturbations of the Self-Stressed Shape

By imposing small perturbations on the self-stressed reference shape one can obtain a response like that of inhomogeneous and anisotropic media. This kind of inhomogeneity (and anisotropy) was discussed in Chapter 8. Here, we illustrate this phenomenon by representations of the perturbed fields introduced in Chapter 3, Section 3.11.4.

According to Chapter 3, formula (3.57), the perturbed Cauchy stresses can be expressed by components, represented in matrix form as follows:

$$[\boldsymbol{T}_\varepsilon] = [\boldsymbol{T}_0] + [\boldsymbol{R}].[\tilde{\boldsymbol{u}}'_\varepsilon] + [\boldsymbol{L}_\gamma].[\widetilde{\boldsymbol{E}}_\varepsilon]\,,$$

where $[\boldsymbol{T}_0]$ is a matrix that corresponds to the initial stress tensor, $[\boldsymbol{R}]$ is the matrix of the fourth-order tensor that represents rotations, and $[\boldsymbol{L}_\gamma]$ is the elasticity tensor matrix:

$$[\boldsymbol{T}_0] = \operatorname{diag}[\mathscr{T}, \mathscr{P}, \mathscr{P}]\,, \quad [\boldsymbol{R}] = \operatorname{diag}[0,0,0,0,\mathscr{R},\mathscr{R}]\,,$$

$$[\boldsymbol{L}_\gamma] = \begin{pmatrix} \mathscr{A} & \mathscr{C} & \mathscr{C} & 0 & 0 & 0 \\ \mathscr{H} & \mathscr{B} & \mathscr{D} & 0 & 0 & 0 \\ \mathscr{H} & \mathscr{D} & \mathscr{B} & 0 & 0 & 0 \\ 0 & 0 & 0 & \mathscr{F} & 0 & 0 \\ 0 & 0 & 0 & 0 & \mathscr{G} & 0 \\ 0 & 0 & 0 & 0 & 0 & \mathscr{G} \end{pmatrix}.$$

Applying formulae from Chapter 3, one can find exact expressions for the tangent elastic moduli $\mathscr{A}$, $\mathscr{B}$, $\mathscr{C}$, $\mathscr{D}$, $\mathscr{F}$, $\mathscr{G}$, $\mathscr{H}$, $\mathscr{P}$, $\mathscr{R}$, and $\mathscr{T}$, related with functions $A$ and $a$:

$$\mathscr{A} = (\zeta(A^2 - 7a^2) + \xi(5a^2 + A^2) + 3\gamma)/9\,,$$
$$\mathscr{B} = (\zeta(a^2 - 13A^2) + \xi(7a^2 + 5A^2) + 6\gamma)/18\,,$$

$$
\begin{aligned}
\mathscr{C} &= (\zeta(4A^2 - a^2) + \xi(2a^2 - 5A^2) + 3\gamma)/9\,,\\
\mathscr{D} &= (\zeta(5A^2 + a^2) + \xi(5A^2 - 11a^2) + 6\gamma)/18\,,\\
\mathscr{F} &= (a^2\xi - A^2\zeta)/2\,,\\
\mathscr{G} &= (a^2 + A^2)(\xi - \zeta)/4\,,\\
\mathscr{H} &= (\zeta(7a^2 - A^2) + \xi(-5a^2 - A^2) + 6\gamma)/18\,,\\
\mathscr{R} &= (A^2 - a^2)(\xi - \zeta)/4\,,\\
\mathscr{T} &= ((1 - a^{-2}A^{-1})\gamma + (a^2 - A^2)(\zeta - \xi))/3\,,\\
\mathscr{P} &= (2(1 - a^{-2}A^{-1})\gamma - (a^2 - A^2)(\zeta - \xi))/6\,,
\end{aligned}
$$

where

$$
a = \frac{f(R)}{R}\,, \quad A = f'(R)\,, \quad \gamma = 3a^2 Ad\,, \quad \zeta = \frac{C_2}{a^3A^2}\left(\frac{a}{A}\right)^{1/3}\,, \quad \xi = \frac{C_1}{a^3A^2}\left(\frac{A}{a}\right)^{1/3}.
$$

The plots in Figure 11.13 illustrate the dependencies of tangent elastic modules on the initial distortions determined by values of $A$ and $a$. The intersection of dashed lines corresponds to a non-stressed reference shape, and, consequently, to a conventional elastic solid

Applying the same formulae, one can obtain the distributions of tangent moduli in solids with Euclidean non-connected Euclidean shapes. The comparison of tangent moduli distributions in such layered structures with their continuous counterparts is shown in Figures 11.14 and 11.15 and allows one to estimate the convergency numerically. Solid lines correspond to distributions in solids obtained by the continuation procedure. The computation is performed in dimensionless variables. The initial radius is taken to be 1, while the total thickness of the unstressed layers is equal to 0.6.

## 11.3 Bending of Rectangular Blocks

### 11.3.1 Deformations of a Single Block

The reference positions of material points $\mathfrak{X} \in \mathfrak{B}$ of a body $\mathfrak{B}$ are given by scalar fields $\mathfrak{X} \mapsto X^k(\mathfrak{X})$, $k = 1, 2, 3$, which determine the reference configuration:

$$
\varkappa_R : \mathfrak{B} \to \mathscr{E}\,, \quad \varkappa_R(\mathfrak{X}) = o + X^k(\mathfrak{X})\boldsymbol{i}_k\,.
$$

Here, $X^k = X^k(\mathfrak{X})$ are simply the Cartesian coordinates of the reference position of the material point $\mathfrak{X}$.

Consider the universal deformation that belongs to Family 1:

$$
\boldsymbol{x}(\boldsymbol{X}) = \sqrt{2\frac{\boldsymbol{i}_1 \cdot \boldsymbol{X} + a}{b}}\,(\boldsymbol{i}\,\cos(b\,\boldsymbol{i}_2 \cdot \boldsymbol{X}) + \boldsymbol{i}_2 \sin(b\,\boldsymbol{i}_2 \cdot \boldsymbol{X})) + \boldsymbol{i}_3 \otimes \boldsymbol{i}_3 \cdot \boldsymbol{X} \tag{11.48}
$$

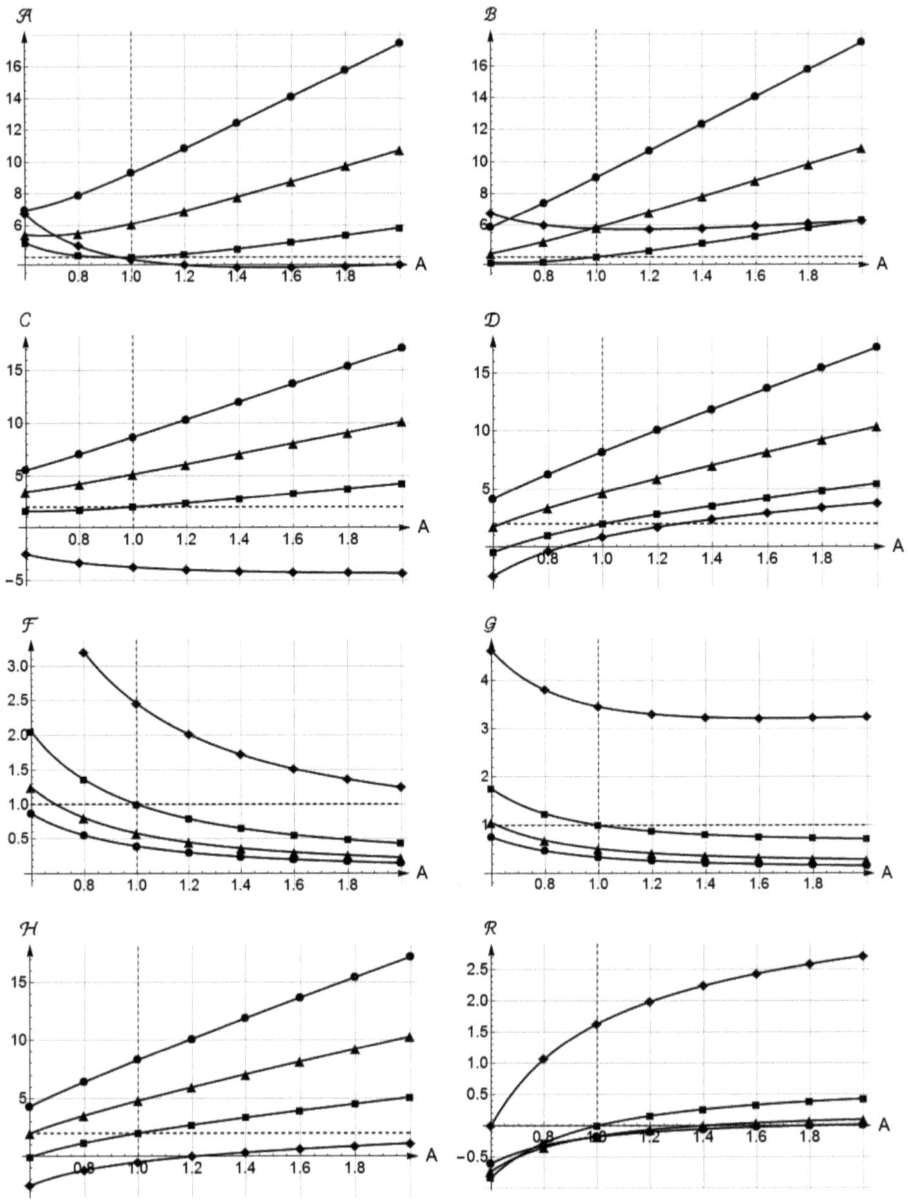


**Fig. 11.13:** Graphs of the dependencies of tangent elastic modules on initial distortions.

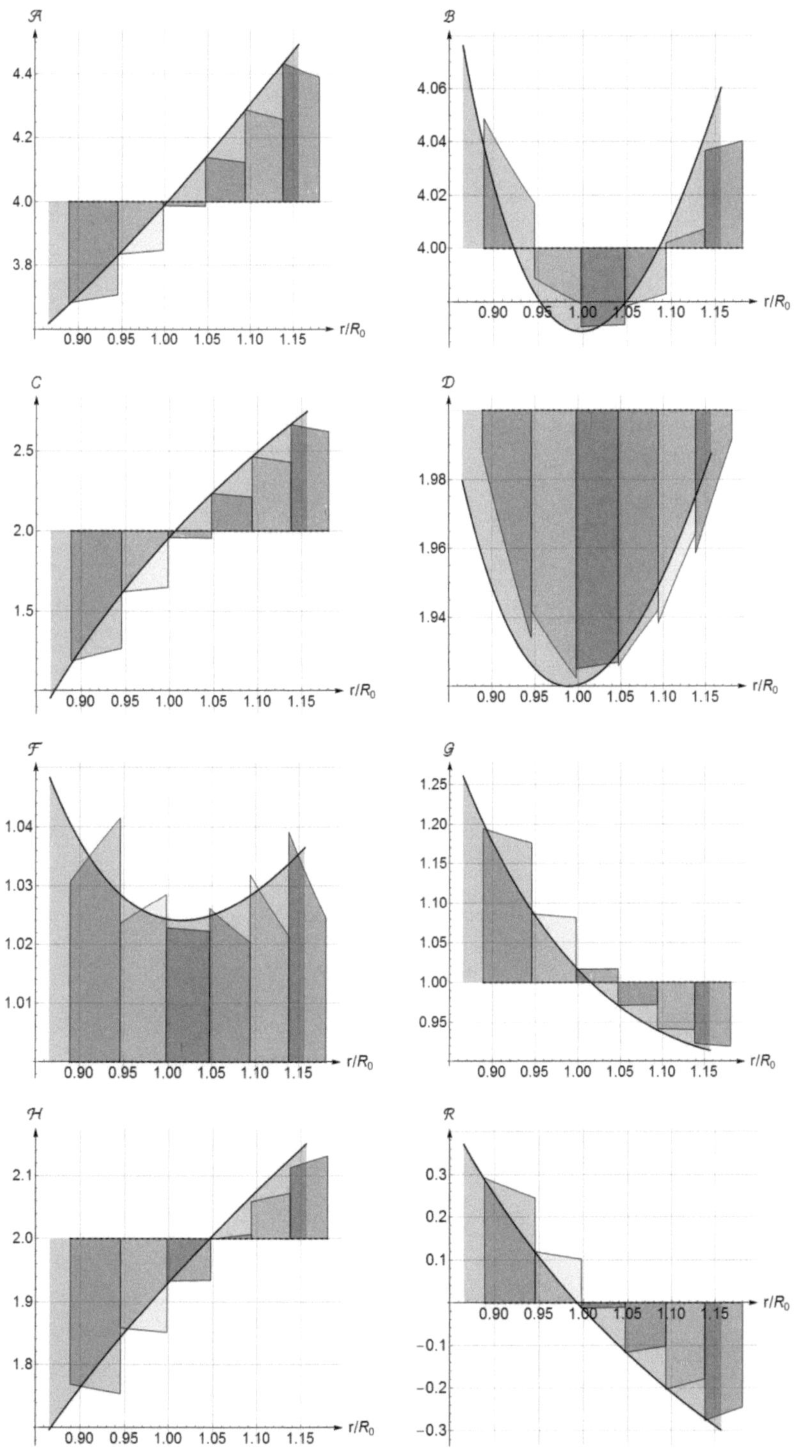


**Fig. 11.14:** Distributions of tangent elastic modules in a self-stressed layered structure for the assembly of six layers and in their continualizations.

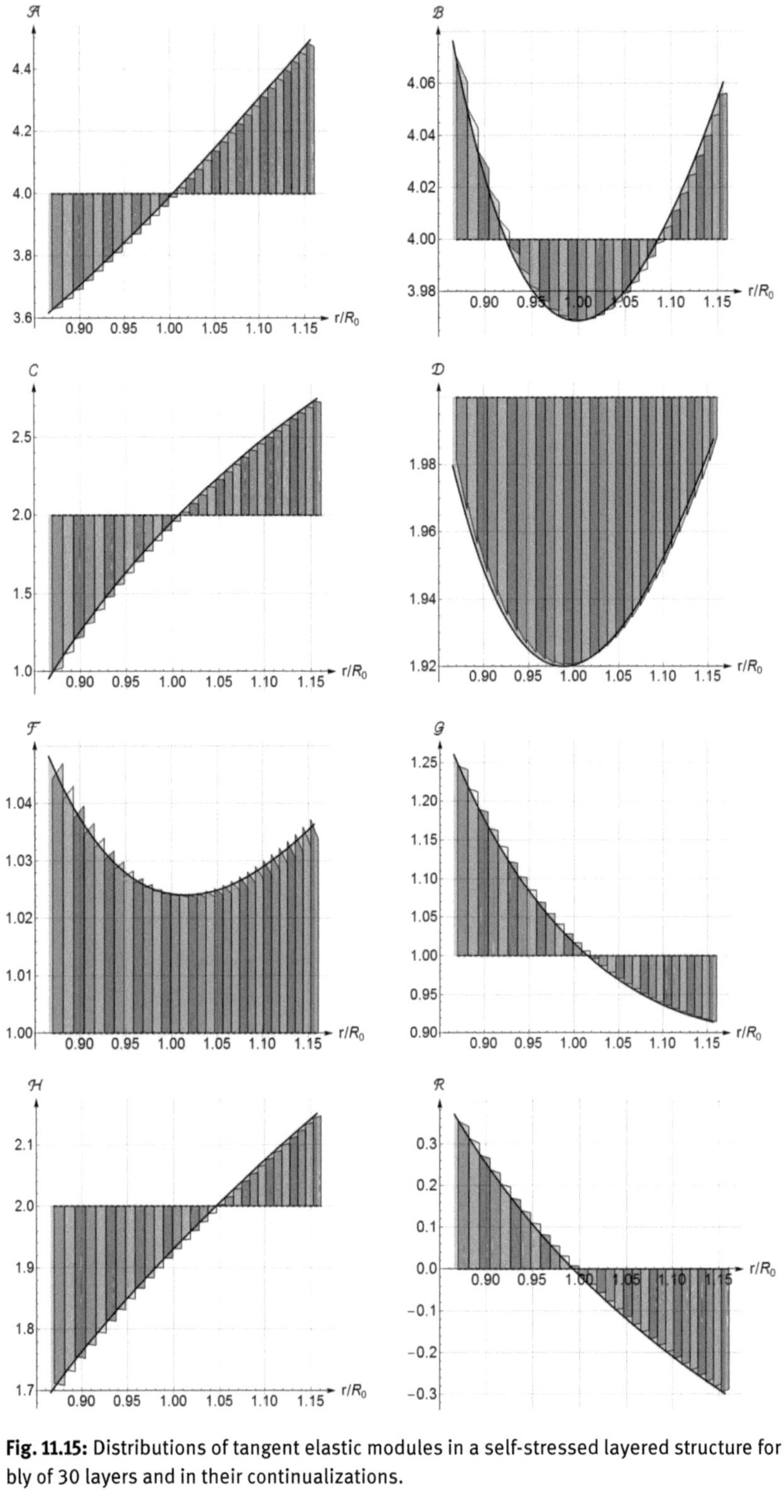


**Fig. 11.15:** Distributions of tangent elastic modules in a self-stressed layered structure for the assembly of 30 layers and in their continualizations.

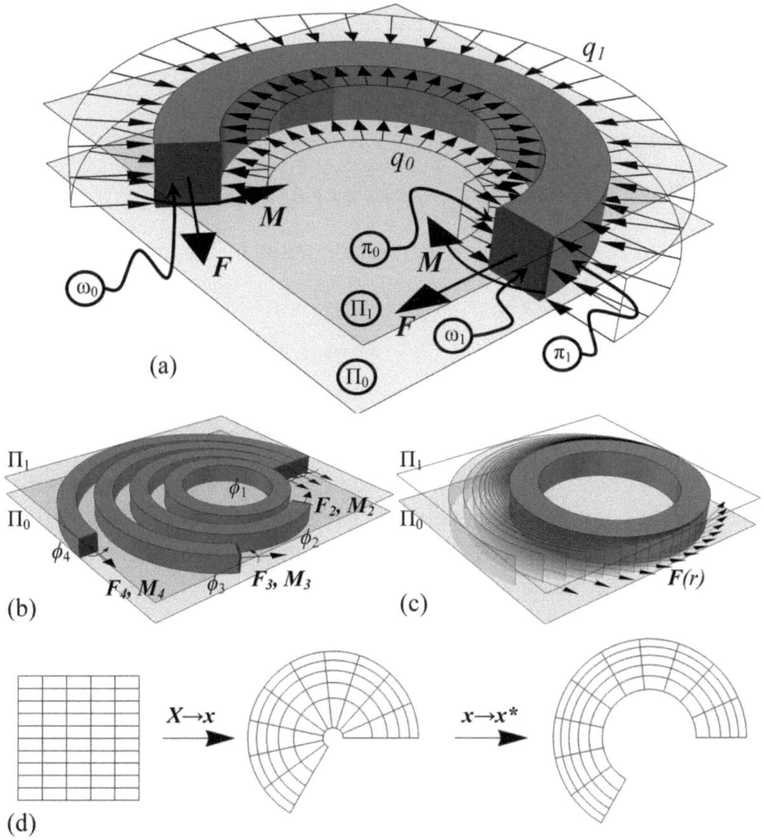


**Fig. 11.16:** Combining the slab assembly.

or, in component form,

$$x = \sqrt{2\frac{X+a}{b}}\cos bY\,, \qquad y = \sqrt{2\frac{X+a}{b}}\sin bY\,, \qquad z = Z\,.$$

Here, $a$ and $b$ are parameters of the deformation (11.48). In the local basis $(\boldsymbol{e}_r, \boldsymbol{e}_\theta, \boldsymbol{e}_z)$ of cylindrical coordinates, the position field has the form

$$\boldsymbol{x} = r\boldsymbol{e}_r + z\boldsymbol{e}_z\,, \qquad r = \sqrt{2\frac{X+a}{b}}\,, \qquad \theta = bY\,.$$

**Remark 11.4.** One can regard (11.48) as a composition bending of a parallelepiped (block) into a sector of a circular cylinder (cylindrical panel), Figure 11.16.

The deformation gradient $\boldsymbol{F}$ of (11.48) can be represented in the following form:

$$\boldsymbol{F} = \frac{1}{b\,r}\,\boldsymbol{e}_r \otimes \boldsymbol{i}^1 + b\boldsymbol{e}_\theta \otimes \boldsymbol{i}^2 + \boldsymbol{e}_z \otimes \boldsymbol{i}^3 \tag{11.49}$$

The corresponding left Cauchy–Green tensor can be represented by

$$\boldsymbol{B} = \boldsymbol{F} \cdot \boldsymbol{F}^T = \frac{1}{b^2 r^2}\, \boldsymbol{e}_r \otimes \boldsymbol{e}_r + b^2 \boldsymbol{e}_\theta \otimes \boldsymbol{e}_\theta + \boldsymbol{e}_z \otimes \boldsymbol{e}_z\,, \tag{11.50}$$

while its inverse is determined as

$$\boldsymbol{B}^{-1} = b^2 r^2\, \boldsymbol{e}_r \otimes \boldsymbol{e}_r + \frac{1}{b^2 r^4}\, \boldsymbol{e}_\theta \otimes \boldsymbol{e}_\theta + \boldsymbol{e}_z \otimes \boldsymbol{e}_z\,.$$

The invariants $I_1$, $I_2$, and $I_3$ of the tensor $\boldsymbol{B}$ can be calculated by the formulae

$$I_1 = \boldsymbol{I} \cdot \boldsymbol{B} = 1 + \frac{1}{b^2 r^2} + b^2 r^2\,, \quad I_2 = \boldsymbol{I} \cdot \boldsymbol{B}^{-1} = I_1\,, \quad I_3 = 1 \tag{11.51}$$

### 11.3.2 Stresses

Suppose that the material of a block is isotropic, homogeneous, and incompressible. Moreover, we assume that, apart from the layered structure, the block is undistorted. Taking into account the expressions (11.50) and (11.51), one can obtain the following decomposition for the Cauchy stress tensor field:

$$\begin{aligned}
\boldsymbol{T} &= T^{rr} \boldsymbol{e}_r \otimes \boldsymbol{e}_r + T^{\theta\theta} \boldsymbol{e}_\theta \otimes \boldsymbol{e}_\theta + T^{zz} \boldsymbol{e}_z \otimes \boldsymbol{e}_z\,,\\
T^{rr} &= -p + \frac{\beth_1}{b^2 r^2} + \beth_{-1} b^2 r^2\,, \quad T^{\theta\theta} = -\frac{p}{r^2} + \beth_1 b^2 + \frac{\beth_{-1}}{b^2 r^4}\,,\\
T^{rr} &= -p + \beth_1 + \beth_{-1}\,.
\end{aligned} \tag{11.52}$$

The hydrostatic component $p$ can be determined from the equilibrium equation $\nabla \cdot \boldsymbol{T} = \boldsymbol{0}$. This vector equation in cylindrical coordinates can be reduced to three equations:

$$\frac{T^{rr}}{r} - r T^{\theta\theta} + \frac{\partial T^{rr}}{\partial r} = 0\,, \quad \frac{T^{\theta\theta}}{\partial \theta} = 0\,, \quad \frac{T^{zz}}{\partial z} = 0\,.$$

It follows from the last two equations that $\boldsymbol{T} = \boldsymbol{T}(r)$, hence the solution of the first equation can be written as

$$T^{rr} = \int \left( r T^{\theta\theta} - \frac{T^{rr}}{r} \right) dr\,.$$

With (11.48), the integrand becomes

$$T^{rr} = \int (\beth_1 - \beth_{-1}) \left( b^2 r - \frac{1}{b^2 r^3} \right) dr\,.$$

Note that the above expressions for the stress remain true if the parameter $b$ is assumed to be a function of the variable $r$. However, if $b = \text{const.}$, then with equality $I_1 = I_2$, which follows from (11.51), one can obtain

$$\begin{aligned}
\frac{dW}{dr} &= \frac{\partial W}{\partial I_1} \frac{\partial I_1}{\partial r} + \frac{\partial W}{\partial I_2} \frac{\partial I_2}{\partial r} = 2 \left( \frac{\partial W}{\partial I_1} + \frac{\partial W}{\partial I_2} \right) \left( b^2 r - \frac{1}{b^2 r^3} \right)\\
&= (\beth_1 - \beth_{-1}) \left( b^2 r - \frac{1}{b^2 r^3} \right).
\end{aligned}$$

Therefore, $T^{rr} = W + p_0$, where the constant of integration $p_0$ defines the invariable part of the hydrostatic component to be defined from boundary conditions. Consequently,

$$p = \beth_1 \frac{1}{b^2 r^2} + \beth_{-1} b^2 r^2 - W - p_0 \,.$$

With this relation taken into account, expressions (11.52) can be converted to

$$T^{rr} = W + p_0 \,, \quad T^{\theta\theta} = \frac{W + p_0}{r^2} + (\beth_1 - \beth_{-1}) \left( b^2 - \frac{1}{b^2 r^4} \right) ,$$
$$T^{zz} = W + p_0 + \beth_1 \left( 1 - \frac{1}{b^2 r^2} \right) + \beth_{-1} \left( 1 - b^2 r^2 \right) .$$

In order to perform calculations numerically, specify the potential $W(I_1, I_2)$ to be the two-constant Mooney–Rivlin potential

$$W = C_1(I_1 - 3) + C_2(I_2 - 3) = \frac{\mu}{2} \frac{(b^2 r^2 - 1)^2}{b^2 r^2} \,.$$

The corresponding scalar reactions can be expressed as

$$\beth_1 = 2\frac{\partial W}{\partial I_1} = 2C_1 = \frac{1}{2}\mu(1 + \beta) \,, \quad \beth_{-1} = -2\frac{\partial W}{\partial I_2} = -2C_2 = -\frac{1}{2}\mu(1 - \beta) \,.$$

Then,

$$T^{rr} = W + p_0 \,, \quad T^{\theta\theta} = \frac{\mu}{2} \left( 3b^2 - \frac{1}{b^2 r^4} \right) + \frac{p_0}{r^2} \,,$$
$$T^{zz} = p_0 + \mu b^2 r^2 - \frac{\mu\beta(b^2 r^2 - 1)^2}{2 b^2 r^2} \,.$$

### 11.3.3 Forces on Boundary Surfaces

The universal deformations (11.48) are controllable, *i.e.*, they can be achieved by specific surface traction applied to the boundary surfaces only. Since deformation (11.48) contains constants $a$, and $b$, and the stress field is determined up to a constant $p_0$, it is possible to satisfy the prescribed boundary conditions pointwise on the cylindrical surfaces $\pi_0$, $\pi_1$ (Figure 11.16 (a)) and integrally on flat surfaces, in such a way that resultant force and moment on these surfaces will be equal to the prescribed values:

$$\boldsymbol{T} \cdot \boldsymbol{e}^r \Big|_{r=r_0} = q_0 \, \boldsymbol{e}_r \,, \quad \boldsymbol{T} \cdot \boldsymbol{e}^r \Big|_{r=r_1} = q_1 \, \boldsymbol{e}_r \,.$$

That is,

$$W \Big|_{r=\sqrt{2a/b}} = q_0 - p_0 \,, \quad W \Big|_{r=\sqrt{2(a+\Delta)/b}} = q_1 - p_0 \qquad (11.53)$$

If the scalar reaction coefficients $\beth_1$, and $\beth_{-1}$ explicitly depend on the principal invariants $I_1$ and $I_2$, then system (11.53) can be solved numerically only. However, if the reaction coefficients are constant, then it becomes possible to find the solution of (11.53) in closed form. So, (11.53) becomes

$$\begin{aligned} 2ba + \frac{1}{2ba} &= \frac{2}{\mu}(q_0 - p_0)\,, \\ 2b(a+\Delta) + \frac{1}{2b(a+\Delta)} &= \frac{2}{\mu}(q_0 - p_0) + 2\delta\,, \end{aligned} \tag{11.54}$$

where $\delta = (q_1 - q_0)/\mu$. Eliminating $p_0$, one can reduce system (11.54) to the following quadratic equation with respect to $a$:

$$a^2 + \Delta a - \frac{\Delta}{4b(\Delta b - \delta)} = 0\,. \tag{11.55}$$

If $\delta < b\Delta$, then the discriminant of (11.55) is positive. Out of the two roots of (11.55) we have to choose the positive one, because the negative root corresponds to the case where part of the body penetrates itself:

$$a = \frac{\Delta}{2}\left(\sqrt{1 + \frac{1}{\Delta b(\Delta b - \delta)}} - 1\right). \tag{11.56}$$

A new value of $p_0$ can be calculated from the formula

$$p_0 = q_0 - \mu\left(ba + \frac{1}{4ba}\right). \tag{11.57}$$

With the found parameters, $a$ and $p_0$, one can determine the radii of the cylindrical boundary surfaces in the actual configuration:

$$\begin{aligned} r_0 &= \frac{1}{b\sqrt{D_0}}\,, \quad r_1 = \frac{\sqrt{D_0}}{b\sqrt{1 - \dfrac{\delta}{\Delta b}}}\,, \\ D_0 &= \Delta b - \delta + \sqrt{(\Delta b - \delta)^2 + 1 - \frac{\delta}{\Delta b}}\,. \end{aligned} \tag{11.58}$$

If the cylindrical boundary surfaces are traction free, *i.e.*, $\delta = q_0 = 0$, then the formulae for $a$ and $p_0$, become simpler:

$$a = \frac{1}{2}\left(\sqrt{\Delta^2 + b^{-2}} - \Delta\right), \quad p_0 = -\mu\sqrt{b^2\Delta^2 + 1}\,.$$

In this case, the radius of the inner boundary cylindrical surface $\pi_0$ is equal to

$$r_0 = \left(b\sqrt{\Delta b + \sqrt{\Delta^2 b^2 + 1}}\right)^{-1} = b^{-1}\sqrt{\sqrt{\Delta^2 b^2 + 1} - \Delta b}\,.$$

and the radius of the surface $\pi_1$ is determined by the expression

$$r_1 = b^{-1}\sqrt{\Delta b + \sqrt{\Delta^2 b^2 + 1}}\,, \quad r_0 r_1 = b^{-2}$$

The distribution of the traction on the surfaces $\omega_0$ and $\omega_1$ (Figure 11.16 (a)) cannot be constant and depends on the explicit expression for universal deformation (11.48), because on any surface $\theta$ = const.,

$$\boldsymbol{e}_{\langle\theta\rangle} \cdot \boldsymbol{T} = \boldsymbol{e}_{\langle\theta\rangle}\left(p_0 + \frac{\mu}{2}\left(3b^2r^2 - \frac{1}{b^2r^2}\right)\right), \quad \boldsymbol{e}_{\langle\theta\rangle} = r\boldsymbol{e}_\theta$$

At the same time, they also depend on the parameter $b$ and on the choice of which one can met the boundary conditions on $\omega_0$ and $\omega_1$ integrally. To do this, calculate the resultant force:

$$\begin{aligned}\boldsymbol{F} &= \boldsymbol{e}_{\langle\theta\rangle}\int_0^h\int_{r_0}^{r_1}\left(p_0 + \frac{\mu}{2}\left(3b^2r^2 - \frac{1}{b^2r^2}\right)\right) dr\, dz \\ &= \boldsymbol{e}_{\langle\theta\rangle} h\left(p_0\,(r_1 - r_0) + \frac{\mu}{2}\left(b^2\left(r_1^3 - r_0^3\right) + \frac{1}{b^2r_1} - \frac{1}{b^2r_0}\right)\right),\end{aligned} \tag{11.59}$$

and the resultant moment

$$\begin{aligned}\boldsymbol{M} &= \boldsymbol{e}_z\int_0^h\int_{r_0}^{r_1}\left(p_0 + \frac{\mu}{2}\left(3b^2r^2 - \frac{1}{b^2r^2}\right)\right)\left(r - \frac{r_0 + r_1}{2}\right) dr\, dz \\ &= \boldsymbol{e}_z\frac{\mu h}{2}\left(\frac{r_1^4 - r_0^4}{4}b^2 + \frac{r_1^2 - r_0^2}{2}\left(\frac{1}{b^2r_0r_1} - b^2r_0r_1\right) - \frac{1}{b^2}\ln\frac{r_1}{r_0}\right).\end{aligned}$$

Note that if the cylindrical surfaces $\pi_0$ and $\pi_1$ are traction free, *i.e.*, $\delta = q_0 = 0$, then

$$\boldsymbol{F} = \boldsymbol{0}, \quad \boldsymbol{M} = \boldsymbol{e}_z\frac{\mu h}{2b^2}\left(\Delta b\sqrt{\Delta^2b^2 + 1} - \operatorname{arsh}(\Delta b)\right).$$

### 11.3.4 Thin Layers

To model the processes of layerwise accretion, it is necessary to consider the case of thin layers (the case of small $\Delta$). Herewith,

$$\delta = k\Delta + o(\Delta), \quad k = \lim_{\Delta\to 0}\frac{\delta}{\Delta},$$

$$\boldsymbol{F} = \boldsymbol{e}_{\langle\theta\rangle}\mu h\left(\frac{f_0\sqrt{b(b-k)} + k}{b^{3/4}\sqrt[4]{b-k}}\Delta + O(\Delta^2)\right), \quad k = \frac{\delta}{\Delta} + O(\Delta^2), \quad f_0 = \frac{q_0}{\mu},$$

$$r_0 = \frac{1}{b\sqrt[4]{1-k/b}} - \frac{1}{2}\sqrt[4]{1-k/b}\,\Delta + \frac{b}{8}\left(1 - \frac{k}{b}\right)^{3/4}\Delta^2 + O(\Delta^3),$$

$$r_1 = \frac{1}{b\sqrt[4]{1-k/b}} + \frac{1}{2}\sqrt[4]{1-k/b}\,\Delta + \frac{b}{8}\left(1 - \frac{k}{b}\right)^{3/4}\Delta^2 + O(\Delta^3),$$

and, hence,

$$r_0 \simeq r_1 \simeq r = \frac{1}{b\sqrt[4]{1-k/b}} + O(\Delta), \quad \Delta_r = r_1 - r_0 = \sqrt[4]{1-k/b}\,\Delta + O(\Delta^3).$$

Thus, the deformation of a thin annular layer for a fixed value of the parameter $b$ can be controlled in the following two ways: **(i)** prescribe the resultant force $\boldsymbol{F}$ and then calculate the corresponding radius of the base or **(ii)** prescribe the base radius and then calculate the resultant force $\boldsymbol{F}$, which must be applied to the surfaces $\omega_0$ and $\omega_1$ to produce the required deformation of the layer. Accordingly, if the base radii $r$ and $b$ are given, then

$$k = b - \frac{1}{r^4 b^3} + O(\Delta)\,,$$

$$\boldsymbol{F} = \boldsymbol{e}_{\langle\theta\rangle}\left(\frac{\mu h}{b^3 r^3}\left(b^4 r^4 + b^2 r^2 f_0 - 1\right)\Delta + O(\Delta^2)\right).$$

### 11.3.5 Discrete Accretion

Consider finite sets $\Omega$ of bodies. In the unstressed configuration, the shapes of these bodies are parallelepipeds $\iota^k \subset \mathscr{E}$, $k = 1, 2, \ldots, n$. These shapes can be deformed by (11.48) into annual domains $\phi^k \subset \mathscr{E}$, *i.e.*, hollow circular cylinders whose bases lie in the planes $\Pi_0$ and $\Pi_1$ (Figure 11.16 (b)). Assume that the sides of the bases $\Delta^k$ and $h^k$ of the parallelepipeds $\iota^k$ are the same, *i.e.*, $\Delta^k = \Delta$, $h^k = h$, but their thicknesses $H^k$, in general, differ. Moreover, we suppose that the images $\phi^k$ of the actual configurations of elements of $\Omega$ do not intersect pairwise and can be composed with each other without gaps and overlaps, *i.e.*, their union $\bigcup_{k=1}^{n} \phi^k$ constitutes the annular domain.

The following types of accretion will be taken into account:

1. *Free accretion.* Each layer during the accretion is free from stresses.
2. *Accretion with a given reference geometry.* The thicknesses of all layers $H^k$ are given in the reference configuration.
3. *Accretion with a given actual geometry.* The position of the body boundary (surface of growth) is given in the actual state.
4. *Accretion with a given tension.* The resultant forces (or resultant moments of forces) acting on accreted layers in the actual configuration are given.

**Type 1. Free accretion.** The thickness $H^1$ of the first element of $\Omega$, and, hence, the value of the parameter $b^1 = 2\pi/H^1$, are given. The quantities $H^k$, $k = 2, \ldots, n$ are determined by the contact conditions:

$$r_0^k = r_1^{k-1}\,, \quad k = 2, 3, \ldots, n\,, \tag{11.60}$$

where $r_0^k$ is the radius of the inner cylindrical boundary surface $\pi_0^k$ of the domain $\phi^k$, and $r_1^{k-1}$ is the radius of the outer cylindrical boundary surface $\pi_0^{k-1}$ of the domain $\phi^{k-1}$, which can be calculated as

$$r_0^k = \frac{1}{b^k}\sqrt{\sqrt{\Delta^2 (b^k)^2 + 1} - \Delta b^k}\,, \quad r_1^{k-1} = \frac{1}{b^{k-1}}\sqrt{\sqrt{\Delta^2 (b^{k-1})^2 + 1} + \Delta b^{k-1}}\,.$$

Here, $b^k = 2\pi/H_1^k$ and $b^{k-1} = 2\pi/H_1^{k-1}$. After simple rearrangements, one arrives at the system consisting of $(n-1)$ equations for $z^k = 1/b^k$, $k = 2, 3, \ldots, n$:

$$z^k\left(\sqrt{\Delta^2 + (z^k)^2} - \Delta\right) = z^{k-1}\left(\sqrt{\Delta^2 + (z^{k-1})^2} - \Delta\right), \quad k = 2, 3, \ldots, n .$$

This system is a sequence of cubic equations and admits the recursive solution

$$z^k = \frac{1}{3}\left(z^{k-1} + \frac{2(z^{k-1})^2}{\sqrt[3]{D_1^{k-1}}} - \sqrt[3]{D_1^{k-1}}\right),$$

$$D_1^{k-1} = z^{k-1}\left(\sqrt{8(z^{k-1})^4 + \left(D_2^{k-1}\right)^2} - D_2^{k-1}\right),$$

$$D_2^{k-1} = 10(z^{k-1})^2 + 27\Delta\left(\Delta + \sqrt{(z^{k-1})^2 + \Delta^2}\right).$$

Thus, all parameters $b^k = 1/z^k$ and, hence, $H^k = 2\pi z^k$, $k = 1, 2, \ldots, n$, are determined. The corresponding values of the parameter $a^k$ are calculated by relation (11.56) for $\delta = 0$.

**Type 2. Accretion with a given reference geometry.** The layer heights $H^1, H^2, \ldots, H^n$ in the reference state are given, and, hence, $b^1 = 2\pi/H^1, \ldots, b^n = 2\pi/H^n$. One has to find the values $a^k$, $k = 1, 2, \ldots, n$. The contact condition (11.60) yields

$$\delta^1 + \delta^2 + \cdots + \delta^n = \delta ,$$

where $\delta = \left(q_1^n - q_0^1\right)/\mu$, $q_0^1$ is the pressure on the surface $\pi_0^1$, and $q_1^n$ is that on the surface $\pi_1^n$. If these surfaces are free from stresses, then $\delta = 0$. The expressions for the contact surface radii in (11.60) are determined by the relations (11.58). Thus, one obtains a system of $n$ equations for the parameters $\delta^n$:

$$\sum_{k=1}^{n} \delta^k = \delta , \tag{11.61}$$

$$\frac{\Delta b^k - \delta^k + \sqrt{(\Delta b^k - \delta^k)^2 + 1 - \delta^k/(\Delta b^k)}}{(b^k)^2 - b^k\delta^k/\Delta}$$

$$= \frac{\Delta b^{k+1} - \delta^{k+1} + \sqrt{(\Delta b^{k+1} - \delta^{k|!+1})^2 + 1 - \delta^{k+1}/(\Delta b^{k+1})}}{(b^{k+1})^2 - b^{k+1}\delta^{k+1}/\Delta}, \quad k = 1, \ldots, n-1 .$$

**Type 3. Accretion with a given actual geometry.** The values of the parameters $b^k$ and $a^k$ are unknown. The radius $R$ of the outer boundary surface of the extreme layer is given. Consider a sequence of assemblies of one, two, and more elements of the set $\Omega$, *i.e.*, of the domain

$$\hat{\phi}^1 = \phi^1 , \quad \hat{\phi}^2 = \phi^1 \cup \phi^1 , \quad \ldots \quad \hat{\phi}^m = \phi^1 \cup \phi^1 \cup \cdots \cup \phi^m .$$

Each $m$-assembly is associated with a system of equations for $m+1$ unknown variables $\{b^m, \delta^1, \delta^2, \dots, \delta^m\}$:

$$\sum_{k=1}^{n} \delta^n = \delta \,, \tag{11.62}$$

$$\frac{\Delta b^k - \delta^k + \sqrt{(\Delta b^k - \delta^k)^2 + 1 - \delta^k/(\Delta b^k)}}{(b^k)^2 - b^k\delta^k/\Delta}$$

$$= \frac{\Delta b^{k+1} - \delta^{k+1} + \sqrt{(\Delta b^{k+1} - \delta^{k+1})^2 + 1 - \delta^{k+1}/(\Delta b^{k+1})}}{(b^{k+1})^2 - b^{k+1}\delta^{k+1}/\Delta} \,, \quad k = 1, \dots, m-1 \,,$$

$$R^2 = \frac{\Delta b^m - \delta^m + \sqrt{(\Delta b^m - \delta^m)^2 + 1 - \delta^m/(\Delta b^m)}}{(b^m)^2 - b^m\delta^m/\Delta} \,.$$

It should be noted that the system for the first assembly $\hat{\phi}^1$, which consists of two equations, the first of which is trivial, determines the width of the first block that can be deformed into an annular domain with a given radius of the outer cylindrical boundary. The systems for succeeding assemblies can be solved recursively.

**Type 4. Accretion with a given tension.** The values of the parameters $b^k$ and $a^k$ are unknown. The resulting force $\boldsymbol{F}$ (or the bending moment $M$) acting on the accreting layer is given. Consider the sequence of assemblies $\hat{\phi}^m$. Each $m$-assembly is associated with a system of equations for $m+1$ unknowns $\{b^m, \delta^1, \delta^2, \dots \delta^m\}$:

$$\sum_{k=1}^{n} \delta^n = \delta \,, \tag{11.63}$$

$$\frac{\Delta b^k - \delta^k + \sqrt{(\Delta b^k - \delta^k)^2 + 1 - \delta^k/(\Delta b^k)}}{(b^k)^2 - b^k\delta^k/\Delta}$$

$$= \frac{\Delta b^{k+1} - \delta^{k+1} + \sqrt{(\Delta b^{k+1} - \delta^{k+1})^2 + 1 - \delta^{k+1}/(\Delta b^{k+1})}}{(b^{k+1})^2 - b^{k+1}\delta^{k+1}/\Delta} \,, \quad k = 1, \dots, m-1 \,,$$

$$F = \mathscr{F}(b^m, \delta^m)$$

where the symbol $\mathscr{F}$ denotes the expression resulting from substituting (11.58), (11.57), and (11.56) into (11.59) (the formula obtained is rather cumbersome and, therefore, is not written out here). The algorithm for solving the sequence of systems is the same as above.

### 11.3.6 Continuous Accretion

The parameter $b$ is replaced by a continuous function $b = b(r)$, which will be called the distortion parameter. This allows us to describe continuous accretion realized as a continuous process of adding material surfaces (Figure 11.16 (c)). The deformation gradient $\boldsymbol{F}$ is replaced by smooth tensor field

$$\mathscr{F} = \frac{1}{b(r)r}\,\boldsymbol{e}_r \otimes \boldsymbol{i}^1 + b(r)\boldsymbol{e}_\theta \otimes \boldsymbol{i}^2 + \boldsymbol{e}_z \otimes \boldsymbol{i}^3 \,. \tag{11.64}$$

The corresponding left Cauchy–Green strain is given by the decomposition

$$\mathscr{B} = \frac{1}{b(r)^2 r^2}\,\boldsymbol{e}_r \otimes \boldsymbol{e}_r + b(r)^2 \boldsymbol{e}_\theta \otimes \boldsymbol{e}_\theta + \boldsymbol{e}_z \otimes \boldsymbol{e}_z \,. \tag{11.65}$$

Its invariants have forms similar to (11.51)

$$I_1 = I_2 = 1 + \frac{1}{b(r)^2 r^2} + b(r)^2 r^2 \,, \quad I_3 = 1$$

However, now the invariants are functionally dependent on the variable $r$ in a more complicated way. The stress tensor components can be calculated as

$$T^{rr} = -p + \frac{\mu(1+\beta)}{2b(r)^2 r^2} - \frac{\mu(1-\beta)}{2} b(r)^2 r^2 \,,$$

$$T^{\theta\theta} = -\frac{p}{r^2} + \frac{\mu(1+\beta)}{2} b(r)^2 - \frac{\mu(1-\beta)}{2b(r)^2 r^4} \,, \quad T^{rr} = -p + \mu\beta \,,$$

where the hydrostatic component $p$ is determined by the integral:

$$p(r) = \mu \left[ \frac{1+\beta}{2b^2(r) r^2} - \frac{1-\beta}{2} b(r)^2 r^2 - \int_{r_0}^{r} \left( b^2(\zeta)\zeta - \frac{1}{b^2(\zeta)\zeta^3} \right) d\zeta \right] + p_0 \,.$$

Analyze the types of continuous accretion that are similar to the types of discrete accretion considered above.

**Type 1. Free accretion.** The distortion parameter is $b = 1/r$. Then,

$$\mathscr{B} = \mathscr{I} \,, \quad I_1 = I_2 = 3 \,, \quad \mathscr{P} = p_0 \mathscr{I} \,, \quad p_0 = \text{const} \,.$$

As a result of such an accretion, one obtains a body in the undistorted configuration, and if $p_0 = 0$, then this configuration is free from stresses.

**Type 2. Accretion with a given distortion.** The distortion parameter is given as a function $b = b(r)$, and it is assumed that it is not equal to $1/r$. The volume of the accreted material $V$ and the respective pressures $q_0$ and $q_1$ on the cylindrical boundary surfaces $\pi_0$ and $\pi_1$ are given. The unknowns are the inner radius $r_0$, the outer radius $r_1$, and the constant hydrostatic pressure component $p_0$. They can be found by solving the system of equations

$$T^{rr}\Big|_{r=r_0} = q_0 \,, \quad T^{rr}\Big|_{r=r_1} = q_1 \,, \quad \pi h \left( r_1^2 - r_0^2 \right) = V \,, \tag{11.66}$$

which can be rewritten as the non-linear integral equation

$$p_0 = q_0 \,, \quad \mu \int_{r_0}^{r_1} \left( b^2(\zeta)\zeta - \frac{1}{b^2(\zeta)\zeta^3} \right) d\zeta + p_0 = q_1 \,, \quad \pi h \left( r_1^2 - r_0^2 \right) = V \,.$$

Discretizing this equation, one arrives at a system of equations similar to (11.61).

**Type 3. Accretion with a given position of the growth surface.** The volume of the accreted material is given as a function of a time-like parameter $\tau \in [0, t[$, *i.e.*, $V = V(\tau)$. The position of the growth boundary $r_1(\tau)$ is given. Just as in the preceding case, the pressures $q_0(\tau)$ and $q_1(\tau)$ on the boundary surfaces $\pi_0$ and $\pi_1$ are given. One has to determine the distortion function $b(r)$, and together with the functions $r_0(\tau)$ and $p_0(\tau)$. They can be found from the system (11.66), which is reduced to the non-linear integral equation

$$\mu \int\limits_{\sqrt{r_1(\tau)^2 - V/(\pi h)}}^{r_1(\tau)} \left( b^2(\zeta)\zeta - \frac{1}{b^2(\zeta)\zeta^3} \right) d\zeta + q_0(\tau) = q_1(\tau) .$$

Discretizing this equation, one obtains a system of equations similar to (11.62).

**Type 4. Accretion with a given tension.** Just as above, the volume of the accreted material $V = V(\tau)$, loadings $q_0(\tau)$ and $q_1(\tau)$ on the boundary surfaces $\pi_0$ and $\pi_1$ and the surface stresses (tension) $g(\tau)$ on the growth boundary are given. It is required to find the distortion function $b(r)$ and the functions $r_0(\tau)$, $r_1(\tau)$ and $p_0(\tau)$. They are determined by the equations

$$T^{rr}\Big|_{r=r_0} = q_0(\tau) , \qquad T^{rr}\Big|_{r=r_1} = q_1(\tau) ,$$
$$r^2 T^{\theta\theta}\Big|_{r=r_1} = g(\tau) , \qquad \pi h \left( r_1^2 - r_0^2 \right) = V(\tau) ,$$

which can now be reduced to the system of algebraic and non-linear integral equations

$$q_1 + \mu \left( b^2 r^2 - \frac{1}{b^2 r^2} \right) = g(\tau) ,$$
$$\mu \int\limits_{\sqrt{r_1(\tau)^2 - V/(\pi h)}}^{r_1(\tau)} \left( b^2(\zeta)\zeta - \frac{1}{b^2(\zeta)\zeta^3} \right) d\zeta + q_0(\tau) = q_1(\tau) .$$

Discretizing these equations, one obtains a system of equations similar to (11.63).

The results of the numerical computations for discrete and continuous accretion are shown in Figure 11.17. The graphs corresponding to continuous accretion are labeled with small stars. The thickness of all layers $h$ is assumed to be equal to 1.

Figure 11.17 (a) displays the distribution of the dimensionless stress $T^{rr}/\mu$ for free discrete accretion (type 1). The thickness of all layers except the first one is assumed to be equal to $\Delta = 0.1$ cm, while the thickness of the first is assumed to be $H^1 = \pi/5$ cm. The number of layers $N = 10$.

Figure 11.17 (b) shows the graphs of $T^{rr}(r)/\mu$ in the case of accretion with a given reference configuration (type 2) for various numbers of layers. The layer thicknesses $\Delta$ were chosen so that the total thickness of the whole layered structure is equal to $0.1$ cm in the reference configuration. The outer boundaries of the accreted body were assumed to be traction free ($\delta = 0$). Note that when the number of layers increases, the distribution of $T^{rr}/\mu$ approaches the corresponding distribution for a continuously accreted body.

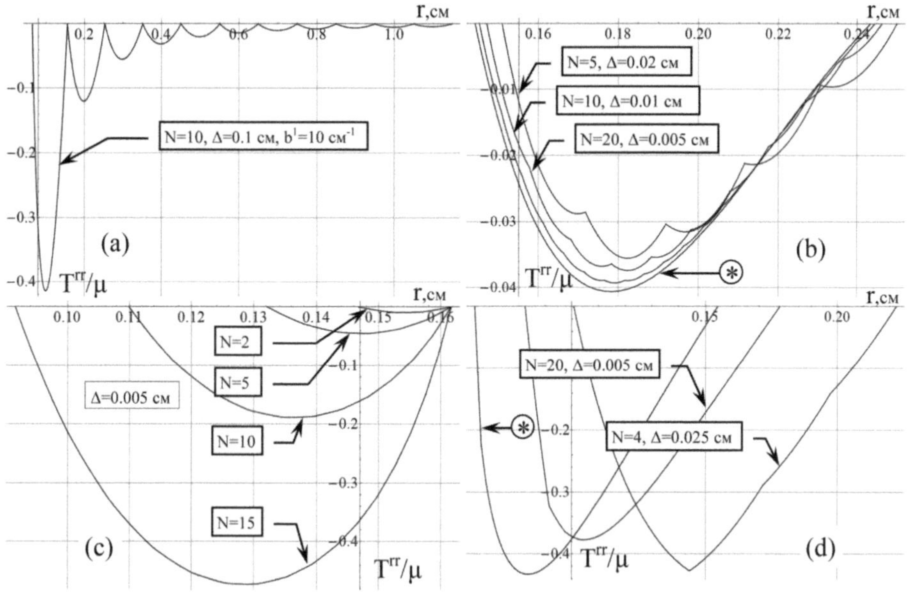


**Fig. 11.17:** Distributions for dimensionless stresses.

Figure 11.17 (c) illustrates the stress distribution in the case of discrete accretion with fixed position of the growth surface (type 3), and in Figure 11.17 (d), for discrete and continuous accretion with given tension (type 4).

# 12 Algebraic Structures

## 12.1 Preliminary Comments on the Use of Sets

In this book, we systematically use sets, the axiomatic treatment of which can be found in [142]. Below, we give remarks on notations related with sets.

The notation $x \in X$ means that the object $x$ is an element of the set $X$. If a set $Y$ is a subset of a set $X$, then we write $Y \subset X$.

The empty set is denoted by $\emptyset$. The set defined by a list of objects is denoted by listing the objects between braces "{" and "}". If $X$ is a set and $P(x)$ is a sentence that is either true of false whenever $x$ belongs to $X$, then the unique set of all $x \in X$ for which $P(x)$ is true is denoted by $\{x \in X \mid P(x)\}$.

The union of two sets $X$ and $Y$ is denoted by $X \cup Y$ and their intersection by $X \cap Y$. The symbol $X \setminus Y$ denotes the difference between $X$ and $Y$.

The symbols $\mathbb{N}$, $\mathbb{Z}$, $\mathbb{Q}$, and $\mathbb{R}$ denote, respectively, the set of natural numbers, the set of integers, the set of rational numbers, and the set of real numbers. If $r$ and $s$ are integers such that $r \leq s$, then the set $\{r, \ldots, s\} \subset \mathbb{Z}$ is defined by

$$\{r, \ldots, s\} := \{k \in \mathbb{Z} \mid r \leq k \leq s\} .$$

If $a < b$ are real numbers, then the open interval is denoted by $]a, b[$ and the closed interval by $[a, b]$. Half-open intervals are denoted, respectively, by $[a, b[$ and $]a, b]$.

## 12.2 Ordered Pairs. Cartesian Products. Relations

**Ordered pairs.** Let $x$ and $y$ be sets. One can construct a new set $(x, y)$, namely, an *ordered pair of x and y*, as follows [142]:

$$(x, y) := \{\{x\}, \{x, y\}\} .$$

In contrast to the set consisting of $x$ and $y$, *i.e.*, $\{x, y\}$, the ordered pair $(x, y)$ contains the information about the order of its elements. Indeed, for arbitrary $x, y, z, w$,

$$(x, y) = (z, w) \quad \text{if and only if} \quad (x = z) \wedge (y = w) .$$

The procedure of ordered pair construction can be repeated recursively. For sets $x, y, z$, define the ordered triple by $(x, y, z) := ((x, y), z)$. In general, an ordered $k$-tuple is defined by

$$(x_1, x_2, \ldots, x_{k-1}, x_k) := ((x_1, x_2, \ldots, x_{k-1}), x_k) .$$

**The Cartesian product.** Let $X$ and $Y$ be sets. The *Cartesian product* $X \times Y$ is the set of all possible ordered pairs $(x, y)$, where $x \in X$ and $y \in Y$:

$$X \times Y := \{(x, y) \mid x \in X \wedge y \in Y\} .$$

https://doi.org/10.1515/9783110563214-012

Recursively, suppose that $X$, $Y$, and $Z$ are sets. Then, $X \times Y \times Z := (X \times Y) \times Z$. In general, for an arbitrary collection of sets $X_1, \ldots, X_k$, one can define their Cartesian product as follows:

$$\begin{aligned} X_1 \times \cdots \times X_k &:= (X_1 \times \cdots \times X_{k-1}) \times X_k \\ &= \{(x_1, x_2, \ldots, x_{k-1}, x_k) \mid x_i \in X_i\,, i = 1, \ldots, k\}\,. \end{aligned}$$

**Relations.** Any subset $R \subset X \times Y$ of a Cartesian product is called a *(binary) relation.* If $R$ is a relation, then we occasionally write $xRy$ instead of $(x, y) \in R$. Now, let $P \subset X \times Y$ and $R \subset Y \times Z$. The relation

$$P^{-1} := \{(y, x) \in Y \times X \mid xRy\}$$

is called the *inverse of P*. The relation

$$R \circ P := \{(x, z) \in X \times Z \mid \exists y \in Y: \quad (xPy \wedge yRz)\}$$

is called the *composition of P and R*.

**Equivalence relation.** Let $X$ be a set, while a relation $\sim$ is defined on $X$ ($\sim \subset X \times X$). The relation $\sim$ is an *equivalence relation on X*, if it satisfies the following axioms:
(i) $\forall x \in X: \;\; x \sim x$;
(ii) $\forall x, y \in X: \;\; (x \sim y) \Rightarrow (y \sim x)$;
(iii) $\forall x, y, z \in X: \;\; (x \sim y) \wedge (y \sim z) \Rightarrow (x \sim z)$.

If $x \in X$, then the set

$$[x] := \{y \in X \mid y \sim x\}$$

is called the *equivalence class of x with respect to* $\sim$. Equivalence classes satisfy the following property: for every $x, y \in X$, either $[x] = [y]$ or $[x] \cap [y] = \emptyset$. The collection of all equivalence classes, *i.e.*,

$$X/\sim := \{[x] \mid x \in X\}\,,$$

is called the *quotient set of X with respect to* $\sim$.

## 12.3 Functions

Let $X$ and $Y$ be sets, while $R \subset X \times Y$ is a relation. This relation is called a *function* [142] if

$$\forall x \quad \forall y_1 \quad \forall y_2: \quad (xRy_1 \wedge xRy_2 \Rightarrow y_1 = y_2)\,.$$

Suppose that $R \subset X \times Y$ is a function. If the set of all first elements of pairs that form $R$ coincides with $X$, *i.e.*,

$$X = \{x \in X \mid \exists y \in Y: \;\; xRy\}\,,$$

then the ordered triple[1] $f = (X, R, Y)$ is called a *map (function, transformation) from X to Y* and is denoted by $f : X \to Y$. Let $Y^X$ denote the set of all mappings from $X$ to $Y$. If $f \in Y^X$, $f = (X, R, Y)$ and $x \in X$, then the definition of $f$ implies that there exists a unique $y \in Y$ such that $xRy$. This element $y$ is called the *value of f at x* and is denoted by $f(x)$. Thus, the equality $y = f(x)$ is no more than an alternative notation for the relation $xRy$. If $f_1, f_2 \in Y^X$ are mappings, then

$$(f_1 = f_2) \Leftrightarrow \forall x \in X: \quad f_1(x) = f_2(x) .$$

For $f : X \to Y$, $A \subset X$, and $B \subset Y$ define the sets

$$\begin{aligned} f(A) &:= \{y \in Y \mid \exists x \in X: \ \ y = f(x)\} = \{f(x) \mid x \in X\} , \\ f^{-1}(B) &= \{x \in X \mid f(x) \in B\} . \end{aligned}$$

These sets, $f(A)$ and $f^{-1}(B)$, are called the *image of A* and the *pre-image of B* with respect to $f$.

One has the following classification of functions. A function $f : X \to Y$ is called:

- a *surjective* (onto) if $f(X) = Y$;
- an *injective* if
$$\forall x, y \in X: \quad (f(x) = f(y)) \Rightarrow (x = y) ;$$
- a *bijection* if $f$ is both surjective and injective.

Suppose that $f = (X, R, Y)$ is a bijective function from $X$ into $Y$. Then, $R^{-1}$ is a function and $f^{-1} := (Y, R^{-1}, X)$ is a bijective function from $Y$ to $X$. The function $f^{-1} : Y \to X$ is called the *inverse of f*.

Let $f \in Y^X$, $f = (X, R, Y)$ and $g \in Z^Y$, $g = (Y, S, Z)$. Then, $S \circ R$ is a function and $g \circ f := (X, S \circ R, Z)$ is a function from $X$ to $Z$. Moreover,

$$\forall x \in X: (g \circ f)(x) = g(f(x)) .$$

The function $g \circ f : X \to Z$ is called the *composition of functions f and g*. The composition is an associative operation: if $h \in Y^X$, $g \in Z^Y$ and $f \in W^Z$, then $(f \circ g) \circ h = f \circ (g \circ h)$. In this regard, parentheses can be omitted, and the sequential composition can be written as $f \circ g \circ h$.

For any set $X$, one can define the *identity map of X*, $\mathrm{Id}_X : X \to X$, as

$$\forall x \in X: \mathrm{Id}_X(x) := x . \tag{12.1}$$

The identity map satisfies the equality $f \circ \mathrm{Id}_X = \mathrm{Id}_Y \circ f$ for each $f \in Y^X$. Moreover, let $f \in Y^X$ and $g \in X^Y$. Then, $g = f^{-1}$ *iff* both $g \circ f = \mathrm{Id}_X$ and $f \circ g = \mathrm{Id}_Y$ hold.

---

1 Giving the definition of a mapping from $X$ to $Y$ as an ordered triple we follow [143].

Suppose that $A \subset X$ and $f \in Y^X$. Then, the map $f|_A : A \to Y$, which is defined by

$$\forall x \in X : f|_A(x) := f(x) ,$$

is called the *restriction of f to A*.

## 12.4 Some Algebraic Structures

### 12.4.1 Groups

**Definition.** A *group* is a pair $(G, \top)$, where $G$ is a non-empty set and

$$\top : G \times G \to G , \quad \top(x, y) = x \top y ,$$

is a mapping (binary operation) that satisfies the following axioms:
$(G_1)$ **Associativity:** $\forall x, y, z \in G : (x \top y) \top z = x \top (y \top z)$;
$(G_2)$ **Existence of identity:** $\exists e \in G \quad \forall x \in G : x \top e = e \top x = x$;
$(G_3)$ **Existence of inverses:** $\forall x \in G \quad \exists x^{-1} \in G : x \top x^{-1} = x^{-1} \top x = e$.

It can be proved that the identity $e$ is a unique element, and for all $x \in G$, the declared element $x^{-1}$ is unique [144]. A group $(G, \top)$ is called *commutative* if

$$\forall x, y \in G : x \top y = y \top x .$$

If such property holds for $(G, \top)$, then the latter is called the *commutative (abelian) group*.

**Homomorphism.** Let $(G_1, \top_1)$ and $(G_2, \top_2)$ be groups. A map $h : G_1 \to G_2$, which preserves group operation, *i.e.*,

$$\forall g_1, g_2 \in G_1 : h(g_1 \top_1 g_2) = h(g_1) \top_2 h(g_2) ,$$

is called a *homomorphism*.

**Isomorphism.** A bijective homomorphism is called an *isomorphism*.

### 12.4.2 Ring

A ring is a triple $(R, \top, \bot)$, where $R$ is a non-empty set and

$$\top : R \times R \to R , \quad \top(x, y) = x \top y ,$$
$$\bot : R \times R \to R , \quad \bot(x, y) = x \bot y ,$$

are operations that satisfy the following axioms:

$R_1$) $(R, \top)$ is abelian group;
$R_2$) $\forall x, y, z \in R\colon (x \perp y) \perp z = x \perp (y \perp z)$;
$R_3$) the following distributivity relations hold:

$$\forall x, y, z \in R\colon (x \top y) \perp z = x \perp z \top y \perp z\,,$$
$$\forall x, y, z \in R\colon z \perp (x \top y) = z \perp x \top z \perp y\,.$$

A ring $(R, \top, \perp)$ is called *commutative* if

$$\forall x, y \in R\colon x \perp y = y \perp x\,.$$

A ring $(R, \top, \perp)$ is a *ring with identity* if

$$\exists e \in G\, \forall x \in G\colon x \perp e = e \perp x = x\,.$$

The operations $\top$ and $\perp$ are often denoted by $+$ and $\cdot$ (juxtaposition).

### 12.4.3 Module

Let $(R, \top, \perp)$ be a commutative ring with the identity $e$. Then, one can define an $R$-*module* [42] as a non-empty set $\mathfrak{M}$ endowed with the binary operations

$$+\colon \mathfrak{M} \times \mathfrak{M} \to \mathfrak{M}\,, \quad (x, y) \mapsto x + y\,,$$
$$\cdot\colon R \times \mathfrak{M} \to \mathfrak{M}\,, \quad (r, y) \mapsto r \cdot y = ry\,,$$

such that
(i) $(\mathfrak{M}, +)$ is an abelian group;
(ii) for all $r, q \in R$ and for all $x, y \in \mathfrak{M}$ the following equalities hold:

$$r(x + y) = rx + ry\,, \quad (r \top q)x = rx + ry\,,$$
$$(r \perp q)x = r(qx)\,, \quad ex = x\,.$$

## 12.5 Linear Spaces and Mappings

### 12.5.1 Vector Space over $\mathbb{R}$

**Definition.** Recall that the set of real numbers, equipped with the arithmetic operations $+$ and $\cdot$, constitutes a field. A *vector (linear) space* $\mathscr{V}$ *over* $\mathbb{R}$ is a module over the field $\mathbb{R}$. Elements of $\mathbb{R}$ are referred to as *scalars*, and elements of $\mathscr{V}$ are referred to as *vectors*. By the definition, the set $\mathscr{V}$ is endowed with two operations: *addition*,

$$+\colon \mathscr{V} \times \mathscr{V} \to \mathscr{V}\,, \quad +(u, v) = u + v\,,$$

and *scalar multiplication*,

$$\cdot\colon \mathbb{R} \times \mathscr{V} \to \mathscr{V}\,, \quad \cdot(\lambda, v) = \lambda \cdot u\,.$$

Occasionally, the dot can be omitted, and the product $\lambda \cdot u$ is written by juxtaposition, $\lambda u$. By virtue of the associativity of +, one can construct linear combinations as follows:

$$\lambda_1 u_1 + \cdots + \lambda_k u_k\,, \quad \text{where} \quad \lambda_i \in \mathbb{R} \text{ and } u_i \in \mathscr{V}, i = 1, \ldots, k\,.$$

**Subspace.** Let $\mathscr{V}$ be a vector space. A subset $\mathscr{U} \subset \mathscr{V}$, with the restrictions of + and · operations to it, is called a *subspace of* $\mathscr{V}$ if it is closed with respect to these restrictions, *i.e.*,

$$\forall u, v \in \mathscr{U} : u + v \in \mathscr{U}\,; \qquad \forall \lambda \in \mathbb{R} \quad u \in \mathscr{U} : \lambda u \in \mathscr{U}\,.$$

**External direct sum.** Let $\mathscr{V}_1, \ldots, \mathscr{V}_k$ be vector spaces. The Cartesian product $\mathscr{V}_1 \times \cdots \times \mathscr{V}_k$ can be endowed with a vector space structure by the following componentwise operations:

$$(x_1, \ldots, x_k) + (y_1, \ldots, y_k) := (x_1 + y_1, \ldots, x_k + y_k)\,,$$
$$\lambda(x_1, \ldots, x_k) := (\lambda x_1, \ldots, \lambda x_k)\,,$$

where $x_i, y_i \in \mathscr{V}_i, i = 1, \ldots, k$ and $\lambda \in \mathbb{R}$. Determined in such a way, the vector space is called the *external direct sum of* $\mathscr{V}_1, \ldots, \mathscr{V}_k$ and is denoted by $\mathscr{V}_1 \oplus \cdots \oplus \mathscr{V}_k$, or, $\bigoplus_{i=1}^k \mathscr{V}_i$.

**Internal direct sum.** Let $\mathscr{V}$ be a vector space and $\mathscr{U}_1, \ldots, \mathscr{U}_k$ be its subspaces. Suppose that for each vector $v \in \mathscr{V}$, there exists the unique collection of vectors $u_i \in \mathscr{U}_i$, $i = 1, \ldots, k$, such that

$$v = u_1 + \cdots + u_k\,.$$

In this case, the vector space $\mathscr{V}$ is called the *(internal) direct sum of* $\mathscr{U}_1, \ldots, \mathscr{U}_k$, i.e.,

$$\mathscr{V} = \mathscr{U}_1 \oplus \cdots \oplus \mathscr{U}_k\,, \quad \text{or}\,, \quad \mathscr{V} = \bigoplus_{i=1}^{k} \mathscr{U}_i\,.$$

**Basis and dimension.** Let $\mathscr{V}$ be a vector space. A set $\{v_1, \ldots, v_k\} \subset \mathscr{V}$ of vectors is called *linearly dependent* if there exists a linear combination $\lambda_1 u_1 + \cdots + \lambda_k u_k = 0$, for some $\lambda_i \in \mathbb{R}$, $i = 1, \ldots, k$, such that $\sum_{i=1}^k \lambda_i^2 \neq 0$. If a set $\{v_1, \ldots, v_k\} \subset \mathscr{V}$ is not linearly dependent, then it is called *linearly independent.*

Let $\mathscr{V}$ be a vector space. This space is called *finite dimensional* if it is either zero space $\{0\}$ or there exists a number $n \in \mathbb{N}$, such that

(i) one can find a linearly independent set $\{v_1, \ldots, v_n\} \subset \mathscr{V}$;

(ii) any set $\{v_1, \ldots, v_m\} \subset \mathscr{V}$ with $m > n$ is linearly dependent.

The number $n$ is called the *dimension of vector space* $\mathscr{V}$ and is denoted by $\dim \mathscr{V}$. By convention, $\dim\{0\} := 0$. If for any $n \in \mathbb{N}$, one can find $n+1$ linearly independent sets of vectors, then $\mathscr{V}$ is called *infinite dimensional.*

Let $\mathscr{V}$ be a vector space. An ordered $n$-tuple $(e_1, \ldots, e_n)$ of vectors from $\mathscr{V}$ is called a *basis for* $\mathscr{V}$ if the set $\{e_1, \ldots, e_n\} \subset \mathscr{V}$ is linearly independent, and for any vector $v \in \mathscr{V}$, there exist real numbers $\lambda_i \in \mathbb{R}$, $i = 1, \ldots, n$, such that

$$v = \lambda_1 e_1 + \cdots + \lambda_n e_n \,.$$

From the linear independence property, it follows that such numbers $\lambda_i$ are unique. For a vector $v \in \mathscr{V}$, the corresponding numbers $\lambda_i$ are called *components* and are denoted by $v^i$. Using the Einstein summation convention, one can write $v = v^i e_i$. The number $n$ of vectors that constitute a basis does not depend on the basis and is equal to the dimension of $\mathscr{V}$: $n = \dim \mathscr{V}$. Any linearly independent set of $n = \dim \mathscr{V}$ vectors forms a basis for $\mathscr{V}$ [42].

### 12.5.2 Linear and $k$-Linear Mappings

**Vector space of linear mappings.** Linear mappings between vector spaces preserve the linear relations between arguments. More precisely, the definition of linear mappings can be formulated as follows. Let $\mathscr{V}$, $\mathscr{W}$ be vector spaces. A mapping $\mathscr{L}: \mathscr{V} \to \mathscr{W}$ is called a *linear mapping (transformation)* if
$L_1$) (**additivity**) $\forall u, v \in \mathscr{V}$: $\mathscr{L}(u + v) = \mathscr{L}(u) + \mathscr{L}(v)$;
$L_2$) (**homogeneity**) $\forall u \in \mathscr{V}\ \forall \lambda \in \mathbb{R}$: $\mathscr{L}(\lambda u) = \lambda \mathscr{L}(u)$.

The set of all linear mappings from $\mathscr{V}$ to $\mathscr{W}$ can be endowed with a vector space structure. To this end, define the linear mappings on $\mathscr{V}$ by the relations

$$(\mathscr{L} + \mathscr{M}): \mathscr{V} \to \mathscr{W}\,, \quad (\mathscr{L} + \mathscr{M})(u) := \mathscr{L}(u) + \mathscr{M}(u)\,,$$
$$(\lambda \mathscr{L}): \mathscr{V} \to \mathscr{W}\,, \quad (\lambda \mathscr{L})(u) := \lambda \mathscr{L}(u)\,,$$

where $\mathscr{L}, \mathscr{M} \in \mathscr{W}^{\mathscr{V}}$ are linear mappings, and $\lambda \in \mathbb{R}$ is a scalar. This turns the set of all linear mappings from $\mathscr{V}$ to $\mathscr{W}$ together with the operations defined above into a vector space denoted by $\mathrm{Lin}(\mathscr{V}; \mathscr{W})$. Note that if $\dim \mathscr{V} = n$ and $\dim \mathscr{W} = m$, then $\dim \mathrm{Lin}(\mathscr{V}; \mathscr{W}) = mn$ [42].

If $\mathscr{L} \in \mathrm{Lin}(\mathscr{V}; \mathscr{W})$ is a bijective mapping, then its inverse, $\mathscr{L}^{-1}$, is similarly linear; $\mathscr{L}^{-1} \in \mathrm{Lin}(\mathscr{W}; \mathscr{V})$. In this book, the set of all invertible linear mappings from $\mathscr{V}$ to $\mathscr{W}$ is denoted by $\mathrm{Inv}(\mathscr{V}; \mathscr{W})$.

Let $\mathscr{V}$, $\mathscr{U}$ and $\mathscr{W}$ be vector spaces. If $\mathscr{L} \in \mathrm{Lin}(\mathscr{V}; \mathscr{U})$ and $\mathscr{M} \in \mathrm{Lin}(\mathscr{U}; \mathscr{W})$, then $\mathscr{M} \circ \mathscr{L} \in \mathrm{Lin}(\mathscr{V}; \mathscr{W})$. Often, the composition $\mathscr{M} \circ \mathscr{L}$ of linear mappings is simply denoted by $\mathscr{M}\mathscr{L}$.

**Isomorphism.** Let $\mathscr{V}$ and $\mathscr{W}$ be vector spaces. A bijective mapping $\mathscr{L} \in \mathrm{Lin}(\mathscr{V}; \mathscr{W})$ is called an *isomorphism*. If for $\mathscr{V}$ and $\mathscr{W}$ at least one isomorphism exists, then these vector spaces are called *isomorphic*. In this case, the following designation is used:

$\mathscr{V} \cong \mathscr{W}$. For finite dimensional spaces $\mathscr{V}$ and $\mathscr{W}$, the following necessary and sufficient condition holds: $\mathscr{V} \cong \mathscr{W}$ *iff* $\dim \mathscr{V} = \dim \mathscr{W}$ [42]. In particular, for any $n$-dimensional vector space $\mathscr{V}$, the following isomorphism $\Phi\colon \mathscr{V} \to \mathbb{R}^n$ exists. Indeed, let $(e_i)_{i=1}^n$ be a basis for $\mathscr{V}$. Then define $\Phi$ as

$$\Phi\colon \mathscr{V} \to \mathbb{R}^n\,, \quad \Phi(v) := (v^1, \ldots, v^n)\,, \quad \text{for} \quad v = v^i e_i\,.$$

Such a mapping is an isomorphism, and one has that $\mathscr{V} \cong \mathbb{R}^n$. This isomorphism depends on a particular choice of basis.

Any isomorphism transforms a basis into another basis. Let $\mathscr{V}, \mathscr{W}$ be finite dimensional isomorphic vector spaces. Suppose that $\dim \mathscr{V} = n$. If $\mathscr{L} \in \mathrm{Lin}(\mathscr{V}; \mathscr{W})$ is the corresponding isomorphism, and $(e_i)_{i=1}^n$ is a basis for $\mathscr{V}$, then $(\mathscr{L}(e_i))_{i=1}^n$ is a basis for $\mathscr{W}$. Thus, for any vector $w \in \mathscr{W}$ one has the unique decomposition $w = w^i \mathscr{L}(e_i)$.

**Dual vector space.** Suppose that $\mathscr{V}$ is a vector space. Any linear transformation $\nu \in \mathrm{Lin}(\mathscr{V}; \mathbb{R})$ is called a *linear functional (covector) on* $\mathscr{V}$. The set of all linear functionals on $\mathscr{V}$ is denoted by $\mathscr{V}^*$, *i.e.*, $\mathscr{V}^* := \mathrm{Lin}(\mathscr{V}; \mathbb{R})$. The vector space $\mathscr{V}^*$ is called the *dual space of* $\mathscr{V}$.

The action of a linear functional $\nu \in \mathscr{V}^*$ on a vector $v \in \mathscr{V}$ is also denoted by angular brackets $\langle \nu, v \rangle$. Here,

$$\langle \cdot, \cdot \rangle\colon \mathscr{V}^* \times \mathscr{V} \to \mathbb{R}\,, \quad \langle \nu, v \rangle := \nu(v)\,.$$

Let $\mathscr{V}$ be an $n$-dimensional vector space and $(e_i)_{i=1}^n$ be some basis for $\mathscr{V}$. Consider the $n$-tuple $(e^i)_{i=1}^n$ of covectors $e^i \in \mathscr{V}^*$, $i = 1, \ldots, n$, such that

$$\left\langle e^i, e_j \right\rangle = \delta^i_j\,, \quad i, j = 1, \ldots, n\,.$$

The $n$-tuple $(e^i)_{i=1}^n$ forms a basis for $\mathscr{V}^*$ and is called the *basis for the dual space* $\mathscr{V}^*$, *dual with respect to the basis for* $\mathscr{V}$, or, for short, the *dual (covector) basis to* $(e_i)_{i=1}^n$ [42]. It follows that $\dim \mathscr{V}^* = n$. If $v \in \mathscr{V}$ and $v = v^i e_i$, then

$$v^i = \langle e^i, v \rangle\,, \quad i = 1, \ldots, n\,.$$

These equalities mean that the dual basis represents the machinery that reads off the components of any vector $v$ from $\mathscr{V}$. Consequently, the components for a covector $\nu \in \mathscr{V}^*$ can be expressed as expansion over $(e_i)_{i=1}^n$ as follows ($\nu = \nu_i e^i$):

$$\nu_i = \langle \nu, e_i \rangle\,, \quad i = 1, \ldots, n\,.$$

Let $\mathscr{V}$ be an $n$-dimensional vector space. Since $\dim \mathscr{V}^* = n$, one has that $\mathscr{V} \cong \mathscr{V}^*$. The corresponding isomorphism depends on a particular choice of basis.

**Reflexivity.** The dual to the dual space is denoted by $\mathscr{V}^{**} := (\mathscr{V}^*)^*$. There is a *natural* isomorphism between vector spaces $\mathscr{V}$ and $\mathscr{V}^{**}$. This isomorphism is represented by *canonical map*

$$\varepsilon\colon \mathscr{V} \to \mathscr{V}^{**}\,, \quad \varepsilon(v) = \varepsilon_v\,,$$

where $\varepsilon_v : \mathscr{V}^* \to \mathbb{R}$ is the mapping that acts on a covector $v$ as follows: $\varepsilon_v(v) := v(v)$. By the definition, $\varepsilon_v \in \mathscr{V}^{**}$ and $\varepsilon$ is an isomorphism [42] that does not depend on the choice of basis. In this regard, a vector $v$ is often identified with the object $\varepsilon_v$, and one writes $v(v)$ instead of $\varepsilon_v(v)$. This property is referred to as *reflexivity*. Note that the infinite dimensional spaces may not be reflexible.

***k*-linear mappings.** The generalization of a linear mapping on the functions of several variables is given by the notion of $k$-linear mapping. Let $\mathscr{V}_1, \ldots, \mathscr{V}_k$ and $\mathscr{W}$ be vector spaces and $\mathscr{L} \in \mathscr{W}^{\mathscr{V}_1 \times \cdots \times \mathscr{V}_k}$ be a mapping. Fix $k$ vectors $v_i \in \mathscr{V}_i$, $i = 1, \ldots, k$ and define *partial mappings* as follows:

$$\mathscr{L}(v_1, \ldots, v_{i-1}, \cdot, v_{i+1}, \ldots, v_k) \colon \mathscr{V}_i \to \mathscr{W} ,$$
$$\mathscr{L}(v_1, \ldots, v_{i-1}, \cdot, v_{i+1}, \ldots, v_k) \colon x \mapsto \mathscr{L}(v_1, \ldots, v_{i-1}, x, v_{i+1}, \ldots, v_k) .$$

A mapping $\mathscr{L} \in \mathscr{W}^{\mathscr{V}_1 \times \cdots \times \mathscr{V}_k}$ is called a *k-linear mapping* if for all vectors $v_i \in \mathscr{V}_i$, $i = 1, \ldots, k$, and for any $i = 1, \ldots, k$, the partial mapping $\mathscr{L}(v_1, \ldots, v_{i-1}, \cdot, v_{i+1}, \ldots, v_k) \colon \mathscr{V}_i \to \mathscr{W}$ is linear. The set of all $k$-linear mappings from $\mathscr{V}_1 \times \cdots \times \mathscr{V}_k$ to $\mathscr{W}$ can be endowed with vector space structure in a similar manner as was done for linear mappings. The resulting vector space is denoted by $\mathrm{Lin}_k(\mathscr{V}_1, \ldots, \mathscr{V}_k; \mathscr{W})$.

**Algebra.** An *algebra over* $\mathbb{R}$ is a real vector space $\mathscr{V}$ endowed with a bilinear (two-linear) map $\mathscr{L} \colon \mathscr{V} \times \mathscr{V} \to \mathscr{V}$ [42]. This map is called a product map, and its image on elements is often written as a juxtaposition. That is, the set $\mathscr{V}$ is endowed with three operations:

$$+\colon \mathscr{V} \times \mathscr{V} \to \mathscr{V} , \qquad +(u, v) = u + v ,$$
$$\cdot\colon \mathbb{R} \times \mathscr{V} \to \mathscr{V} , \qquad \cdot(\lambda, v) = \lambda \cdot u ,$$
$$\mathscr{L}\colon \mathscr{V} \times \mathscr{V} \to \mathscr{V} , \qquad \mathscr{L}(u, v) = uv .$$

If $\mathscr{L}$ satisfies the property

$$\forall u, v \in \mathscr{V} \colon uv = vu ,$$

then the algebra $\mathscr{V}$ is called *commutative*. If the property

$$\forall u, v, w \in \mathscr{V} \colon (uv)w = u(vw) ,$$

holds for $\mathscr{L}$, then the algebra $\mathscr{V}$ is called *associative*.

### 12.5.3 Tensor Products of Vector Spaces

The ideas of the abstract tensor product are briefly introduced in the present section. We follow [1, 3, 145]. Apparently, for the first time, such a construction appeared in the treatise of N. Bourbaki [144].

**Quotient vector space.** Let $\mathscr{U}$ be a subspace of a vector space $\mathscr{V}$. Consider the relation $\sim_{\mathscr{U}}$ between vectors from $\mathscr{V}$ [42]:

$$\forall u, v \in \mathscr{V} : (u \sim_{\mathscr{U}} v) \Leftrightarrow (u - v \in \mathscr{U}) .$$

It is easy to show that this relation is an equivalence relation on $\mathscr{V}$. Let $x \in \mathscr{V}$ and $[x]$ be the corresponding equivalence class. It can be shown that $[x] = \{x + u \mid u \in \mathscr{U}\}$. Indeed,

$$\begin{aligned}[x] &= \{v \in \mathscr{V} \mid v \sim_{\mathscr{U}} x\} = \{v \in \mathscr{V} \mid v - x \in \mathscr{U}\} \\ &= \{v \in \mathscr{V} \mid \exists u \in \mathscr{U} : v = x + u\} = \{x + u \mid u \in \mathscr{U}\} .\end{aligned}$$

The quotient set of $\mathscr{V}$ with respect to $\sim_{\mathscr{U}}$ is denoted by $\mathscr{V}/\mathscr{U}$. It can be endowed with a real vector space structure as follows. Let $a \in \mathscr{V}/\mathscr{U}$ be an equivalence class. If $\lambda \in \mathbb{R}$ and $u, v \in a$, then $\lambda u - \lambda v \in \mathscr{U}$ and $\lambda u \sim_{\mathscr{U}} \lambda v$. Taking this into account, one can define $\lambda a$ as

$$\lambda a := [\lambda u] ,$$

for some $u \in a$. Thus, the scalar multiplication

$$\cdot : \mathbb{R} \times \mathscr{V}/\mathscr{U} \to \mathscr{V}/\mathscr{U} , \quad (\lambda, a) \mapsto \lambda a ,$$

is defined. Let $a, b \in \mathscr{V}/\mathscr{U}$ be an equivalence classes. Take representatives $x_1, x_2 \in a$, $y_1, y_2 \in b$. One can obtain that $x_1 + y_1 \sim_{\mathscr{U}} x_2 + y_2$. Thus, the definition

$$a + b := [x + y] ,$$

where $x \in a$ and $y \in b$, is correct. The addition operation is defined as follows:

$$+ : \mathscr{V}/\mathscr{U} \times \mathscr{V}/\mathscr{U} \to \mathscr{V}/\mathscr{U} , \quad (a, b) \mapsto a + b .$$

With addition and scalar multiplication operations, the set $\mathscr{V}/\mathscr{U}$ becomes a vector space. This is called the *quotient space of* $\mathscr{V}$ *modulo* $\mathscr{U}$.

**Formal linear combinations.** The notion of the tensor product is based on the idea of a formal linear combination. Such an idea allows one to define finite sums like $\sum_{i=1}^{m} a_i x_i$, where $a_i$ are real numbers, and $x_i$ are elements of a set $X$.

Suppose that $X$ is a set. A *formal linear combination of elements of* $X$ [3] is a scalar-valued function $f : X \to \mathbb{R}$, such that the set

$$K_f := \{x \in X \mid f(x) \neq 0\} = f^{-1}(\mathbb{R} \setminus \{0\})$$

of all elements $x \in X$ that do not transform to zero by $f$ is a finite set. The set $M(X)$ of all such functions is a real vector space with the typical operations of pointwise addition and scalar multiplication. Its basis is formed by the functions $(\delta_x)_{x \in X}$ from $M(X)$, which can be defined as follows:

$$\delta_x(y) := \begin{cases} 1, & y = x , \\ 0, & y \neq x . \end{cases}$$

Each element $f \in M(X)$ has the unique decomposition $f = \sum_{i=1}^{k} a_i \delta_{x_i}$, where $x_1, \ldots, x_k$ are the elements of $X$, for which $f(x) \neq 0$ and $a_i = f(x_i)$. By the definition of $K_f$, the sum is finite. Typically, the function $\delta_x$ is *identified* with $x$, and the set $X$ is thought of as the subset of $M(X)$. Thus, one writes $f = \sum_{i=1}^{k} a_i x_i$.

**Tensor product.** Suppose that $\mathscr{V}_1$ and $\mathscr{V}_2$ are real vector spaces. The set $\mathscr{V}_1 \times \mathscr{V}_2$ consists of all possible pairs $(v_1, v_2)$ and $v_i \in \mathscr{V}_i$, and can be endowed with a vector space structure by coordinatewise addition and coordinatewise scalar multiplication. Then, the expressions (where $v_i, v_i', v_i'' \in \mathscr{V}_i$, $\lambda \in \mathbb{R}$)

$$\begin{aligned} (v_1' + v_1'', v_2) - (v_1', v_2) - (v_1'', v_2)\,, &\qquad (v_1, v_2' + v_2'') - (v_1, v_2') - (v_1, v_2'')\,, \\ \lambda(v_1, v_2) - (\lambda v_1, v_2)\,, &\qquad \lambda(v_1, v_2) - (v_1, \lambda v_2)\,, \end{aligned} \tag{12.2}$$

are not equal to zero. Consider the space $M(\mathscr{V}_1 \times \mathscr{V}_2)$ of all formal linear combinations. Let $\mathscr{N}$ be its subspace that spans over all possible differences (12.2). The *tensor product* $\mathscr{V}_1 \otimes \mathscr{V}_2$ [144] is the quotient vector space

$$\mathscr{V}_1 \otimes \mathscr{V}_2 := M(\mathscr{V}_1 \times \mathscr{V}_2)/\mathscr{N}\,.$$

The equivalence class of an element $(v_1, v_2)$ is denoted by

$$v_1 \otimes v_2 := [(v_1, v_2)]\,,$$

and is called the *tensor product of* $v_1$ *and* $v_2$ [3]. The tensor product $\otimes$ satisfies

$$\begin{aligned} (v_1' + v_1'') \otimes v_2 = v_1' \otimes v_2 + v_1'' \otimes v_2\,, &\qquad v_1 \otimes (v_2' + v_2'') = v_1 \otimes v_2' + v_1 \otimes v_2''\,, \\ \lambda(v_1 \otimes v_2) = (\lambda v_1) \otimes v_2\,, &\qquad \lambda(v_1 \otimes v_2) = v_1 \otimes (\lambda v_2)\,. \end{aligned}$$

**Tensor product and associativity.** Let $\mathscr{U}, \mathscr{V}, \mathscr{W}$ be three vector spaces. Since $(\mathscr{U} \otimes \mathscr{V}) \otimes \mathscr{W}$ is naturally isomorphic to $\mathscr{U} \otimes (\mathscr{V} \otimes \mathscr{W})$, one can simply write $\mathscr{U} \otimes \mathscr{V} \otimes \mathscr{W}$ for these sets and $u \otimes v \otimes w$ for their elements. Thus, one inductively comes to the products $\mathscr{V}_1 \otimes \cdots \otimes \mathscr{V}_k$ and $v_1 \otimes \cdots \otimes v_k$ for $k$ vector spaces and their elements.

**Basis and dimension of tensor product.** Suppose that $\mathscr{V}_1$ and $\mathscr{V}_2$ are real vector spaces with dimensions $n_1$ and $n_2$. If $(e_j^{(i)})_{j=1}^{n_i}$ are bases for $\mathscr{V}_i$, $i = 1, 2$, then the collection

$$\left(e_{i_1}^{(1)} \otimes e_{i_2}^{(2)}\right)_{1 \le i_1 \le n_1, 1 \le i_2 \le n_2}\,,$$

is the basis for $\mathscr{V}_1 \otimes \mathscr{V}_2$. Thus, $\dim(\mathscr{V}_1 \otimes \mathscr{V}_2) = n_1 n_2$ [3].

**Tensor product and linear mappings.** Suppose that $\mathscr{V}_1, \ldots, \mathscr{V}_k$ are finite-dimensional vector spaces. It can be proven that the following natural isomorphisms exist:

$$\begin{aligned} \mathscr{V}_1 \otimes \cdots \otimes \mathscr{V}_k &\cong \mathrm{Lin}_k\left(\mathscr{V}_1^*, \ldots, \mathscr{V}_k^*; \mathbb{R}\right)\,, \\ \mathscr{V}_1^* \otimes \cdots \otimes \mathscr{V}_k^* &\cong \mathrm{Lin}_k(\mathscr{V}_1, \ldots, \mathscr{V}_k; \mathbb{R})\,. \end{aligned}$$

That is, the tensor product of $k$ vector spaces is naturally isomorphic to the vector space of $k$-linear mappings.

### 12.5.4 Vectors and Linear Mappings in Euclidean Space

**Inner (scalar) product.** The vector space structure does not presuppose the presence of a "measuring tool" that allows one to measure the lengths of vectors and angles between them. Such a tool is given by an additional structure: the Euclidean space structure. This structure is generated by the inner product.

Let $\mathscr{V}$ be a vector space. An *inner (scalar) product on* $\mathscr{V}$ is a mapping $(\cdot) : \mathscr{V} \times \mathscr{V} \to \mathbb{R}$ with the following properties:

(i) *Symmetry.* $\forall u, v \in \mathscr{V} : u \cdot v = v \cdot u$.

(ii) *Linearity in the first coordinate.*

$$\forall v_1, v_2, w \in \mathscr{V}\ \forall \lambda_1, \lambda_2 \in \mathbb{R} : (\lambda_1 v_1 + \lambda_2 v_2) \cdot w = \lambda_1 (v_1 \cdot w) + \lambda_2 (v_2 \cdot w) .$$

(iii) *Positive definiteness.* $\forall v \in \mathscr{V} : v \cdot v \geq 0$. Moreover,

$$\forall v \in \mathscr{V} : (v \cdot v = 0 \Leftrightarrow v = 0) .$$

In this case, $(\mathscr{V}, \cdot)$ is called a *Euclidean vector space*, vectors from $\mathscr{V}$ are denoted by Latin boldface minuscules, and linear mappings between Euclidean vector spaces are denoted by Latin boldface majuscules. That is, $\boldsymbol{v} \in \mathscr{V}$, $\boldsymbol{L} \in \mathrm{Lin}(\mathscr{V}; \mathscr{W})$. Such notations are used in classical works in mechanics [37, 47, 59, 104, 110, 146] and seem to be a tribute to the notation.

**Dual vector basis.** Suppose that $(\mathscr{V}, \cdot)$ is an $n$-dimensional Euclidean vector space. Let $(\boldsymbol{e}_i)_{i=1}^n$ be a basis for $\mathscr{V}$. In conventional mechanics, one uses the *dual vector basis* $(\boldsymbol{e}^i)_{i=1}^n$ [37]. This basis is defined by the relations

$$\boldsymbol{e}^i \cdot \boldsymbol{e}_j = \delta^i_j , \quad i, j = 1, \ldots, n .$$

Denote by $g_{ij} = \boldsymbol{e}_i \cdot \boldsymbol{e}_j$, $i, j = 1, \ldots, n$, the components of the bilinear form that defines the inner product $(\cdot)$. Note that $g_{ij}$ are also called the components of the metric tensor, but we reserve this term for the corresponding field (see Chapter 2, Section 2.2.5). The $n \times n$ matrix $[g_{ij}]$ is symmetric, non-degenerate, and such that $g = \det[g_{ij}] > 0$. The elements of the dual basis $(\boldsymbol{e}^i)_{i=1}^n$ can be decomposed over the basis $(\boldsymbol{e}_i)_{i=1}^n$:

$$\boldsymbol{e}^i = g^{ij} \boldsymbol{e}_j , \quad i = 1, \ldots, n ,$$

where $[g^{ij}]$ is the inverse matrix: $g^{ik} g_{kj} = \delta^i_j$. A vector $\boldsymbol{v} \in \mathscr{V}$ can be decomposed either over the basis $(\boldsymbol{e}_i)_{i=1}^n$ or over the dual basis $(\boldsymbol{e}^i)_{i=1}^n$. In doing so, $\boldsymbol{v} = v^i \boldsymbol{e}_i = v_i \boldsymbol{e}^i$, where $v^i = \boldsymbol{v} \cdot \boldsymbol{e}^i$ are contravariant components of $\boldsymbol{v}$, while $v_i = \boldsymbol{v} \cdot \boldsymbol{e}_i$ are covariant components of $\boldsymbol{v}$, $i = 1, \ldots, n$. Therefore, the action of elements of the dual vector basis by scalar multiplication is similar to the direct action of the dual covector basis: both read off the components of a vector. Note that covariant and contravariant components of a vector $\boldsymbol{v}$ are related to each other via contractions with matrix $[g_{ij}]$ or its inverse:

$$v_i = g_{ij}v^j \quad \text{and} \quad v^i = g^{ij}v_j\,, \qquad i = 1, \dots, n\,.$$

It can be proven that in $n$-dimensional Euclidean vector space $(\mathscr{V}, \cdot)$, there exists an orthonormal basis $(\boldsymbol{i}_k)_{k=1}^n$, $\boldsymbol{i}_k \cdot \boldsymbol{i}_s = \delta_{ks}$ (it can be obtained by Gram–Schmidt orthogonalization [42]). The corresponding dual vector basis coincides with $(\boldsymbol{i}_k)_{k=1}^n$, but formally we will use the upper index notation $(\boldsymbol{i}^k)_{k=1}^n$ even in this case, for consistency.

**Theorem** (The Riesz representation theorem). *Let $(\mathscr{V}, \cdot)$ be a finite dimensional Euclidean vector space. For any linear functional $\nu \in \mathscr{V}^*$, there exists a unique vector $\boldsymbol{w} \in \mathscr{V}$ for which*

$$\forall \boldsymbol{v} \in \mathscr{V}: \nu(\boldsymbol{v}) = \boldsymbol{v} \cdot \boldsymbol{w}\,.$$

Due to the importance of this theorem, we list one of the possible proofs below.

*Proof.* **1. Uniqueness.** Let $\nu \in \mathscr{V}^*$ be a covector. Suppose that there exist at least two vectors $\boldsymbol{w}_1, \boldsymbol{w}_2 \in \mathscr{V}$, such that $\nu(\boldsymbol{v}) = \boldsymbol{v} \cdot \boldsymbol{w}_1 = \boldsymbol{v} \cdot \boldsymbol{w}_2$, for any vector $\boldsymbol{v} \in \mathscr{V}$. Using the linearity property of the inner product, one obtains that $\boldsymbol{v} \cdot (\boldsymbol{w}_1 - \boldsymbol{w}_2) = 0$ for an arbitrary vector $\boldsymbol{v} \in \mathscr{V}$. In particular, this equality holds for $\boldsymbol{v} = \boldsymbol{w}_1 - \boldsymbol{w}_2$, and the positive definiteness property implies that $\boldsymbol{w}_1 = \boldsymbol{w}_2$. The uniqueness is proven.

**2. Existence.** The existence can be proven in the following manner. Suppose that $\dim \mathscr{V} = n$. Choose a basis $(\boldsymbol{e}_k)_{k=1}^n$ for $\mathscr{V}$ and the corresponding dual bases $(e^k)_{k=1}^n$ for $\mathscr{V}^*$ and $(\boldsymbol{e}^k)_{k=1}^n$ for $\mathscr{V}$. For a covector $\nu \in \mathscr{V}^*$, one has $\nu = \nu_k e^k$. If $\boldsymbol{v} \in \mathscr{V}$, then $\nu(\boldsymbol{v}) = v^k \nu_k$, where $\boldsymbol{v} = v^k \boldsymbol{e}_k$. Set $\boldsymbol{w} := \nu_k \boldsymbol{e}^k$. Then, $\boldsymbol{v} \cdot \boldsymbol{w} = v^k \nu_k$, and $\boldsymbol{w}$ is exactly the appropriate vector. □

**Musical isomorphisms.** The Riesz theorem allows us to establish mutually inverse correspondences between vectors and covectors. Indeed, if $\boldsymbol{w} \in \mathscr{V}$, then one can define a linear functional $\nu_{\boldsymbol{w}} \in \mathscr{V}^*$, such that for all $\boldsymbol{v} \in \mathscr{V}$, we have $\nu_{\boldsymbol{w}}(\boldsymbol{v}) := \boldsymbol{v} \cdot \boldsymbol{w}$. The mapping $(\cdot)^\flat : \mathscr{V} \to \mathscr{V}^*$, defined as $\boldsymbol{v} \mapsto \boldsymbol{v}^\flat := \nu_{\boldsymbol{v}}$, is an isomorphism. It is linear by construction. The injectivity is proven similarly as in the first part of the proof of the Riesz theorem, and the surjectivity follows directly from the Riesz theorem.

Denote by $\boldsymbol{w}_\nu$ the vector dual (by the Riesz theorem) to a covector $\nu \in \mathscr{V}^*$. That is, for all $\boldsymbol{v} \in \mathscr{V}$, we have $\nu(\boldsymbol{v}) := \boldsymbol{v} \cdot \boldsymbol{w}_\nu$. The mapping $(\cdot)^\sharp : \mathscr{V}^* \to \mathscr{V}$, which is defined as $\nu \mapsto \nu^\sharp := \boldsymbol{w}_\nu$, is similarly an isomorphism. For this, it is sufficient to see that $(\cdot)^\sharp$ and $(\cdot)^\flat$ are mutually inverse.

Mutually inverse isomorphisms

$$(\cdot)^\flat : \mathscr{V} \to \mathscr{V}^* \quad \text{and} \quad (\cdot)^\sharp : \mathscr{V}^* \to \mathscr{V}\,,$$

are called *musical isomorphisms* [3]. They establish the relations between the bilinear mapping $\langle \cdot, \cdot \rangle$ and the inner product $(\cdot)$:

$$\forall \boldsymbol{v}, \boldsymbol{w} \in \mathscr{V}: \langle \boldsymbol{v}^\flat, \boldsymbol{w} \rangle = \boldsymbol{v} \cdot \boldsymbol{w}\,,$$
$$\forall \boldsymbol{w} \in \mathscr{V} \quad \nu \in \mathscr{V}^*: \nu^\sharp \cdot \boldsymbol{w} = \langle \nu, \boldsymbol{w} \rangle\,.$$

Consequently, if $(e^k)_{k=1}^n$ is a covector basis, dual to $(\boldsymbol{e}_k)_{k=1}^n$, then

$$\boldsymbol{e}^i = (e^i)^\sharp \quad \text{and} \quad e^i = (\boldsymbol{e}^i)^\flat\,, \quad i = 1, \ldots, n\,.$$

**Transposition.** Let $(\mathscr{V}, \cdot_{\mathscr{V}})$ and $(\mathscr{W}, \cdot_{\mathscr{W}})$ be Euclidean spaces with scalar products $\cdot_{\mathscr{V}}$ and $\cdot_{\mathscr{W}}$, respectively. For a linear map $\boldsymbol{L} \in \operatorname{Lin}(\mathscr{V}; \mathscr{W})$, there exists a unique map $\boldsymbol{L}^{\mathrm{T}} \in \operatorname{Lin}(\mathscr{W}; \mathscr{V})$ such that

$$\forall \boldsymbol{u} \in \mathscr{V} \quad \forall \boldsymbol{v} \in \mathscr{W}: \boldsymbol{v} \cdot_{\mathscr{W}} \boldsymbol{L}[\boldsymbol{u}] = \boldsymbol{u} \cdot_{\mathscr{V}} \boldsymbol{L}^{\mathrm{T}}[\boldsymbol{v}]\,. \tag{12.3}$$

This map is called the *transpose of* $\boldsymbol{L}$.

If $\boldsymbol{L} \in \operatorname{Lin}(\mathscr{V}; \mathscr{V})$ is a linear map with the property $\boldsymbol{L}^{\mathrm{T}} = \boldsymbol{L}$, then it is called *symmetric*. The set of all symmetric linear maps from $\mathscr{V}$ to $\mathscr{V}$ is denoted by $\operatorname{Sym}(\mathscr{V})$.

**Inner product in the space of linear mappings.** The inner (scalar) product on $\mathscr{V}$ induces the inner product

$$(:) : \operatorname{Lin}(\mathscr{V}; \mathscr{V}) \times \operatorname{Lin}(\mathscr{V}; \mathscr{V}) \to \mathbb{R}\,,$$

on the space $\operatorname{Lin}(\mathscr{V}; \mathscr{V})$ of linear mappings in the following way:

$$\forall \boldsymbol{L}, \boldsymbol{M} \in \operatorname{Lin}(\mathscr{V}; \mathscr{V}): \boldsymbol{L}:\boldsymbol{M} = \operatorname{tr}(\boldsymbol{L}^{\mathrm{T}}\boldsymbol{M})\,. \tag{12.4}$$

It is easy to check that the so-defined mapping (:) satisfies all axioms for the inner product.

Applying the Riesz theorem, we obtain the representation of a linear functional $\varepsilon \in \operatorname{Lin}(\mathscr{V}; \mathscr{V})^*$, *i.e.*, $\varepsilon : \operatorname{Lin}(\mathscr{V}; \mathscr{V}) \to \mathbb{R}$:

$$\forall \boldsymbol{L} \in \operatorname{Lin}(\mathscr{V}; \mathscr{V}): \varepsilon(\boldsymbol{L}) = \boldsymbol{T}_\varepsilon:\boldsymbol{L}\,,$$

where $\boldsymbol{T}_\varepsilon \in \operatorname{Lin}(\mathscr{V}; \mathscr{V})$ is a uniquely determined element of $\operatorname{Lin}(\mathscr{V}; \mathscr{V})$.

**Orthogonal operator.** Consider the special type of linear transformation, an orthogonal operator. The linear transformation $\boldsymbol{Q} \in \operatorname{Lin}(\mathscr{V}; \mathscr{V})$ is called the *orthogonal operator* if it preserves the inner product:

$$\forall \boldsymbol{u}, \boldsymbol{v} \in \mathscr{V}: \boldsymbol{Q}[\boldsymbol{u}] \cdot \boldsymbol{Q}[\boldsymbol{v}] = \boldsymbol{u} \cdot \boldsymbol{v}\,. \tag{12.5}$$

By definition, an orthogonal operator maps orthonormal bases to orthonormal bases. It is an invertible operator, and

$$\boldsymbol{Q}\boldsymbol{Q}^{\mathrm{T}} = \boldsymbol{I}\,.$$

Conversely, any linear operator with this property is orthogonal [42, 82].

**Positive definiteness and square root.** A linear map $\boldsymbol{L}$ is called *positive definite* if

$$\forall \boldsymbol{u} \in \mathscr{V}: (\boldsymbol{u} \neq \boldsymbol{0}) \Rightarrow \boldsymbol{u} \cdot \boldsymbol{L}[\boldsymbol{u}] > 0\,.$$

For a positive definite operator $\boldsymbol{L}$, the notion of *square root* is defined as follows [82]: there exists an operator $\boldsymbol{M} \in \operatorname{Lin}(\mathscr{V}; \mathscr{V})$ such that $\boldsymbol{M}^2 = \boldsymbol{L}$. This operator is unique and is denoted by $\sqrt{\boldsymbol{L}}$.

**Norm.** Let $\mathscr{V}$ be a vector space. A *norm* on $\mathscr{V}$ is a function $\|\cdot\|\colon \mathscr{V} \to \mathbb{R}$ with the properties:
(i) *Non-negativeness.* $\forall v \in \mathscr{V}: \|v\| \geq 0$.
(ii) *Definiteness.* If $\|v\| = 0$ then $v = 0$.
(iii) *Homogeneity.* $\forall v \in \mathscr{V}\ \forall \lambda \in \mathbb{R}: \|\lambda v\| = |\lambda|\|v\|$.
(iv) *Triangle inequality.* $\forall v, w \in \mathscr{V}: \|v + w\| \leq \|v\| + \|w\|$.

Suppose that $(\mathscr{V}, \cdot)$ is a Euclidean vector space. The scalar products $(\cdot)$ and $(:)$ on $\mathscr{V}$ and $\mathrm{Lin}(\mathscr{V};\mathscr{V})$ generate norms on these spaces as follows:

$$\|\cdot\|_{\mathscr{V}}: \mathscr{V} \to \mathbb{R}\,, \qquad \|\boldsymbol{u}\|^2_{\mathscr{V}} := \boldsymbol{u}\cdot\boldsymbol{u}\,,$$
$$\|\cdot\|_{\mathrm{Lin}(\mathscr{V};\mathscr{V})}: \mathrm{Lin}(\mathscr{V};\mathscr{V}) \to \mathbb{R}\,, \qquad \|\boldsymbol{L}\|^2_{\mathrm{Lin}(\mathscr{V};\mathscr{V})} := \boldsymbol{L}{:}\boldsymbol{L}\,.$$

## 12.6 Linear Groups

**The general linear groups** Let $s \in \mathbb{N}$. Consider the set of all invertible $s \times s$ matrices with real elements. This set, together with the ordinary matrix multiplication operation, forms a group denoted by $\mathrm{GL}(s;\mathbb{R})$. This group is called the *general linear group of degree s.*

**The automorphism group.** Let $\mathscr{V}$ be an $s$-dimensional real vector space. The set of all possible isomorphisms from $\mathscr{V}$ onto $\mathscr{V}$ (automorphisms) together with the composition defines a group $\mathrm{GL}(\mathscr{V})$. This group is called the *automorphism group.*

**Isomorphism $\mathrm{GL}(s;\mathbb{R}) \cong \mathrm{GL}(\mathscr{V})$.** If two groups are isomorphic, then one of them may be treated as a representation of the other. Consider the groups $\mathrm{GL}(\mathscr{V})$ and $\mathrm{GL}(s;\mathbb{R})$, where $\mathscr{V}$ is an $s$-dimensional real vector space. These groups are isomorphic. Indeed, if one fixes some basis $(e_i)_{i=1}^s$ for $\mathscr{V}$, then for every automorphism $\mathscr{L} \in \mathrm{GL}(\mathscr{V})$, one has

$$\mathscr{L}(e_j) = \Omega^i{}_j e_i\,, \quad i, j = 1, \ldots, s\,.$$

The correspondence $\mathscr{L} \mapsto [\Omega^i{}_j]$ is the desired isomorphism. Thus, the group $\mathrm{GL}(s;\mathbb{R})$ can be considered as a "coordinate" representation of the group $\mathrm{GL}(\mathscr{V})$.

**The orthogonal group.** Let $\mathscr{V}$ be an $s$-dimensional Euclidean vector space. The orthogonal operators $\boldsymbol{T} \in \mathrm{Lin}(\mathscr{V};\mathscr{V})$ constitute an *orthogonal group* under the composition operation. This group is denoted by $\mathrm{O}(s)$ and is isomorphic to the group of $s \times s$ orthogonal matrices[2].

2 A square matrix $Q$ is called orthogonal if $QQ^{\mathrm{T}} = Q^{\mathrm{T}}Q = I$.

## 12.7 Affine Space

*Affine space (over $\mathbb{R}$)* is a triple $(\mathscr{A}, \mathscr{V}, \psi)$ such that the following axioms are satisfied:

$\mathscr{A}_1$) $\mathscr{A}$ is a non-empty set, whose elements are referred to as *points*.

$\mathscr{A}_2$) $\mathscr{V}$ is a vector space over $\mathbb{R}$, whose elements are referred to as *translation vectors*.

$\mathscr{A}_3$) $\psi\colon \mathscr{A} \times \mathscr{A} \to \mathscr{V}$ is a mapping, which assigns to each ordered pair $(a, b) \in \mathscr{A} \times \mathscr{A}$ some vector from $\mathscr{V}$, denoted by an arrow over concatenated symbols of points, $\overrightarrow{ab}$, or by formal subtraction, $b - a$. The mapping $\psi$ satisfies the following Weyl axioms [35]:

$W_1$) for every points $a, b, c \in \mathscr{A}$ the following *Chasles' relation* holds:

$$\overrightarrow{ab} + \overrightarrow{bc} + \overrightarrow{ca} = 0 \in \mathscr{V} .$$

$W_2$) for any point $a \in \mathscr{A}$ and for any vector $v \in \mathscr{V}$, there exists a unique point $b \in \mathscr{A}$, such that $\overrightarrow{ab} = v$.

Consider the elementary consequences from the definition. Taking $a = b = c$ in the Weyl axiom $W_1$) implies that $\overrightarrow{aa} = 0$. If one takes in $W_1$) $a = c$, then it leads to the relation $\overrightarrow{ab} = -\overrightarrow{ba}$. Using the latter equality, the Chasles relation can be rewritten in the form

$$\overrightarrow{ab} + \overrightarrow{bc} = \overrightarrow{ac} .$$

According to the Weyl axiom $W_2$), for any fixed point $a \in \mathscr{A}$, the mapping

$$\psi_a\colon \mathscr{A} \to \mathscr{V} , \quad \psi_a(x) := \overrightarrow{ax} ,$$

is a bijection. This allows one to define the external operation

$$+\colon \mathscr{A} \times \mathscr{V} \to \mathscr{A} , \quad +\colon (a, v) \mapsto a + v := \psi_a^{-1}(v) ,$$

which assigns to each ordered pair $(a, v)$ a unique point $b \in \mathscr{A}$, such that $\overrightarrow{ab} = v$. From the definition of $(+)$, it follows that

$(+_1)$ if $a, b \in \mathscr{A}$ are two arbitrary points, then there exists a unique vector $u \in \mathscr{V}$, such that $a + u = b$;

$(+_2)$ for any point $a \in \mathscr{A}$, the following identity holds: $a + 0 = a$;

$(+_3)$ for any point $a \in \mathscr{A}$ and all vectors $u, v \in \mathscr{V}$, one has $(a + u) + v = a + (u + v)$.

**Remark 12.1.** Properties $(+_1)$–$(+_3)$ of the operation $+$ are often stated as affine space axioms [144]. In this case, the affine space is defined as the ordered triple $(\mathscr{A}, \mathscr{V}, +)$, where $\mathscr{A}$ is a non-empty set, $\mathscr{V}$ is a real vector space, and

$$+\colon \mathscr{A} \times \mathscr{V} \to \mathscr{A} , \quad +\colon (a, v) \mapsto a + v ,$$

is an external binary operation that satisfies the conditions $(+_1)$–$(+_3)$.

Let $\overrightarrow{ab}$ denote such a unique vector $u \in \mathscr{V}$, that $a + u = b$. For the mapping

$$\mathscr{A} \times \mathscr{A} \to \mathscr{V}\,, \quad (a, b) \mapsto \overrightarrow{ab}\,,$$

the Weyl axioms $W_1$) and $W_2$) follow from these conditions.

The particular case of affine space is represented by the triple $(\mathscr{V}, \mathscr{V}, \psi)$, where $\mathscr{V}$ is a real vector space and $\psi\colon \mathscr{V} \times \mathscr{V} \to \mathscr{V}$ is a mapping such that $\psi\colon (a, b) \mapsto b - a$. In this case, $\mathscr{A} = \mathscr{V}$, and vectors from $\mathscr{V}$ are "points" of the affine space.

The *dimension* of affine space $(\mathscr{A}, \mathscr{V}, \psi)$ is the dimension of the space $\mathscr{V}$: $\dim \mathscr{A} := \dim \mathscr{V}$. In this book, we restrict ourselves to finite dimensional affine spaces.

In the affine space $(\mathscr{A}, \mathscr{V}, \psi)$, $\dim \mathscr{A} = n$, one can naturally introduce the notion of coordinates. Let $(e_i)_{i=1}^n$ be a basis for $\mathscr{V}$ and let $o \in \mathscr{A}$ be a fixed point (origin). The ordered pair $(o, (e_i)_{i=1}^n)$ is called an *affine coordinate system*. If $x \in \mathscr{A}$ is arbitrary point, then its *coordinates* with respect to $(o, (e_i)_{i=1}^n)$ are components of the vector $\overrightarrow{ox}$ with respect to the basis $(e_i)_{i=1}^n$. The properties of the mapping $\psi$ imply that any point $x \in \mathscr{A}$ is uniquely characterized by its coordinates $(x^i)_{i=1}^n$ and

$$x = o + x^i e_i\,.$$

Suppose that $a, b \in \mathscr{A}$ are two points that are represented in $(o, (e_i)_{i=1}^n)$ by coordinates $(a^i)_{i=1}^n$ and $(b^i)_{i=1}^n$, respectively. Then, $\overrightarrow{ab} = \overrightarrow{ao} + \overrightarrow{ob} = (b^i - a^i)e_i$, and the vector $\overrightarrow{ab}$ has the components $(b^i - a^i)_{i=1}^n$.

The previous reasonings show that the affine space reflects the classical notions of analytical geometry. In particular, it supports the ideas of a vector and a coordinate. If one chooses a coordinate system, then it can consider points as ordered tuples. The difference between tuples gives the components of a vector that connects the corresponding points. Affine space results in all possible translations from a fixed origin.

Let $(\mathscr{A}, \mathscr{V}, \psi)$ be an affine space with $\dim \mathscr{A} = n$. Suppose that $\mathscr{U} \subset \mathscr{V}$ is a subspace of $\mathscr{V}$ and $a \in \mathscr{A}$ is some point. Then, the set

$$\mathscr{F} := a + \mathscr{U} = \{a + h \mid h \in \mathscr{U}\}\,,$$

is called the *affine subspace* of dimension $\dim \mathscr{U}$ [147]. If $\dim \mathscr{F} = 1$, then $\mathscr{F}$ is called an *affine line*. In this case, the subspace $\mathscr{U}$ is generated by a non-zero vector $h \in \mathscr{U}$, and one can write

$$\mathscr{F} = \{a + \lambda h \mid \lambda \in \mathbb{R}\}\,.$$

An affine subspace of dimension 2 is called an *affine plane*. If $\dim \mathscr{U} = n - 1$, then the affine subspace is called an *affine hyperplane*.

Two affine subspaces $\mathscr{F}_1 = a_1 + \mathscr{U}_1$ and $\mathscr{F}_2 = a_2 + \mathscr{U}_2$ are called *strongly parallel*, if $\mathscr{U}_1 = \mathscr{U}_2$. They are called *weakly parallel* if $\mathscr{U}_1 \subsetneq \mathscr{U}_2$. Thus, affine space is an absolute world of lines and planes; the world that is represented in classical synthetic geometry. At the same time, the definition of affine space does not presuppose any metric data. In particular, the notion of an angle is undefined. To build the analog of synthetic Euclidean space, one needs an additional tool. In Chapter 2, such tool is provided by a scalar product.

# 13 Review of Smooth Manifolds and Vector Bundles

## 13.1 Smooth Manifolds

### 13.1.1 Topological Spaces

**Topology.** Let $\mathfrak{M}$ be a set. A collection $\mathfrak{S}$ of its subsets is called a *topology on* $\mathfrak{M}$, if it satisfies the following axioms:
($\mathfrak{S}_1$) $\mathfrak{M}$ and $\emptyset$ are elements of $\mathfrak{S}$.
($\mathfrak{S}_2$) If $\{O_\alpha\}_{\alpha\in I}$ is any family of elements of $\mathfrak{S}$, then its union, $\bigcup_{\alpha\in I} O_\alpha$, is an element of $\mathfrak{S}$.
($\mathfrak{S}_3$) If $O_1, \dots, O_k$ are elements of $\mathfrak{S}$, then their intersection, $\bigcap_{i=1}^k O_i$, is an element of $\mathfrak{S}$.

Each element $O$ of the set $\mathfrak{S}$ is called the *open subset of* $\mathfrak{M}$. If $F \subset \mathfrak{M}$ is such a subset of $\mathfrak{M}$ that $\mathfrak{M} \setminus F \in \mathfrak{S}$, then $F$ is called a *closed subset of* $\mathfrak{M}$. The ordered pair $(\mathfrak{M}, \mathfrak{S})$ is called a *topological space* [64, 148, 149], and the elements of $\mathfrak{M}$ are called *points*. In the entire book, if this does not lead to any misunderstanding, we omit the symbol $\mathfrak{S}$ and denote the topological space $(\mathfrak{M}, \mathfrak{S})$ simply by $\mathfrak{M}$.

**Metric space.** Consider the particular example of a topological space. Let $\mathfrak{M}$ be a set. A *metric on* $\mathfrak{M}$ is a mapping $d\colon \mathfrak{M} \times \mathfrak{M} \to \mathbb{R}$ that satisfies the following axioms:
(i) *Positivity.* $\forall p, q \in \mathfrak{M}\colon d(p, q) \geq 0$. Moreover, $d(p, q) = 0$ *iff* $p = q$.
(ii) *Symmetry.* $\forall p, q \in \mathfrak{M}\colon d(p, q) = d(q, p)$.
(iii) *Triangle inequality.* $\forall p, q, r \in \mathfrak{M}\colon d(p, r) \leq d(p, q) + d(q, r)$.

The ordered pair $(\mathfrak{M}, d)$ is called a *metric space* [64, 148]. The topology on the metric space can be induced by the metric. Indeed, let $p \in \mathfrak{M}$ be a point and $r > 0$ be a number. Then the *open ball* $B_r(p)$ *of radius* $r$ *around* $p$ is the set

$$B_r(p) := \{x \in \mathfrak{M} \mid d(x, p) < r\}\,.$$

Let $\mathfrak{S}_d \subset 2^{\mathfrak{M}}$ be the following collection of subsets $O \subset \mathfrak{M}$:

$$\mathfrak{S}_d := \{O \in 2^{\mathfrak{M}} \mid \forall p \in O\ \exists r > 0\colon B_r(p) \subset O\}\,.$$

The collection $\mathfrak{S}_d$ satisfies axioms ($\mathfrak{S}_1$)–($\mathfrak{S}_3$). Thus, $\mathfrak{S}_d$ is a topology on $\mathfrak{M}$, called the *metric topology*.

A vector space $\mathscr{V}$ with norm $\|\cdot\|$ gives the example of metric space, in which the metric is defined by $d(u, v) := \|u - v\|$.

**The space $\mathbb{R}^s$.** The particular case of topological space is one of the main actors in manifold theory; the vector space $\mathbb{R}^s$, $s \in \mathbb{N}$. It is equipped with a norm $\|\cdot\|$, such that $\|(x^1, \dots, x^s)\| := \sqrt{(x^1)^2 + \cdots + (x^s)^2}$. As for any metric space, the topology on $\mathbb{R}^s$ is

https://doi.org/10.1515/9783110563214-013

based on the notion of the open ball: if $p \in \mathbb{R}^s$ and $r > 0$, then the open ball $B_r(p)$ is defined by

$$B_r(p) = \{x \in \mathbb{R}^s \mid \|x - p\| < r\} .$$

Thus, a set $O \subset \mathbb{R}^s$ is open *iff* for any point $p \in O$, there exists such $r > 0$, that $B_r(p) \subset O$.

**Neighborhoods.** If $\mathfrak{M}$ is a topological space and $p \in \mathfrak{M}$, then by a *neighborhood of* $p$ we mean any open set $O$ that contains $p$ [64]. Often, the notion of neighborhood is introduced in a more general sense [148, 149]. That is, $V \subset \mathfrak{M}$ is a neighborhood of $p$, if $p \in V$ and there exists such an open subset $O$ of $\mathfrak{M}$ that $p \in O \subset \mathfrak{M}$. We do not use such a generalization.

**Subspace topology.** Suppose that $(\mathfrak{M}, \mathfrak{S})$ is a topological space and $\mathfrak{U} \subset \mathfrak{M}$ is a subset. The topology $\mathfrak{S}$ defined on $\mathfrak{M}$ induces the topology $\mathfrak{S}_{\mathfrak{U}}$ on $\mathfrak{U}$ as follows:

$$\mathfrak{S}_{\mathfrak{U}} := \{V \subset \mathfrak{U} \mid \exists O \in \mathfrak{S} \colon V = \mathfrak{U} \cap O\} .$$

The topology $\mathfrak{S}_{\mathfrak{U}}$ is called the *subspace topology*, and the space $(\mathfrak{U}, \mathfrak{S}_{\mathfrak{U}})$ is called the *subspace of* $(\mathfrak{M}, \mathfrak{S})$. The elementary properties of the subspace follow from its definition:

i) A subset $V \subset \mathfrak{U}$ is closed in $\mathfrak{U}$ if and only if there exists a closed subset $F$ of $\mathfrak{M}$, such that $V = \mathfrak{U} \cap F$,
ii) If $V$ is open in $\mathfrak{U}$ and $\mathfrak{U}$ is open in $\mathfrak{M}$, then $V$ is open in $\mathfrak{M}$. The statement remains true if "open" is changed everywhere into "closed".
iii) If $V \subset \mathfrak{U}$ is open in $\mathfrak{M}$, then it is open in $\mathfrak{U}$. The case of closed subset is similar.

We use these properties in the definition of the joining operation $\vee$ (see Chapter 8, Section 8.5.1).

**Interior. Closure. Boundary.** Let $(\mathfrak{M}, \mathfrak{S})$ be a topological space and $A \subset \mathfrak{M}$. Its *interior in* $\mathfrak{M}$, denoted by $\operatorname{Int} A$, is the maximal open subset of $\mathfrak{M}$ that contains $A$:

$$\operatorname{Int} A := \bigcup \{O \in \mathfrak{S} \mid O \subset A\} .$$

It follows that $\operatorname{Int} A \in \mathfrak{S}$ and $\operatorname{Int} A \subset A$. The *closure of* $A$ *in* $\mathfrak{M}$, denoted by $\overline{A}$, is the minimal closed set that contains $A$:

$$\overline{A} := \bigcap \{O \in \mathfrak{S} \mid A \subset \mathfrak{M} \setminus O\} .$$

The *exterior* of $A$, denoted by $\operatorname{Ext} A$, is the interior if its complement:

$$\operatorname{Ext} A := \operatorname{Int} (\mathfrak{M} \setminus A) .$$

The *boundary* of $A$, denoted by $\partial A$, is defined as follows:

$$\partial A := \mathfrak{M} \setminus (\operatorname{Int} A \cup \operatorname{Ext} A) .$$

Note that the interior, exterior, closure, and boundary of a set can be defined in an alternative way. Particularly, the definitions of these notions in terms of the metric are given in [148].

**Continuous mappings.** The topological structures over two sets allow us to define the notion of continuous mapping between them without referring to the particular properties of $\mathbb{R}$, which may not even be used in the definition of topology on these sets. Such formalism is applicable to situations when topology is not generated by any metric. Let $(\mathfrak{M}, \mathfrak{S}_{\mathfrak{M}})$ and $(\mathfrak{N}, \mathfrak{S}_{\mathfrak{N}})$ be two topological spaces, and $f: \mathfrak{M} \to \mathfrak{N}$ be a mapping. Such a mapping is called *continuous* [64], if for any open set $O \subset \mathfrak{N}$ its pre-image, $f^{-1}(O)$, is open in $\mathfrak{M}$. Denoting the set of all continuous mappings with the domain $\mathfrak{M}$ and the codomain $\mathfrak{N}$ by $C^0(\mathfrak{M}; \mathfrak{N})$, one arrives at the following definition:

$$f \in C^0(\mathfrak{M}; \mathfrak{N}) \quad \Leftrightarrow \quad \left(\forall O \in \mathfrak{S}_{\mathfrak{N}} : f^{-1}(O) \in \mathfrak{S}_{\mathfrak{M}}\right) .$$

**Inclusion map.** Suppose that $(\mathfrak{M}, \mathfrak{S})$ is a topological space and $\mathfrak{U} \subset \mathfrak{M}$. The mapping $\iota_{\mathfrak{U}}: \mathfrak{U} \to \mathfrak{M}$, which is defined by

$$\forall p \in \mathfrak{U}: \iota_{\mathfrak{U}}(p) := p , \tag{13.1}$$

is called the *inclusion map of* $\mathfrak{U}$ *in* $\mathfrak{M}$. The inclusion map is a continuous mapping between topological spaces $(\mathfrak{U}, \mathfrak{S}_{\mathfrak{U}})$ and $(\mathfrak{M}, \mathfrak{S})$ [64].

**Homeomorphism.** If $f: \mathfrak{M} \to \mathfrak{N}$ is bijective, and both $f, f^{-1}$ are continuous, then $f$ is called a *homeomorphism*.

**Hausdorff space.** Conventional analysis of $\mathbb{R}^s$ assumes that any two distinct points possess disjoint neighborhoods. The definition of a topological space given by axioms $(\mathfrak{S}_1)$–$(\mathfrak{S}_3)$ is so wide that this assumption, in general, is no longer valid. This, in particular, leads to a loss of uniqueness of the convergent sequences' limits. To avoid such a situation one needs to restrict the set of all possible topologies on $\mathfrak{M}$ by the following condition: for each pair of distinct points $p, q \in \mathfrak{M}$, there exist disjoint neighborhoods $U$ of $p$ and $V$ of $q$, i.e., $U \cap V = \emptyset$. The space $(\mathfrak{M}, \mathfrak{S})$, where the topology $\mathfrak{S}$ satisfies this condition, is called the *Hausdorff space* [64].

**Basis of topology.** Through the analogy for the decomposition of an arbitrary vector over some basis, chosen in vector space, one can try to find the subset $\mathscr{B}$ of the topology $\mathfrak{S}$, which would be much "narrower" than $\mathfrak{S}$ and permits one to represent any open subset as a union of the elements from $\mathscr{B}$. More precisely, if $\mathfrak{S}$ is a topology on $\mathfrak{M}$, then a collection $\mathscr{B}$ of some subsets of $\mathfrak{M}$ is called a *basis for the* $\mathfrak{S}$, if
($\mathscr{B}_1$) $\mathscr{B} \subset \mathfrak{S}$;
($\mathscr{B}_2$) for every open subset $O \in \mathfrak{S}$ there exists such a family $\{U_\alpha\}_{\alpha \in I}$ of elements from $\mathscr{B}$, that $\bigcup_{\alpha \in I} U_\alpha = O$.

For example, the set $\mathscr{B} = \{B_r(x^1, \ldots, x^s) \mid (r \in \mathbb{Q}) \wedge (\forall i \in \{1, \ldots, s\}: x^i \in \mathbb{Q})\}$ is a basis for the topology on $\mathbb{R}^s$.

**Second countable space.** A topological space $(\mathfrak{M}, \mathfrak{S})$ is *second countable* if a countable basis for its topology exists. Let $\mathscr{U}$ be a collection of subsets of $\mathfrak{M}$. This collection is a *cover of* $\mathfrak{M}$, if $\bigcup_{U\in\mathscr{U}} U = \mathfrak{M}$. The cover is called open, if all its elements are open sets. Any subset $\mathscr{U}' \subset \mathscr{U}$, which is a cover of $\mathfrak{M}$ itself, is called a *subcover of* $\mathscr{U}$. According to Lindelöf's theorem, the second countability allows one to extract a countable subcover of any open cover of $\mathfrak{M}$. The advantage of such an assertion appears in manifold theory; one can choose a countable set of charts for a complete coordinate description of a manifold.

**Topological manifolds.** The formal definition of topological manifold bears on the idea of covering a set, underlying the manifold, by a system of local coordinates. More strictly, a topological space $(\mathfrak{M}, \mathfrak{S})$ is an *s–dimensional topological manifold*[1], $s \in \mathbb{N}$, [64] if:

($M_1$) $(\mathfrak{M}, \mathfrak{S})$ is a Hausdorff space with countable base;

($M_2$) for each point $p \in \mathfrak{M}$ one can find such a neighborhood $U \in \mathfrak{S}$ and such an open set $O \subset \mathbb{R}^s$ that there exists a homeomorphism $\varphi\colon U \to O$ between them.

The axiom ($M_2$) asserts that for $\mathfrak{M}$, there exists a set $A = \{(U_\alpha, \varphi_\alpha)\}_{\alpha\in I}$, such that

($A_1$) for each $\alpha \in I$, the ordered pair $(U_\alpha, \varphi_\alpha)$ consists of a non-empty open subset $U_\alpha$ of $\mathfrak{M}$ and a homeomorphism $\varphi_\alpha\colon U_\alpha \to O_\alpha$ between $U_\alpha$ and an open set $O_\alpha \subset \mathbb{R}^s$;

($A_2$) $\bigcup_{\alpha\in I} U_\alpha = \mathfrak{M}$.

Such a set $A$ is called an *atlas on* $\mathfrak{M}$. The natural number $\dim \mathfrak{M} := s$ defines the *dimension of* $\mathfrak{M}$, each ordered pair $(U_\alpha, \varphi_\alpha)$ specifies a *(coordinate) chart on* $\mathfrak{M}$. We refer to $U_\alpha$ as the *(coordinate) domain* and to $\varphi_\alpha$ as the *coordinate map*. Since, due to the axiom ($M_1$), the space $(\mathfrak{M}, \mathfrak{S})$ is second countable, one can choose a countable index set $I$.

**Transition map.** The atlas $A$ defines a collection of local coordinates on the manifold. Each point $p \in \mathfrak{M}$ is contained in the domain $U_\alpha$ of some chart $(U_\alpha, \varphi_\alpha)$. The coordinate map $\varphi_\alpha$ endows $p$ with an $s$-tuple $(x^1, \dots, x^s) = \varphi_\alpha(p)$, which represents the local coordinates of the point $p$. Suppose that for some distinct $\alpha, \beta \in I$, the coordinate domains $U_\alpha$ and $U_\beta$ have a non-empty intersection: $U_\alpha \cap U_\beta \neq \emptyset$. The points $p$ from the coordinate domain $U_\alpha$ have local coordinates $(x^1, \dots, x^s) = \varphi_\alpha(p)$, while the points $p$ from the coordinate domain $U_\beta$ have local coordinates $(\tilde{x}^1, \dots, \tilde{x}^s) = \varphi_\beta(p)$. In the intersection $U_\alpha \cap U_\beta$, one has the homeomorphism

$$\varphi_\beta \circ \varphi_\alpha^{-1}|_{\varphi_\alpha(U_\alpha \cap U_\beta)}\colon \varphi_\alpha(U_\alpha \cap U_\beta) \to \varphi_\beta(U_\alpha \cap U_\beta)\,, \tag{13.2}$$

1 The definition of a topological manifold makes sense even in the case $s = 0$, but we do not consider it.

which is called a *transition map*. This homeomorphism determines the coordinate transformation $(x^1, \dots, x^s) \mapsto \tilde{x}^i$, $i = 1, \dots, s$, which associates $s$-tuple $(x^1, \dots, x^s)$ with $(\tilde{x}^1, \dots, \tilde{x}^s)$, which gives different coordinate representations for a point $p \in U_\alpha \cap U_\beta$.

**Open submanifolds.** In some cases, one needs to consider mappings which domains are open subsets of manifolds. Mappings that are defined on shapes of the body may serve as an example (we assume here that the dimensions of the body and the physical space coincide). Suppose that $(\mathfrak{M}, \mathfrak{S})$ is a topological $s$-dimensional manifold with atlas $A = \{(U_\alpha, \varphi_\alpha)\}_{\alpha \in I}$, and that $\mathfrak{U} \subset \mathfrak{M}$ is an open set. Consider it with the induced topology. The topological space $(\mathfrak{U}, \mathfrak{S}_\mathfrak{U})$ is Hausdorff, with a countable base. Let $J \subset I$ be a set with the property:

$$\forall \alpha \in J\colon \mathfrak{U} \cap U_\alpha \neq \emptyset\,.$$

Since each point of $\mathfrak{U}$ is contained in a coordinate domain of some chart from $A$, the set $J$ is not empty. Define the set $A_\mathfrak{U} = \{(\mathfrak{U} \cap U_\alpha, \varphi_\alpha|_{\mathfrak{U} \cap U_\alpha})\}_{\alpha \in J}$. For each $\alpha \in J$, the set $\mathfrak{U} \cap U_\alpha$ is open relative to the topology $\mathfrak{S}_\mathfrak{U}$, and the mapping $\varphi_\alpha|_{\mathfrak{U} \cap U_\alpha}$ is a homeomorphism onto its image. Thus, $A_\mathfrak{U}$ is an atlas on $\mathfrak{U}$, and the topological space $(\mathfrak{U}, \mathfrak{S}_\mathfrak{U})$ is the topological $s$-dimensional manifold. We refer to such a space as the *open submanifold* of $\mathfrak{M}$ [64].

**Product manifolds.** In this book, we consider mappings that may be defined on products of topological manifolds. The material velocity gives an example of such a mapping. The topological structure can be defined on the product of topological manifolds as follows. Suppose that $(\mathfrak{M}_1, \mathfrak{S}_1)$ and $(\mathfrak{M}_2, \mathfrak{S}_2)$ are topological manifolds with atlases $A_1 = \{(U_\alpha, \varphi_\alpha)\}_{\alpha \in I}$, $A_2 = \{(V_\beta, \psi_\beta)\}_{\beta \in J}$ and dimensions $s_1$ and $s_2$, respectively. Then, the set

$$\mathscr{B} = \{O_1 \times O_2 \mid (O_1 \in \mathfrak{S}_1) \wedge (O_2 \in \mathfrak{S}_2)\}$$

generates the unique topology $\mathfrak{S}$ on $\mathfrak{M}_1 \times \mathfrak{M}_2$ (called the *product topology*), for which $\mathscr{B}$ is the base [64]. Every set $O \in \mathfrak{S}$ is the union $O = \bigcup_{U \in \mathscr{C}_O} U$ of sets (they are Cartesian products) from some subset $\mathscr{C}_O \subset \mathscr{B}$. It is easy to check that the so-defined topological space $(\mathfrak{M}_1 \times \mathfrak{M}_2, \mathfrak{S})$ is a Hausdorff space with a countable base. One can define an atlas $A = \{(U_\alpha \times V_\beta, \varphi_\alpha \times \psi_\beta)\}_{\alpha \in I, \beta \in J}$ on $\mathfrak{M}_1 \times \mathfrak{M}_2$, where

$$\varphi_\alpha \times \psi_\beta\colon U_\alpha \times V_\beta \to \mathbb{R}^{s_1 + s_2}\,, \qquad \varphi_\alpha \times \psi_\beta(p, q) := (\varphi_\alpha(p), \psi_\beta(q))$$

are homeomorphisms onto their images [64]. Thus, $(\mathfrak{M}_1 \times \mathfrak{M}_2, \mathfrak{S})$ is a $s_1 + s_2$-dimensional topological manifold.

**Topological manifolds with boundary.** The definition of a topological manifold is useful for the description of a body and a physical space as open sets [37]. However, the formalization of the part of a body requires a more complicated construction. It must contain the information about the interior of the part, treated as a subbody, as well as about the boundary, which is the arena of interaction between the part and the rest of the body. Such a construction consists of two topological manifolds, where one describes the interior and another describes the boundary. To this end, introduce the set $\mathbb{H}^s$, that is, the *closed s-dimensional upper half-space*:

$$\mathbb{H}^s = \{(x^1, \ldots, x^s) \in \mathbb{R}^s \mid x^s \geq 0\} .$$

Consider $\mathbb{H}^s$ as the topological space with the topology induced from $\mathbb{R}^s$.

A topological space $(\mathfrak{M}, \mathfrak{S})$ is an *s-dimensional topological manifold with boundary*, $s \in \mathbb{N}$, [64] if:

$(MB_1)$ $(\mathfrak{M}, \mathfrak{S})$ is a Hausdorff space with a countable base;

$(MB_2)$ for each point $p \in \mathfrak{M}$, one can find such a neighborhood $U \in \mathfrak{S}$ and such a set $O$, which is either an open subset of $\mathbb{R}^s$ or an open subset of $\mathbb{H}^s$, that there exists a homeomorphism $\varphi\colon U \to O$ between them.

The axiom $(MB_2)$, like the corresponding $(M_2)$, asserts that for $\mathfrak{M}$, there exists a set $A = \{(U_\alpha, \varphi_\alpha)\}_{\alpha \in I}$, where

$(AB_1)$ for each $\alpha \in I$, the ordered pair $(U_\alpha, \varphi_\alpha)$ is such that $U_\alpha$ is an open subset of $\mathfrak{M}$ and $\varphi_\alpha\colon U_\alpha \to O_\alpha$ is a mapping between $U_\alpha$ and a set $O_\alpha$. Such a set is either an open subset of $\mathbb{R}^s$ or an open subset of $\mathbb{H}^s$. The mapping $\varphi_\alpha$ is a homeomorphism;

$(AB_2)$ $\bigcup_{\alpha \in I} U_\alpha = \mathfrak{M}$.

The definition for *atlas, chart, etc.*, can be repeated here unchanged. A chart $(U_\alpha, \varphi_\alpha)$ is said to be an *interior chart*, if $\varphi_\alpha(U_\alpha)$ is an open subset of $\mathbb{R}^s$, and it is said to be a *boundary chart*, if $\varphi_\alpha(U_\alpha)$ is an open subset of $\mathbb{H}^s$, such that $\varphi_\alpha(U_\alpha) \cap \partial\mathbb{H}^s \neq \emptyset$. The latter condition specifies that each point from a boundary chart has a local coordinate $x^s = 0$. Such a distinction in charts results in a distinction in points; a point from $\mathfrak{M}$ is called an *interior point of* $\mathfrak{M}$, if it belongs to the domain of some interior chart, and such a point is called a *boundary point of* $\mathfrak{M}$, if it is an element of the domain of a boundary chart. Such a definition implies the representation of $\mathfrak{M}$ as a disjoint union

$$\mathfrak{M} = \operatorname{Int}\mathfrak{M} \sqcup \partial\mathfrak{M} ,$$

in which $\operatorname{Int}\mathfrak{M}$ is the set of all interior points, the *interior of* $\mathfrak{M}$, and $\partial\mathfrak{M}$ is the set of all boundary points, the *boundary*[2] *of* $\mathfrak{M}$ [64].

Note that the topological space $\mathfrak{M}$, which is endowed with the structure prescribed by axioms $(M_1)$–$(M_2)$, is a specific case of a manifold with boundary; all its points are interior ones, that is, $\partial\mathfrak{M} = \emptyset$. We will refer to such a manifold as an "s-dimensional topological manifold", while in the common case ($\partial\mathfrak{M} \neq \emptyset$), we will use the term "s-dimensional topological manifold with boundary".

2 Note that such notions are not equivalent to the topological notions of interior and boundary.

Suppose that $\mathfrak{M}$ is an $s$-dimensional topological manifold with boundary. The interior, Int $\mathfrak{M}$, is an open subset of $\mathfrak{M}$, and it is an $s$-dimensional topological manifold. The boundary, $\partial\mathfrak{M}$, is a closed subset of $\mathfrak{M}$, and it is also a topological manifold, but its dimension is less per unit, *i.e.*, $\dim \partial\mathfrak{M} = s - 1$ [64].

### 13.1.2 Smooth Structure

**Smooth compatibility.** The notion of a topological manifold allows one to introduce local coordinates in some neighborhood of any point. Such a notion provides the analytical description for points in a body and a physical space. The structure of the topological manifold, however, is not sufficient for the correct definition of the smoothness of the mappings, specified on its subsets. Therefore, the manifold must be endowed with an additional structure, namely a smooth structure. This allows one to define the notion of a smooth mapping on the manifold independently to a particular choice of coordinates, only relative to the common properties of all atlases in this structure.

Suppose that $\mathfrak{M}$ is a topological manifold with atlas $A = \{(U_\alpha, \varphi_\alpha)\}_{\alpha\in I}$.

Charts $(U_\alpha, \varphi_\alpha)$, $(U_\beta, \varphi_\beta)$ on $\mathfrak{M}$ are called *smoothly compatible* [3], if one of the following holds:

1) $U_\alpha \cap U_\beta = \emptyset$;
2) $U_\alpha \cap U_\beta \neq \emptyset$, and the transition map (13.2) is a $C^\infty$-diffeomorphism.

An atlas $A$ in which any two charts are smoothly compatible with each other is *called a smooth atlas.*

The case of manifolds with boundary is considered in the analogous way. However, some words should be mentioned about transition maps. In the case of manifold without boundary, the transition map $\varphi_\beta \circ \varphi_\alpha^{-1}$ acts between open subsets of $\mathbb{R}^s$ and its smoothness is understood in the standard sense of calculus (all its component functions have continuous partial derivatives of all orders). In the case of manifold with boundary, the set $\varphi_\alpha(U_\alpha \cap U_\beta)$ can be open in $\mathbb{H}^s$, but not in $\mathbb{R}^s$. Thus, one requires to generalize the notion of smoothness for $\varphi_\beta \circ \varphi_\alpha^{-1}$. Suppose that $O$ is an open subset of $\mathbb{H}^s$ and $f\colon O \to \mathbb{R}^k$ is some mapping. This mapping is called to be of class $C^\infty$, if for any point $x \in O$ there exists an open subset $U \subset \mathbb{R}^s$, that contains $x$, and a smooth function $\tilde{f}\colon U \to \mathbb{R}^k$, such that $\tilde{f}|_{U\cap\mathbb{H}^s} = f|_{U\cap\mathbb{H}^s}$. In other words, $f$ is smooth, if locally it can be extended to a mapping that is smooth in the ordinary sense.

**Equivalence relation between smooth atlases.** Let $\mathfrak{M}$ be a topological manifold with or without boundary. Consider the set of all possible atlases on $\mathfrak{M}$. Two smooth atlases $A = \{(U_\alpha, \varphi_\alpha)\}_{\alpha\in I}$, $B = \{(V_\beta, \psi_\beta)\}_{\beta\in J}$ on $\mathfrak{M}$ are equivalent ($A \sim B$), if their union, $A \cup B$, results in a smooth atlas. This relation satisfies all conditions for equivalent relations for sets. The collection of all smooth atlases can be split into equivalence classes $D = [A]$. These equivalence classes are called *smooth structures on the manifold* $\mathfrak{M}$ [49, 111].

Suppose that $D$ is a smooth structure. The *maximal atlas* $A_{\max}$, which corresponds to $D$, is the union of all smooth atlases that constitute $D$. Any chart from the maximal atlas is called a *smooth chart*.

The ordered triple $(\mathfrak{M}, \mathfrak{S}, D)$ is called a *smooth manifold* if $\partial\mathfrak{M} = \emptyset$ and a *smooth manifold with boundary* in the opposite case.

For the sake of brevity, we suppress the symbols $\mathfrak{S}$, $D$ in $(\mathfrak{M}, \mathfrak{S}, D)$ and write "$\mathfrak{M}$ is a smooth manifold" (or, respectively, a "smooth manifold with boundary").

### 13.1.3 Smooth Mappings

Smooth mappings between smooth manifolds play a primary role in the present book. Configurations, deformations, and energy density distributions are only a few examples of such mappings. Bearing on the notion of smooth structure over a manifold, one can obtain a very common definition for the smooth mappings between them [3, 119].

**The real algebra $C^\infty(\mathfrak{M})$ of smooth functions.** Let $(\mathfrak{M}, \mathfrak{S}, [A])$ be an $s$-dimensional smooth manifold (with or without boundary). Consider a function $f: \mathfrak{M} \to \mathbb{R}$. Choose some chart $(U, \varphi) \in A$. The mapping

$$\tilde{f}_\varphi = f \circ \varphi^{-1} : \varphi(U) \to \mathbb{R} ,$$

acts on the set $\varphi(U) \subset \mathbb{R}^s$, which is open either in $\mathbb{R}^s$ or in $\mathbb{H}^s$. We refer to the map $\tilde{f}_\varphi$ as the *coordinate (local) representation of $f$ with respect to the chart* $(U, \varphi)$. Thus, each function $f: \mathfrak{M} \to \mathbb{R}$ can be represented locally as a function defined on an open subset of $\mathbb{R}^s$ or $\mathbb{H}^s$. The notion of smoothness for such functions was defined in conventional calculus [44]. Taking this into account one can define the notion of a smooth function on a smooth manifold.

A function $f: \mathfrak{M} \to \mathbb{R}$ is called a *smooth function*, or of a *class* $C^\infty$, if for every $p \in \mathfrak{M}$, there exists a chart $(U, \varphi) \in A$ such that $p \in U$, and the coordinate representation $\tilde{f}_\varphi$ is smooth on the set $\varphi(U)$. Denote the set of all smooth functions $f: \mathfrak{M} \to \mathbb{R}$ by $C^\infty(\mathfrak{M})$. Such a set forms a real vector space, if one defines addition and multiplication on scalars pointwise:

$$(f+g): \mathfrak{M} \to \mathbb{R} , \quad (f+g)(p) := f(p) + g(p) ,$$
$$(\lambda f): \mathfrak{M} \to \mathbb{R} , \quad (\lambda f)(p) := \lambda f(p) ,$$

where $f, g \in C^\infty(\mathfrak{M})$, $\lambda \in \mathbb{R}$. Pointwise multiplication

$$(fg): \mathfrak{M} \to \mathbb{R} , \quad (fg)(p) := f(p)g(p) ,$$

turns $C^\infty(\mathfrak{M})$ into commutative and associative algebra[3] over $\mathbb{R}$.

---

3 The definition of algebra is given in Chapter 12, Section 12.5.2.

Let $(\mathfrak{M}, \mathfrak{S}, D)$ be an $s$-dimensional smooth manifold. If $\mathfrak{U} \subset \mathfrak{M}$ is an open subset of $\mathfrak{M}$, then, considering it as the open submanifold of $\mathfrak{M}$, one can use the notion $C^\infty(\mathfrak{U})$ for the set of all smooth functions $f\colon \mathfrak{U} \to \mathbb{R}$. Choose some smooth chart $(U, \varphi)$ from the maximal atlas of $\mathfrak{M}$. For each $i = 1, \ldots, s$ define the projection map

$$\pi_s^i\colon \mathbb{R}^s \to \mathbb{R}\,, \quad \pi_s^i(x^1, \ldots, x^s) := x^i\,.$$

Choose one of $i \in \{1, \ldots, s\}$ and construct the scalar-valued function

$$\varphi^i := \pi_s^i \circ \varphi\colon U \to \mathbb{R}\,. \tag{13.3}$$

The coordinate representation of this function is $\widetilde{\varphi}^i = \varphi^i \circ \varphi^{-1}\colon \varphi(U) \to \mathbb{R}$. Obviously, $\widetilde{\varphi}^i = \pi_s^i|_{\varphi(U)}$. Since $\pi_s^i$ is smooth,

$$\forall i = 1, \ldots, s\colon \varphi^i \in C^\infty(U)\,.$$

**Smooth mappings between manifolds.** Suppose that $(\mathfrak{M}, \mathfrak{S}_\mathfrak{M}, [A])$ and $(\mathfrak{N}, \mathfrak{S}_\mathfrak{N}, [B])$ are smooth manifolds (with or without boundary) with dimensions $\dim \mathfrak{M} = s$ and $\dim \mathfrak{N} = k$. Consider a continuous mapping $\varkappa\colon \mathfrak{M} \to \mathfrak{N}$. Choose some chart $(V, \psi) \in B$. One can find at least one chart $(U, \varphi) \in A$, such that $U \cap \varkappa^{-1}(V) \neq \emptyset$. Due to the continuity of $\varkappa$ the set $U \cap \varkappa^{-1}(V)$ is open in $U$ and, hence, the set $\varphi(U \cap \varkappa^{-1}(V))$ is open either in $\mathbb{R}^s$ or in $\mathbb{H}^s$. The mapping

$$\widetilde{\varkappa}_{\varphi,\psi} = \psi \circ \varkappa \circ \varphi^{-1}\colon \varphi(U \cap \varkappa^{-1}(V)) \to \psi(V)$$

is called the *coordinate (local) representation of* $\varkappa$ *with respect to the charts* $(U, \varphi)$ *and* $(V, \psi)$. It is a map between subsets of $\mathbb{R}^s$ and $\mathbb{R}^k$. Let $(x^j)$ denote local coordinates for $\mathfrak{M}$ and $(y^i)$ denote local coordinates for $\mathfrak{N}$. Then, the map $\widetilde{\varkappa}_{\varphi,\psi}$ can be represented by $k$ mappings

$$y^i\colon (x^1, \ldots, x^s) \mapsto y^i(x^1, \ldots, x^s)\,, \quad i = 1, \ldots, k\,.$$

A mapping $\varkappa\colon \mathfrak{M} \to \mathfrak{N}$ is called a *smooth mapping* or of a *class* $C^\infty$ if it is continuous and for every $p \in \mathfrak{M}$, there exist charts $(U, \varphi) \in A$ and $(V, \psi) \in B$, such that $p \in U$, $U \cap \varkappa^{-1}(V) \neq \emptyset$ and the coordinate representation $\widetilde{\varkappa}_{\varphi,\psi}$ is of class $C^\infty$. In this book, we denote the set of all smooth mappings $\varkappa\colon \mathfrak{M} \to \mathfrak{N}$ by $C^\infty(\mathfrak{M}; \mathfrak{N})$.

**Diffeomorphism.** Let $(\mathfrak{M}, \mathfrak{S}_\mathfrak{M}, [A])$, and $(\mathfrak{N}, \mathfrak{S}_\mathfrak{N}, [B])$ be smooth manifolds (with or without boundary). A mapping $\varkappa\colon \mathfrak{M} \to \mathfrak{N}$ is called a $C^\infty$*-diffeomorphism* (or just a *diffeomorphism*) if it is bijective, and both $\varkappa$, $\varkappa^{-1}$ are of class $C^\infty$.

Consider an example. Suppose that $(\mathfrak{M}, \mathfrak{S}, [A])$ is a smooth $s$-dimensional manifold. Choose some chart $(U, \varphi) \in A$. The mappings $\varphi^i$ (13.3), $i = 1, \ldots s$, are smooth. Now consider the mapping $\varphi\colon U \to O$. Here, $O = \varphi(U)$ is an open submanifold of $\mathbb{R}^s$ with the chart $(O, \mathrm{Id}_O)$. The coordinate representation of $\varphi$ has the form $\widetilde{\varphi}_{\varphi,\mathrm{Id}_O} = \mathrm{Id}_O \circ \varphi \circ \varphi^{-1} = \mathrm{Id}_O$. This mapping is of class $C^\infty$, and we have that $\varphi \in C^\infty(U; O)$. Since the mapping $\varphi$ is a homeomorphism, the inverse map $\varphi^{-1}\colon O \to U$ exists. Its coordinate representation is similar to $\mathrm{Id}_O$. Thus, $\varphi\colon U \to O$ is a diffeomorphism between smooth manifolds that are represented by $U$ and $O$.

**Immersions and embeddings.** Let $(\mathfrak{M}, \mathfrak{S}_{\mathfrak{M}}, [A])$ and $(\mathfrak{N}, \mathfrak{S}_{\mathfrak{N}}, [B])$ be smooth manifolds with dimensions $\dim \mathfrak{M} = s$ and $\dim \mathfrak{N} = k$. Among smooth mappings from $C^\infty(\mathfrak{M}; \mathfrak{N})$, there are mappings with specific properties that play an important role in the definitions of configurations and deformations. We will analyze them in detail.

Suppose that $\varkappa \in C^\infty(\mathfrak{M}; \mathfrak{N})$ and $p \in \mathfrak{M}$. Choose some charts $(U, \varphi) \in A$, $(V, \psi) \in B$, such that $p \in U$, $U \cap \varkappa^{-1}(V) \neq \emptyset$, and consider the map

$$\widetilde{\varkappa}_{\varphi,\psi}\colon (x^1, \dots, x^s) \mapsto (y^1, \dots, y^k)\,.$$

One can calculate the Jacobian matrix of this map at point $\varphi(p)$, that is, the matrix with elements

$$\left.\frac{\partial y^i}{\partial x^j}\right|_{\varphi(p)}, \quad i = 1, \dots, k, \quad j = 1, \dots, s\,.$$

Next, one can calculate the number of linearly independent rows (or columns) in this matrix. This number is called the *rank of $\varkappa$ at $p$* and is denoted by $\operatorname{rank}|_p \varkappa$.

We restrict ourselves to the case $s \le k$. Then, a map $\varkappa \in C^\infty(\mathfrak{M}; \mathfrak{N})$ is called a *smooth immersion* if its rank is constant and is equal to $\operatorname{rank}|_p \varkappa = s$ for every point $p \in \mathfrak{M}$. If, in addition, $\varkappa$ is a homeomorphism onto its image, $\varkappa(\mathfrak{M})$, then $\varkappa$ is called a *smooth embedding*.

**Partitions of unity.** For a global definition of the function on a manifold that is covered by several maps, one has to apply some technique for gluing coordinate representations in different charts. This technique can be developed bearing on the topological concept of the partition of unity. In particular, it allows one to define the integration over an arbitrary subset of a manifold.

Suppose that $\mathfrak{M}$ is a smooth manifold with or without boundary. The support $\operatorname{supp} f$ of a mapping $f\colon \mathfrak{M} \to \mathbb{R}$ is the closure of the set $f^{-1}(\mathbb{R} \setminus \{0\})$:

$$\operatorname{supp} f = \overline{f^{-1}(\mathbb{R} \setminus \{0\})}\,.$$

*A partition of unity on a smooth manifold* $\mathfrak{M}$ [123] is a family $\{(U_\alpha, f_\alpha)\}_{\alpha \in I}$ that satisfies the following conditions:

i) The family $\{U_\alpha\}_{\alpha \in I}$ is a locally finite open cover of[4] $\mathfrak{M}$.
ii) For each $\alpha \in I$, one has that $f_\alpha \in C^\infty(\mathfrak{M})$ and $\operatorname{supp} f_\alpha \subset U_\alpha$.
iii) For all $\alpha \in I$ and for each $x \in \mathfrak{M}$: $f_\alpha(x) \ge 0$.
iv) for all $x \in \mathfrak{M}$, one has $\sum_{\alpha \in I} f_\alpha(x) = 1$.

By i) the sum in iv) is finite. If $A = \{(V_\alpha, \varphi_\alpha)\}_{\alpha \in I}$ is an atlas on $\mathfrak{M}$, then a *partition of unity, subordinated to $A$*, is such a partition of unity $\{(U_\beta, f_\beta)\}_{\beta \in J}$, that for any $\beta \in J$ there exists $\alpha \in I$ with the property $U_\beta \subset V_\alpha$.

4 That is, any point from $\mathfrak{M}$ has a neighborhood that intersects only with a finite number of $U_\alpha$.

### 13.1.4 Embedded Submanifolds

The embedding of a smooth manifold into another smooth manifold, probably, with higher dimension, defines a subset of it. This subset can be endowed with the structure of a smooth manifold also. If this structure is aligned with the structure of an enveloping manifold, then the image of the embedded manifold becomes the submanifold of the enveloping one. Such a way to define submanifolds is intensively used in continuum mechanics. For example, the shape of a body is, on the one hand, an image of a material manifold and, on the other hand, a submanifold of the physical space. Another example is given by the definition of the body part, which is a smooth manifold with boundary, aligned with a smooth structure of the whole body (material manifold). Below, we intend to briefly treat submanifolds by following [3].

**Definition of a submanifold.** Let $\mathfrak{M}$ be a smooth manifold with or without boundary. A smooth manifold (with or without boundary) $\mathfrak{N} \subset \mathfrak{M}$ is called an *embedded submanifold of* $\mathfrak{M}$ if

(i) $\mathfrak{N}$ is endowed with the subspace topology;
(ii) the smooth structure of $\mathfrak{N}$ is such that the inclusion map $\iota_{\mathfrak{N}}\colon \mathfrak{N} \to \mathfrak{M}$ (13.1) is a smooth embedding.

An open submanifold is an example of an embedded submanifold. Conversely, if $\mathfrak{M}$ is a smooth manifold, and $\mathfrak{N} \subset \mathfrak{M}$ is an embedded submanifold of $\mathfrak{M}$ such that $\dim \mathfrak{N} = \dim \mathfrak{M}$, then $\mathfrak{N}$ is an open subset of $\mathfrak{M}$.

**An image as a submanifold.** Consider two smooth manifolds $\mathfrak{M}$, $\mathfrak{N}$ with dimensions $\dim \mathfrak{M} \le \dim \mathfrak{N}$. Let $\varkappa\colon \mathfrak{M} \to \mathfrak{N}$ be a smooth embedding. The set $\varkappa(\mathfrak{M})$ can be endowed with topological and smooth structures as follows.

The topology $\mathfrak{S}_{\varkappa(\mathfrak{M})}$ on $\varkappa(\mathfrak{M})$ is induced from $\mathfrak{N}$. Let $A = \{(U_\alpha, \varphi_\alpha)\}_{\alpha\in I}$ be an atlas from the smooth structure of $\mathfrak{M}$. Then, the family $\{\varkappa(U_\alpha)\}_{\alpha\in I}$ is an open cover of $\varkappa(\mathfrak{M})$ (in the induced topology of $\varkappa(\mathfrak{M})$). Since $\varkappa$ is a smooth embedding, the mapping

$$\hat{\varkappa}\colon \mathfrak{M} \to \varkappa(\mathfrak{M}) , \quad \hat{\varkappa}(p) := \varkappa(p) ,$$

is a homeomorphism. For $\alpha \in I$ one has that $\varphi_\alpha\colon U_\alpha \to O_\alpha$ is a homeomorphism and in this regard, $\psi_\alpha = \varphi_\alpha \circ \hat{\varkappa}^{-1}\colon \varkappa(U_\alpha) \to O_\alpha$ is a homeomorphism. Suppose that $\alpha, \beta \in I$ are distinct indices with $\varkappa(U_\alpha) \cap \varkappa(U_\beta) \neq \emptyset$. Then, $U_\alpha \cap U_\beta \neq \emptyset$ and

$$\varphi_\beta \circ \varphi_\alpha^{-1}|_{\varphi_\alpha(U_\alpha\cap U_\beta)}\colon \varphi_\alpha(U_\alpha \cap U_\beta) \to \varphi_\beta(U_\alpha \cap U_\beta) ,$$

is a $C^\infty$-diffeomorphism. Hence,

$$\psi_\beta \circ \psi_\alpha^{-1} = \varphi_\beta \circ \hat{\varkappa}^{-1} \circ \hat{\varkappa} \circ \varphi_\alpha^{-1} ,$$

is also a $C^\infty$-diffeomorphism. It proves that $A_{\varkappa(\mathfrak{M})} = \{(\varkappa(U_\alpha), \psi_\alpha)\}_{\alpha\in I}$ is a smooth atlas on $\varkappa(\mathfrak{M})$. Therefore, $(\varkappa(\mathfrak{M}), \mathfrak{S}_{\varkappa(\mathfrak{M})}, [A_{\varkappa(\mathfrak{M})}])$ is a smooth manifold.

Consider the mapping $\widehat{\varkappa}\colon \mathfrak{M} \to \varkappa(\mathfrak{M})$. In the appropriate charts $(U_\alpha, \varphi_\alpha)$ and $(\varkappa(U_\alpha), \psi_\alpha)$, one has

$$\widetilde{\widehat{\varkappa}} = \psi_\alpha \circ \widehat{\varkappa} \circ \varphi_\alpha^{-1} = \mathrm{Id}_{O_\alpha} .$$

Similarly, for $\widehat{\varkappa}^{-1}$, one obtains

$$\widetilde{\widehat{\varkappa}^{-1}} = \varphi_\alpha \circ \widehat{\varkappa}^{-1} \circ \psi_\alpha^{-1} = \mathrm{Id}_{O_\alpha} .$$

Hence, $\widehat{\varkappa}$ is a $C^\infty$-diffeomorphism. Moreover, the inclusion map $\iota_{\varkappa(\mathfrak{M})}\colon \varkappa(\mathfrak{M}) \to \mathfrak{N}$ is a composition

$$\iota_{\varkappa(\mathfrak{M})} = \varkappa \circ \widehat{\varkappa}^{-1} ,$$

of a diffeomorphism and a smooth embedding and, consequently, $\iota_{\varkappa(\mathfrak{M})}$ is a $C^\infty$-embedding. As a result, we conclude that triple $(\varkappa(\mathfrak{M}), \mathfrak{S}_{\varkappa(\mathfrak{M})}, [A_{\varkappa(\mathfrak{M})}])$ defines an embedded submanifold of $\mathfrak{N}$.

**Restricting maps to submanifolds.** Suppose that $\mathfrak{M}$, $\mathfrak{N}$ are smooth manifolds, and $\varkappa\colon \mathfrak{M} \to \mathfrak{N}$ is a smooth mapping. If $\mathfrak{U} \subset \mathfrak{M}$ is an embedded submanifold, then $\varkappa|_{\mathfrak{U}} = \varkappa \circ \iota_{\mathfrak{U}}\colon \mathfrak{U} \to \mathfrak{N}$ is a composition of smooth mappings. Therefore, the restriction $\varkappa|_{\mathfrak{U}}\colon \mathfrak{U} \to \mathfrak{N}$ is smooth.

## 13.2 The "Tower" of Tensor Spaces

One can use Whitney's embedding theorem and consider the embedding of a smooth manifold (a body or a physical space) in some ambient Euclidean space of higher dimension. The representation of a smooth manifold as a hypersurface in a Euclidean space allows one to introduce tangent vectors on it as directed line segments by appealing to the generalization of conventional Euclidean geometry on higher dimensions. However, such a direct approach turns out to be cumbersome because of the higher dimension of the ambient space. At the same time, the description of geometric objects by methods of internal geometry has always determined the elegance of the geometric theory. In this regard, it seems more convenient to define the tangent vectors, covectors, *etc.*, as intrinsic objects on smooth manifolds.

### 13.2.1 Tangent Space to a Smooth Manifold

Suppose that $\mathfrak{M}$ is a smooth $s$-dimensional manifold and $p \in \mathfrak{M}$ is a point. Intuitively, a tangent vector to $\mathfrak{M}$ at $p$ represents an "infinitesimal", or, in other words, "local", object. Such an object must be introduced in an intrinsic manner, that is, in terms of the "inhabitants" of a smooth manifold; the *smooth curves* and *smooth functions* defined on it. One can formalize the notion of locality by dealing with smooth curves and smooth functions as follows.

**Tangent vectors as an equivalence class of curves.** A *smooth curve on* $\mathfrak{M}$ is a mapping $\chi \in C^\infty(\mathbb{J};\mathfrak{M})$, where $\mathbb{J} \subset \mathbb{R}$ is a real line interval (open or closed). A curve $\chi \in C^\infty(\mathbb{J};\mathfrak{M})$ *passes through the point* $p \in \mathfrak{M}$, if there exists $c \in \mathbb{J}$, such that $\chi(c) = p$. Consider the set $\mathfrak{Cu}_p(\mathfrak{M})$ of all smooth curves that satisfy the following conditions: $\chi \in \mathfrak{Cu}_p(\mathfrak{M})$ *iff*
($\mathfrak{Cu}_1$) $\chi$ is a smooth curve on $\mathfrak{M}$ that is defined on an open interval containing 0.
($\mathfrak{Cu}_2$) $\chi(0) = p$. Moreover, $\chi^{-1}(\{p\}) = \{0\}$.

The condition $\chi^{-1}(\{p\}) = \{0\}$ requires that $0 \in \mathbb{R}$ is the unique point that transforms to $p$. Let $(U, \varphi)$ be a chart from the maximal atlas of $\mathfrak{M}$ that contains $p$. Introduce the following equivalence relation $\sim_p$ [119]:

$$\forall \chi_1, \chi_2 \in \mathfrak{Cu}_p(\mathfrak{M}) \colon (\chi_1 \sim_p \chi_2) \Leftrightarrow \left( \frac{d(\varphi \circ \chi_1)(t)}{dt}\bigg|_{t=0} = \frac{d(\varphi \circ \chi_2)(t)}{dt}\bigg|_{t=0} \right).$$

Note that $\varphi \circ \chi_i$, $i = 1, 2$, are smooth mappings from symmetric intervals to $\mathbb{R}^s$, and the derivative $d/dt$ is understood in the sense of conventional calculus [44]. If $(V, \psi)$ is another chart that contains $p$, then by the chain rule one could be assured that if $\chi_1 \sim_p \chi_2$ relatively to the chart $(U, \varphi)$, then $\chi_1 \sim_p \chi_2$ relatively to $(V, \psi)$. Thus, the equivalence relation does not depend on a particular chart. Each equivalence class $u = [\chi]_p$ is called a *tangent vector at p*, and the quotient set

$$T_p\mathfrak{M} := \mathfrak{Cu}_p(\mathfrak{M})/ \sim_p$$

is called *the tangent space to* $\mathfrak{M}$ *at p* [119].

**The vector structure of the tangent space.** Every chart $(U, \varphi)$ from the maximal atlas of $\mathfrak{M}$, which contains $p$, generates the mapping

$$\mathfrak{c}_p^\varphi \colon T_p\mathfrak{M} \to \mathbb{R}^s, \quad \mathfrak{c}_p^\varphi \colon u = [\chi]_p \mapsto \frac{d(\varphi \circ \chi)}{dt}\bigg|_{t=0} = (u^1, \dots, u^s). \tag{13.4}$$

By the definition of the equivalence relation $\sim_p$, such a mapping is well defined, *i.e.*, its value does not depend on a representative $\chi \in u$. Moreover, the mapping $\mathfrak{c}_p^\varphi$ is a bijection. This can be easily verified by means of the following arguments.

Suppose that $u_1, u_2 \in T_p\mathfrak{M}$ and $u_1 \neq u_2$. If one assumes that $\mathfrak{c}_p^\varphi(u_1) = \mathfrak{c}_p^\varphi(u_2)$ then, choosing representatives $\chi_1 \in u_1$ and $\chi_2 \in u_2$, one obtains that $\chi_1 \sim_p \chi_2$. This implies a contradiction with $u_1 \neq u_2$. Thus, $\mathfrak{c}_p^\varphi$ is injective.

Now, let $(u^1, \dots, u^s) \in \mathbb{R}^s$ be an arbitrary $s$-tuple and $(p^1, \dots, p^s) = \varphi(p)$. Choose $a > 0$, such that the image $\tilde{\chi}(]-a, a[)$ of the curve,

$$\tilde{\chi} \in C^\infty(]-a, a[; \mathbb{R}^s), \quad \tilde{\chi}(t) := (p^1 + u^1 t, \dots, p^s + u^s t),$$

is contained in $\varphi(U)$. Taking $\chi = \varphi^{-1} \circ \tilde{\chi}$, one obtains the curve $\chi \in \mathfrak{Cu}_p(\mathfrak{M})$, such that $\mathfrak{c}_p^\varphi([\chi]_p) = (u^1, \dots, u^s)$. Thus, $\mathfrak{c}_p^\varphi$ is surjective. We have arrived at the desired conclusion: $\mathfrak{c}_p^\varphi$ is bijective.

By definition, the mapping $\mathfrak{c}_p^{\varphi}$ depends on the chart $(U, \varphi)$. Let $(U, \varphi)$ and $(V, \psi)$ be two charts from the maximal atlas that contain $p$. The chain rule implies that

$$\mathfrak{c}_p^{\psi} = D_{\varphi(p)}(\psi \circ \varphi^{-1}) \circ \mathfrak{c}_p^{\varphi} , \tag{13.5}$$

where $D_{\varphi(p)}(\psi \circ \varphi^{-1}) \in \mathrm{Lin}(\mathbb{R}^s; \mathbb{R}^s)$ is the total derivative of the transition map $\psi \circ \varphi^{-1}$ at $\varphi(p)$.

Since $\mathfrak{c}_p^{\varphi} : T_p\mathfrak{M} \to \mathbb{R}^s$ is a bijective mapping with values on the $s$-dimensional vector space $\mathbb{R}^s$, one can induce the real vector space structure of the latter into $T_p\mathfrak{M}$ by means of $\mathfrak{c}_p^{\varphi}$. To this end, define the mappings

$$\begin{aligned} +&: T_p\mathfrak{M} \times T_p\mathfrak{M} \to T_p\mathfrak{M} , \quad (u + v) := \left(\mathfrak{c}_p^{\varphi}\right)^{-1} \left(\mathfrak{c}_p^{\varphi}(u) + \mathfrak{c}_p^{\varphi}(v)\right) , \\ \cdot&: \mathbb{R} \times T_p\mathfrak{M} \to T_p\mathfrak{M} , \quad (\lambda \cdot u) := \left(\mathfrak{c}_p^{\varphi}\right)^{-1} \left(\lambda \mathfrak{c}_p^{\varphi}(u)\right) , \end{aligned}$$

where $u, v \in T_p\mathfrak{M}$ and $\lambda \in \mathbb{R}$. Although the mapping $\mathfrak{c}_p^{\varphi}$ depends on the chart $(U, \varphi)$, direct calculations taking the equality (13.5) into account show that the introduced mappings $(+)$ and $(\cdot)$ are chart-independent. Thus, the mappings $(+)$ and $(\cdot)$ are a well-defined real vector space structure on $T_p\mathfrak{M}$, and $\mathfrak{c}_p^{\varphi}$ is the vector space isomorphism relative to such a structure. This implies that $\dim T_p\mathfrak{M} = s$.

**Coordinate basis.** The standard basis of $\mathbb{R}^s$ is formed by the collection of the $s$-tuples $(\delta_i^1, \ldots, \delta_i^s)$, where 1 stands in the $i$-th position and 0 stands in others. Choose some smooth chart $(U, \varphi)$, which contains $p$. Since $\mathfrak{c}_p^{\varphi}$ is the vector space isomorphism, the collection $(e_i, \ldots, e_s)$ of tangent vectors

$$e_i = \left(\mathfrak{c}_p^{\varphi}\right)^{-1} \left(\delta_i^1, \ldots, \delta_i^s\right) \in T_p\mathfrak{M} , \quad i = 1, \ldots, s , \tag{13.6}$$

forms a basis in $T_p\mathfrak{M}$. Each vector $e_i$ is an equivalence class of the curve $\chi_i$, such that

$$\varphi \circ \chi_i(t) = \left(p^1 + \delta_i^1 t, \ldots, p^s + \delta_i^s t\right) , \tag{13.7}$$

where $(p^1, \ldots, p^s) = \varphi(p)$. We refer to the basis $(e_i, \ldots, e_s)$ as *coordinate basis*. Such a basis depends on the chart $(U, \varphi)$. Each tangent vector $u \in T_p\mathfrak{M}$ has a decomposition $u = u^i e_i$, where $(u^1, \ldots, u^s) = \mathfrak{c}_p^{\varphi}(u)$.

Suppose that $(U, \varphi)$ and $(V, \psi)$ are two smooth charts that contain $p$ and establish the local coordinates $(x^i)$, $(\tilde{x}^i)$. If $(u^i)$ and $(\tilde{u}^i)$ are components of the same vector $u \in T_p\mathfrak{M}$ in these charts, then the equality (13.5) implies that

$$\tilde{u}^i = \left.\frac{\partial \tilde{x}^i}{\partial x^j}\right|_{\varphi(p)} u^j , \quad i = 1, \ldots, s . \tag{13.8}$$

Here, $\psi \circ \varphi^{-1} : (x^1, \ldots, x^s) \mapsto (\tilde{x}^1, \ldots, \tilde{x}^s)$ is a transition map (13.2).

**The tangent vector as a derivation of the space of germs.** Let $U \subset \mathfrak{M}$ be an open set and $f \in C^\infty(U; \mathbb{R})$ be a smooth function. The ordered pair $(f, U)$ is called a *smooth function element* on $\mathfrak{M}$ [3].

Denote the set of all smooth function elements $(f, U)$ on $\mathfrak{M}$, where $p \in U$, by $\mathrm{Fel}_p(\mathfrak{M})$. Define the equivalence relation $\sim_p$ on $\mathrm{Fel}_p(\mathfrak{M})$ by the following condition: $(f, U) \sim_p (g, V)$, if there exists an open set $W \subset U \cap V$ containing $p$ such that $f|_W = g|_W$. The equivalence class $\mathsf{f} = [(f, U)]_p$ of a function element $(f, U)$ is called the *germ of f at p* [3], and the quotient set

$$C^\infty(\mathfrak{M}) := \mathrm{Fel}_p(\mathfrak{M})/\sim_p$$

represents the set of all germs of smooth functions at $p$. Note that if $U$ and $V$ are neighborhoods of $p$, $V \subset U$ and $(f, U) \in \mathrm{Fel}_p(\mathfrak{M})$, then $(f|_V, V) \in [(f, U)]_p$. In this regard, so that there will not be any confusion, we will write $[f]_p$ instead of $[(f, U)]_p$.

The set $C^\infty(\mathfrak{M})$ is endowed by the real vector space structure as follows. Define the binary operations

$$+\colon C^\infty(\mathfrak{M}) \times C^\infty(\mathfrak{M}) \to C^\infty(\mathfrak{M})\,, \qquad \mathsf{f} + \mathsf{g} := [(f + g, U \cap V)]_p\,,$$
$$\cdot\colon \mathbb{R} \times C^\infty(\mathfrak{M}) \to C^\infty(\mathfrak{M})\,, \qquad \lambda\mathsf{f} := [(\lambda f + g, U)]_p\,,$$

where $(f, U) \in \mathsf{f}$, $(g, V) \in \mathsf{g}$ and $\lambda \in \mathbb{R}$. It is easy to check that these operations are well defined and bring $C^\infty(\mathfrak{M})$ to the real vector space. Moreover, the operation

$$\cdot\colon C^\infty(\mathfrak{M}) \times C^\infty(\mathfrak{M}) \to C^\infty(\mathfrak{M})\,, \quad \mathsf{f} + \mathsf{g} := [(fg, U \cap V)]_p\,,$$

turns $C^\infty(\mathfrak{M})$ into real associative algebra.

Since for arbitrary $(f, U), (g, V) \in \mathsf{f}$, the equality $f(p) = g(p)$ holds, the notion $\mathsf{f}(p) := f(p)$ is well defined.

A *derivation of* $C^\infty(\mathfrak{M})$ [3, 48, 132] is a linear map

$$v\colon C^\infty(\mathfrak{M}) \to \mathbb{R}$$

that satisfies the following *product rule*:

$$\forall \mathsf{f}, \mathsf{g} \in C^\infty(\mathfrak{M})\colon v(\mathsf{f}\mathsf{g}) = \mathsf{f}(p)v(\mathsf{g}) + \mathsf{g}(p)v(\mathsf{f})\,.$$

Denote the set of all derivations by $D_p\mathfrak{M}$. Each element of $D_p\mathfrak{M}$ is called a *tangent vector at p* [3, 48, 132]. This gives an alternative definition of a tangent vector.

Suppose that $u, v \in D_p\mathfrak{M}$ and $\lambda \in \mathbb{R}$. Since $u$ and $v$ are linear mappings, the sum $(u + v)\colon C^\infty(\mathfrak{M}) \to \mathbb{R}$ and the product $(\lambda u)\colon C^\infty(\mathfrak{M}) \to \mathbb{R}$ are also linear mappings. Direct calculations show that they satisfy the product rule. Hence, the set $D_p\mathfrak{M}$ is a real vector space.

Let $v \in D_p\mathfrak{M}$. The following elementary properties follow directly from the definition of the derivation:

- If $\mathsf{f}$ is a germ of a constant function then[5] $v(\mathsf{f}) = 0$.
- If $\mathsf{f}(p) = \mathsf{g}(p) = 0$ then $v(\mathsf{f}\mathsf{g}) = 0$.

5 If $f\colon U \to \mathbb{R}$ is such that $f(U) = \{c\}$, then $[f]_p = c[e]_p$, where $e \in C^\infty(\mathfrak{M})$ is such function that $e(\mathfrak{M}) = \{1\}$. Since $v([f]_p) = cv([e]_p)$, it is sufficient to prove that $v([e]_p) = 0$. Using the equality $[e]_p[e]_p = [e]_p$ and the product rule one obtains that $v([e]_p) = v([e]_p[e]_p) = 2v([e]_p)$. This implies the desired equality.

**The isomorphism $T_p\mathfrak{M} \cong D_p\mathfrak{M}$.** Both equivalence classes of curves and derivations of germs are no more than alternative formalizations for the idea of locality. Despite the fact that such formalizations are constructed with different argumentations, they represent a unified object (up to natural isomorphism). Let $u \in T_p\mathfrak{M}$ be a tangent vector. It generates the mapping $\mathscr{L}_u : C^\infty(\mathfrak{M}) \to \mathbb{R}$,

$$\mathscr{L}_u(\mathsf{f}) := \left.\frac{d(f \circ \chi)}{dt}\right|_{t=0}, \tag{13.9}$$

where $\chi \in u$ and $(f, U) \in \mathsf{f}$. The chain rule and the definitions of $\mathsf{f}$ and $u$ imply that the value of $\mathscr{L}_u$ does not depend on representatives from $u$ and $\mathsf{f}$. It follows from the definition that (13.9) does not require any chart. Thus, $\mathscr{L}_u$ is well defined. Since $d/dt$ is linear and satisfies the Leibniz rule, $\mathscr{L}_u \in D_p\mathfrak{M}$.

Consider the mapping $\mathscr{L} : T_p\mathfrak{M} \to D_p\mathfrak{M}$, $\mathscr{L} : u \mapsto \mathscr{L}_u$, where $\mathscr{L}_u$ is defined by (13.9). Such a mapping is the desired isomorphism, because

- *The mapping $\mathscr{L}$ is linear.* It follows from the linearity of $\mathfrak{c}_p^\varphi$.
- *The mapping $\mathscr{L}$ is injective.* Suppose that $\mathscr{L}_{u_1} = \mathscr{L}_{u_2}$. Choose a smooth chart $(U, \varphi)$ that contains $p$. Take $\varphi^i := \pi_s^i \circ \varphi$ (13.3), where

$$\pi_s^i : \mathbb{R}^s \to \mathbb{R}, \quad \pi_s^i(x^1, \dots, x^s) = x^i, \quad i = 1, \dots, s.$$

  Then one has $\mathscr{L}_{u_1}([\varphi^i]_p) = \mathscr{L}_{u_2}([\varphi^i]_p)$, for all $i = 1, \dots, s$. Choosing representatives $\chi_1 \in u_1, \chi_2 \in u_2$, one obtains

$$\left.\frac{d\left(\pi_s^i \circ \varphi \circ \chi_1\right)}{dt}\right|_{t=0} = \left.\frac{d\left(\pi_s^i \circ \varphi \circ \chi_2\right)}{dt}\right|_{t=0}, \quad i = 1, \dots, s,$$

  and this implies $u_1 = u_2$.
- *The mapping $\mathscr{L}$ is surjective.* Suppose that $v \in D_p\mathfrak{M}$. One needs to obtain such tangent vector $u \in T_p\mathfrak{M}$, for which $\mathscr{L}_u = v$. Choose a smooth chart $(U, \varphi)$, that contains $p$ and denote $\varphi^i = \pi_s^i \circ \varphi$ and $(p^1, \dots, p^s) = \varphi(p)$.
  Take a germ $\mathsf{f} = [(f, \widetilde{U})]_p \in C^\infty(\mathfrak{M})$. Without any loss of generality, one can assume that $\widetilde{U} \subset U$. Using Taylor's formula, one obtains the following decomposition on $\varphi(\widetilde{U})$:

$$\begin{aligned} f \circ \varphi^{-1}(x^1, \dots, x^s) = f(p) + \left.\frac{\partial(f \circ \varphi^{-1})}{\partial x^i}\right|_{\varphi(p)} (x^i - p^i) \\ + f_{ij}(x^1, \dots, x^s)(x^i - p^i)(x^j - p^j), \end{aligned} \tag{13.10}$$

  where $f_{ij}$ is a smooth function (the remainder term) on $\varphi(\widetilde{U})$. Identifying $f(p)$ and $p^i$ with the corresponding constant mappings on $\widetilde{U}$, one can rewrite (13.10) on $\widetilde{U}$:

$$f = f(p) + \left.\frac{\partial(f \circ \varphi^{-1})}{\partial x^i}\right|_{\varphi(p)} (\varphi^i - p^i) + \tilde{f}_{ij}(\varphi^i - p^i)(\varphi^j - p^j), \tag{13.11}$$

where $\widetilde{f}_{ij} = f_{ij} \circ \varphi$. Using the algebraic structure of germs, the properties of $v$, and the relation (13.11), one obtains

$$v([f]_p) = \left.\frac{\partial(f \circ \varphi^{-1})}{\partial x^i}\right|_{\varphi(p)} v\left([\varphi^i]_p\right) . \tag{13.12}$$

The formula (13.12) was derived for an arbitrary germ $[f]_p$.
Define a smooth curve $\chi : t \mapsto \chi(t)$ by the relation

$$\chi(t) = \varphi^{-1}\left(p^1 + tv\left([\varphi^1]_p\right), \dots, p^s + tv\left([\varphi^s]_p\right)\right) .$$

Here, $(U, \varphi)$ and $p^i$ are the same as above. Such a curve belongs to $\mathfrak{Cu}_p(\mathfrak{M})$. Set $u := [\chi]_p \in T_p\mathfrak{M}$. Choose a germ $[f]_p \in C^\infty(\mathfrak{M})$. Direct calculations show that

$$\mathscr{L}_u([f]_p) = \left.\frac{d(f \circ \chi)}{dt}\right|_{t=0} = D_{\varphi(p)}(f \circ \varphi^{-1}) \left.\frac{d(\varphi \circ \chi)}{dt}\right|_{t=0} = \left.\frac{\partial(f \circ \varphi^{-1})}{\partial x^i}\right|_{\varphi(p)} v\left([\varphi^i]_p\right) .$$

One obtains that $v(\mathrm{f}) = \mathscr{L}_u(\mathrm{f})$ for each germ $\mathrm{f} \in C^\infty(\mathfrak{M})$. Thus, $u$ is the desired vector.

The above reasonings show that $\mathscr{L}$ is a natural isomorphism. Finally, we can claim that $T_p\mathfrak{M} \cong D_p\mathfrak{M}$.

Suppose that $(U, \varphi)$ is a smooth chart that contains $p$, and $(e_1, \dots, e_s)$ is the corresponding coordinate basis (13.6) of $T_p\mathfrak{M}$, which is generated by curves (13.7). Since $\mathscr{L}$ is the isomorphism, then $(\mathscr{L}_{e_1}, \dots, \mathscr{L}_{e_s})$ is the basis in $D_p\mathfrak{M}$. Thus, if $u = u^i e_i$, then $\mathscr{L}_u = u^i \mathscr{L}_{e_i}$. Let $[f]_p \in C^\infty(\mathfrak{M})$. Then,

$$\mathscr{L}_{e_i}([f]_p) = \left.\frac{d(f \circ \chi)}{dt}\right|_{t=0} = D_{\varphi(p)}(f \circ \varphi^{-1})\left(\delta_i^1, \dots, \delta_i^s\right) = \left.\frac{\partial(f \circ \varphi^{-1})}{\partial x^i}\right|_{\varphi(p)} .$$

The derivation $\mathscr{L}_{e_i}$ acts like a partial derivative. This motivates the following notation:

$$\partial_i|_p := \mathscr{L}_{e_i} , \quad i = 1, \dots, s . \tag{13.13}$$

Then, for a tangent vector $u \in T_p\mathfrak{M}$, one has $\mathscr{L}_u = u^i \partial_i|_p$. In the following, we identify $u \in T_p\mathfrak{M}$ with its image $\mathscr{L}_u \in D_p\mathfrak{M}$ by virtue of the isomorphism $\mathscr{L}$. Thus, $(e_i)_{i=1}^s$ is identified with $(\partial_i|_p)_{i=1}^s$ and one obtains the following expansion:

$$u = u^i \partial_i|_p$$

for a tangent vector $u \in T_p\mathfrak{M}$.

Suppose that $(U, \varphi)$ and $(V, \psi)$ are two smooth charts that contain $p$ and establish local coordinates $(x^i)$, $(\tilde{x}^i)$. The relations (13.8) imply that

$$\partial_{x^i}|_p = \left.\frac{\partial \tilde{x}^j}{\partial x^i}\right|_{\varphi(p)} \partial_{\tilde{x}^j}|_p , \quad i = 1, \dots, s , \tag{13.14}$$

where $(x^1, \dots, x^s) \mapsto (\tilde{x}^1, \dots, \tilde{x}^s)$ is the transition map $\psi \circ \varphi^{-1}$. Symbols $x$ and $\tilde{x}$ in $\partial_{x^i}|_p$ and $\partial_{\tilde{x}^i}|_p$ indicate that we are dealing with different coordinate bases. In this book, we use such notations if there is any danger of confusion.

An equivalence class of a curve or, similarly, a derivation of germs, represents an "infinitesimal" object at $p$. The set of all such objects, the tangent space $T_p\mathfrak{M}$, is the formalization of the idea of an "infinitesimal" neighborhood of $p$.

**The tangent vector as a derivation of the algebra $C^\infty(\mathfrak{M})$.** A *derivation at* $p$ is a linear map $v\colon C^\infty(\mathfrak{M}) \to \mathbb{R}$ that satisfies the "product rule" relation [3, 119]:

$$\forall f, g \in C^\infty(\mathfrak{M})\colon v(fg) = f(p)v(g) + g(p)v(f) .$$

The set of all derivations at point $p$ can be endowed with the structure of a vector space by means of the operations of addition and scalar multiplication, the same as defined previously for linear mappings. Denote this space by $V_p\mathfrak{M}$. For any derivation $v \in V_p\mathfrak{M}$, the following elementary properties hold:

- If $f \in C^\infty(\mathfrak{M})$ is a constant function then $v(f) = 0$.
- If $f(p) = g(p) = 0$ then $v(fg) = 0$.

Suppose that $f, g \in C^\infty(\mathfrak{M})$ are such mappings that $f|_W = g|_W$ for some neighborhood $W \subset \mathfrak{M}$ of $p$. Then, for any $v \in C^\infty(\mathfrak{M})$, we have $v(f) = v(g)$ just by the definition of a germ at $p$. A similar locality property still holds for any $v \in V_p\mathfrak{M}$: $v(f) = v(g)$, but contrary to the case of germ, this property needs to be proved. The proof of such a property based on the construction of a function $h\colon \mathfrak{M} \to \mathbb{R}$ such that $h(p) = 1$ and $h|_{\mathfrak{M}\setminus W} = 0$ is given in [119].

Any vector $u \in T_p\mathfrak{M}$ defines the derivation $\mathscr{L}_u\colon C^\infty(\mathfrak{M}) \to \mathbb{R}$ in much the same way as was the case for germs (see (13.9)):

$$\mathscr{L}_u(f) := \left.\frac{d(f \circ \chi)}{dt}\right|_{t=0} ,$$

where $\chi \in u$. Reasonings similar to those for constructing the natural isomorphism $T_p\mathfrak{M} \cong D_p\mathfrak{M}$ allow one to construct the natural isomorphism $T_p\mathfrak{M} \cong V_p\mathfrak{M}$.

### 13.2.2 The tangent Map at a Point

Suppose that $\mathfrak{M}$ and $\mathfrak{N}$ are $s$ and $k$-dimensional smooth manifolds, respectively. Let $p \in \mathfrak{M}$ be a point and $\varkappa\colon \mathfrak{M} \to \mathfrak{N}$ be a smooth mapping. This mapping transforms all points of $\mathfrak{M}$ into points of $\mathfrak{N}$. The idea of the tangent map is to define the linear transformation of $T_p\mathfrak{M}$ to $T_{\varkappa(p)}\mathfrak{N}$, which is aligned with given $\varkappa$. Below, two alternative definitions are given [3, 123].

- **The first definition.** *Tangent map to* $\varkappa$ *at point* $p$ is the mapping

$$T_p\varkappa\colon T_p\mathfrak{M} \to T_{\varkappa(p)}\mathfrak{N} ,$$
$$\forall u \in T_p\mathfrak{M}\colon T_p\varkappa(u) := [\varkappa \circ \chi]_{\varkappa(p)} ,$$

where $\chi \in u$. The chain rule implies that such a definition does not depend on the representative of $u$. Thus, $T_p\varkappa$ transforms the equivalence class of a curve into the equivalence class of the image of this curve.

- **The second definition.** *The tangent map to $\varkappa$ at point $p$* is the mapping

$$T_p\varkappa\colon D_p\mathfrak{M} \to D_{\varkappa(p)}\mathfrak{N}\,,$$
$$\forall v \in T_p\mathfrak{M}\colon\ T_p\varkappa(v)(\mathrm{f}) := v([f \circ \varkappa]_p)\,,$$

where $f \in \mathrm{f} \in C^\infty(\mathfrak{N})$.

The so-defined tangent map satisfies the following properties [3]. Let $\mathfrak{M}$, $\mathfrak{N}$, $\mathfrak{P}$ be smooth manifolds, $\varkappa \in C^\infty(\mathfrak{M};\mathfrak{N})$, $\gamma \in C^\infty(\mathfrak{N};\mathfrak{P})$ are smooth mappings, and $p \in \mathfrak{M}$. Then:

i) The mapping $T_p\varkappa\colon T_p\mathfrak{M} \to T_{\varkappa(p)}\mathfrak{N}$ is linear.

ii) The tangent map to a composition is a composition of tangent maps:

$$T_p(\gamma \circ \varkappa) = T_{\varkappa(p)}\gamma \circ T_p\varkappa\colon T_p\mathfrak{M} \to T_{\gamma\circ\varkappa(p)}\mathfrak{P}\,.$$

iii) $T_p\mathrm{Id}_{\mathfrak{M}} = \mathrm{Id}_{T_p\mathfrak{M}}$.

iv) If $\varkappa$ is a diffeomorphism then $T_p\varkappa\colon T_p\mathfrak{M} \to T_{\varkappa(p)}\mathfrak{N}$ is isomorphism and $(T_p\varkappa)^{-1} = T_{\varkappa(p)}\varkappa^{-1}$.

Such properties are similar to the corresponding properties of the total derivative $D_xf$ of a smooth mapping $f\colon \mathbb{R}^s \to \mathbb{R}^k$ at point $x$. The property i) follows from the chain rule applied to total derivative and from the definition of $\mathfrak{c}_p^\varphi$. The properties ii) and iii) follow from the definition of the tangent map. Finally, the property iv) follows from ii) and iii).

### 13.2.3 Cotangent Space to a Smooth Manifold

Let $\mathfrak{M}$ be a smooth $s$-dimensional manifold and $p \in \mathfrak{M}$ be a point. A real vector space of linear functionals (covectors) over $T_p\mathfrak{M}$ (the dual vector space) is denoted by $T_p^*\mathfrak{M} := (T_p\mathfrak{M})^*$ and is called the *cotangent space to $\mathfrak{M}$ at $p$*. Like in Chapter 12, Section 12.5.2, we define the bilinear operation

$$\langle\cdot,\cdot\rangle_p\colon T_p^*\mathfrak{M} \times T_p\mathfrak{M} \to \mathbb{R}\,, \quad \langle v, u\rangle_p := v(u)\,.$$

It follows from the natural isomorphism between $T_p\mathfrak{M}$ and $(T_p\mathfrak{M})^{**}$ that $\langle\cdot,\cdot\rangle_p$ is symmetric:

$$\forall v \in T_p^*\mathfrak{M}\ \forall u \in T_p\mathfrak{M}\colon\ \langle v, u\rangle_p = \langle u, v\rangle_p\,.$$

**The coordinate dual basis.** Choose a smooth chart $(U, \varphi)$, such that $p \in U$. To this chart there corresponds the coordinate basis $(\partial_i|_p)_{i=1}^s$. The mapping $\mathfrak{c}_p^\varphi$ (13.4) generates

$s$ linear functionals

$$dx^i|_p = \pi^i_s \circ c^{\varphi}_p : T_p\mathfrak{M} \to \mathbb{R}\,, \quad i = 1, \dots, s\,, \tag{13.15}$$

that satisfy the following conditions: if $u = u^i \partial_i|_p \in T_p\mathfrak{M}$, then $dx^i|_p(u) = u^i$. Thus,

$$\left\langle dx^i|_p, \partial_j|_p \right\rangle_p = \delta^i_j\,, \quad i, j = 1, \dots, s\,,$$

and $(dx^i|_p)^s_{i=1}$ is a dual basis. We refer to it as the *coordinate dual basis.*

Each covector $v \in T^*_p\mathfrak{M}$ can be uniquely decomposed as $v = v_i dx^i|_p$. Suppose that $(U, \varphi)$ and $(V, \psi)$ are two smooth charts with local coordinates $(x^i)$, $(\tilde{x}^i)$ that contain $p$. Therefore, an arbitrary covector $v \in T^*_p\mathfrak{M}$ can be decomposed in two ways: $v = v_i dx^i|_p = \tilde{v}_i d\tilde{x}^i|_p$. The formula (13.14) implies that

$$v_i = \left.\frac{\partial \tilde{x}^j}{\partial x^i}\right|_{\varphi(p)} \tilde{v}_j\,, \quad i = 1, \dots, s\,. \tag{13.16}$$

**The differential.** The differential of a smooth function provides an example of a covector. Suppose that $f \in C^\infty(\mathfrak{M})$. The *differential of $f$ at $p$* [3] is a mapping $df_p : T_p\mathfrak{M} \to \mathbb{R}$ that acts as follows:

$$\forall u \in T_p\mathfrak{M} : df_p(u) := u([f]_p)\,, \tag{13.17}$$

where $[f]_p$ is the germ of $f$ at $p$. It is a linear mapping and, thus, the element of $T^*_p\mathfrak{M}$. Using the dual basis (13.15) one obtains

$$df_p = \left.\frac{\partial(f \circ \varphi^{-1})}{\partial x^i}\right|_{\varphi(p)} dx^i|_p\,.$$

Let $(U, \varphi)$ be a smooth chart. For $i = 1, \dots, s$, consider the differential $d\varphi^i_p$ of the mapping $\varphi^i$ (13.3) at some point $p \in U$. Since $\varphi^i \circ \varphi^{-1} = \pi^i_s$, one has

$$d\varphi^i_p = \left.\frac{\partial \pi^i_s}{\partial x^j}\right|_{\varphi(p)} dx^j|_p = \delta^i_j dx^j|_p\,.$$

Hence, $d\varphi^i_p = dx^i|_p$. This formula means that the covector $dx^i|_p$ can be considered as the result of applying of the operation $d$ to the map $\varphi^i$. For a point $x$, the latter returns its $i$-th coordinate $x^i$. Such an observation motivates the use of the notation $dx^i|_p$ for covectors which constitute the coordinate dual basis.

### 13.2.4 Remarks on the Tangent Spaces

**The tangent space to an open submanifold.** Suppose that $\mathfrak{M}$ is a smooth manifold and $\mathfrak{U} \subset \mathfrak{M}$ is an open subset. Let $p \in \mathfrak{U}$ be a point. One has two tangent spaces: $T_p\mathfrak{M}$ and $T_p\mathfrak{U}$, where $\mathfrak{U}$ is considered as an open submanifold of $\mathfrak{M}$. The tangent map, $T_p\iota_{\mathfrak{U}} : T_p\mathfrak{U} \to T_p\mathfrak{M}$ to the inclusion map $\iota_{\mathfrak{U}} : \mathfrak{U} \to \mathfrak{M}$ (13.1) is a vector space isomorphism [3]. Since this isomorphism is basis and chart independent, it is natural, and one can identify any vector $u \in T_p\mathfrak{U}$ with its image, $T_p\iota_{\mathfrak{U}}(u)$, in $T_p\mathfrak{M}$. In this entire book we make such an identification.

**The tangent space to the product manifold.** Suppose that $\mathfrak{M}_1$ and $\mathfrak{M}_2$ are smooth manifolds. Then, their direct product, $\mathfrak{M}_1 \times \mathfrak{M}_2$, is a smooth manifold, similarly. Let $(p_1, p_2) \in \mathfrak{M}_1 \times \mathfrak{M}_2$ be a point and $T_{(p_1,p_2)}\mathfrak{M}_1 \times \mathfrak{M}_2$ be the tangent space at $(p_1, p_2)$, defined in the usual manner. Let

$$\mathrm{pr}_i \colon \mathfrak{M}_1 \times \mathfrak{M}_2 \to \mathfrak{M}_i\,, \quad \mathrm{pr}_i(q_1, q_2) := q_i\,, \quad i = 1, 2\,,$$

be projections. Then the map

$$\pi \colon T_{(p_1,p_2)}\mathfrak{M}_1 \times \mathfrak{M}_2 \to T_{p_1}\mathfrak{M}_1 \oplus T_{p_2}\mathfrak{M}_2\,,$$
$$\forall u \in T_{(p_1,p_2)}\mathfrak{M}_1 \times \mathfrak{M}_2 \colon \pi(u) := (T_{(p_1,p_2)}\mathrm{pr}_1(u), T_{(p_1,p_2)}\mathrm{pr}_2(u))\,,$$

is a natural isomorphism [3]. This observation allows us to consider the tangent vectors to a product manifold simply as ordered pairs of tangent vectors to each multiplier.

**Velocity vectors of curves.** Any tangent vector represents an equivalence class of smooth curves. In this regard, it can be interpreted as a generalized velocity vector along a curve. Let $\mathfrak{M}$ be a smooth manifold. Suppose that $\chi \colon \mathbb{J} \to \mathfrak{M}$ is a smooth curve that is defined on an open interval $\mathbb{J} \subset \mathbb{R}$. Let $t_0 \in \mathbb{J}$ be a point. The space $\mathbb{J}$ is a one-dimensional smooth manifold, and, thus, the coordinate basis of $T_{t_0}\mathbb{J}$ consists of one vector $\partial_t|_{t_0}$. The *velocity of* $\chi$ *at* $t_0$ is the vector $\chi'(t_0) \in T_{\chi(t_0)}\mathfrak{M}$,

$$\chi'(t_0) := T_{t_0}\chi(\partial_t|_{t_0})\,. \tag{13.18}$$

Choose some smooth chart $(U, \varphi)$ on $\mathfrak{M}$ and the corresponding coordinate basis $(\partial_i|_{\chi(t_0)})_{i=1}^s$, where $s = \dim \mathfrak{M}$. Then, the coordinate representation of $\chi'(t_0)$ is the following:

$$\chi'(t_0) = \left.\frac{d\chi^i(t)}{dt}\right|_{t_0} \partial_i|_{\chi(t_0)}\,.$$

Here, $\chi^i$, $i = 1, \dots, s$, is the coordinate representation of $\chi$.

**The tangent space to a smooth manifold with boundary.** Suppose that $\mathfrak{M}$ is a smooth $s$-dimensional manifold with boundary. Let $p \in \mathfrak{M}$. The space $C^\infty(\mathfrak{M})$ of germs at $p$ is defined in a similar manner as for the smooth manifolds without boundary. It does not matter whether $p$ is an interior point or a boundary point. The tangent space $T_p\mathfrak{M}$ is then defined as the vector space of all derivations on $C^\infty(\mathfrak{M})$. It has dimension $s$ everywhere [48]. One can consider derivations of the algebra $C^\infty(\mathfrak{M})$ similarly [3]. At the same time, the approach of the equivalence class of curves requires a minor modification. In the case of a boundary point, curves should be defined not only on open intervals but also on the intervals of the form $[0, \varepsilon[$ or $]-\varepsilon, 0]$ [3]. For every point $p \in \mathfrak{M}$, one has the coordinate basis $(\partial_i|_p)_{i=1}^s$, for $T_p\mathfrak{M}$. Note that if $p \in \partial\mathfrak{M}$ then, since the boundary $\partial\mathfrak{M}$ is a smooth $(s-1)$-dimensional manifold (submanifold of $\mathfrak{M}$), one can consider either the vector space $T_p\mathfrak{M}$ of dimension $s$ or the vector space $T_p\partial\mathfrak{M}$ of dimension $s-1$.

The definitions of the tangent map and the cotangent space are repeated unchanged. We adopt the convention of identifying $T_p\partial\mathfrak{M}$ with its image under the injective map $T_p\iota_{\partial\mathfrak{M}} : T_p\partial\mathfrak{M} \to T_p\mathfrak{M}$, which is tangent to the inclusion map $\iota_{\partial\mathfrak{M}} : \partial\mathfrak{M} \to \mathfrak{M}$.

**Inward–outward pointing vectors.** Suppose that $\mathfrak{M}$ is a smooth manifold with a boundary and $p \in \partial\mathfrak{M}$. Vectors of $T_p\mathfrak{M}$ can be divided into three classes: vectors that are tangent to $T_p\partial\mathfrak{M}$, vectors, that are inward-pointing, and vectors that are outward-pointing. The latter two classes are defined as follows [3]. A vector $u \in T_p\mathfrak{M} \setminus T_p\partial\mathfrak{M}$ is called *inward-pointing*, if there exists a smooth curve $\chi : [0, \varepsilon[\to \mathfrak{M}$, $\varepsilon > 0$, such that $\chi(0) = p$ and $\chi'(0) = u$. A vector $u \in T_p\mathfrak{M} \setminus T_p\partial\mathfrak{M}$ is called *outward-pointing*, if there exists a smooth curve $\chi : ]-\varepsilon, 0] \to \mathfrak{M}$, $\varepsilon > 0$, such that $\chi(0) = p$ and $\chi'(0) = u$.

### 13.2.5 The "Tower"

Suppose that $\mathfrak{M}$ is an $s$-dimensional smooth manifold (with or without boundary). The tangent, $T_p\mathfrak{M}$, and the cotangent, $T_p^*\mathfrak{M}$, spaces are associated with each point $p \in \mathfrak{M}$. They represent the "upper" level of the "tower" and are not sufficient to construct mathematical models of continuum mechanics. Other "levels" are constructed by means of the "upper one" and the additional "building technique", the abstract tensor product of vector spaces introduced in Chapter 12, Section 12.5.3. Such a formalism does not need a Euclidean structure and provides an effective way to construct tensors in non-Euclidean spaces.

Let $p \in \mathfrak{M}$. One can generate all possible finite tensor products from copies of $T_p\mathfrak{M}$ and $T_p^*\mathfrak{M}$ by using the notion of the abstract tensor product. Among them are ($k \in \mathbb{N}$):

- Contravariant tensors of rank $k$:

$$T^k(T_p\mathfrak{M}) := \underbrace{T_p\mathfrak{M} \otimes \cdots \otimes T_p\mathfrak{M}}_{k \text{ copies}} ,$$

  which can be decomposed over the basis

$$(e_{i_1} \otimes \cdots \otimes e_{i_k})_{1 \le i_1, \ldots, i_k \le s} .$$

  Here, $(e_i)_{i=1}^s$ is a basis for $T_p\mathfrak{M}$.
- Covariant tensors of rank $k$:

$$T^k\left(T_p^*\mathfrak{M}\right) := \underbrace{T_p^*\mathfrak{M} \otimes \cdots \otimes T_p^*\mathfrak{M}}_{k \text{ copies}} ,$$

  which can be decomposed over the basis

$$(\vartheta^{i_1} \otimes \cdots \otimes \vartheta^{i_k})_{1 \le i_1, \ldots, i_k \le s} .$$

  Here, $(\vartheta^i)_{i=1}^s$ is a basis for $T_p^*\mathfrak{M}$.

- Mixed tensors of type $(k, l)$:

$$T^{(k,l)}(T_p\mathfrak{M}) := \underbrace{T_p\mathfrak{M} \otimes \cdots \otimes T_p\mathfrak{M}}_{k \text{ copies}} \otimes \underbrace{T_p^*\mathfrak{M} \otimes \cdots \otimes T_p^*\mathfrak{M}}_{l \text{ copies}} .$$

The basis in this space of mixed tensors can be combined as follows:

$$\left(e_{i_1} \otimes \cdots \otimes e_{i_k} \otimes \vartheta^{j_1} \otimes \cdots \otimes \vartheta^{j_l}\right)_{1 \le i_1,\dots,i_k \le s, \quad 1 \le j_1,\dots,j_l \le s} ,$$

where $(e_i)_{i=1}^s$ is a basis for $T_p\mathfrak{M}$ and $(\vartheta^i)_{i=1}^s$ is the dual basis for $T_p^*\mathfrak{M}$.

We use the standard agreements:

$$T^0(T_p\mathfrak{M}) = T^0\left(T_p^*\mathfrak{M}\right) := \mathbb{R} ,$$
$$T^{(0,k)}(T_p\mathfrak{M}) = T^k\left(T_p^*\mathfrak{M}\right) , \quad T^{(k,0)}(T_p\mathfrak{M}) = T^k(T_p\mathfrak{M}) , \quad \dots$$

The dimensions of the spaces introduced above are: $\dim T^k(T_p\mathfrak{M}) = \dim T^k(T_p^*\mathfrak{M}) = s^k$ and $\dim T^{(k,l)}(T_p\mathfrak{M}) = s^{k+l}$. All such spaces can be ordered by the levels of the symbolic “tower”, as is shown below:

$$\begin{array}{cccc} \dots & \dots & \dots & \dots \\ \\ T_p\mathfrak{M} \otimes T_p\mathfrak{M} & T_p\mathfrak{M} \otimes T_p^*\mathfrak{M} & \overset{\cong}{\longleftrightarrow} \; T_p^*\mathfrak{M} \otimes T_p\mathfrak{M} & T_p^*\mathfrak{M} \otimes T_p^*\mathfrak{M} \\ \\ & T_p\mathfrak{M} & T_p^*\mathfrak{M} & \end{array}$$

### 13.2.6 Exterior Forms

The calculus of exterior forms was developed in the papers of F. Klein [41], H. Weyl [35], and E. Cartan [150] for a mathematically rigorous definition of the operations of integration on manifolds that are of general structure. Since the mid-twentieth century, this calculus has been intensively used in theoretical physics to formalize the fields defined in the curved space-time [34, 151]. In continuum mechanics, the calculus of exterior forms is used rather rarely: most of the problems are stated in Euclidean space, and there is no need to complicate the traditional apparatus of integration. At the same time, the present book goes beyond traditional Euclidean statements, and it seems most concise and convenient to use the elements of the calculus of exterior forms to determine the fields that represent the densities of physical quantities: mass forces, stresses, *etc.*

**Exterior forms: definition.** Let $\mathfrak{M}$ be a smooth $s$-dimensional manifold. By virtue of the natural isomorphism it is convenient to consider each element $\mathscr{T} \in T^k(T_p^*\mathfrak{M})$ as a

$k$-linear mapping

$$\mathcal{T}\colon \underbrace{T_p\mathfrak{M} \times \cdots \times T_p\mathfrak{M}}_{k \text{ times}} \to \mathbb{R}\,.$$

A mapping $\omega \in T^k(T_p^*\mathfrak{M})$ is called an *exterior k-form* (or antisymmetric tensor) [3] if for all tangent vectors $u_1, \dots, u_k \in T_p\mathfrak{M}$ and for each pair of distinct indices $i, j \in \{1, \dots, k\}$, one has

$$\omega(u_1, \dots, u_i, \dots, u_j, \dots, u_k) = -\omega(u_1, \dots, u_j, \dots, u_i, \dots, u_k)\,.$$

The set $\Lambda^k(T_p^*\mathfrak{M}) \subset T^k(T_p^*\mathfrak{M})$ of all exterior forms is the subspace of $T^k(T_p^*\mathfrak{M})$.

**The wedge product.** The tensor product $\otimes$, applied to exterior forms, in general does not lead beyond the set of exterior forms, but a new operation, namely, the wedge (exterior) product can be defined in such a way.

Recall that a permutation is a bijection $\sigma\colon \{1, \dots, k\} \to \{1, \dots, k\}$, $k \in \mathbb{N}$. Let $S_k$ be the set of all permutations. This set, being endowed with a binary operation of the composition $\circ$, is a group. For any tensor $\mathcal{T} \in T^k(T_p^*\mathfrak{M})$, a permutation $\sigma \in S_k$ generates a new tensor ${}^\sigma\mathcal{T} \in T^k(T_p^*\mathfrak{M})$:

$$\forall u_1, \dots, u_k \in T_p\mathfrak{M}\colon {}^\sigma\mathcal{T}(u_1, \dots, u_k) := \mathcal{T}(u_{\sigma(1)}, \dots, u_{\sigma(k)})\,.$$

For a permutation $\sigma \in S_k$, its *sign* is the number

$$\operatorname{sgn}\sigma := \begin{cases} 1, & \text{if } \sigma \text{ is even}\,, \\ -1, & \text{if } \sigma \text{ is odd}\,. \end{cases}$$

One has the equivalent definition for the exterior form as follows: $\omega \in \Lambda^k(T_p^*\mathfrak{M})$ iff $\omega \in T^k(T_p^*\mathfrak{M})$, and for all tangent vectors $u_1, \dots, u_k \in T_p\mathfrak{M}$, for every permutation $\sigma \in S_k$, one has

$$\omega(u_{\sigma(1)}, \dots, u_{\sigma(k)}) = (\operatorname{sgn}\sigma)\omega(u_1, \dots, u_k)\,.$$

The mapping $\operatorname{Alt}\colon T^k(T_p^*\mathfrak{M}) \to \Lambda^k(T_p^*\mathfrak{M})$,

$$\operatorname{Alt}\mathcal{T} := \frac{1}{k!}\sum_{\sigma\in S_k}(\operatorname{sgn}\sigma)({}^\sigma\mathcal{T})\,,$$

is called an *alternation* [3]. In the explicit form, for all $u_1, \dots, u_k \in T_p\mathfrak{M}$, one has

$$\operatorname{Alt}\mathcal{T}(u_1, \dots, u_k) = \frac{1}{k!}\sum_{\sigma\in S_k}(\operatorname{sgn}\sigma)\mathcal{T}(u_1, \dots, u_k)\,.$$

Let $\omega \in \Lambda^k(T_p^*\mathfrak{M})$ and $\eta \in \Lambda^l(T_p^*\mathfrak{M})$. There are several definitions for the *wedge product*. The first one is $\omega\wedge\eta := \operatorname{Alt}(\omega\otimes\eta)$. Such a definition is used in [145]. We prefer to use another definition:

$$\omega \wedge \eta := \frac{(k+l)!}{k!\,l!}\operatorname{Alt}(\omega \otimes \eta)\,. \tag{13.19}$$

**Remark 13.1.** In particular, for covectors $\alpha$ and $\beta$, one has

$$\alpha \wedge \beta = \alpha \otimes \beta - \beta \otimes \alpha .$$

Hence, $\wedge \colon \Lambda^k(T_p^*\mathfrak{M}) \times \Lambda^l(T_p^*\mathfrak{M}) \to \Lambda^{k+l}(T_p^*\mathfrak{M})$. We refer to such a mapping as the *wedge product* or *exterior product*. It satisfies the following properties [3]:

$\wedge_1$) **Bilinearity.** For $\omega, \omega' \in \Lambda^k(T_p^*\mathfrak{M})$, $\eta, \eta' \in \Lambda^l(T_p^*\mathfrak{M})$ and $a, a' \in \mathbb{R}$,

$$(a\omega + a'\omega') \wedge \eta = a(\omega \wedge \eta) + a'(\omega' \wedge \eta) ,$$
$$\omega \wedge (a\eta + a'\eta') = a(\omega \wedge \eta) + a'(\omega \wedge \eta') .$$

$\wedge_2$) **Associativity.** For $\omega \in \Lambda^k(T_p^*\mathfrak{M})$, $\eta \in \Lambda^l(T_p^*\mathfrak{M})$ and $\mu \in \Lambda^r(T_p^*\mathfrak{M})$,

$$\omega \wedge (\eta \wedge \mu) = (\omega \wedge \eta) \wedge \mu .$$

$\wedge_3$) **Anticommutativity.** For $\omega \in \Lambda^k(T_p^*\mathfrak{M})$, $\eta \in \Lambda^l(T_p^*\mathfrak{M})$,

$$\omega \wedge \eta = (-1)^{kl} \eta \wedge \omega .$$

$\wedge_4$) If $v^1, \dots, v^k \in T_p^*\mathfrak{M}$ and $u_1, \dots, u_k \in T_p\mathfrak{M}$, then

$$v^1 \wedge \cdots \wedge v^k(u_1, \dots, u_k) = \det[v^i(u_j)] .$$

The factor $(k+l)!/(k!l!)$ is essential for $\wedge_4$). In this regard, in [3] the term “determinant convention” is used for (13.19).

Suppose that $(\vartheta^i)_{i=1}^s$ is a basis for $T_p^*\mathfrak{M}$. The collection

$$(\vartheta^{i_1} \wedge \cdots \wedge \vartheta^{i_k})_{1 \le i_1 < i_2 < \cdots < i_k \le s}$$

represents a basis for $\Lambda^k(T_p^*\mathfrak{M})$; for each exterior $k$-form $\omega$, the following decomposition holds:

$$\omega = \sum_{1 \le i_1 < i_2 < \cdots < i_k \le s} \omega_{i_1 i_2 \dots i_k} \vartheta^{i_1} \wedge \cdots \wedge \vartheta^{i_k} ,$$

where $\omega_{i_1 i_2 \dots i_k} = \omega(e_{i_1}, \dots, e_{i_k})$. Therefore, $\dim \Lambda^k(T_p^*\mathfrak{M}) = \binom{s}{k}$ [3].

**Remark 13.2.** For $\omega \in \Lambda^k(T_p^*\mathfrak{M})$, and vectors $u_1, \dots, u_k \in T_p\mathfrak{M}$ there is a relation between the exterior $k$-form and a determinant:

$$\omega(u_1, \dots, u_k) = \sum_{1 \le i_1 < i_2 < \cdots < i_k \le s} \omega_{i_1 i_2 \dots i_k} \begin{vmatrix} u_1^{i_1} & \dots & u_1^{i_k} \\ \dots & \dots & \dots \\ u_k^{i_1} & \dots & u_k^{i_k} \end{vmatrix} ,$$

where $\vartheta^{i_j}(u_l) = u_l^j$. In particular, for $\omega \in \Lambda^s(T_p^*\mathfrak{M})$, and vectors $u_1, \dots, u_s \in T_p\mathfrak{M}$ one obtains

$$\omega(u_1, \dots, u_s) = \omega_{1 \dots s} \begin{vmatrix} u_1^1 & \dots & u_1^s \\ \dots & \dots & \dots \\ u_s^1 & \dots & u_s^s \end{vmatrix} .$$

Thus, the formalism of the exterior product, as well as the calculus of exterior forms as a whole, are closely related to the theory of determinants, the development of which led to the appearance of geometric concepts of oriented area and volume [41].

The collection $(\Lambda^k(T_p^*\mathfrak{M}))_{k=0}^s$ generates the vector space $\Lambda(T_p^*\mathfrak{M})$, which is the external direct sum[6]

$$\Lambda\left(T_p^*\mathfrak{M}\right) = \bigoplus_{k=0}^{s} \Lambda^k\left(T_p^*\mathfrak{M}\right) .$$

Its dimension is $2^s$. The operation $\wedge$ turns $\Lambda(T_p^*\mathfrak{M})$ into associative and anticommutative graded algebra. Such algebra is called the *Grassmann algebra* of $T_p\mathfrak{M}$ [3].

## 13.3 Vector Bundles and Their Sections

Tensors in their algebraic skins of linear operators or multilinear forms, characterize physical processes locally, in the infinitesimal neighborhood of some spatial or material point. In order to describe the process globally, one has to broaden the algebraic notion of a tensor to the notion of tensor field. The formalization of a tensor field in the framework of smooth manifold theory is carried out through the geometrical concept of a section over a vector bundle. Apparently, for the first time an extensive exposition of this theory was given by N. Steenrod in [152]. In the present book, we follow [3, 123].

### 13.3.1 Smooth Vector Bundles of Rank *k*

Suppose that $\mathfrak{M}$ is a smooth manifold. A *smooth real vector bundle of rank k over* $\mathfrak{M}$ [3] is a triple $(\mathfrak{E}, \mathfrak{M}, \pi)$, in which $\mathfrak{E}$ is a smooth manifold, and $\pi\colon \mathfrak{E} \to \mathfrak{M}$ is a surjective smooth mapping such that

$\mathfrak{E}_1$) For each $p \in \mathfrak{M}$, the set (which is called *fiber over p*) $\mathfrak{E}_p := \pi^{-1}(\{p\})$ is endowed with a $k$-dimensional real vector space structure.

$\mathfrak{E}_2$) For each $p \in \mathfrak{M}$, there exist a neighborhood $U \subset \mathfrak{M}$ of $p$ and a diffeomorphism

$$\Phi\colon \pi^{-1}(U) \to U \times \mathbb{R}^k ,$$

which is called a *smooth local trivialization of* $\mathfrak{E}$ *over* $U$, and such that

- $\Phi_1$) $\mathrm{pr}_1 \circ \Phi = \pi|_{\pi^{-1}(U)}$, where $\mathrm{pr}_1 : U \times \mathbb{R}^k \to U$ is the projection;
- $\Phi_2$) for each $q \in U$ the mapping $\Phi|_{\mathfrak{E}_q}$ is a vector space isomorphism from $\mathfrak{E}_q$ to $\{q\} \times \mathbb{R}^k \cong \mathbb{R}^k$.

In the following, for the sake of simplicity, we use the notation $\mathrm{rank}(\mathfrak{E})$ for a rank of a smooth vector bundle $(\mathfrak{E}, \mathfrak{M}, \pi)$.

6 See Chapter 12, Section 12.5.1.

Generally, vector bundles are introduced for topological spaces [153], but we restrict our attention to smooth manifolds. Since we consider only real numbers, the word “real” in the definition of a vector bundle is further omitted. If $(\mathfrak{E}, \mathfrak{M}, \pi)$ is a smooth vector bundle of rank $k$, the following terminology is used: the space $\mathfrak{E}$ is called the *total space of the bundle*, $\mathfrak{M}$ is called its *base*, and $\pi$ is its *projection*. If this does not lead to confusion, we follow the standard practice of shortenings and write: “$\mathfrak{E}$ is a vector bundle over $\mathfrak{M}$”, or “$\pi\colon \mathfrak{E} \to \mathfrak{M}$ is a vector bundle”. If $\Phi$ is a global trivialization, that is, $U = \mathfrak{M}$, then $(\mathfrak{E}, \mathfrak{M}, \pi)$ is called a *trivial bundle*. In this case, $\mathfrak{E}$ is diffeomorphic to $\mathfrak{M} \times \mathbb{R}^k$.

Let us make some comments on the definition of vector bundles. Since the mapping $\pi\colon \mathfrak{E} \to \mathfrak{M}$ is surjective, the set $\mathfrak{E}$ can be represented as the disjoint union of $k$-dimensional vector spaces: $\mathfrak{E} = \bigsqcup_{p\in\mathfrak{M}} \mathfrak{E}_p$. Locally, such a union is diffeomorphic to a Cartesian product. That is, if $U$ is a neighborhood from $\mathfrak{E}_2$), then $\pi^{-1}(U) = \bigsqcup_{q\in U} \mathfrak{E}_q$ is diffeomorphic to the smooth manifold $U \times \mathbb{R}^k$. The diffeomorphism $\Phi$ preserves the fiber structure; the condition $\Phi_1$) is equivalent to the following: for each $q \in U$, one has $\Phi(\mathfrak{E}_q) \subset \{q\} \times \mathbb{R}^k$. Here, $\{q\} \times \mathbb{R}^k$ can be considered as the “fiber” of $U \times \mathbb{R}^k = \bigsqcup_{q\in U}\left(\{q\} \times \mathbb{R}^k\right)$. In addition, vector structures of $\mathfrak{E}_q$ and $\{q\} \times \mathbb{R}^k$ are aligned by $\Phi_2$.

A point $p$ may have two neighborhoods $U$ and $V$ with smooth local trivializations

$$\Phi\colon \pi^{-1}(U) \to U \times \mathbb{R}^k\,, \quad \Psi\colon \pi^{-1}(V) \to V \times \mathbb{R}^k\,.$$

As is shown in [3], there exists a smooth mapping

$$\tau\colon U \cap V \to \mathrm{GL}(k; \mathbb{R})\,,$$

such that the composition $\Phi \circ \Psi^{-1}\colon (U \cap V) \times \mathbb{R}^k \to (U \cap V) \times \mathbb{R}^k$ has the form

$$\Phi \circ \Psi^{-1}(p, v) = (p, \tau(p)v)\,.$$

Here, $\tau(p)v$ is the action of the $k \times k$ matrix on a vector $v \in \mathbb{R}^k$. The mapping $\tau$ is traditionally called the *transition function* [3].

### 13.3.2 Tangent and Cotangent Bundles

A smooth manifold $\mathfrak{M}$ is endowed with tensorial “towers” over each of its points. The “first” floor gives the families $\{T_p\mathfrak{M}\}_{p\in\mathfrak{M}}$, $\{T_p^*\mathfrak{M}\}_{p\in\mathfrak{M}}$ of tangent and cotangent spaces. These families, together with transformation laws of vectors and covectors, are enough to construct vector bundle structures. Our purpose is to give a sketch of how it can be done. Details and proofs can be found in [3, 119].

**The disjoint union.** First, let us introduce the notion of the disjoint union. Assume that one has a family $\{X_i\}_{i\in I}$ of non-empty sets. These sets are not supposed to be contained in an ambient set, nor to be pairwise disjoint. The set

$$\bigsqcup_{i\in I} X_i := \bigcup_{i\in I}(\{i\} \times X_i) = \{(i, x) \mid (i \in I) \wedge (x \in X_i)\}$$

is called the *disjoint union of the family* $\{X_i\}_{i\in I}$. For each $i \in I$, the inclusion map

$$\iota_i\colon X_i \to \bigsqcup_{i\in I} X_i\,, \quad \iota_i(x) = (i, x)$$

is defined. Denote $X_i' = \iota_i(X_i) = \{i\} \times X_i$. By virtue of such a constriction, if $i \neq j$ then $X_i'$ is disjoint with $X_j'$, and the set $X = \bigsqcup_{i\in I} X_i$ is a disjoint union in the usual sense, $X = \bigsqcup_{i\in I} X_i'$. The set $X_i'$ is a "copy" of $X_i$ in the disjoint union, and the index $i$ in $(i, x)$ serves as a "tag" that indicates where $x$ came from.

**Construction of smooth vector bundles.** Suppose that $\mathfrak{M}$ is a smooth manifold (with or without boundary), and $\{\mathfrak{E}_p\}_{p\in\mathfrak{M}}$ is a family of $k$-dimensional real vector spaces. Let

$$\mathfrak{E} = \bigsqcup_{p\in\mathfrak{M}} \mathfrak{E}_p = \bigcup_{p\in\mathfrak{M}} (\{p\} \times \mathfrak{E}_p)$$

and $\pi\colon \mathfrak{E} \to \mathfrak{M}$ be a mapping $(p, u) \mapsto \pi(p, u) = p$. Furthermore, let a family $\{(U_i, \Phi_i)\}_{i\in I}$ be given. Here, $\{U_i\}_{i\in I}$ is an open cover of $\mathfrak{M}$, and for each $i \in I$, the $\Phi_i$ is the bijective mapping

$$\Phi_i\colon \pi^{-1}(U_i) \to U_i \times \mathbb{R}^k\,,$$

such that $\Phi_i|_{\mathfrak{E}_p}\colon \mathfrak{E}_p \to \{p\}\times\mathbb{R}^k$ is a vector space isomorphism for each $p \in U_i$. Suppose that the family $\{(U_i, \Phi_i)\}_{i\in I}$ satisfies the following condition. If $i, j \in I$ are such indices that $U_i \cap U_j \neq \emptyset$, then there exists a smooth mapping

$$\tau_{ij}\colon U_i \cap U_j \to \mathrm{GL}(k; \mathbb{R})\,,$$

such that the composition $\Phi_i \circ \Phi_j^{-1}\colon (U_i \cap U_j) \times \mathbb{R}^k \to (U_i \cap U_j) \times \mathbb{R}^k$ gives

$$\Phi_i \circ \Phi_j^{-1}(p, v) = (p, \tau_{ij}(p)v)\,.$$

With these conditions, the set $\mathfrak{E}$ has a unique topology and smooth structure, that turn it into a smooth manifold (with or without boundary). The triple $(\mathfrak{E}, \mathfrak{M}, \pi)$ turns into a smooth vector bundle of rank $k$ over $\mathfrak{M}$ with $\pi$ as the projection and $\{\Phi_i\}_{i\in I}$ as smooth local trivializations.

The proof of this claim is given in [3].

**Tangent and cotangent bundles.** Suppose that $\mathfrak{M}$ is a smooth $s$-dimensional manifold with or without boundary. One has the family $\{T_p\mathfrak{M}\}_{p\in\mathfrak{M}}$ and $\{T_p^*\mathfrak{M}\}_{p\in\mathfrak{M}}$ of tangent and cotangent spaces. Denote by $T\mathfrak{M}$ and $T^*\mathfrak{M}$ the following disjoint unions:

$$T\mathfrak{M} := \bigsqcup_{p\in\mathfrak{M}} T_p\mathfrak{M}\,, \quad T^*\mathfrak{M} := \bigsqcup_{p\in\mathfrak{M}} T_p^*\mathfrak{M}$$

and define the mappings $\pi_t\colon T\mathfrak{M} \to \mathfrak{M}$ and $\pi_c\colon T^*\mathfrak{M} \to \mathfrak{M}$ that both act like $(p, v) \mapsto p$.

Choose a smooth atlas $\{(U_\alpha, \varphi_\alpha)\}_{\alpha\in I}$ from the smooth structure of $\mathfrak{M}$ and define the families $\{\Phi_{t;\alpha}\}_{\alpha\in I}$, $\{\Phi_{c;\alpha}\}_{\alpha\in I}$ of mappings

$$\Phi_{t;\alpha}\colon \pi_t^{-1}(U_\alpha) \to U_\alpha \times \mathbb{R}^s\,, \qquad \Phi_{t;\alpha}\left(p, u^i\partial_i|_p\right) := (p, (u^1, \ldots, u^s))\,,$$
$$\Phi_{c;\alpha}\colon \pi_c^{-1}(U_\alpha) \to U_\alpha \times \mathbb{R}^s\,, \qquad \Phi_{c;\alpha}\left(p, v_i dx^i|_p\right) := (p, (v_1, \ldots, v_s))\,.$$

Suppose that one has distinct indices $\alpha, \beta \in I$, such that $U_\alpha \cap U_\beta \neq \emptyset$. The equalities (13.8) and (13.16), which represent the transformation laws of vector and covector, allow us to write the following relations for $(p, u) \in \pi_t^{-1}(U_\alpha \cap U_\beta)$ and $(p, v) \in \pi_c^{-1}(U_\alpha \cap U_\beta)$:

$$\Phi_{t;\alpha} \circ \Phi_{t;\beta}^{-1}(p, (u^1, \ldots, u^s)) = (p, \tau_{t;\alpha\beta}(p)(u^1, \ldots, u^s))\,,$$
$$\Phi_{c;\alpha} \circ \Phi_{c;\beta}^{-1}(p, (v_1, \ldots, v_s)) = (p, \tau_{c;\alpha\beta}(p)(v_1, \ldots, v_s))\,,$$

where the smooth mappings $\tau_{t;\alpha\beta}$ and $\tau_{c;\alpha\beta}$ acting from $U_\alpha \cap U_\beta$, result in values from $\mathrm{GL}(s; \mathbb{R})$ and can be defined as follows:

$$\tau_{t;\alpha\beta}(p) := D_{\varphi_\beta(p)}\left(\varphi_\alpha \circ \varphi_\beta^{-1}\right)\,, \quad \tau_{c;\alpha\beta}(p) := \left(D_{\varphi_\beta(p)}\left(\varphi_\alpha \circ \varphi_\beta^{-1}\right)\right)^{-\mathrm{T}}\,.$$

Hence, by the smooth vector bundle construction theorem, $(T\mathfrak{M}, \mathfrak{M}, \pi_t)$ and $(T^*\mathfrak{M}, \mathfrak{M}, \pi_c)$ are smooth vector bundles of rank $s$. The local charts for $T\mathfrak{M}$ and $T^*\mathfrak{M}$ are represented by the pairs $(\pi_t^{-1}(U_\alpha), \widetilde{\varphi}_{t;\alpha})$, $(\pi_c^{-1}(U_\alpha), \widetilde{\varphi}_{c;\alpha})$, where

$$\widetilde{\varphi}_{t;\alpha}\colon \pi_t^{-1}(U_\alpha) \to \mathbb{R}^{2s}\,, \qquad \widetilde{\varphi}_{t;\alpha}(p, u) := (p^1, \ldots, p^s, u^1, \ldots, u^s)\,,$$
$$\widetilde{\varphi}_{c;\alpha}\colon \pi_c^{-1}(U_\alpha) \to \mathbb{R}^{2s}\,, \qquad \widetilde{\varphi}_{c;\alpha}(p, v) := (p^1, \ldots, p^s, v_1, \ldots, v_s)\,,$$

for $(p^1, \ldots, p^s) = \varphi_\alpha(p)$, $u = u^i\partial_i|_p$ and $v = v_i dx^i|_p$. Thus, $T\mathfrak{M}$ and $T^*\mathfrak{M}$ are smooth two-manifolds. The $2s$-tuples $(x^i, u^i)$ and $(x^i, v^i)$ are called *natural coordinates* on $T\mathfrak{M}$ and $T^*\mathfrak{M}$, respectively [3].

### 13.3.3 Operations on Vector Bundles

Previously, we introduced a sketch of reasonings on constructing tangent and cotangent bundles. Similar reasonings may be provided for vector bundles, whose fibers are arbitrary tensor products of tangent and cotangent spaces. Another way of reasoning may be provided by constructing only a tangent bundle. Then, one can construct its *dual vector bundle*, that is, the cotangent bundle. Using such two bundles one can construct their *tensor products* and *direct sums*. Such methodology is useful in the theory of connections. If one introduces a connection on some vector bundle, then vector bundle operations allow one to induce this connection on the dual bundle, and so on. Below, we give the brief descriptions of vector bundle operations. For a comprehensive treatment, we refer the reader to the book [119].

**The dual bundle.** Suppose that $(\mathfrak{E}, \mathfrak{M}, \pi)$ is a smooth vector bundle of rank $k$. Denote

$$\mathfrak{E}^* := \bigsqcup_{p\in\mathfrak{M}} \mathfrak{E}_p^* ,$$

where $\mathfrak{E}_p^*$ is the dual vector space to $\mathfrak{E}_p$. Define the mapping $\pi^* : \mathfrak{E}^* \to \mathfrak{M}$ by the rule $(p, v) \mapsto p$. The triple $(\mathfrak{E}^*, \mathfrak{M}, \pi^*)$ is a smooth vector bundle of rank $k$. The transition function $\tau^*$ of such a structure is equal to $\tau^*(p) = (\tau(p))^{-T}$, where $\tau$ is a transition function of $(\mathfrak{E}, \mathfrak{M}, \pi)$.

According to the definition of the dual vector bundle, the cotangent bundle is just dual for the tangent bundle.

**The Whitney sum.** Suppose that $(\mathfrak{E}_1, \mathfrak{M}, \pi_1)$ and $(\mathfrak{E}_2, \mathfrak{M}, \pi_2)$ are smooth vector bundles of rank $k_1$ and $k_2$ over the same base $\mathfrak{M}$. Define $\mathfrak{E}_1 \oplus \mathfrak{E}_2$ as

$$\mathfrak{E}_1 \oplus \mathfrak{E}_2 := \bigsqcup_{p\in\mathfrak{M}} (\mathfrak{E}_{1;p} \oplus \mathfrak{E}_{2;p}) ,$$

where $\mathfrak{E}_{i;p} = \pi_i^{-1}(\{p\})$. The sign $\oplus$ in $\mathfrak{E}_{1;p}\oplus\mathfrak{E}_{2;p}$ represents the external direct sum; each element of $\mathfrak{E}_{1;p} \oplus \mathfrak{E}_{2;p}$ is an ordered pair $(v_1, v_2)$, where $v_i \in \mathfrak{E}_{i;p}$. Define the mapping $\pi\colon \mathfrak{E}_1 \oplus \mathfrak{E}_2 \to \mathfrak{M}$ by the rule $(p, v) \mapsto p$. The triple $(\mathfrak{E}_1 \oplus \mathfrak{E}_2, \mathfrak{M}, \pi)$ is a smooth vector bundle of rank $k_1 + k_2$.

**The tensor product.** Suppose that $(\mathfrak{E}_1, \mathfrak{M}, \pi_1)$ and $(\mathfrak{E}_2, \mathfrak{M}, \pi_2)$ are smooth vector bundles of rank $k_1$ and $k_2$ over the same base $\mathfrak{M}$. Denote

$$\mathfrak{E}_1 \otimes \mathfrak{E}_2 := \bigsqcup_{p\in\mathfrak{M}} (\mathfrak{E}_{1;p} \otimes \mathfrak{E}_{2;p}) ,$$

where $\mathfrak{E}_{i;p} = \pi_i^{-1}(\{p\})$. Define the mapping $\pi\colon \mathfrak{E}_1 \otimes \mathfrak{E}_2 \to \mathfrak{M}$ as in the previous cases. The triple $(\mathfrak{E}_1 \otimes \mathfrak{E}_2, \mathfrak{M}, \pi)$ is a smooth vector bundle of rank $k_1 k_2$.

### 13.3.4 Vector Bundles of Higher Rank

Using the tensor product of vector bundles and the tangent constructed above, $T\mathfrak{M}$, and the cotangent, $T^*\mathfrak{M}$, bundles, one can generate new bundles from the collection of "towers" over the smooth $s$-dimensional manifold $\mathfrak{M}$ (with or without boundary). Among such vector bundles are [3]:

- The bundle of contravariant $k$-tensors with total space

$$T^k(T\mathfrak{M}) = \bigsqcup_{p\in\mathfrak{M}} T^k(T_p\mathfrak{M}) .$$

- The bundle of covariant $k$-tensors with total space

$$T^k(T^*\mathfrak{M}) = \bigsqcup_{p\in\mathfrak{M}} T^k\left(T_p^*\mathfrak{M}\right) .$$

- The bundle of mixed $(k, l)$-tensors with total space

$$T^{(k,l)}(T\mathfrak{M}) = \bigsqcup_{p\in\mathfrak{M}} T^{(k,l)}(T_p\mathfrak{M}) .$$

- The bundle of exterior $k$-forms with total space

$$\Lambda^k(T^*\mathfrak{M}) = \bigsqcup_{p\in\mathfrak{M}} \Lambda^k\left(T_p^*\mathfrak{M}\right) .$$

### 13.3.5 Sections of Vector Bundles

Vector and tensor fields can be formalized as sections of the corresponding vector bundles. Suppose that $(\mathfrak{E}, \mathfrak{M}, \pi)$ is a smooth vector bundle. A *local section of* $\mathfrak{E}$ is a continuous map $\sigma\colon U \to \mathfrak{E}$ that is defined on an open set $U \subset \mathfrak{M}$, satisfying $\pi \circ \sigma = \mathrm{Id}_U$. The latter equality is equivalent to the following:

$$\forall p \in U\colon \sigma(p) \in \mathfrak{E}_p .$$

If $U = \mathfrak{M}$, then $\sigma$ is a *global section of* $\mathfrak{E}$ [3]. We are dealing with *smooth* local and global sections. The “smoothness” is understood in the general sense of $C^\infty$-mappings between smooth manifolds (see Section 13.1.3).

The set containing all smooth global sections of a smooth vector bundle $(\mathfrak{E}, \mathfrak{M}, \pi)$ can be endowed with the real vector space structure by a pointwise addition and a scalar multiplication defined as

$$\begin{aligned}(\sigma_1 + \sigma_2)(p) &:= \sigma_1(p) + \sigma_2(p) ,\\ (\lambda\sigma_1)(p) &:= \lambda\sigma_1(p) ,\end{aligned}$$

where $\sigma_1, \sigma_2$ are smooth global sections of $\mathfrak{E}$ and $\lambda \in \mathbb{R}$. The vector space obtained is denoted by $\mathrm{Sec}(\mathfrak{E})$. Its null element is represented by the null section $p \mapsto 0 \in \mathfrak{E}_p$.

With arbitrary sections $\sigma \in \mathrm{Sec}(\mathfrak{E})$ and $f \in C^\infty(\mathfrak{M})$, one can define their multiplication

$$f\sigma\colon \mathfrak{M} \to \mathfrak{E} , \quad (f\sigma)(p) := f(p)\sigma(p) .$$

This operation turns the vector space $\mathrm{Sec}(\mathfrak{E})$ into a module over $C^\infty(\mathfrak{M})$ [3]. In this book, we often use the notation $\sigma_p$ for the value of a section $\sigma$ at a point $p$, that is, $\sigma_p := \sigma(p)$.

Smooth sections of the tangent $(T\mathfrak{M}, \mathfrak{M}, \pi_t)$ and the cotangent $(T^*\mathfrak{M}, \mathfrak{M}, \pi_c)$ bundles are particular cases of the constructions discussed above. We use these notations for them:

$$\mathrm{Vec}(\mathfrak{M}) := \mathrm{Sec}(T\mathfrak{M}) , \quad \mathrm{CVec}(\mathfrak{M}) := \mathrm{Sec}(T^*\mathfrak{M}) .$$

Following classical terminology, we refer to the elements of $\mathrm{Vec}(\mathfrak{M})$ as *vector fields* and to the elements of $\mathrm{CVec}(\mathfrak{M})$ to *covector fields*. Consider the case of $\mathrm{Vec}(\mathfrak{M})$ in more detail. Each vector field $u \in \mathrm{Vec}(\mathfrak{M})$ and a function $f \in C^\infty(\mathfrak{M})$ generates a new mapping $u(f) \in C^\infty(\mathfrak{M})$:

$$u(f)(p) := u_p([f]_p) .$$

Here, $[f]_p$ is the germ of $f$ at $p$ (see Section 13.2.1). Such a definition allows one to introduce the bracket operation $[\cdot,\cdot]\colon \mathrm{Vec}(\mathfrak{M}) \times \mathrm{Vec}(\mathfrak{M}) \to \mathrm{Vec}(\mathfrak{M})$ by the relation

$$[u, v]f := u(vf) - v(uf) . \tag{13.20}$$

This operation is called the *Lie bracket* [3]. It has the following properties:

i) **Bilinearity.** For $\alpha, \beta \in \mathbb{R}$ and $u, v, w \in \mathrm{Vec}(\mathfrak{M})$, one has

$$[\alpha u + \beta v, w] = \alpha[u, w] + \beta[v, w] ,$$
$$[u, \alpha v + \beta w] = \alpha[u, v] + \beta[u, w] .$$

ii) **Antisymmetry.** For $u, v \in \mathrm{Vec}(\mathfrak{M})$:

$$[u, v] = -[v, u] .$$

iii) **Jacobi identity.** For $u, v, w \in \mathrm{Vec}(\mathfrak{M})$:

$$[u, [v, w]] + [v, [w, u]] + [w, [u, v]] = 0 .$$

iv) For $f, g \in C^\infty(\mathfrak{M})$ and $u, v \in \mathrm{Vec}(\mathfrak{M})$, one has

$$[fu, gv] = fg[u, v] + fu(g)v - gv(f)u . \tag{13.21}$$

**Remark 13.3.** If $h \in C^\infty(\mathfrak{M})$ is an arbitrary scalar function, then we have

$$[fu, gv] = fg(u(v(h)) - v(u(h))) + fu(g)vh - gv(f)uh = (fg[u, v] + (fu(g)v - gv(f)u)h .$$

This relation with the (13.20) implies (13.21).

The operation $\langle\cdot,\cdot\rangle_p$ that was introduced for the elements of tangent and cotangent spaces (see Section 13.2.3) can be generalized to sections pointwise. To do this, define the bilinear mapping $\langle\cdot,\cdot\rangle\colon \mathrm{CVec}(\mathfrak{M}) \times \mathrm{Vec}(\mathfrak{M}) \to \mathbb{R}$:

$$\langle v, u\rangle := \langle v_p, u_p\rangle_p .$$

In the same manner, we can give the definitions of bilinear mappings $\llcorner$, $\lrcorner$:

$$\llcorner\colon (u \otimes \vartheta, v) \mapsto u \otimes \vartheta \llcorner v = \langle \vartheta, v\rangle u , \quad \lrcorner\colon (v, \vartheta \otimes u) \mapsto v \lrcorner \vartheta \otimes u = \langle \vartheta, v\rangle u .$$

Here, $u$ is a smooth section of a vector bundle, and $v$ and $\vartheta$ are vector and covector fields.

**The wedge product of differential forms.** A smooth section of the bundle $\Lambda^k(T^*\mathfrak{M})$ of exterior $k$-forms is called a *differential k-form* or, simply, a *k-form* [3]. The space of all such sections is denoted by

$$\Omega^k(\mathfrak{M}) := \mathrm{Sec}(\Lambda^k(T^*\mathfrak{M})) .$$

The wedge product of two differential forms is defined as $(\omega \wedge \eta)_p := \omega_p \wedge \eta_p$. On the right-hand side of this definition, the symbol $\wedge$ represents the operation (13.19), which acts between exterior forms, while on the left-hand side, the symbol $\wedge$ defines a new operation between differential forms.

**Interior multiplication of differential forms.** If $\omega \in \Omega^k(\mathfrak{M})$ and $u \in \mathrm{Vec}(\mathfrak{M})$, then one can define the $(k-1)$-form $u \lrcorner \omega \in \Omega^k(\mathfrak{M})$ by

$$p \mapsto (u \lrcorner \omega)_p , \quad (u \lrcorner \omega)_p(v_1, \ldots, v_{k-1}) := \omega_p(u_p, v_1, \ldots, v_{k-1}) .$$

The differential form is called the *interior multiplication of $\omega$ by $u$.*

### 13.3.6 Vector Bundle Homomorphisms

**Definition.** Suppose that $(\mathfrak{E}_1, \mathfrak{M}_1, \pi_1)$ and $(\mathfrak{E}_2, \mathfrak{M}_2, \pi_2)$ are two smooth vector bundles. Consider the mappings from $\mathfrak{E}_1$ to $\mathfrak{E}_2$ that transform fibers to fibers. Since fibers are vector spaces, one expects that such mappings, restricted to fibers, be linear. At the same time, to define a mapping that transforms a fiber $\mathfrak{E}_{1;p}$ to a fiber $\mathfrak{E}_{2;q}$, one requires additional information to identify the point $q$ from a given $p$. Such reasonings lead to the following definition. A continuous mapping $F\colon \mathfrak{E}_1 \to \mathfrak{E}_2$ is called a *vector bundle homomorphism* [3], if there exists a map $f\colon \mathfrak{M}_1 \to \mathfrak{M}_2$ satisfying

$$\pi_2 \circ F = f \circ \pi_1 ,$$

with the property: for each $p \in \mathfrak{M}_1$, the restriction $F|_{\mathfrak{E}_{1;p}} : \mathfrak{E}_{1;p} \to \mathfrak{E}_{2;f(p)}$ is linear. The mapping $f$ in this definition is unique and continuous.

**Remark 13.4.** Assume that $f_1, f_2$ are mappings that correspond to a bundle homomorphism $F$. Choose some section $\sigma\colon \mathfrak{M}_1 \to \mathfrak{E}_1$ (for example, the zero section). Then, $\pi_2 \circ F = f_i \circ \pi_1$ and $\pi_1 \circ \sigma = \mathrm{Id}_{\mathfrak{M}_1}$ imply $f_i = \pi_2 \circ F \circ \sigma$, $i = 1, 2$. This, in turn, implies both continuity and the uniqueness of $f$.

In this book, we are dealing with smooth bundle homomorphisms only; a bundle homomorphism $F$ is a smooth mapping between smooth manifolds $\mathfrak{E}_1$ and $\mathfrak{E}_2$.

**A global tangent map.** Suppose that $\mathfrak{M}$ and $\mathfrak{N}$ are smooth manifolds, and $f \in C^\infty(\mathfrak{M}; \mathfrak{N})$ is a smooth mapping. One has the family $\{T_p f\}_{p\in\mathfrak{M}}$ of tangent maps (see Section 13.2.2) $T_p f\colon T_p\mathfrak{M} \to T_p\mathfrak{N}$. A *global tangent map* [3] is the map $Tf\colon T\mathfrak{M} \to T\mathfrak{N}$

between tangent bundles, whose restriction to each tangent space $T_p\mathfrak{M}$ is $T_pf$. The mapping $Tf$ is smooth since in natural coordinates, one has

$$\widetilde{Tf}(x^i, u^i) = \left(y^i, u^j \partial_{x^j} y^i\right) ,$$

where $y^i(x^j)$ is the coordinate representation of $f$. By definition, the map $Tf$ is a tangent bundle homomorphism.

### 13.3.7 Pullback and Pushforward

Suppose that $\mathfrak{M}$ and $\mathfrak{N}$ are smooth manifolds, and $f : \mathfrak{M} \to \mathfrak{N}$ is a smooth mapping. For example, in the framework of continuum mechanics, $\mathfrak{M}$ can represent a body, $\mathfrak{N}$ can represent a physical space, while $f$ defines a configuration. The definition of desired fields on $\mathfrak{M}$ gives the so-called material description, while their definition on $\mathfrak{N}$ gives rise to the spatial one. Pullback and pushforward define a class of transformations from the spatial description to the material and vice versa [60].

**The case of a vector field.** Suppose that $f$ is a diffeomorphism. Define the mappings

$$\begin{aligned} f^* &: \mathrm{Vec}(\mathfrak{N}) \to \mathrm{Vec}(\mathfrak{M}) , \\ f_* &: \mathrm{Vec}(\mathfrak{M}) \to \mathrm{Vec}(\mathfrak{N}) . \end{aligned}$$

The first mapping, $f^*$, acts on a vector field $u \in \mathrm{Vec}(\mathfrak{N})$ and gives

$$f^*u := (Tf^{-1}) \circ u \circ f \in \mathrm{Vec}(\mathfrak{M}) .$$

The field $f^*u$ is called the *pullback of $u$ by $f$*. The second mapping, $f_*$, assigns to each vector field $u \in \mathrm{Vec}(\mathfrak{M})$ the vector field

$$f_*u := (Tf) \circ u \circ f^{-1} \in \mathrm{Vec}(\mathfrak{N}) .$$

This is called the *pushforward of $u$ by $f$*. These definitions are illustrated by the commutative diagrams:

$$\begin{array}{ccc} \mathfrak{M} & \xrightarrow{\;f\;} & \mathfrak{N} \\ \downarrow{\scriptstyle f^*u} & & \downarrow{\scriptstyle u} \\ T\mathfrak{M} & \xleftarrow{\;Tf^{-1}\;} & T\mathfrak{N} \end{array} \qquad \begin{array}{ccc} \mathfrak{M} & \xleftarrow{\;f^{-1}\;} & \mathfrak{N} \\ \downarrow{\scriptstyle u} & & \downarrow{\scriptstyle f_*u} \\ T\mathfrak{M} & \xrightarrow{\;Tf\;} & T\mathfrak{N} \end{array}$$

By the definition, $f^*$ and $f_*$ are isomorphisms, and $(f^*)^{-1} = f_*$.

**The case of a covector field.** Define the mapping

$$f^* : \mathrm{CVec}(\mathfrak{N}) \to \mathrm{CVec}(\mathfrak{M}) \,,$$

which assigns to each covector field $v \in \mathrm{CVec}(\mathfrak{N})$ the covector field $f^* v \in \mathrm{CVec}(\mathfrak{M})$, which acts on a tangent vector $u \in T\mathfrak{M}$ and results in $f^* v(u) := v \llcorner Tf(u)$. The covector field $f^* v$ is called the *pullback of* $v$ *by* $f$.

Suppose that $f$ is a diffeomorphism. The mapping

$$f_* : \mathrm{CVec}(\mathfrak{M}) \to \mathrm{CVec}(\mathfrak{N}) \,,$$

acts as an inverse mapping: for each covector field $v \in \mathrm{CVec}(\mathfrak{M})$, one has $f_* v := (f^{-1})^* v$. The covector field $f_* v$ is called the *pushforward of* $v$ *by* $f$.

**The case of a $k$-form.** Define the mapping

$$f^* : \Omega^k(\mathfrak{N}) \to \Omega^k(\mathfrak{M}) \,,$$

that acts on a $k$-form $\omega \in \Omega^k(\mathfrak{N})$ as follows. For a point $p \in \mathfrak{M}$, the value of $(f^* \omega)_p$ on tangent vectors $u_1, \ldots, u_k \in T_p\mathfrak{M}$ is

$$(f^* \omega)_p(u_1, \ldots, u_k) := \omega_{f(p)}(Tf(u_1), \ldots, Tf(u_k)) \,.$$

The $k$-form $f^* \omega$ is called the *pullback of* $\omega$ *by* $f$.

Like in the previous paragraph suppose that $f$ is a diffeomorphism. Then the mapping

$$f_* : \Omega^k(\mathfrak{M}) \to \Omega^k(\mathfrak{N}) \,,$$

for each $k$-form $\omega \in \Omega^k(\mathfrak{M})$ gives $f_* \omega := (f^{-1})^* \omega$. The $k$-form $f_* \omega$ is called the *pushforward of* $\omega$ *by* $f$.

**The case of a mixed $(k, l)$-tensor field.** Suppose that $f$ is a diffeomorphism. The pullback and pushforward are generalized as follows. Define the mapping

$$f_* \colon \mathrm{Sec}(T^{(k,l)}(T\mathfrak{M})) \to \mathrm{Sec}(T^{(k,l)}(T\mathfrak{N})) \,,$$

which acts on a tensor field $\mathcal{T} \in \mathrm{Sec}(T^{(k,l)}(T\mathfrak{M}))$. For a point $p \in \mathfrak{N}$, the value of the image, $f_* \mathcal{T}$, is equal to

$$(f_* \mathcal{T})_p(v_1, \ldots, v_k; u_1, \ldots, u_l) := \mathcal{T}_{f^{-1}(p)}\left(f^* v_1, \ldots, f^* v_k; f^* u_1, \ldots, f^* u_l\right) \,,$$

for all $u_1, \ldots, u_l \in T\mathfrak{N}$, $v_1, \ldots, v_k \in T^*\mathfrak{N}$, $f^* u_i = (Tf^{-1})u_i$, $(f^* v_j)v = v_j(Tfv)$. The mixed tensor field $f_* \mathcal{T}$ is called the *pushforward of* $\mathcal{T}$ *by* $f$.

The mapping

$$f^* \colon \mathrm{Sec}(T^{(k,l)}(T\mathfrak{N})) \to \mathrm{Sec}(T^{(k,l)}(T\mathfrak{M})) \,,$$

acts on each tensor field $\mathcal{T} \in \mathrm{Sec}(T^{(k,l)}(T\mathfrak{N}))$ by $f^* \mathcal{T} = (f^{-1})_* \mathcal{T}$. The mixed tensor field $f^* \mathcal{T}$ is called the *pullback of* $\mathcal{T}$ *by* $f$.

**Remark 13.5.** Note that in the definition of pushforward, we take tangent vectors $u_i \in T\mathfrak{M}$ and identify them with constant fields. Thus,

$$f^* u_i|_{f^{-1}(p)} = ((Tf^{-1}) \circ u_i \circ f)(f^{-1}(p)) = (Tf^{-1})u_i\,, \quad i = 1, \dots, k\,.$$

### 13.3.8 Smooth Frames

Suppose that $(\mathfrak{E}, \mathfrak{M}, \pi)$ is a smooth vector bundle of rank $k$, and $U$ is an open subset of $\mathfrak{M}$. A $k$-tuple $(\varepsilon_1, \dots, \varepsilon_k)$ of smooth[7] local sections is called a *local frame for* $\mathfrak{E}$ *over* $U$ [3], if for each $p \in U$, their values form a basis $(\varepsilon_1(p), \dots, \varepsilon_k(p))$ of the $k$-dimensional vector space $\mathfrak{E}_p$.

Let $\sigma$ be a local section defined in the open set $U$. For an arbitrary $p \in U$, one has $\sigma(p) = \sigma^i(p)\varepsilon_i(p)$, where $k$-tuple $(\sigma^1(p), \dots, \sigma^k(p))$ of real numbers is uniquely defined. Thus, one has $k$-mappings $\sigma^i : U \to \mathbb{R}$, which are called the *component functions of* $\sigma$ with respect to the local frame $(\varepsilon_1, \dots, \varepsilon_k)$. The smoothness of $\sigma$ is equivalent to the smoothness of each component function [3].

Consider local frames for standard vector bundles. Let $\mathfrak{M}$ be a smooth $s$-dimensional manifold. Suppose that $U$ is a coordinate neighborhood and $(x^i)$ are local coordinates. Then one can consider the following structures:

- **The tangent bundle,** $T\mathfrak{M}$**.** The frame $(\partial_i)_{i=1}^s$, which is defined on $U$ by $p \mapsto \partial_i|_p$, is called a *coordinate frame*. Here, $\partial_i|_p$ are the symbols defined by (13.13). For a vector field $u \in \mathrm{Vec}(U)$, one has $u = u^i\partial_i$.
- **The cotangent bundle,** $T^*\mathfrak{M}$**.** The frame $(dx^i)_{i=1}^s$, which is defined on $U$ by $p \mapsto dx^i|_p$, is called a *coordinate coframe*. Here, $dx^i|_p$ are the symbols defined by (13.15). For a covector field $v \in \mathrm{CVec}(U)$, one has $v = v_i dx^i$.
- **The bundle of exterior** $k$**-forms,** $\Lambda^k(T^*\mathfrak{M})$**.** Coordinates $(x^i)$ generate the local frame $(dx^{i_1} \wedge \dots \wedge dx^{i_k})_{1 \le i_1 < i_2 < \dots < i_k \le s}$ over $U$. This frame is defined as

  $$p \mapsto dx^{i_1}|_p \wedge \dots \wedge dx^{i_k}|_p\,, \quad 1 \le i_1 < i_2 < \dots < i_k \le s\,.$$

  For a $k$-form $\omega \in \Omega^k(U)$, one obtains

  $$\omega = \sum_{1 \le i_1 < i_2 < \dots < i_k \le s} \omega_{i_1 i_2 \dots i_k} dx^{i_1} \wedge \dots \wedge dx^{i_k}\,.$$

- **The bundle of mixed** $(k, l)$**-tensors,** $T^{(k,l)}(T\mathfrak{M})$**.** The local frame over $U$, which corresponds to $(x^i)$, is represented by the collection

  $$\left(\partial_{i_1} \otimes \dots \otimes \partial_{i_k} \otimes dx^{j_1} \otimes \dots \otimes dx^{j_l}\right)_{1 \le i_1, \dots, i_k \le s\,, 1 \le j_1, \dots, j_l \le s}\,,$$

7 We are dealing with smooth fields only and thus we suppose that elements of the ordered tuple $(\varepsilon_1, \dots, \varepsilon_k)$ are smooth.

where each element is the mapping

$$p \mapsto \partial_{i_1}|_p \otimes \cdots \otimes \partial_{i_k}|_p \otimes dx^{j_1}|_p \otimes \cdots \otimes dx^{j_l}|_p \,.$$

For a mixed tensor $\mathcal{T} \in \mathrm{Sec}(T^{(k,l)}(TU))$, one has

$$\mathcal{T} = T^{i_1 \dots i_k}{}_{j_1 \dots j_l} \partial_{i_1} \otimes \cdots \otimes \partial_{i_k} \otimes dx^{j_1} \otimes \cdots \otimes dx^{j_l} \,.$$

If $u \in \mathrm{Vec}(\mathfrak{M})$ and $v \in \mathrm{CVec}(\mathfrak{M})$, then we use the symbolic equalities $u = u^i \partial_i$ and $v = v_i dx^i$. These equalities acquire an exact meaning if one considers them within a coordinate neighborhood. If the equalities are considered in common points of overlapping coordinate neighborhoods, then component functions are related via (13.8) and (13.16). A similar situation holds for an arbitrary tensor field.

Suppose that there exists a frame over the whole smooth manifold $\mathfrak{M}$. We refer to such a frame as a *global frame* [3]. The particular case is a global frame for tangent bundle $T\mathfrak{M}$. If such a frame exists, then the manifold is called *parallelizable*. The necessary and sufficient condition for parallelizability is the triviality of $T\mathfrak{M}$.

### 13.3.9 Exterior Differentiation

For any $f \in C^\infty(\mathfrak{M})$, the formula (13.17) defines the mapping $df : p \mapsto df_p$ or, similarly, the element of $\mathrm{CVec}(\mathfrak{M})$. Thus, one obtains the operation

$$d\colon \Omega^0(\mathfrak{M}) \to \Omega^1(\mathfrak{M}) \,, \quad f \mapsto df \,,$$

where $\Omega^0(\mathfrak{M}) := C^\infty(\mathfrak{M})$ and $\Omega^1(\mathfrak{M}) := \mathrm{CVec}(\mathfrak{M})$.

The operation $d$ can be extended to arbitrary $k$-forms in the following way [3]. Let $\mathfrak{M}$ be a smooth $s$-dimensional manifold with or without boundary. There exist the unique operations

$$d\colon \Omega^k(\mathfrak{M}) \to \Omega^{k+1}(\mathfrak{M}) \,, \quad k = 0, 1, \dots, s-1 \,,$$

called *exterior differentiation*, which satisfy the properties:

i) $d(a\alpha + b\beta) = a d\alpha + b d\beta$, where $\alpha, \beta \in \Omega^k(\mathfrak{M})$ and $a, b \in \mathbb{R}$;
ii) for $\alpha \in \Omega^0(\mathfrak{M})$ the field $d\alpha\colon p \mapsto d\alpha_p$ is defined by the formula (13.17);
iii) $d(\alpha \wedge \beta) = d\alpha \wedge \beta + (-1)^k \alpha \wedge d\beta$, for $\alpha \in \Omega^k(\mathfrak{M})$ and $\beta \in \Omega^l(\mathfrak{M})$;
iv) $d^2\alpha = d(d\alpha) = 0$, for all $\alpha \in \Omega^k(\mathfrak{M})$.

For a $k$-form $\alpha$, the $k+1$-form $d\alpha$ is called the *exterior derivative of* $\alpha$.

**Remark 13.6.** The exterior derivative of a $k$-form

$$\alpha = \sum_{1 \le i_1 < i_2 < \cdots < i_k \le s} \alpha_{i_1 i_2 \dots i_k} dx^{i_1} \wedge \cdots \wedge dx^{i_k} \,,$$

according to the properties i)–iv) has the form

$$d\alpha = \sum_{1\leq i_1<i_2<\cdots<i_k\leq s} d\alpha_{i_1 i_2 \ldots i_k} \wedge dx^{i_1} \wedge \cdots \wedge dx^{i_k} \,,$$

where $d\alpha_{i_1 i_2 \ldots i_k} = \sum_{i=1}^{s} \partial_i \alpha_{i_1 i_2 \ldots i_k} dx^i$. The equivalent representation holds [3]:

$$d\alpha = \sum_{1\leq j_1<\cdots<j_{k+1}\leq s} \left( \sum_{a=1}^{k+1} (-1)^{a+1} \frac{\partial \alpha_{j_1 \ldots \widehat{j_a} \ldots j_{k+1}}}{\partial x^{j_a}} \right) dx^{j_1} \wedge \cdots \wedge dx^{j_{k+1}} \,,$$

where the sign ^ indicates that the corresponding index is omitted.

### 13.3.10 The Riemannian Metric and Musical Isomorphisms

Suppose that $\mathfrak{M}$ is a smooth $s$-dimensional manifold with or without boundary. The conventional geometric tool that allows one to measure lengths of vectors from a tangent space $T_p\mathfrak{M}$ and angles between them is provided by an inner product. However, at different points of $\mathfrak{M}$, this tool defines differently, and in order to introduce it on the whole manifold we ought to use some functions. Since we are dealing with smooth manifolds and smooth mappings only, it is reasonable to consider these functions to be smooth. This results in the field of inner products that is a Riemannian metric.

By the formal definition, a *Riemannian metric on* $\mathfrak{M}$ [3] is a smooth section $g \in \mathrm{Sec}(T^*\mathfrak{M} \otimes T^*\mathfrak{M})$, such that for every point $p \in \mathfrak{M}$, the following axioms hold:

$g_1$) $\forall u, v \in T_p\mathfrak{M}\colon\ g_p(u, v) = g_p(v, u)$;
$g_2$) $\forall u \in T_p\mathfrak{M}\colon\ g_p(u, u) \geq 0$;
$g_3$) $\forall u \in T_p\mathfrak{M}\colon\ (g_p(u, u) = 0, \text{ iff } u = 0)$.

**Remark 13.7.** Each smooth $s$-dimensional manifold $\mathfrak{M}$ (with or without boundary) admits a Riemannian metric [3]. The reasonings about the existence of a Riemannian metric are based on two observations. The first one is the Euclidean metric $g^{\mathrm{Eucl}} = \delta_{ij} dx^i \otimes dx^j$ on $\mathbb{R}^s$, written in Cartesian coordinates. Choose some atlas $A = \{(U_\alpha, \varphi_\alpha)\}_{\alpha\in I}$ of the smooth structure of $\mathfrak{M}$. This atlas generates the family $\{g_\alpha\}_{\alpha\in I}$ of pullbacks $g_\alpha := \varphi_\alpha^* g^{\mathrm{Eucl}}$; each of these is a Riemannian metric on the corresponding coordinate domain $U_\alpha$. The second observation is based on the conception of a smooth partition of unity $\{f_\alpha\}_{\alpha\in I}$, which is subordinate to the cover $\{U_\alpha\}_{\alpha\in I}$. It allows one to "glue" the so-obtained Riemannian metrics into a global Riemannian metric $g = \sum_{\alpha\in I} f_\alpha g_\alpha$ on $\mathfrak{M}$.

Since the Riemannian metric provides a tool for geometrical measurements, the choice of a certain one from all possible Riemannian metrics on $\mathfrak{M}$ is not arbitrary in physical applications. Such a choice is motivated by physical arguments. The construction of the material metric (see Chapter 8, Section 8.4) may serve as an example of such an argument.

For a fixed $p \in \mathfrak{M}$, the mapping $u \mapsto g_p(u, \cdot)$ is a vector space isomorphism between $T_p\mathfrak{M}$ and $T_p^*\mathfrak{M}$. It allows us to extend the definitions of *musical isomorphisms* given for vectors in Chapter 12, Section 12.5.4, to sections. That is, the mappings $(\cdot)^\flat : T\mathfrak{M} \to T^*\mathfrak{M}$ and $(\cdot)^\sharp : T^*\mathfrak{M} \to T\mathfrak{M}$ are defined by

$$\forall u \in T\mathfrak{M} \quad u \mapsto u^\flat \in T^*_{\pi_t(u)}\mathfrak{M} : \langle u^\flat, \cdot \rangle_{\pi_t(u)} = g_{\pi_t(u)}(u, \cdot) \,;$$
$$\forall v \in T^*\mathfrak{M} \quad v \mapsto v^\sharp \in T_{\pi_c(v)}\mathfrak{M} : g_{\pi_c(v)}(v^\sharp, \cdot) = \langle v, \cdot \rangle_{\pi_c(v)} \,.$$

The following diagrams illustrate relations between them:

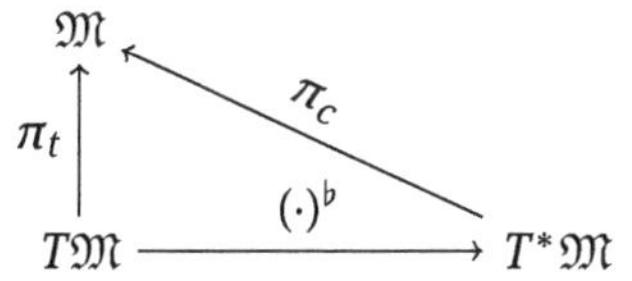


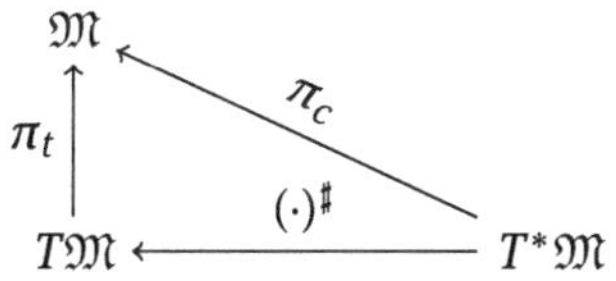


In terms of the coordinate frame $(\partial_i)_{i=1}^s$ and the corresponding coframe $(dx^i)_{i=1}^s$, one has the following representations:

$$u^\flat = g_{ij} u^j dx^i \,, \quad v^\sharp = g^{ij} v_j \partial_i \,,$$

where $g_{ij} = g(\partial_i, \partial_j)$ and $[g^{ij}]$ is the inverse to $[g_{ij}]$.

If $v = g(u, \cdot)$, then $u^\flat = v$, $v^\sharp = u$ and $v_i = g_{ij}u^j$, $u^i = g^{ij}v_j$. Further, we denote $v_i = v_i$. Thus, in coordinates, musical isomorphisms can be interpreted as rising and lowering index operations.

Musical isomorphisms can be generalized to tensor spaces as follows. Operation $(\cdot)^{k\flat}$, $k \in \mathbb{N}$ is defined by the relation

$$(\cdot)^{k\flat} : S_1 \otimes \ldots S_{k-1} \otimes T\mathfrak{M} \otimes S_{k+1} \otimes S_r \to S_1 \otimes \ldots S_{k-1} \otimes T^*\mathfrak{M} \otimes S_{k+1} \otimes S_r \,,$$

that is, $\mathcal{T}^{k\flat}(f_1, \ldots, f_{k-1}, u, f_{k+1}, \ldots, f_r) = \mathcal{T}(f_1, \ldots, f_{k-1}, u^\flat, f_{k+1}, \ldots, f_r)$, and operation $(\cdot)^{k\sharp}$, $k \in \mathbb{N}$ is defined as follows:

$$(\cdot)^{k\sharp} : S_1 \otimes \ldots S_{k-1} \otimes T^*\mathfrak{M} \otimes S_{k+1} \otimes S_r \to S_1 \otimes \ldots S_{k-1} \otimes T\mathfrak{M} \otimes S_{k+1} \otimes S_r \,,$$

that is, $\mathcal{T}^{k\sharp}(f_1, \ldots, f_{k-1}, v, f_{k+1}, \ldots, f_r) = \mathcal{T}(f_1, \ldots, f_{k-1}, v^\sharp, f_{k+1}, \ldots, f_r)$. Here, $S_j$ denotes either $T\mathfrak{M}$ or $T^*\mathfrak{M}$.

In the particular case of second-rank tensor fields, one has the following decompositions:

- The case $\mathcal{T} = T^{ij}\partial_i \otimes \partial_j$:

$$\mathcal{T}^{1\flat} = g_{ki} T^{ij} dx^k \otimes \partial_j \,, \quad T_k{}^j = g_{ki} T^{ij} \,.$$

- The case $\mathcal{T} = T_{ij} dx^i \otimes dx^j$:

$$\mathcal{T}^{1\sharp} = g^{ki} T_{ij} \partial_k \otimes dx^j \,, \quad T^k{}_j = g^{ki} T_{ij} \,.$$

- The case $\mathscr{T} = T_{ij}dx^i \otimes dx^j$:

$$\mathscr{T}^{2\sharp} = g^{kj}T_{ij}dx^i \otimes \partial_k\,, \quad T_i^{\ k} = g^{kj}T_{ij}\,.$$

The Riemannian metric $\mathscr{g}$ induces a field of inner products between elements of $T^*\mathfrak{M}$ in the following way. Denote $\mathscr{g}_{(1)} := \mathscr{g}^{1\sharp 2\sharp}$. Thus, $\mathscr{g}_{(1)} = g^{ij}\partial_i \otimes \partial_j$. For any covectors $\mu, \nu \in T^*\mathfrak{M}$, one obtains $\mathscr{g}_{(1)}(\mu, \nu) = \mu_i \nu_j g^{ij}$. In general, the Riemannian metric $\mathscr{g}$ generates a field $\mathscr{g}_{(k)}$ of inner products between elements of $\Lambda^k(T^*\mathfrak{M})$. Let $\alpha = \sum_{i_1<\cdots<i_k} \alpha_{i_1\ldots i_k} e^{i_1} \wedge \cdots \wedge e^{i_k}$ and $\beta = \sum_{j_1<\cdots<j_k} \beta_{j_1\ldots j_k} e^{j_1} \wedge \cdots \wedge e^{j_k}$ be exterior $k$-forms. Then, [123]

$$\mathscr{g}_{(k)}(\alpha, \beta) := \sum_{\substack{1\le i_1<\cdots<i_k\le s \\ j_1,\ldots,j_k\in\{1,\ldots,s\}}} g^{i_1 j_1} \ldots g^{i_k j_k} \alpha_{i_1\ldots i_k} \beta_{j_1\ldots j_k}\,.$$

## 13.4 Orientation and Integration on Manifolds

### 13.4.1 The Volume Form and Orientation of Smooth Manifolds

Suppose that $\mathfrak{M}$ is an $s$-dimensional smooth manifold (with or without boundary). For each point $p \in \mathfrak{M}$, the tangent space $T_p\mathfrak{M}$ is an $s$-dimensional vector space, and, thus, the classical notion of orientation is defined for it. An orientation is one of two equivalence classes of *ordered* bases. Recall that two ordered bases are equivalent if their transition matrix has a positive determinant. The notion of determinant is intimately related with the notion of the exterior $s$-form, and this implies that any non-zero exterior form $\mu \in \Lambda^s(T_p^*\mathfrak{M})$ determines an orientation on $T_p\mathfrak{M}$ [3]. Choosing an orientation for each space $T_p\mathfrak{M}$, one arrives at a field of orientations, or, equivalently, at a field of non-zero $s$-forms. With the requirement of smoothness, one obtains the notion of volume form.

**Volume form.** A *volume form* is an $s$-form $\mu \in \Omega^s(\mathfrak{M})$, such that $\mu(p) \neq 0$ for any $p \in \mathfrak{M}$. If a volume form exists for the manifold, then the latter is called *orientable*.

Suppose that $\mathfrak{M}$ is a smooth $s$-dimensional orientable manifold with or without boundary. Two volume forms $\mu_1, \mu_2 \in \Omega^s(\mathfrak{M})$ are called equivalent, if there exists a function $f \in C^\infty(\mathfrak{M})$, such that $f(p) > 0$ for every $p \in \mathfrak{M}$, and $\mu_1 = f\mu_2$. It can be proved that this is an equivalence relation [123]; an *orientation* of $\mathfrak{M}$ is any equivalence class $[\mu]$. Then, the oriented manifold can be considered as an *ordered pair* $(\mathfrak{M}, [\mu])$.

For an oriented manifold $(\mathfrak{M}, [\mu])$, one can introduce the orientation for each tangent space. That is, $(b_i)_{i=1}^s$ in $T_p\mathfrak{M}$ is positively (negatively) oriented respectively to $\mu$, if $\mu_p(b_1, \ldots, b_s) > 0$ ($< 0$).

**Orientation on the boundary.** Suppose that the orientation on a smooth manifold $\mathfrak{M}$ with boundary is fixed, *i.e.*, a volume form $\mu$ is chosen. The boundary $\partial\mathfrak{M}$ can be oriented in such a way that this orientation is aligned with $[\mu]$. Choose a smooth outward-pointing vector field $p \mapsto N(p) \in T_p\mathfrak{M} \setminus T_p\partial\mathfrak{M}$. Then, the $(n-1)$-form $\iota^*_{\partial\mathfrak{M}}(N \lrcorner \mu)$ is a volume form on $\partial\mathfrak{M}$, and it results in the desired orientation [3].

**The Riemannian volume form.** Assume that the Riemannian metric $\mathcal{g}$ is chosen on $\mathfrak{M}$. At each tangent space $T_p\mathfrak{M}$, one can get an orthonormal frame by the Gram–Schmidt orthonormalization procedure. Let $\mathfrak{M}$ be oriented. Then, in the chosen orientation, there exists a unique volume form $\mu$, such that $\mu_p(b_1, \dots, b_s) = 1$ for each positively oriented orthonormal frame for $T_p\mathfrak{M}$ [3]. In local coordinates,

$$\mu = \sqrt{g}\, dx^1 \wedge \dots \wedge dx^s\,, \quad g = \det[g_{ij}]\,.$$

We denote this differential form by $dV$ and call it the *Riemannian volume form.*

### 13.4.2 The Hodge Star Operator

Assume that $\mathfrak{M}$ is oriented and the volume form $dV$ is generated by the Riemannian metric $\mathcal{g}$. Fix some orientation for $T_p\mathfrak{M}$ relative to $dV$ and define the *Hodge star* isomorphism [123]

$$*: \Lambda^k\left(T_p^*\mathfrak{M}\right) \to \Lambda^{s-k}\left(T_p^*\mathfrak{M}\right)\,,$$
$$\forall \alpha, \beta \in \Lambda^k\left(T_p^*\mathfrak{M}\right) : \alpha \wedge *\beta = \mathcal{g}_{(k)}(\alpha, \beta) dV\,.$$

If $(e_1, \dots, e_s)$ is a positively oriented $\mathcal{g}$-orthonormal frame of $T_p\mathfrak{M}$, and $(e^1, \dots, e^s)$ is the corresponding coframe, then

$$*(e^{\sigma(1)} \wedge \dots \wedge e^{\sigma(k)}) = (\operatorname{sgn} \sigma) e^{\sigma(k+1)} \wedge \dots \wedge e^{\sigma(s)}\,,$$

for a permutation $\sigma \in S_s$, such that $\sigma(1) < \dots < \sigma(k)$ and $\sigma(k+1) < \dots < \sigma(s)$.

**Remark 13.8.** In particular, when $n = 2$,

$$*e^1 = e^2\,, \quad *e^2 = -e^1\,,$$

and when $n = 3$,

$$*e^1 = e^2 \wedge e^3\,, \quad *(e^1 \wedge e^3) = -e^2\,.$$

It can be shown that the Hodge star operator obeys the following properties [123]:
(i) $\forall \alpha, \beta \in \Lambda^k(T_p^*\mathfrak{M}) \quad \alpha \wedge *\beta = \beta \wedge *\alpha$;
(ii) $*1 = dV$, $*dV = 1$;
(iii) $\forall \alpha \in \Lambda^k(T_p^*\mathfrak{M}) \quad * *\alpha = (-1)^{k(s-1)}\alpha$;
(iv) $\forall \alpha, \beta \in \Lambda^k(T_p^*\mathfrak{M}) \quad \mathcal{g}_{(k)}(\alpha, \beta) = \mathcal{g}_{(s-k)}(*\alpha, *\beta)$.

In the frame $(e^{i_1}, \dots, e^{i_s})$, dual to the oriented frame, one has

$$*(e^{i_1} \wedge \dots \wedge e^{i_k}) = \sqrt{g} \sum_{\substack{j_1 < \dots < j_k \\ j_{k+1} < \dots < j_s}} \varepsilon_{(j_1, \dots, j_s)} g^{i_1 j_1} \dots g^{i_k j_k} e^{j_{k+1}} \wedge \dots \wedge e^{j_s}\,,$$

where $g = \det[g_{ij}]$. In particular, for $k = 1$, the following expression holds:

$$* e^i = \sqrt{g} \sum_{j=1}^{s} (-1)^{j-1} g^{ij} e^1 \wedge \cdots \wedge \hat{e}^j \wedge \cdots \wedge e^s , \qquad (13.22)$$

where $\hat{e}^j$ denotes that the $e^j$ should be omitted.

For $\alpha = \sum_{i_1<\cdots<i_k} \alpha_{i_1\dots i_k} e^{i_1} \wedge \cdots \wedge e^{i_k}$, one has

$$*\alpha = \sum_{i_1<\cdots<i_k} \sum_{\substack{j_1<\cdots<j_k \\ j_{k+1}<\cdots<j_s}} \varepsilon_{(j_1,\dots,j_s)} \alpha_{i_1\dots i_k} g^{i_1 j_1} \dots g^{i_k j_k} e^{j_{k+1}} \wedge \cdots \wedge e^{j_s} .$$

Finally, we note that the Hodge star operator acts pointwise on differential forms:

$$(*\alpha)_p = *(\alpha_p) .$$

### 13.4.3 Integration of Differential Forms and Stokes' Theorem

In the framework of the general geometric considerations developed in this book, the balance equations of continuum mechanics can be formulated in terms of integrals on differential forms over smooth manifolds. Such integrals are constructed consequentially, from integrals over open sets of $\mathbb{R}^s$ to integrals over manifolds. Below we show the sketch of such construction [3, 123].

**Integration over open set in $\mathbb{R}^s$.** Suppose that $U \subset \mathbb{R}^s$ is an open set, and a differential form $\omega \in \Omega^s(U)$ has a compact support[8]. If relatively to the natural frame in $\mathbb{R}^s$, $\omega = \omega_{1\dots s} dx^1 \wedge \cdots \wedge dx^s$, then integral of the $\omega$ over the set $U$ is defined as follows:

$$\int_U \omega := \int_U \omega_{1\dots s}(x) dx^1 \dots dx^s .$$

The integral on the right-hand side is simply a multiple integral.

**Integration over a single chart.** Let $\mathfrak{M}$ be a smooth $s$-dimensional manifold and $(U, \varphi)$ be a smooth chart. If $\omega \in \Omega^s(\mathfrak{M})$ and $\operatorname{supp} \omega \subset U$, then $\omega|_U$ has the same support. Hence, $\varphi^*(\omega|_U)$ has compact support, if so has $\omega$.

Let $\mathfrak{M}$ be a smooth and oriented manifold and the exterior form $\omega \in \Omega^s(U)$ have a compact support $\operatorname{supp} \omega \subset U$, where $(U, \varphi)$ is a smooth positively oriented chart. Then, one can define

$$\int_U \omega := \int_{\varphi(U)} (\varphi^{-1})^*(\omega|_U) .$$

8 Support for a differential form is defined analogously to the support of a function [123].

If $\operatorname{supp}\omega \subset U \cap V$, where $(U, \varphi)$ and $(V, \psi)$ are positively oriented charts, then

$$\int_U \omega = \int_V \omega .$$

**Integration over manifolds.** Above we defined the integration over a single chart. To extend this definition to the whole manifold, the partition of unity can be used. Let $\mathfrak{M}$ be an oriented smooth manifold, $A$ be an atlas with positively oriented charts, and $\{(U_\alpha, f_\alpha)\}_{\alpha\in I}$ be a partition of unity, subordinated to $A$. For $\omega \in \Omega^s(\mathfrak{M})$, define $\omega_\alpha := f_\alpha\omega$ (that is, $\omega_\alpha$ has a compact support in $U_\alpha$). Define the integral as

$$\int_{\mathfrak{M}} \alpha := \sum_{\alpha\in A} \int_{U_\alpha} \omega_\alpha .$$

The right-hand side of the equality contains only a finite amount of non-zero terms, and its value does not depend on the choice of atlas and the corresponding partition of unity.

**Theorem** (Stokes' theorem). *Let $\mathfrak{M}$ be an oriented smooth manifold with boundary and $\omega \in \Omega^{s-1}(\mathfrak{M})$ be a $(s-1)$-form with a compact support. Then the Stokes' theorem [3] asserts that*

$$\int_{\mathfrak{M}} d\omega = \int_{\partial\mathfrak{M}} \iota^*_{\partial\mathfrak{M}}\omega , \tag{13.23}$$

*where $\iota_{\partial\mathfrak{M}} : \partial\mathfrak{M} \to \mathfrak{M}$ is the inclusion map (13.1). For short, to avoid confusion, we write*

$$\int_{\mathfrak{M}} d\omega = \int_{\partial\mathfrak{M}} \omega .$$

# 14 Connections on Principal Bundles

In this chapter, we sketch the elements of the theory of connections on principal bundles that generalize all notions of connection mentioned in the previous sections. The comprehensive treatment of the theory of connections can be found in [145, 154]. Below, we only give a minimal set of definitions and reasonings, which allow us to interpret the concept of the material connection in terms of general geometrical considerations. This approach seems to be very useful because it allows us to look over the theory of incompatible finite deformations, and, probably, the theories of defects, growth, crystal plasticity, shrinkage, *etc.* from a very common standpoint. Moreover, this general geometrical treatment is really applicable in a number of branches of theoretical physics, which illuminates the universality of geometrical methods in physics and the mechanics of solids.

## 14.1 Lie Groups and Lie Algebras

### 14.1.1 Lie Groups and Homomorphisms

**Lie groups.** A *Lie group* [119] $G$ is a set that is endowed with a structure of a smooth manifold and a structure of a group with the group operation

$$\top\colon G \times G \to G\,,$$

and the inversion map

$$\mathrm{inv}\colon G \to G\,,$$

which are smooth mappings with respect to the smooth structure of the manifold. The general linear group $\mathrm{GL}(s;\mathbb{R})$ and the automorphism group $\mathrm{GL}(\mathscr{V})$ may serve as examples of Lie groups.

**Homomorphisms of Lie groups.** Suppose that $G_1$ and $G_2$ are Lie groups. A *Lie group homomorphism from* $G_1$ *to* $G_2$ is a smooth map $h\colon G_1 \to G_2$, such that

$$\forall a, b \in G\colon\ h(a \top_1 b) = h(a) \top_2 h(b)\,.$$

Here, $\top_i$ is a group operation of $G_i$, $i = 1, 2$. If there exists such a diffeomorphism $h\colon G_1 \to G_2$, for which both $h$ and $h^{-1}$ are homomorphisms, then $G_1$ and $G_2$ are called *isomorphic Lie groups*. The mapping $h$ is then called a *Lie group isomorphism*.

### 14.1.2 Group Action

Let $(G, \top)$ be a group with identity $e$ and $\mathfrak{M}$ be a set.

https://doi.org/10.1515/9783110563214-014

**Left and right actions.** A *left action of* $G$ *on* $\mathfrak{M}$ is a map

$$\triangleright\colon G \times \mathfrak{M} \to \mathfrak{M}\,, \quad \triangleright(g, x) = g \triangleright x\,,$$

such that

$LA_1)$ $\forall g_1, g_2 \in G \quad \forall x \in \mathfrak{M}\colon g_1 \triangleright (g_2 \triangleright x) = (g_1 \top g_2) \triangleright x$;

$LA_2)$ $\forall x \in \mathfrak{M}\colon e \triangleright x = x$.

A *right action of* $G$ *on* $\mathfrak{M}$ is a map

$$\triangleleft\colon \mathfrak{M} \times G \to \mathfrak{M}\,, \quad \triangleleft(x, g) = x \triangleleft g\,,$$

such that

$RA_1)$ $\forall g_1, g_2 \in G\ \forall x \in \mathfrak{M}\colon (x \triangleleft g_1) \triangleleft g_2 = x \triangleleft (g_1 \top g_2)$;

$RA_2)$ $\forall x \in \mathfrak{M}\colon x \triangleleft e = x$.

**Remark 14.1.** For a given left action $\triangleright$, one can construct the right action $\triangleleft$ by the formula:

$$x \triangleleft g := g^{-1} \triangleright x\,.$$

Indeed, $x \triangleleft e = e \triangleright x = x$. Then, the sequence of equalities

$$\begin{aligned}(x \triangleleft g_1) \triangleleft g_2 &= (g_1^{-1} \triangleright x) \triangleleft g_2 = g_2^{-1} \triangleright (g_1^{-1} \triangleright x)\\ &= (g_2^{-1} \top g_1^{-1}) \triangleleft x = (g_1 \top g_2)^{-1} \triangleright x = x \triangleleft (g_1 \top g_2)\,,\end{aligned}$$

proves that the so-defined operation $\triangleleft$ satisfies the conditions stated above for the right action.

**Orbit.** Let $\triangleright\colon G \times \mathfrak{M} \to \mathfrak{M}$ be a left action. For any $x \in \mathfrak{M}$, one can define the *orbit of* $x$ as the set

$$O_x^{\triangleright} := \{y \in \mathfrak{M} \mid \exists g \in G\colon g \triangleright x = y\}\,.$$

Thus, $O_x^{\triangleright}$ is simply the image of the partial map $\triangleright(\cdot, x)\colon G \to \mathfrak{M}$. Any orbit is nonempty set because $x \in O_x^{\triangleright}$ always.

For a right action $\triangleleft\colon \mathfrak{M} \times G \to \mathfrak{M}$, the orbit of $x \in \mathfrak{M}$ is defined in a similar way as the image of the partial map $\triangleleft(x, \cdot)\colon G \to \mathfrak{M}$, *i.e.*,

$$O_x^{\triangleleft} := \{y \in \mathfrak{M} \mid \exists g \in G\colon x \triangleleft g = y\}\,.$$

**Orbit space.** Given that the left action $\triangleright\colon G \times \mathfrak{M} \to \mathfrak{M}$ induces an equivalence relation $\sim$ on $\mathfrak{M}$:

$$\forall x, y \in \mathfrak{M}\colon (x \sim y) \Leftrightarrow (\exists g \in G\colon y = g \triangleright x)\,.$$

By virtue of this relation, for each $x \in \mathfrak{M}$, one has $[x] = O_x^{\triangleright}$. Thus, $\mathfrak{M}$ can be split into orbits:

$$\mathfrak{M} = \bigcup_{x \in \mathfrak{M}} O_x^{\triangleright}\,.$$

The quotient set is denoted by $\mathfrak{M}/G$ and is called the *orbit space*. The case of right action can be considered similarly.

**Stabilizer.** For any given $x \in \mathfrak{M}$, one can define the subset of $G$ that consists of elements that do not change $x$, *i.e.*,

$$S_x^{\triangleright} := \{g \in G \mid g \triangleright x = x\} ,$$

for left action $\triangleright$, and

$$S_x^{\triangleleft} := \{g \in G \mid x \triangleleft g = x\} ,$$

for right action $\triangleleft$. The set $S_x^{\triangleright}$ (or $S_x^{\triangleleft}$) is called the *stabilizer of* $x$. Any stabilizer is nonempty; at least it consists of the identity $e$.

**Free action.** An action $\triangleright$ (or $\triangleleft$) is *free* if

$$\forall x \in \mathfrak{M}: S_x = \{e\} .$$

**Smooth action.** If $\mathfrak{M}$ is a smooth manifold, $G$ is a Lie group and $\triangleright: G \times \mathfrak{M} \to \mathfrak{M}$ (or $\triangleleft: \mathfrak{M} \times G \to \mathfrak{M}$) is a smooth mapping, then the corresponding action is called a *smooth action.*

### 14.1.3 Lie Algebra of the Lie Group

**Lie algebra.** A *(real) Lie algebra* is a real vector space $A$ that is endowed with a map

$$[\cdot, \cdot]: A \times A \to A , \quad (x, y) \mapsto [x, y] ,$$

called the *bracket*, such that the following conditions hold

i) **Bilinearity.** For $\alpha, \beta \in \mathbb{R}$ and $x, y, z \in A$, one has

$$[\alpha x + \beta y, z] = \alpha[x, z] + \beta[y, z] ,$$
$$[x, \alpha y + \beta z] = \alpha[x, y] + \beta[x, z] .$$

ii) **Antisymmetry.** For $x, y \in A$:

$$[x, y] = -[y, x] .$$

iii) **Jacobi identity.** For $x, y, z \in A$:

$$[x, [y, z]] + [y, [z, x]] + [z, [x, y]] = 0 .$$

The space $\mathrm{Vec}(\mathfrak{M})$ of smooth vector fields defined on a smooth manifold $\mathfrak{M}$ is as an example of Lie algebra. In this case, brackets are defined by (13.20).

**Objects of anholonomity.** In Chapter 9, Section 9.2.6, we considered the decompositions (9.14) of Lie brackets applied to a global frame. Coefficients $c_{ij}{}^m$ of such decompositions are called objects of anholonomity. This construction is intimately related with the following consideration. Let $A$ be the Lie algebra, $\dim A = s$. Choose some basis $(e_i)_{i=1}^s$ for $A$. Then, one has decompositions of brackets:

$$[e_i, e_j] = c_{ij}{}^k e_k \,, \quad i, j = 1, \dots, s \,. \tag{14.1}$$

Note the difference between relations (9.14) and (14.1). The former relation is between frames, *i.e.*, sections, while the latter is a relation between constant vectors. One can consider (14.1) as (9.14) written pointwise. The minus sign in (14.1) has been omitted. The defining properties of Lie algebra, *i.e.*, antisymmetry and the Jacobi identity, imply

$$c_{ij}{}^k + c_{ji}{}^k = 0 \,,$$
$$c_{jk}{}^l c_{il}{}^r + c_{ki}{}^l c_{jl}{}^r + c_{ij}{}^l c_{kl}{}^r = 0 \,.$$

Thus, coefficients $c_{ij}{}^k$ are not arbitrary and need to obey the relations obtained.

**Lie algebra homomorphism.** If $A_1$ and $A_2$ are Lie algebras, then a linear map $L\colon A_1 \to A_2$ is called a *Lie algebra homomorphism* if

$$\forall x, y \in A_1 \colon L([x, y]) = [L(x), L(y)] \,.$$

If $L\colon A_1 \to A_2$ is a bijection such that both $L$ and $L^{-1}$ are Lie algebra homomorphisms, then $L$ is called a *Lie algebra isomorphism*. In this case, $A_1$ and $A_2$ are called *isomorphic Lie algebras*.

**Translations and invariance.** Suppose that $G$ is a Lie group. Each $a \in G$ induces the following mappings:

$$L_a\colon G \to G \,, \qquad L_a(g) := a \top g \,,$$
$$R_a\colon G \to G \,, \qquad R_a(g) := g \top a \,.$$

These mappings are called *left translation* and *right translation*, respectively [119]. They satisfy the following relations:

$$L_{g_1} \circ L_{g_2} = L_{g_1 \top g_2} \,, \quad R_{g_1} \circ R_{g_2} = R_{g_2 \top g_1} \,, \quad L_{g_1} \circ R_{g_2} = R_{g_2} \circ L_{g_1} \,,$$

for all $g_1, g_2 \in G$. These relations show that $L_a$ and $R_b$ are diffeomorphisms with inverse $L_{a^{-1}}$ and $R_{a^{-1}}$, respectively.

Consider sections from $\mathrm{Vec}(G)$. A vector field $u \in \mathrm{Vec}(G)$ is called *left-invariant* [119] if

$$\forall a \in G \colon (L_a)_* u = u \,.$$

Recall that lower the asterisk $*$ denotes the pushforward (Chapter 13, Section 13.3.7). Let $a \in G$ be a point. Then, $TL_a\colon TG \to TG$. For a left-invariant vector field $u \in \mathrm{Vec}(G)$ and $g \in G$, one has

$$TL_a u|_{a^{-1} \top g} = u_g \,.$$

This equality implies the following relation:

$$\forall g_1, g_2 \in G \colon TL_{g_1} u|_{g_2} = u_{g_1 \top g_2} \,.$$

**Lie subalgebra.** If $A$ is a Lie algebra, then a linear subspace $A_1 \subset A$ is called a *Lie subalgebra of A* if it is closed under brackets. Thus, $A_1$ is algebra itself with the restriction $[\cdot, \cdot]|_{A_1 \times A_1}$. If $G$ is a Lie group, then all left-invariant vector fields on $G$ form Lie subalgebra Lie($G$) of Vec($G$) [3, 119]. Therefore, Lie($G$) is a Lie algebra. It is called the *Lie algebra of G*. The mapping

$$L\colon \mathrm{Lie}(G) \to T_e G\,, \quad L(u) := u_e\,,$$

is a vector space isomorphism [3]. Thus, dim Lie($G$) = dim $G$. Moreover, it follows from the definition of $L$ that left-invariant vector fields are uniquely determined by their values at the identity.

### 14.1.4 Adjoint Representation

Suppose that $G$ is a Lie group and $\mathscr{V}$ is a finite dimensional real vector space. Recall that GL($\mathscr{V}$) is an automorphism group, which is a Lie group. Introduce the following definitions:

- A representation of the Lie group $G$ on $\mathscr{V}$ is a Lie group homomorphism

$$h\colon G \to \mathrm{GL}(\mathscr{V})\,.$$

- A representation of the Lie algebra Lie($G$) on $\mathscr{V}$ is a Lie algebra homomorphism

$$L\colon \mathrm{Lie}(G) \to \mathfrak{gl}(\mathscr{V})\,.$$

Let $(G, \top)$ be a Lie group. For any $a \in G$, define the mapping $C_a\colon G \to G$ by the rule $C_a(b) := a \top b \top a^{-1}$. This mapping is an automorphism of $G$ for each $a$. The assignment $a \mapsto C_a$ defines homomorphism from $G$ to the group of automorphisms of $G$. Define the adjoint representations [119]:

$$\mathrm{Ad}\colon G \to \mathrm{GL}(\mathscr{V})\,, \qquad \mathrm{Ad}(a) := dC_a\,,$$
$$\mathrm{ad}\colon \mathrm{Lie}(G) \to \mathfrak{gl}(\mathscr{V})\,, \qquad \forall X, Y \in G\colon \mathrm{ad}(X)(Y) := [X, Y]\,.$$

### 14.1.5 Exponential Mapping

**The one-parameter subgroup.** Let $G$ be a Lie group. Consider $\mathbb{R}$ as an additive Lie group. Any Lie group homomorphism $\chi\colon \mathbb{R} \to G$ is called a *one-parameter subgroup of G*. In Chapter 6, Section 6.4, we considered integral curves. The notion of the one-parameter subgroup is closely related with the notion of the integral curve. That is, any one-parameter subgroup of $G$ is the integral curve of a left-invariant vector field starting at the identity [3]. In this regard, for a given $u \in$ Lie($G$), the one-parameter subgroup determined by $u$ is called the *one-parameter subgroup generated by u*.

**Exponential map.** Let $G$ be a Lie group and Lie$(G)$ be the corresponding Lie algebra. Define a map

$$\exp\colon \mathrm{Lie}(G) \to G\,, \quad \exp u := \chi(1)\,,$$

where $\chi$ is the one-parameter subgroup generated by $u$. This map is called the *exponential map of G*. It is a smooth map that satisfies the relation [3]

$$\begin{aligned} \forall u \in \mathrm{Lie}(G)\ \forall s, t \in \mathbb{R}\colon \quad & \exp{(s+t)u} = \exp{su}\,\exp{tu}\,, \\ \forall u \in \mathrm{Lie}(G)\colon \quad & (\exp u)^{-1} = \exp{(-u)}\,, \\ \forall u \in \mathrm{Lie}(G)\ \forall n \in \mathbb{Z}\colon \quad & (\exp u)^{n} = \exp{(nu)}\,. \end{aligned}$$

## 14.2 Principal Bundles

### 14.2.1 Bundles

**Bundle.** A *bundle* [153] is a triple $(\mathfrak{E}, \mathfrak{M}, \pi)$, where $\mathfrak{E}$ and $\mathfrak{M}$ are topological spaces and $\pi\colon \mathfrak{E} \to \mathfrak{M}$ is a continuous surjective mapping. This definition includes the definition of the vector bundle (Chapter 13, Section 13.3.1) as a particular case. The space $\mathfrak{E}$ is called the *total space*, $\mathfrak{M}$ is called the *base space*, and $\pi$ is referred to as the *projection*. For any point $p \in \mathfrak{M}$, the set $\mathfrak{E}_p := \pi^{-1}(\{p\})$ is called the *fiber over p*. Thus, $\mathfrak{E}$ is decomposed into a disjoint union of fibers, $\mathfrak{E} = \bigsqcup_{p\in\mathfrak{M}} \mathfrak{E}_p$.

**Product bundle.** A *product bundle* over a topological space $\mathfrak{B}$ with fiber $\mathfrak{F}$ is the triple $(\mathfrak{M} \times \mathfrak{F}, \mathfrak{M}, \mathrm{pr}_1)$, where $\mathrm{pr}_1 : \mathfrak{M} \times \mathfrak{F} \to \mathfrak{M}$ is the projection on the first factor.

**Subbundle.** Suppose that $(\mathfrak{E}, \mathfrak{M}, \pi)$ is a bundle. A triple $(\mathfrak{E}_1, \mathfrak{M}_1, \pi_1)$ is called a *subbundle of* $(\mathfrak{E}, \mathfrak{M}, \pi)$, if $\mathfrak{E}_1$ is a subspace of $\mathfrak{E}$, $\mathfrak{M}_1$ is a subspace of $\mathfrak{M}$ and $\pi_1 = \pi|_{\mathfrak{E}_1} : \mathfrak{E}_1 \to \mathfrak{M}_1$.

**Bundle restriction.** Let $(\mathfrak{E}, \mathfrak{M}, \pi)$ be a bundle and $\mathfrak{N}$ be an open subset of $\mathfrak{M}$. Then, a triple $(\pi^{-1}(\mathfrak{N}), \mathfrak{N}, \pi|_{\pi^{-1}(\mathfrak{N})})$ is called a *restriction of the bundle* $(\mathfrak{E}, \mathfrak{M}, \pi)$ *on* $\mathfrak{N}$. Note that $\pi^{-1}(\mathfrak{N}) = \bigsqcup_{p\in\mathfrak{N}} \mathfrak{E}_p$.

**Sections.** A *section* of a bundle $(\mathfrak{E}, \mathfrak{M}, \pi)$ is a continuous map

$$\sigma\colon \mathfrak{M} \to \mathfrak{E}\,,$$

such that

$$\pi \circ \sigma = \mathrm{Id}_{\mathfrak{M}}\,.$$

In other words, a section is a continuous map $\sigma\colon \mathfrak{M} \to \mathfrak{E}$, such that $\sigma(p) \in \mathfrak{E}_p$, for each $p \in \mathfrak{M}$.

**Bundle morphisms.** Figuratively speaking, a bundle morphism is a map that preserves fibers. Let $(\mathfrak{E}_1, \mathfrak{M}_1, \pi_1)$ and $(\mathfrak{E}_2, \mathfrak{M}_2, \pi_2)$ be two bundles. A *bundle morphism* is a pair $(u, f)$ of continuous maps $u\colon \mathfrak{E}_1 \to \mathfrak{E}_2$ and $f\colon \mathfrak{M}_1 \to \mathfrak{M}_2$, such that $\pi_2 \circ u = f \circ \pi_1$. That is, the following diagram commutes:

$$\begin{array}{ccc} \mathfrak{E}_1 & \xrightarrow{\;u\;} & \mathfrak{E}_2 \\ {\scriptstyle \pi_1}\big\downarrow & & \big\downarrow{\scriptstyle \pi_2} \\ \mathfrak{M}_1 & \xrightarrow[\;f\;]{} & \mathfrak{M}_2 \end{array}$$

This definition includes the definition of the vector bundle homomorphism (Chapter 13, Section 13.3.6) as a particular case. A *bundle isomorphism* is a bundle morphism $(u, f)$ such that the maps $u, f$ are invertible, and the pair $(u^{-1}, f^{-1})$ is also a bundle morphism. The corresponding bundles are called *isomorphic*. If a bundle is isomorphic to some product bundle, then it is called *trivial*.

**Fiber bundles.** Generally, topological structures of fibers at different points may differ one from another. In order to exclude such bundles and consider only bundles with topologically equivalent fibers, we give a more specific definition. Let $\mathfrak{F}$ be a topological space. A *fiber bundle over* $\mathfrak{M}$ *with model fiber* $\mathfrak{F}$ [3] is a bundle $(\mathfrak{E}, \mathfrak{M}, \pi)$ with the following property: for each $p \in \mathfrak{M}$ there exists a neighborhood $U$ of $p$ in $\mathfrak{M}$ and a homeomorphism $\Phi\colon \pi^{-1}(U) \to U \times \mathfrak{F}$ called a *local trivialization of* $\mathfrak{E}$ *over* $U$, such that

$$\mathrm{pr}_1 \circ \Phi = \pi|_{\pi^{-1}(U)} \,.$$

where $\mathrm{pr}_1\colon U \times \mathfrak{F} \to U$ is the projection.

### 14.2.2 Principal Bundles

**Definition of principal bundles.** Let $\mathfrak{M}$ and $\mathfrak{P}$ be smooth manifolds, $\pi\colon \mathfrak{P} \to \mathfrak{M}$ be a smooth surjective mapping, $G$ be a Lie group with group operation $\top$, and $\triangleleft\colon \mathfrak{P} \times G \to \mathfrak{P}$ be a free smooth right action[1]. A *principal bundle* [119] is a 5-tuple $(\mathfrak{P}, \mathfrak{M}, \pi, G, \triangleleft)$ with the property: for every $p \in \mathfrak{M}$ there exist a neighborhood $U$ of $p$ and a diffeomorphism

$$\Phi\colon \pi^{-1}(U) \to U \times G \,,$$

such that

$$\forall q \in \pi^{-1}(U)\colon \Phi(q) = (\pi(q), \varphi(q)) \,,$$

where $\varphi\colon \pi^{-1}(U) \to G$ satisfies the condition

$$\forall q \in \pi^{-1}(U)\; \forall g \in G\colon \varphi(q \triangleleft g) = \varphi(q) \top g \,.$$

1 The case of left action is considered in an analogous way.

Note that the definition of $\Phi$ implies that it is a local trivialization. Thus, $(\mathfrak{P}, \mathfrak{M}, \pi)$ is a fiber bundle over $\mathfrak{M}$ with the model fiber $G$. Next, $\mathfrak{M}$ is diffeomorphic to the orbit space $\mathfrak{P}/G$ [119]. The group $G$ is called the *structure group of* $\mathfrak{P}$. If $U = \mathfrak{M}$, then $(\mathfrak{P}, \mathfrak{M}, \pi, G, \triangleleft)$ is called the *trivial principal bundle.*

The principal bundle is a very general abstract object that covers various kinds of material descriptions for solids. In particular, if we associate a fiber with a manifold that represents local deformations as some general, but not necessarily linear transformations, then we arrive at a generalization of the formal notion for structurally inhomogeneous bodies considered in this book.

**Bundle restriction.** Let $(\mathfrak{P}, \mathfrak{M}, \pi, G, \triangleleft)$ be a principal bundle and $U$ be an open subset of $\mathfrak{M}$. Denote $\mathfrak{P}_U := \pi^{-1}(U)$ and take the restrictions $\pi_U := \pi|_U : \mathfrak{P}_U \to U$, $\triangleleft_U := \triangleleft|_{\mathfrak{P}_U \times G} : \mathfrak{P}_U \times G \to \mathfrak{P}_U$ of $\pi$, $\triangleleft$. The 5-tuple $(\mathfrak{P}_U, U, \pi_U, G, \triangleleft_U)$ is a principal bundle [119]. It is called the *restriction of the bundle* $(\mathfrak{P}, \mathfrak{M}, \pi, G, \triangleleft)$ *on* $U$.

**Sections.** A *section* of a principal bundle $(\mathfrak{P}, \mathfrak{M}, \pi, G, \triangleleft)$ is simply a section of the bundle $(\mathfrak{P}, \mathfrak{M}, \pi)$. A *local section* of $(\mathfrak{P}, \mathfrak{M}, \pi, G, \triangleleft)$ over an open subset $U$ of $\mathfrak{M}$ is a section of the principal bundle $(\mathfrak{P}_U, U, \pi_U, G, \triangleleft_U)$. One has the following criteria: a principal bundle is trivial *iff* it admits a global section. The proof can be found in [119].

**Morphisms.** Let $(\mathfrak{P}_1, \mathfrak{M}_1, \pi_1, G_1, \triangleleft_1)$ and $(\mathfrak{P}_2, \mathfrak{M}_2, \pi_2, G_2, \triangleleft_2)$ be principal fiber bundles. A *morphism* is a pair $(u, f)$ of smooth mappings $u : \mathfrak{P}_1 \to \mathfrak{P}_2$ and $f : G_1 \to G_2$ such that
(i) $f$ a homomorphism of Lie groups;
(ii) for all $g \in G_1$ and all $p \in \mathfrak{P}_1$ one has

$$u(p \triangleleft_1 g) = u(p) \triangleleft_2 f(g) .$$

A morphism $(u, f)$ is called an *isomorphism* if $u$ is a diffeomorphism and $f$ is an isomorphism of Lie groups.

### 14.2.3 Frame Bundles

Suppose that $\mathfrak{M}$ is a smooth $s$-dimensional manifold. For each point $p \in \mathfrak{M}$ denote

$$L_p\mathfrak{M} := \{(e_i)_{i=1}^s \mid (e_i)_{i=1}^s \text{ is a basis for } T_p\mathfrak{M}\} .$$

This is the set of all bases (frames) of $T_p\mathfrak{M}$. Thus, one can define the set of all bases for all tangent spaces:

$$L\mathfrak{M} := \bigsqcup_{p \in \mathfrak{M}} L_p\mathfrak{M} .$$

For a basis $(e_i)_{i=1}^s$ for $T_p\mathfrak{M}$ and an element $[\Omega^i{}_j] \in \mathrm{GL}(s;\mathbb{R})$, one has that $(\Omega^i{}_j e_i)_{j=1}^s$ is a basis for $T_p\mathfrak{M}$ again. The assignment

$$\triangleleft\colon L\mathfrak{M} \times \mathrm{GL}(s;\mathbb{R}) \to L\mathfrak{M}\,, \quad \left((e_i)_{i=1}^s, \left[\Omega^i{}_j\right]\right) \mapsto \left(\Omega^i{}_j e_i\right)_{j=1}^s$$

is a free right action of $\mathrm{GL}(s;\mathbb{R})$ on $L\mathfrak{M}$. Let $p \in \mathfrak{M}$. If $(e_i)_{i=1}^s$ and $(e_i')_{i=1}^s$ are two bases for $T_p\mathfrak{M}$, then the first one can be obtained from the second one by multiplication on some element $\mathrm{GL}(s;\mathbb{R})$. Thus, the orbit $O_p$ is just $L_p\mathfrak{M}$.

Define the mapping

$$\pi\colon L\mathfrak{M} \to \mathfrak{M}\,, \quad \pi\left((p, e_i)_{i=1}^s\right) := p\,.$$

The space $L\mathfrak{M}$ can be given a smooth structure that turns it into a $s + s^2$-dimensional smooth manifold. The 5-tuple $(L\mathfrak{M}, \mathfrak{M}, \pi, \mathrm{GL}(s;\mathbb{R}), \triangleleft)$ is a principal bundle called a *frame bundle* [119].

It should be noted that the structure of the frame bundle corresponds exactly to the representation for structurally inhomogeneous solids as a continuous collection of elementary volumes that do not match one another.

### 14.2.4 Associated Bundles

Let $(\mathfrak{P}, \mathfrak{M}, \pi, G, \triangleleft)$ be a principal bundle, $\mathfrak{F}$ be a smooth manifold, and $\tilde{\triangleleft}\colon \mathfrak{F} \times G \to \mathfrak{F}$ be a right action. Actions $\triangleleft$ and $\tilde{\triangleleft}$ generate the product action

$$\triangleleft \times \tilde{\triangleleft}\colon \mathfrak{P} \times \mathfrak{F} \times G \to \mathfrak{P} \times \mathfrak{F}$$

via the formula

$$(p, f) \triangleleft \times \tilde{\triangleleft} g := (p \triangleleft g, f \tilde{\triangleleft} g)\,.$$

Denote the orbit space $(\mathfrak{P}\times\mathfrak{F})/G$ by $\mathfrak{P}_{\mathfrak{F}} := (\mathfrak{P}\times\mathfrak{F})/G$. It can be endowed with a smooth manifold structure [119].

The natural projection $\mathrm{pr}_1 : \mathfrak{P} \times \mathfrak{F} \to \mathfrak{P}$ induces a smooth surjective mapping (recall that $\mathfrak{P}/G$ is diffeomorphic to $\mathfrak{M}$)

$$\pi_{\mathfrak{F}} : \mathfrak{P}_{\mathfrak{F}} \to \mathfrak{M}\,, \quad \pi_{\mathfrak{F}}([p, f]) := \pi(p)\,.$$

Thus, $(\mathfrak{P}_{\mathfrak{F}}, \mathfrak{M}, \pi_{\mathfrak{F}})$ is a bundle. It can be endowed with a local trivialization, and the triple $(\mathfrak{P}_{\mathfrak{F}}, \mathfrak{M}, \pi_{\mathfrak{F}})$ is a fiber bundle. It is called the *fiber bundle associated to* $(\mathfrak{P}, \mathfrak{M}, \pi, G, \triangleleft)$ *with fiber* $\mathfrak{F}$ [154]. We deal with the particular case, when $\mathfrak{F}$ is a finite-dimensional vector space. In this case, the associated bundle $(\mathfrak{P}_{\mathfrak{F}}, \mathfrak{M}, \pi_{\mathfrak{F}})$ is a vector bundle.

## 14.3 Connections

### 14.3.1 Connections on the Principal Bundle

Let $(\mathfrak{P}, \mathfrak{M}, \pi, G, \lhd)$ be a principal bundle. Each element $A \in \mathrm{Lie}(G)$ induces a vector field on $\mathfrak{P}$:

$$\forall p \in \mathfrak{P}\ \forall f \in C^\infty(\mathfrak{P}) \colon X^A_p f := f(p \lhd \exp(tA)) \, .$$

Define a map $i \colon \mathrm{Lie}(G) \to \mathrm{Vec}(\mathfrak{P})$ that brings $A$ to $X^A$, *i.e.*,

$$i \colon A \mapsto X^A \, .$$

It can be shown that the map $i$ is a Lie algebra homomorphism, *i.e.*,

$$i([A, B]) = [i(A), i(B)] \, .$$

Let $p \in \mathfrak{P}$. Then, one can define *vertical subspace* $V_p\mathfrak{P}$ of tangent space $T_p\mathfrak{P}$ as follows:

$$T_p\mathfrak{P} \supset V_p\mathfrak{P} := \ker(\pi_*) \, ,$$

where $\ker(\pi_*)$ is the kernel of the pushforward of projection $\pi$, *i.e.*,

$$\ker(\pi_*) := \{v \in T_p\mathfrak{P} \mid \pi_*(v) = 0\} \, .$$

Since the curve $t \mapsto \gamma(t) := p \lhd \exp(tA)$ lies entirely in the fiber $\pi^{-1}(\{\pi(p)\})$, then for each $p \in \mathfrak{P}$ one has $X^A_p \in V_p\mathfrak{P}$, or equivalently, $\pi_*(X^A_p) = 0$. This construction is illustrated in Figure 14.1.

A *connection* on a principal bundle $(\mathfrak{P}, \mathfrak{M}, \pi, G, \lhd)$ is a distribution that assigns to every $p \in \mathfrak{P}$ a *horizontal* vector subspace $H_p\mathfrak{P} \subset T_p\mathfrak{P}$, such that [154]

(i) $H_p\mathfrak{P} \oplus V_p\mathfrak{P} = T_p\mathfrak{P}$.
(ii) $(\cdot \lhd g)_*(H_p\mathfrak{P}) = H_{p\lhd g}\mathfrak{P}$.

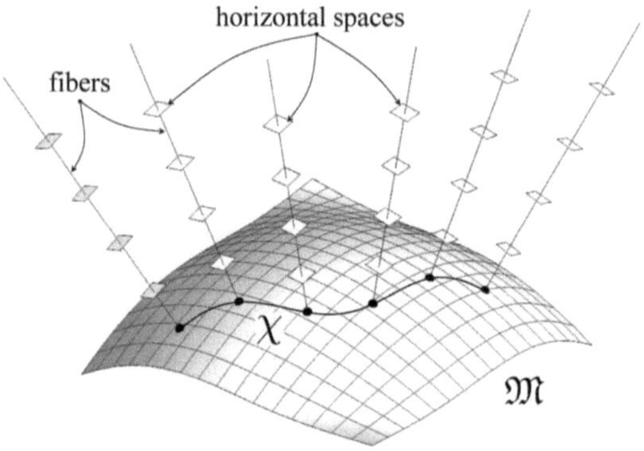


**Fig. 14.1:** Base manifold $\mathfrak{M}$, curve $\chi$ on it, fibers, and horizontal spaces.

(iii) The unique decomposition at point $p$, $u_p = \text{hor}(u_p) + \text{ver}(u_p)$ where $\text{hor}(u_p) \in H_p\mathfrak{P}$, $\text{ver}(u_p) \in V_p\mathfrak{P}$, gives for every smooth $u \in \text{Vec}(\mathfrak{P})$ two smooth vector fields $\text{hor}(u)$ and $\text{ver}(u)$.

This distribution may be represented by the Lie algebra-valued 1-form $\omega\colon p \mapsto \omega_p$,

$$\omega_p \in \text{Lin}(T_p\mathfrak{P}; \text{Lie}(G))\,, \quad u_p \mapsto \omega_p(u_p) := i_p^{-1}(\text{ver}(u))\,.$$

The form $\omega$ is called the *connection 1-form* [154]. In terms of $\omega$, one has that

$$H_p\mathfrak{P} = \ker(\omega_p) := \{v \in T_p\mathfrak{P} \mid \omega_p(v) = 0\}\,.$$

A connection 1-form $\omega$ (with respect to a given connection) has the properties [154]:
(i) $\omega_p(X_p^A) = A$;
(ii) $((\cdot \triangleleft g)^*\omega)_p(u_p) = \text{Ad}_{g^{-1}*}\omega_p(u_p)$;
(iii) $\omega$ is smooth.

### 14.3.2 Local Representation of Connections

Consider an open subset $U$ of $\mathfrak{M}$ and choose a local section $\sigma\colon U \to \mathfrak{P}$. Such a local section induces the field

$$\omega^U\colon \text{Vec}(U) \to \text{Lie}(G)\,, \quad \omega^U = \sigma^*\omega\,,$$

and local trivialization of the principal bundle:

$$h\colon U \times G \to \mathfrak{P}\,, \quad (p, g) \mapsto \sigma(p) \triangleleft g\,.$$

The form $h^*\omega$ is defined on $U \times G$:

$$(h^*\omega)_{(p,g)}\colon T_{(p,g)}(U \times G) \cong T_pU \oplus T_gG \to T\mathfrak{P}\,.$$

The 1-forms $\sigma^*\omega$ and $h^*\omega$ are related as follows:

$$(h^*\omega)_{(p,g)}(v, \gamma) = \text{Ad}_{g^{-1}*}\omega^U(v) + \Xi_g(\gamma)\,,$$

where $\Xi_g\colon T_gG \to \text{Lie}(G)$ is a Maurer–Cartan form, which from left invariant $L_g^A$ recovers the vector $A \in \text{Lie}(G)$, *i.e.*,

$$\Xi_g\colon L_g^A \mapsto A\,.$$

### 14.3.3 Local Representation on Frame Bundles

Let $\mathfrak{M}$ be an $s$-dimensional smooth manifold. Consider a frame bundle $\mathfrak{P} = L\mathfrak{M}$ with group $G = \text{GL}(s; \mathbb{R})$. Any choice of smooth chart $(U, \varphi)$ on $\mathfrak{M}$ induces a section $\sigma\colon \mathfrak{M} \to L\mathfrak{M}$

$$\forall p \in U\colon \sigma(p) := (\partial_1|_p, \ldots, \partial_s|_p)\,.$$

The local representation of a connection $\omega$ is given by a Lie algebra-valued 1-form defined on $U$ as follows:

$$\omega^U := \sigma^* \omega \, .$$

The components of this form in a chart-induced basis are

$$\Gamma^i{}_{j\mu} := (\omega^U)^i{}_{j\mu} = d\varphi^i{}_j \left( \omega^U(\partial_\mu) \right) \, ,$$

where index $\mu$ corresponds to the components of 1-form, while indices $i, j$ correspond to the components of the Lie algebra representation.

The local representation of the Maurer–Cartan form may be obtained as follows. Let $\varphi^i{}_j(g) =: g^i{}_j$ be coordinates on open subset of a connected part of $\mathrm{GL}(s; \mathbb{R})$, say, on $\mathrm{GL}^+(s; \mathbb{R})$, containing $\mathrm{Id}_G$. Consider the left invariant

$$\left( L^A \varphi^i{}_j \right)_g = \left( \varphi^i{}_j(g \top \exp(tA)) \right)' (0) = g^i{}_k A^k{}_i \, .$$

Hence,

$$L^A_g = g^i{}_k A^k{}_j \frac{\partial}{\partial \varphi^i{}_j} |_g \, .$$

One can define the Maurer–Cartan form as follows:

$$(\Xi_g)^i{}_j = (g^{-1})^i{}_k d\varphi^k{}_j \, .$$

Indeed,

$$(\Xi_g)^i{}_j L^A_g = (g^{-1})^i{}_k d\varphi^k{}_j g^p{}_r A^r{}_q \left. \frac{\partial}{\partial \varphi^r{}_q} \right|_g = (g^{-1})^i{}_p g^p{}_r A^r{}_j = A^i{}_j \, .$$

### 14.3.4 Gauge Maps

Consider two local representations $\sigma_1^* \omega$ and $\sigma_2^* \omega$ for a given connection $\omega$, generated by local sections $\sigma_1$ and $\sigma_2$ with intersecting domains $U_1 \cap U_2 \neq \emptyset$. The gauge map $\Omega \colon U_1 \cap U_2 \to G$ defines the unique $\Omega(p)$, such that

$$\sigma_2(p) = \sigma_1(p) \triangleleft \Omega(p) \, .$$

Local representations $\omega^{U_1}$ and $\omega^{U_2}$ are related as follows:

$$\omega^{U_2} = \mathrm{Ad}_{\Omega^{-1} *} \omega^{U_1} + \Omega^* \Xi \, .$$

The coordinate representation of the last term is the following:

$$(\Omega^* \Xi_p)^i{}_j \partial_\mu |_p = (\Omega(p)^{-1})^i{}_k \partial_\mu \Omega(p)^k{}_j dx^\mu =: \Omega^{-1} d\Omega \, .$$

### 14.3.5 Parallel Transport

**Horizontal lift of a curve.** Let $\pi\colon \mathfrak{P} \to \mathfrak{M}$ be a principal bundle, $\omega$ be a connection 1-form, and $\chi\colon [0,1] \to \mathfrak{M}$ be a smooth curve on $\mathfrak{M}$ with $\chi(0) = a$, $\chi(1) = b$. There exists a unique curve $\chi^{\uparrow}\colon [0,1] \to \mathfrak{P}$, which satisfies the following conditions:

(i) $\chi^{\uparrow}$ starts from a given point $p \in \mathfrak{P}$, which lies in the same fiber with $a$, *i.e.*,

$$\chi^{\uparrow}(0) = p \in \pi^{-1}(\{a\}) \subset \mathfrak{P} .$$

(ii) $\chi^{\uparrow}$ projects down to the curve $\chi$, *i.e.*,

$$\pi \circ \chi^{\uparrow} = \chi .$$

(iii) the vertical part of the velocity vector of the lifted curve $\chi^{\uparrow}$ vanishes at each point, *i.e.*,

$$\forall t \in [0,1]\colon \operatorname{ver}((\chi^{\uparrow})'(t)) .$$

(iv) the pushforward of the velocity vector of $\chi^{\uparrow}$ coincides with the velocity vector of the curve $\chi$, *i.e.*,

$$\forall t \in [0,1]\colon \pi_*((\chi^{\uparrow})'(t)) = \chi'(t) .$$

The relation between the base curve and its lifted counterpart is illustrated in Figure 14.2.

In order to find a horizontal lift $\chi^{\uparrow}$ for a given curve $\chi$ on $\mathfrak{M}$ and a given point $p \in \mathfrak{P}$, one has to solve the Cauchy problem for first-order ODE. Actually, if $y\colon [0,1] \to \mathfrak{P}$ is an arbitrary curve projecting to $\chi$ (*i.e.*, $\pi \circ y = \chi$), then $\chi^{\uparrow}$ can be obtained as follows:

$$\chi^{\uparrow}(t) = y(t) \triangleleft g(t) ,$$

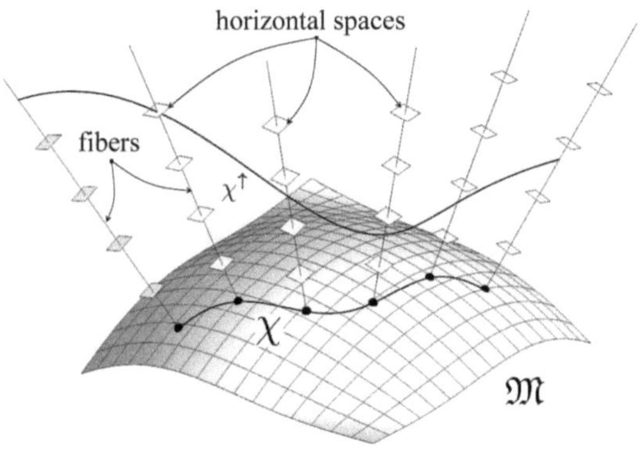


**Fig. 14.2:** Base manifold, the curve on it, fibers, horizontal spaces, and lifted curve.

where $g\colon [0,1] \to G$ is a curve on $G$ that can be obtained as a solution to the Cauchy problem

$$\begin{aligned} \mathrm{Ad}_{g(t)*}\omega_{\gamma(t)}(\gamma'(t)) + \Xi_{g(t)}(g'(t)) &= 0\,, \\ \gamma(0) \triangleleft g(0) &= p\,. \end{aligned} \tag{14.2}$$

If the principal bundle is a frame bundle and, consequently, $G$ is a matrix group, then the ODE (14.2) takes the form

$$g(t)^{-1}\omega_{\gamma(t)}(\gamma'(t))g(t) + g^{-1}(t)\dot{g}(t) = 0\,.$$

Hence,

$$\dot{g}(t) = -\omega_{\gamma(t)}(\gamma'(t))g(t)\,.$$

Let $\sigma$ be a section of the frame bundle. Taking into account the local representation of the connection $\omega$,

$$\omega_{\gamma(t)}(\gamma'(t)) = \omega_{\sigma\circ\chi(t)}(\chi'(t)) = \Gamma^{U,\sigma}(\chi(t))(\chi'(t)) = \Gamma^{U,\sigma}{}_{\mu}(\chi(t))\dot{\chi}^{\mu}(t)\,,$$

where $\Gamma^{U,\sigma} := \sigma^*\omega$, we obtain the following Cauchy problem:

$$\dot{g}(t) = -\Gamma_{\mu}(\chi(t))g(t)\,, \quad g(0) = g_0\,, \tag{14.3}$$

where, for the sake of simplicity, we denote $\Gamma^{U,\sigma}{}_{\mu}$ by $\Gamma_{\mu}$. The solution for (14.3) can be obtained by a path-ordered integral

$$g(t) = \left[P\exp\left(-\int_0^t d\tau\,\Gamma(\tau)\right)\right]g_0\,,$$

where

$$\begin{aligned} P\exp\left(-\int_0^t d\tau\,\Gamma(\tau)\right) := e - \int_0^t d\tau_1\,\Gamma(\tau_1) + \int_0^t d\tau_1 \int_0^{\tau_1} d\tau_2\,\Gamma(\tau_1)\Gamma(\tau_2) \\ - \int_0^t d\tau_1 \int_0^{\tau_1} d\tau_2 \int_0^{\tau_2} d\tau_3\,\Gamma(\tau_1)\Gamma(\tau_2)\Gamma(\tau_3) + \cdots \end{aligned}$$

Finally, the horizontal lift of a curve $\chi\colon [0,1] \to \mathfrak{M}$ is given by the explicit expression

$$\tilde{\chi}^{\uparrow}(t) = \sigma\circ\chi(t) \triangleleft \left[P\exp\left(-\int_0^t d\tau\,\Gamma^{U,\sigma}{}_{\mu}(\chi(\tau))\dot{\chi}^{\mu}(\tau)\right)\right]g_0\,.$$

**Parallel transport map.** Let $\chi\colon [0,1] \to \mathfrak{M}$ be a smooth curve on $\mathfrak{M}$ and $\chi^{\uparrow}\colon [0,1] \to \mathfrak{P}$ be a horizontal lift through $p \in \pi^{-1}(\{\chi(0)\})$. Then, the parallel transport map is defined as follows:

$$T_{\chi}\colon \pi^{-1}(\{\chi(0)\}) \to \pi^{-1}(\{\chi(1)\})\,, \quad p \mapsto \chi_p^{\uparrow}(1)\,.$$

This map is a bijection because $(\triangleleft\, g)_* H_p = H_{p\triangleleft g}$. This construction is illustrated in Figure 14.3.

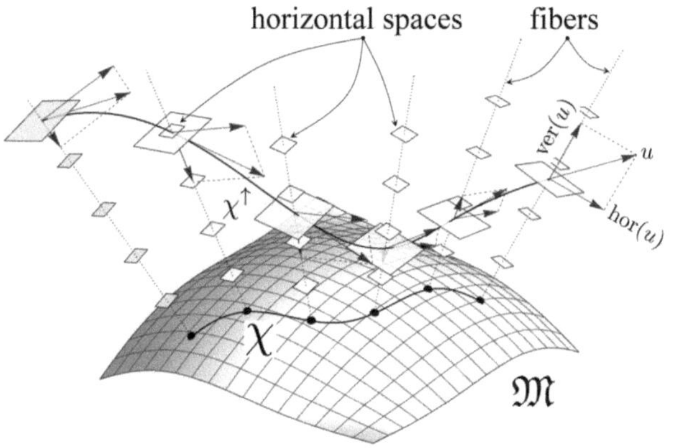


**Fig. 14.3:** Base manifold, the curve on it, fibers, horizontal spaces, lifted curve, and vertical and horizontal components of tangent vector $u$.

**Holonomy group of $\omega$.** Consider special types of curves, namely *loops*, which ends coincide, *i.e.*, $\chi\colon [0,1] \to \mathfrak{M}$ with $\chi(0) = \chi(1)$.

Let $a \in \mathfrak{M}$ and let $A$ be the set of all loops $\chi\colon [0,1] \to \mathfrak{M}$ with $\chi(0) = \chi(1) = a$ (Figure 14.4). Then, the subgroup of $G$,

$$\mathrm{Hol}_a(\omega) = \{g_\chi \in G \mid \exists \chi \in A : \chi_p^{\uparrow}(1)\} ,$$

is called the *holonomy group of $\omega$*.

**Horizontal lift of a curve $\chi$ to the associated bundle.** Let $\pi\colon \mathfrak{P} \to \mathfrak{M}$ be a principal bundle and $\omega$ be a connection 1-form. Let $\pi_{\mathfrak{F}}\colon \mathfrak{P}_{\mathfrak{F}} \to \mathfrak{M}$ be an associated fiber bun-

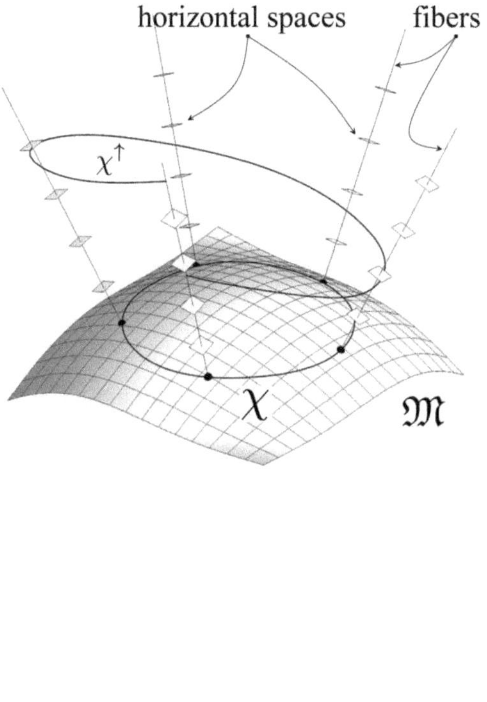


**Fig. 14.4:** The loop.

dle,[2] whose typical fiber $\mathfrak{F}$ acts from the left by $\triangleleft$. Then, the horizontal lift of a curve $\chi$ to the associated bundle that passes through the point $[p, f] \in \mathfrak{P}_{\mathfrak{F}}$ is a curve

$$\chi^{\uparrow \mathfrak{P}_{\mathfrak{F}}} : [0, 1] \to \mathfrak{P}_{\mathfrak{F}} , \quad \chi^{\uparrow \mathfrak{P}_{\mathfrak{F}}}(t) := [\chi^{\uparrow}(t), f] .$$

**Parallel transport on the associated bundle.** Define the mapping

$$T_{\chi}^{\mathfrak{P}_{\mathfrak{F}}} : \pi_{\mathfrak{F}}^{-1}(\{\chi(0)\}) \to \pi^{-1}(\{\chi(1)\}) , \quad [p, f] \mapsto \chi_{[p,f]}^{\uparrow \mathfrak{P}_{\mathfrak{F}}}(1) .$$

This map is a bijection between fibers.

**Parallel transport of a vector field.** If $\mathfrak{F}$ is a real vector space and the action $\triangleright : G \times \mathfrak{F} \to \mathfrak{F}$ is linear, then $\mathfrak{P}_{\mathfrak{F}}$ is a vector bundle. The parallel transport on it is called the *parallel transport of a vector field.*

### 14.3.6 Curvature

**Covariant exterior derivative.** Let $\pi : \mathfrak{P} \to \mathfrak{M}$ be a principal bundle with a connection $\omega$ and let $\varphi$ be an $A$-valued $k$-form. Then,

$$D\varphi : \Omega^k(\mathfrak{P}) \to A ,$$
$$D\varphi(u_1, \ldots, u_k) := d\varphi(\text{hor}(u_1), \ldots, \text{hor}(u_k)) ,$$

is a covariant exterior derivative of $\varphi$.

**Curvature.** The curvature of the connection $\omega$ is the Lie algebra-valued 2-form [154]

$$\Omega : \Omega^2(\mathfrak{P}) \to \text{Lie}(G) , \quad \Omega := D\omega .$$

Curvature 2-form can be expressed as follows:

$$\Omega = d\omega + \omega \,\bar{\wedge}\, \omega , \tag{14.4}$$

where $\omega \,\bar{\wedge}\, \omega(u, v) = [\omega(u), \omega(v)]$.

**Curvature of the connection on the frame bundle.** If $\pi : \mathfrak{P} \to \mathfrak{M}$ is a frame bundle and $G$ is a matrix group, then

$$\Omega^i{}_j = d\omega^i{}_j + \omega^i{}_k \wedge \omega^k{}_j .$$

---

2 Recall that $\mathfrak{P}_{\mathfrak{F}} = \{[p, f], p \in \mathfrak{P}, f \in \mathfrak{F}\}$, where $[p, f]$ is an equivalence class that corresponds to the following equivalence relation:

$$((p, f) \sim (p', f')) \Leftrightarrow (\exists g \in G : p' = p \triangleleft g, f' = g^{-1} \triangleright f) .$$

**Local representation of the curvature.** If a local section $\sigma\colon U \to \mathfrak{P}$ on $\mathfrak{M}$ is chosen, then it becomes possible to obtain a local representation of the connection 1-form, $\sigma^*\omega = \Gamma \in \Omega^1(\mathfrak{M}) \otimes \mathrm{Lie}(G)$ and local representation of the curvature 2-form,

$$\sigma^*\Omega = \mathrm{Riem} \in \Omega^2(\mathfrak{M}) \otimes \mathrm{Lie}(G)\,.$$

Due to the linearity of pullback,

$$\sigma^*\Omega = \sigma^*(d\omega + \omega \,\bar{\bar{\wedge}}\, \omega) = \sigma^*(d\omega) + \sigma^*(\omega) \,\bar{\bar{\wedge}}\, \sigma^*(\omega)\,.$$

Hence,

$$\mathrm{Riem} = d\Gamma + \Gamma \,\bar{\bar{\wedge}}\, \Gamma\,.$$

**Local representation of the curvature on the frame bundle.** Riem is a matrix-valued 2-form

$$\mathrm{Riem}^i{}_{j\mu\nu} = \partial_j \Gamma^i{}_{\mu\nu} - \partial_\mu \Gamma^i{}_{j\nu} + \Gamma^l{}_{\mu\nu}\Gamma^i{}_{jl} - \Gamma^l{}_{j\nu}\Gamma^i{}_{\mu l}\,,$$

where indices $i, j$ correspond to matrix elements, while indices $\mu, \nu$ correspond to 2-form components.

### 14.3.7 Torsion

**Solder form.** Let $V$ be a linear representation space of $G$ with $\dim V = \dim \mathfrak{M}$. Then, the $V$-valued 1-form $\theta$ on $\mathfrak{P}$ is called a solder form, if [154]

(i) $\theta$ is a “vertical” 1-form, *i.e.*,

$$\forall u \in \mathrm{Vec}(\mathfrak{P})\colon \theta(\mathrm{ver}(u)) = 0\,.$$

(ii) The following compatibility condition ($G$-equivariant condition) is satisfied:

$$g \triangleright (\triangleleft\, g)^*\theta = \theta\,.$$

(iii) Tangent bundle $T\mathfrak{M}$ and the associated bundle $\mathfrak{P}_V$ are isomorphic:

$$T\mathfrak{M} \cong \mathfrak{P}_V\,.$$

**Remark 14.2.** The solder form is introduced to provide identification of $V$ with each tangent space to $\mathfrak{M}$.

**Solder form on the frame bundle.** Let $\mathfrak{P} = L\mathfrak{M}$ be a frame bundle. Then, solder form $\theta$ defines the map

$$\theta\colon \mathrm{Vec}(\mathfrak{P}) \to \mathbb{R}^{\dim \mathfrak{M}} \equiv V\,, \quad \theta_e(u) := \left(z_e^{-1} \circ \pi_*\right)(u)\,,$$

where $e$ is a point on $\mathfrak{P}$ that corresponds to some frame, and $z$ is a mapping $z_e \in \mathrm{Lin}(\mathbb{R}^{\dim \mathfrak{M}}; T_{\pi(e)}\mathfrak{M})$,

$$z_e\colon (0, \ldots, 0, 1, 0, \ldots, 0) \mapsto e_i \quad (i\text{-th vector of a frame})\,.$$

Thus,

$$z_e^{-1} : T_{\pi(e)}\mathfrak{M} \to \mathbb{R}^{\dim \mathfrak{M}} , \quad z \mapsto \epsilon(z) ,$$

where $z \in T_{\pi(e)}\mathfrak{M}$ and $\epsilon$ is a frame dual to $e$.

**Torsion form.** Let $\pi: \mathfrak{P} \to \mathfrak{M}$ be a principal bundle, $\omega$ be a connection 1-form, and $\theta$ be a solder 1-form. Then, the torsion is a vector-valued 2-form defined on the principal bundle as follows [154]:

$$\Theta := D\theta \in \Omega^2(\mathfrak{P}) \otimes V .$$

Torsion can be expressed in terms of the exterior derivative

$$\Theta = d\theta + \omega \wedge \theta .$$

If the principal bundle is a frame bundle, and $G$ is matrix group, then the torsion can be expressed in components

$$\Theta^i = d\theta^i + \omega^i{}_k \wedge \theta^k .$$

**Local representation of torsion.** Let $\sigma: U \to \mathfrak{P}$ be a local section. Then, the local representation of the torsion can be obtained as a pullback

$$T = \sigma^* \Theta \in \Omega^2(\mathfrak{M}) \otimes V .$$

### 14.3.8 Bianchi Identities

The first Bianchi identity can be written in the form

$$D\Omega = 0 .$$

We have that $\Omega^3(\mathfrak{P}) \otimes V \ni D\Theta = \Omega \wedge \Theta$. If $\mathfrak{P}$ is a frame bundle, and $G$ is a Lie group, then second Bianchi identity is given by the relation

$$(D\Theta)^i = \Omega^i{}_k \Theta^k .$$

### 14.3.9 Covariant Derivatives on Associated Vector Bundles

Let $\pi: \mathfrak{P} \to \mathfrak{M}$ be a principal bundle with the right action $\triangleleft$ and the connection 1-form $\omega$, and let $\pi_{\mathfrak{F}} : \mathfrak{P}_{\mathfrak{F}} \to \mathfrak{M}$ be the associated vector bundle that is a trivial bundle $G \times \mathfrak{F}$, which typical fiber is a vector space $\mathfrak{F}$. Let the left action $\triangleright: G \times \mathfrak{F} \to \mathfrak{F}$ be defined on $G \times \mathfrak{F}$.

The geometrical idea of a covariant derivative is illustrated in Figure 14.5. Parallel transport along a smooth curve $\chi$, which brings a point $p$ to the point $p'$, is defined on the principal bundle $\mathfrak{P}$ by the horizontal lift $\chi^{\uparrow}$ through $p \in \mathfrak{P}$. Associating the

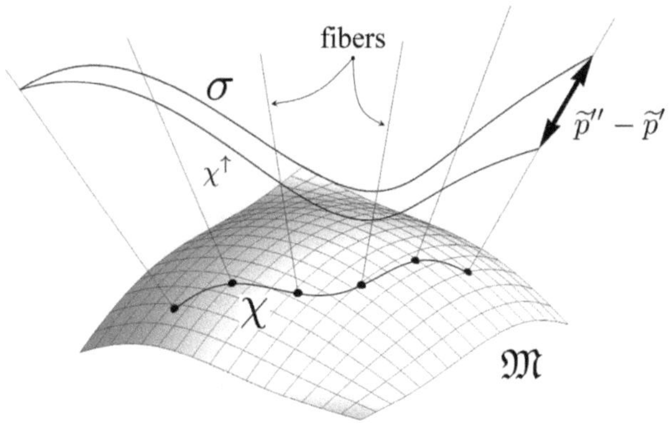


**Fig. 14.5:** The idea of the covariant derivative.

bundles $\pi: \mathfrak{P} \to \mathfrak{M}$ and $\pi_{\mathfrak{F}}: \mathfrak{P}_{\mathfrak{F}} \to \mathfrak{M}$ (provided by the left action $\triangleright$) allows us to determine the associated parallel transport on $\mathfrak{P}_{\mathfrak{F}}$ that brings $\widetilde{p}$ to $\widetilde{p}'$. Let $\sigma: \mathfrak{M} \to \mathfrak{P}_{\mathfrak{F}}$ be a section of the associated bundle that brings point $a$ to $\widetilde{p}$. Then, this section brings point $b$ to some point $\widetilde{p}''$, which, in general, does not coincide with $\widetilde{p}'$. Because the fiber $\mathfrak{F}$ is a vector space it is possible to calculate the difference between two vectors: $\widetilde{p}'' - \widetilde{p}'$ and, passing to the limit, obtain the differential quotient. This differential quotient determines the covariant derivative.

### 14.3.10 Direct Construction of Covariant Derivatives on Principal Bundles

Let $\pi: \mathfrak{P} \to \mathfrak{M}$ be a principal bundle with group $G$ right action $\triangleleft$, and let $\pi_{\mathfrak{F}}: \mathfrak{P}_{\mathfrak{F}} \to \mathfrak{M}$ be the associated vector bundle. Recall that the fiber $\mathfrak{F}$ is a vector space and the left $G$-action is a linear map $\triangleright\, G: \mathfrak{F} \to \mathfrak{F}$. Let $\sigma$ be a section of $\mathfrak{P}_{\mathfrak{F}}$, *i.e.*, $\sigma: \mathfrak{M} \to \mathfrak{P}_{\mathfrak{F}}$, $\sigma \circ \pi_{\mathfrak{F}} = \mathrm{Id}_{\mathfrak{M}}$. Define an $\mathfrak{F}$-valued function

$$\Phi_\sigma: \mathfrak{P} \to \mathfrak{F}\,, \quad \Phi_\sigma(p) := i_p^{-1}(\sigma(\pi(p)))\,, \tag{14.5}$$

where

$$i_p: \mathfrak{F} \to \pi_{\mathfrak{F}}^{-1}(\pi_{\mathfrak{F}})(\{\pi(p)\})\,, \quad i_p(f) := [p, f]\,.$$

The function $\Phi_\sigma$ is $G$-equivariant, *i.e.*, $\Phi_\sigma(p \triangleleft g) = g^{-1} \triangleright \Phi_\sigma(p)$. Indeed,

$$\Phi_\sigma(p \triangleleft g) = i_{p\triangleleft g}^{-1}(\sigma(\pi(p \triangleleft g)))i_p(\Phi_\sigma(p))\,.$$

Taking into account the following relation:

$$i_p(\Phi_\sigma(p)) = [p, \Phi_\sigma(p)] = \left[p \triangleleft g, g^{-1} \triangleright \Phi_\sigma(p)\right] = i_{p\triangleleft g}\left(g^{-1} \triangleright \Phi_\sigma(p)\right)\,,$$

we obtain

$$\Phi_\sigma(p \triangleleft g) = i_{p\triangleleft g}^{-1} i_{p\triangleleft g}\left(g^{-1} \triangleright \Phi_\sigma(p)\right) = g^{-1} \triangleright \Phi_\sigma(p)\,,$$

which proves the property of $G$-equivariance.

Moreover, there is a bijective correspondence between sections $\sigma\colon \mathfrak{M} \to \mathfrak{P}_{\mathfrak{F}}$ and functions $\Phi_\sigma\colon \mathfrak{P} \to \mathfrak{F}$. In order to prove it, let us consider some that $G$-equivariant mapping $\Phi_\sigma$ is given. Then, we can reconstruct the section $\sigma_\Phi\colon \mathfrak{M} \to \mathfrak{P}_{\mathfrak{F}}$ that determines $\Phi$ by (14.5) as the equivalence class

$$\forall x \in \mathfrak{M}\colon \sigma_\Phi(x) := [p, \Phi(p)] \,. \tag{14.6}$$

The section $\sigma_\Phi$ is well defined, because

$$\forall p' \in \pi^{-1}(\{x\}) \quad \exists! g \in G\colon p' = p \triangleleft g \,,$$

and

$$[p', \Phi(p')] = [p \triangleleft g, \Phi(p \triangleleft g)] \,.$$

Taking the equivariance relation into account, we obtain

$$[p', \Phi(p')] = [p \triangleleft g, g^{-1} \triangleright \Phi(p)] = [p, \Phi(p)] \,.$$

Since $\pi_{\mathfrak{F}} \circ \sigma_\Phi = \mathrm{Id}_{\mathfrak{M}}$, $\sigma_\Phi$ is a section on $\mathfrak{P}_{\mathfrak{F}}$. Thus, for a given section $\sigma$ one can define the $G$-equivariant function $\Phi_\sigma$ by (14.5) and for a given $G$-equivariant mapping, one can define the section $\sigma_\Phi$ by (14.6). Moreover, we have

$$\forall x \in \mathfrak{M}\colon \sigma_{\Phi_\sigma(x)} = [p, \Phi_\sigma(p)] = \left[p, i_p^{-1}(\sigma(\pi(p)))\right] = \sigma(\pi(p)) = \sigma(x) \,,$$

and

$$\Phi_{\sigma_\Phi}(p) = i_p^{-1}(\sigma_\Phi(\pi(p))) = i_p^{-1}([p, \Phi(p)]) = \Phi(p) \,.$$

Thus, the correspondence $\sigma_\Phi \leftrightarrow \Phi_\sigma$ is bijective.

If $\Phi\colon \mathfrak{P} \to \mathfrak{F}$ is $G$-equivariant and $A \in \mathrm{Lie}(G)$ is an element of Lie algebra $\mathrm{Lie}(G)$, then

$$\forall p \in \mathfrak{P}\ \forall t \in \mathbb{R}\colon \Phi(p \triangleleft \exp(At)) = \exp(-At) \triangleright \Phi(p) \,.$$

Recall that $\mathfrak{F}$ is a finite-dimensional vector space and $G\ \triangleright\colon \mathfrak{F} \to \mathfrak{F}$ is a linear mapping. Then,

$$\frac{d}{dt}(\Phi(p \triangleleft \exp(At)))\bigg|_{t=0} = \frac{d}{dt}(\exp(-At) \triangleright \Phi(p))\bigg|_{t=0} \,,$$

and

$$d\Phi|_p\left(V_p^A\right) = -A \triangleright \Phi(p) \,,$$

where $V_p^A$ is an induced vector field on $\mathfrak{P}$. Taking into account that $A = \omega(V_p^A)$, where $\omega$ is a connection 1-form, we obtain the corollary from $G$-invariance of $\Phi$ for a linear $\triangleright$:

$$d\Phi(V^A) + \omega(V^A) \triangleright \Phi = 0 \,. \tag{14.7}$$

Now, everything is prepared for constructing the covariant derivative. Recall that $\forall V \in T_p\mathfrak{P}\ D\Phi(V) \in \mathfrak{F}$ and

$$D\Phi := d\Phi \circ \mathrm{hor} \,.$$

It is easy to show that

$$D\Phi(V) = d\Phi(V) + \omega(V) \triangleright \Phi\,. \tag{14.8}$$

Indeed, if $V$ is vertical, then $V = V^A$ and $D\Phi(V) = d\Phi \circ \mathrm{hor}(V) = 0$. On the other hand, keeping in mind (14.7), we obtain $d\Phi(V^A) + \omega(V^A) \triangleright \Phi = 0$. If $V$ is horizontal, then $D\Phi := d\Phi(V)$ and $\omega(V) = 0$, so the equation (14.8) is also satisfied.

Let $V \in T_p\mathfrak{P}$ be an arbitrary vector. Define $D_V\Phi := D\Phi(V)$. In order to locally trivialize a bundle, choose a section $\varphi$ in the principal bundle and construct the pullbacks

$$\begin{aligned} \Phi\colon \mathfrak{P} \to \mathfrak{F} \quad &\text{to} \quad \varphi^*\Phi = \Phi \circ \varphi\,, \\ \omega \in \Omega^1(\mathfrak{P}) \otimes \mathrm{Lie}(G) \quad &\text{to} \quad \omega^{U,\varphi} := \varphi^*\omega \in \Omega^1(\mathfrak{M}) \otimes \mathrm{Lie}(G)\,, \\ D\Phi \in \Omega^1(\mathfrak{P}) \otimes \mathfrak{F} \quad &\text{to} \quad \varphi^* D\Phi \in \Omega^1(\mathfrak{M}) \otimes \mathfrak{P}\,. \end{aligned}$$

Then

$$\begin{aligned} \forall t \in T_x\mathfrak{M}\colon (\varphi^* D\Phi)(t) &= \varphi^*(d\Phi + \omega \triangleright \Phi)(t) \\ &= \varphi^*(d\Phi)(t) + \varphi^*(\omega \triangleright \Phi)(t) \\ &= d(\varphi^*\Phi)(t) + (\varphi^*\omega)(t) \triangleright \varphi^*\Phi \\ &= ds(t) + \omega^{U,\varphi} \triangleright s\,, \end{aligned}$$

where $s\colon U \to \mathfrak{F}$, $s := \varphi^*\Phi$ is a local $\mathfrak{F}$-valued function and $\omega^{U,\varphi} := \varphi^*\omega$. Now, introduce the notation $\nabla_t s$:

$$\nabla_t s := ds(t) + \omega^{U,\varphi} \triangleright s\,. \tag{14.9}$$

The first item on the right-hand side of (14.9) corresponds to the partial derivatives, and the second one corresponds to the connection. It is easy to check that the so-defined mapping $\nabla_t s$ satisfies the following conditions:
(i) $\nabla_{ft+gs}\sigma = f\nabla_t\sigma + g\nabla_s\sigma$;
(ii) $\nabla_t(\sigma + \tau) = \nabla_t\sigma + \nabla_t\tau$;
(iii) $\nabla_t(f\sigma) = (tf)\sigma + f\nabla_t\sigma$.

**Concluding remarks.** From an abstract geometrical point of view, the description of a stress-free shape for a structurally inhomogeneous body is equivalent to choosing a section on a principal bundle. The base of this bundle represents the body, while its fibers represent possible local transformations that bring the neighborhoods of material points into a uniform state. In order to choose a section, one must take into account some extra information about the structure of the body. This information can be obtained from the solution of the evolutionary problem. From general considerations, it is equivalent to some gauge procedure.

# Bibliography


[1] Sternberg S. Lectures on differential geometry. vol. 2. Prentice-Hall Englewood Cliffs, NJ; 1964.

[2] Edelen DGB. Applied Exterior Calculus. Mineola, N.Y.: Dover Publ. Inc.; 2011.

[3] Lee JM. Introduction to Smooth Manifolds. New York: Springer; 2012.

[4] Gibson I, Rosen DW, Stucker B, et al. Additive manufacturing technologies. New York: Springer; 2010.

[5] Epstein M, Elzanowski M. Material inhomogeneities and their evolution: A geometric approach. Berlin: Springer Science & Business Media; 2007.

[6] Epstein M. The Geometrical Language of Continuum Mechanics. Cambridge: Cambridge University Press; 2010.

[7] Steinmann P. Geometrical Foundations of Continuum Mechanics. Springer Berlin Heidelberg; 2015.

[8] Maugin GA. Material inhomogeneities in elasticity. Boca Raton: CRC Press; 1993.

[9] Levi-Civita T. Nozione di parallelismo in una varietà qualunque e conseguente specificazione geometrica della curvatura riemanniana. Rendiconti del Circolo Matematico di Palermo. 1916 dec;42(1):173–204.

[10] Lutzen J. Interactions between mechanics and differential geometry in the 19th century. Arch Hist Exact Sci. 1995;49(1):1–72.

[11] Choquet-Bruhat Y. General Relativity and the Einstein Equations. Oxford: Oxford University Press (OUP); 2008.

[12] Berger M. A panoramic view of Riemannian geometry. Berlin: Springer Science & Business Media; 2012.

[13] Eisenhart LP. Riemannian geometry. Princeton: Princeton university press; 1997.

[14] Eisenhart LP. Non-riemannian geometry. North Chelmsford: Courier Corporation; 2012.

[15] Petersen P. Riemannian geometry. Berlin: Springer Science & Business Media; 2006.

[16] Eckart C. The thermodynamics of irreversible processes. IV. The theory of elasticity and anelasticity. Physical Review. 1948;73(4):373–382.

[17] Kondo K. Non-Riemannian geometry of imperfect crystals from a macroscopic viewpoint. Memoirs of the unifying study of the basic problems in engineering science by means of geometry. 1955;1:6–17.

[18] Kondo K. Geometry of elastic deformation and incompatibility. In: Memoirs of the Unifying Study of the Basic Problems in Engineering Science by Means of Geometry. vol. 1. Tokyo: Gakujutsu Bunken Fukyo-Kai; 1955. p. 5–17.

[19] Kondo K. Non-Riemannian and Finslerian approaches to the theory of yielding. International Journal of Engineering Science. 1963;1(1):71–88.

[20] Kondo K. On the analytical and physical foundations of the theory of dislocations and yielding by the differential geometry of continua. International Journal of Engineering Science. 1964;2(3):219–251.

[21] Weingarten J. Sulle superficie di discontinuita nella teoria della elasticita dei corpi solidi. Rend Reale Accad dei Lincei, classe di sci, fis, mat e nat. 1901;10.1.:57–60.

[22] Volterra V, Volterra E. Sur les distorsions des corps élastiques (théorie et applications). In: Memorial des sciences mathematiques, 147. Gauthier-Villars; 1960. p. 3–117.

[23] Nye JF. Some geometrical relations in dislocated crystals. Acta metallurgica. 1953;1(2):153–162.

[24] Kroner E. Dislocation Theory as a Physical Field Theory. Meccanica. 1996;31:577–587.


https://doi.org/10.1515/9783110563214-015

[25] Yavari A. A geometric theory of growth mechanics. Journal of Nonlinear Science. 2010;20(6):781–830.

[26] Decher G, Schlenoff JB, editors. Multilayer Thin Films. Weinheim: Wiley VCH Verlag GmbH; 2012.

[27] Choy KL. Chemical vapour deposition of coatings. Progress in materials science. 2003;48(2):57–170.

[28] Zubov LM. Nonlinear theory of dislocations and disclinations in elastic bodies. vol. 47. Berlin: Springer Science & Business Media; 1997.

[29] Acharya A. A model of crystal plasticity based on the theory of continuously distributed dislocations. Journal of the Mechanics and Physics of Solids. 2001;49(4):761–784.

[30] Quinn G, Gengnagel C. A review of elastic grid shells, their erection methods and the potential use of pneumatic formwork. In: WIT Transactions on the Built Environment. vol. 136; 2014. p. 129–143.

[31] Sachs R, Wu HH. General Relativity for Mathematicians. New York: Springer-Verlag; 1977.

[32] Poisson E. A Relativist's Toolkit: The Mathematics of Black-Hole Mechanics. Cambridge: Cambridge University Press; 2004.

[33] Kopeikin S, Efroimsky M, Kaplan G. Relativistic Celestial Mechanics of the Solar System. Weinheim: Wiley VCH Verlag GmbH; 2011.

[34] Misner CW, Thorne KS, Wheeler JA, Kaiser DI. Gravitation. Princeton: Princeton University Press; 2017.

[35] Weyl H. Space, Time, Matter. Mineola, N.Y.: Dover Publications Inc.; 1952.

[36] Euclid. Elements. Books I–XIII. New York City: Barnes & Noble; 2006.

[37] Truesdell C, Noll W. The Non-Linear Field Theories of Mechanics. Springer Berlin Heidelberg; 2004.

[38] Chaichian M, Merches I, Tureanu A. Mechanics. An Intensive Course. Berlin Heidelberg: Springer-Verlag; 2012.

[39] Helrich CS. Analytical Mechanics. Switzerland: Springer International Publishing; 2017.

[40] Hestenes D, Sobczyk G. Clifford Algebra to Geometric Calculus: A Unified Language for Mathematics and Physics. Fundamental Theories of Physics. Netherlands: Springer; 2012.

[41] Klein F. Elementary Mathematics from a Higher Standpoint. Volume II: Geometry. Berlin Heidelberg: Springer-Verlag; 2016.

[42] Roman S. Advanced Linear Algebra. New York: Springer-Verlag; 1992.

[43] Jammer M. Concepts of force. Mineola, N.Y.: Dover Publications; 2011.

[44] Schwartz L. Analyse: II, Calcul différentiel et equations différentielles. Collection Enseignement des Sciences. Paris: Hermann; 1992.

[45] Lurie AI. Non-linear theory of elasticity. Amsterdam: Elsevier; 2012.

[46] Gibbs JW. Vector Analysis. New Haven Yale University Press; 1901.

[47] Gurtin ME. The linear theory of elasticity. In: Linear Theories of Elasticity and Thermoelasticity. Springer; 1973. p. 1–295.

[48] Tu LW. An Introduction to Manifolds. New York: Springer-Verlag; 2011.

[49] Nakahara M. Geometry, topology and physics. Boca Raton: CRC Press; 2003.

[50] Jammer M. Concepts of Space: The History of Theories of Space in Physics. Mineola, N.Y.: Dover Publications; 2012.

[51] van Dantzig D. On the phenomenological thermodynamics of moving matter. Physica. 1939;6(7):673–704.

[52] Truesdell CA. A first course in rational continuum mechanics. Volume 1-General concepts. vol. 1. San Diego, CA: Academic Press, Inc.; 1977.

[53] Schuller FP. All spacetimes beyond Einstein (Obergurgl Lectures). arXiv. 2011;p. 1–44.

[54] Logunov AA, Mestvirishvili MA, Petrov VA. How were the Hilbert-Einstein equations discovered? Physics-Uspekhi. 2004;47(6):607.
[55] Stephani H, Kramer D, MacCallum M, Hoenselaers C, Herlt E. Exact Solutions of Einstein's Field Equations. 2nd ed. Cambridge Monographs on Mathematical Physics. Cambridge: Cambridge University Press; 2003.
[56] Boyer RH, Lindquist RW. Maximal Analytic Extension of the Kerr Metric. Journal of Mathematical Physics. 1967;8.
[57] Aldrovandi R, Pereira JG. Teleparallel Gravity. An Introduction. Netherlands: Springer; 2013.
[58] Kellogg OD. Foundations of Potential Theory. Berlin Heidelberg: Springer-Verlag; 1967.
[59] Truesdell C, Toupin R. The classical field theories. Berlin: Springer; 1960.
[60] Marsden JE, Hughes TJ. Mathematical foundations of elasticity. North Chelmsford, MA: Courier Corporation; 1994.
[61] Madsen IH, Tornehave J. From calculus to cohomology: de Rham cohomology and characteristic classes. Cambridge: Cambridge University Press; 1997.
[62] Hatcher A. Algebraic Topology. Algebraic Topology. Cambridge: Cambridge University Press; 2002.
[63] Yavari A. Compatibility Equations of Nonlinear Elasticity for Non-Simply-Connected Bodies. Archive for Rational Mechanics and Analysis. 2013;209(1):237–253.
[64] Lee JM. Introduction to Topological Manifolds. New York: Springer; 2011.
[65] Ciarlet PG, Mardare C, Shen M. Saint Venant compatibility equations in curvilinear coordinates. Analysis and Applications. 2007;5(3):231–251.
[66] de Wit R. Linear theory of static disclinations. In: Simmons JA, de Wit R, Bullough R, editors. Fundamental aspects of dislocation theory. vol. I. (U.S.): National Bureau of Standards; 1970. p. 651–673.
[67] Mindlin RD, Tiersten HF. Effects of couple-stresses in linear elasticity. Arch Rational Mech Anal. 1962;11(1):415–448.
[68] Ryzhak EI. Direct coordinate-free derivation of the compatibility equation for finite strains. Mechanics of Solids. 2014;49(4):382–388.
[69] Maugin GA. Continuum Mechanics Through the Eighteenth and Nineteenth Centuries. Switzerland: Springer International Publishing; 2014.
[70] Truesdell C. Essays in the History of Mechanics. Berlin: Springer Science & Business Media; 1968.
[71] Euler L. Continuation des recherches sur la theorie du mouvement des fluides. Mémoires de l'académie des sciences de Berlin. 1757;p. 316–361.
[72] Euler L. Sectio tertia de motu fluidorum lineari potissimum aquae. Novi Commentarii academiae scientiarum Petrop. 1771;p. 219–360.
[73] Cauchy AL. Recherches sur l'équilibre et le mouvement intŕieur des corps solides ou fluides, élastiques ou non élastiques. Bulletin de la Société philomatique. 1823;p. 9–13.
[74] Cauchy AL. De la pression ou tension dans un corps solide. Ex de Math. 1827;p. 42–56.
[75] Poisson SD. Memoire sur l'equilibre et le mouvement des corps elastiques. Mem Acad Sci Inst France. 1829;8:356.
[76] Capecchi D, Ruta GC. Piola's contribution to continuum mechanics. Archive for History of Exact Sciences. 2007 mar;61(4):303–342. Available from: http://dx.doi.org/10.1007/s00407-007-0002-x.
[77] Devreese JT, Berghe GV. 'Magic is no magic': the wonderful world of Simon Stevin. England: WIT Press; 2008.
[78] Cauchy AL. Sur la condensation et la dilatation des corps solides. Ex de Math. 1827;p. 60–69.

[79] Cauchy AL. Sur les équations qui expriment les conditions d'équilibre, ou les lois du mouvement intérieur d'un corps solide, élastique, ou non élastique. Oeuvres (2) Gauthier–Villars. 1890;8:195–226.
[80] Noll W. The Foundations of Classical Mechanics in the Light of Recent Advances in Continuum Mechanics. In: The Foundations of Mechanics and Thermodynamics. Springer Science & Business Media; 1974. p. 31–47.
[81] Bonet J, Wood RD. Nonlinear Continuum Mechanics for Finite Element Analysis. Cambridge: Cambridge University Press; 2008.
[82] Halmos PR. Finite-Dimensional Vector Spaces. New York: Springer; 1993.
[83] Ericksen JL. Deformations possible in every compressible, isotropic, perfectly elastic material. Journal of Mathematics and Physics. 1955;34:126–128.
[84] Ericksen JL. Deformations possible in every isotropic, incompressible, perfectly elastic body. ZAMP. 1954;5:466–488.
[85] Singh M, Pipkin AC. Note on Ericksen's problem. ZAMP. 1965;16:706–709.
[86] Klingbeil WW, Shield RT. On a class of solutions in plane finite elasticity. ZAMP. 1966;17:489–511.
[87] Shield RT. Deformations possible in every compressible, isotropic, perfectly elastic material. Journal of Elasticity. 1971;1:91–92.
[88] Saccomandi G. Universal Solutions and Relations in Finite Elasticity. In: Hayes M, Saccomandi G, editors. Topics in Finite Elasticity. Vienna: Springer; 2001. p. 95–130.
[89] Green EA, Rivlin SR, Shield TR. General Theory of Small Elastic Deformations Superposed on Finite Elastic Deformations. In: Proceedings of The Royal Society A: Mathematical, Physical and Engineering Sciences. vol. 211; 1952. p. 128–154.
[90] Dehlinger U. Zur Rekristallisation reiner Metalle. Annalen der Physik V Folge. 1929;p. 749–793.
[91] Taylor GI. The Mechanism of Plastic Deformation of Crystals. Part I. Theoretical. Proceedings of the Royal Society A: Mathematical, Physical and Engineering Sciences. 1934 jul;145(855):362–387.
[92] Taylor GI. The Mechanism of Plastic Deformation of Crystals. Part II. Comparison with Observations. Proceedings of the Royal Society A: Mathematical, Physical and Engineering Sciences. 1934 jul;145(855):388–404.
[93] Orowan E. Zur Kristallplastizität. I. Zeitschrift für Physik. 1934 sep;89(9-10):605–613.
[94] Polanyi M. Über eine Art Gitterstörung, die einen Kristall plastisch machen könnte. Zeitschrift für Physik. 1934 sep;89(9-10):660–664.
[95] Frenkel J. Zur Theorie der Elastizitätsgrenze und der Festigkeit kristallinischer Körper. Zeitschrift für Physik. 1926 jul;37(7-8):572–609.
[96] Kröner E. Allgemeine Kontinuumstheorie der Versetzungen und Eigenspannungen. Archive for Rational Mechanics and Analysis. 1959;4:18–334.
[97] Cosserat E, Cosserat F. Théorie des Corps Déformables. Paris: Hermann; 1909.
[98] Mindlin RD. Micro-structure in linear elasticity. Arch Ration Mech and Analysis. 1964;p. 51–78.
[99] Friedel MG. Les États Mésomorphes de la metiére. Annales de Physique. 1922;p. 273–474.
[100] Oseen CW. The theory of liquid crystals. Transactions of the Faraday Society. 1933;29(140):883.
[101] Frank FC. I. Liquid crystals. On the theory of liquid crystals. Discussions of the Faraday Society. 1958;25:19.
[102] Kadic A, Edelen DGB. A Gauge Theory of Dislocations and Disclinations. Berlin Heidelberg: Springer-Verlag; 1983.

[103] Edelen DGB. A four-dimensional formulation of defect dynamics and some of its consequences. International Journal of Engineering Science. 1980;18(9):1095–1116.

[104] Noll W. Materially uniform simple bodies with inhomogeneities. Archive for Rational Mechanics and Analysis. 1967;27(1):1–32.

[105] Huang X, Liu Z, Xie H. Recent progress in residual stress measurement techniques. Acta Mechanica Solida Sinica. 2013;26(6):570–583.

[106] Schajer GS. Practical Residual Stress Measurement Methods. United Kingdom: Wiley; 2013.

[107] Burger DJ. Sphereland: A Fantasy about Curved Spaces and an Expanding Universe. New York City: HarperCollins Publishers; 1983.

[108] Abbott EA. Flatland: A Romance of Many Dimensions. Mineola, N.Y.: Dover; 1992.

[109] Wang CC. On the geometric structures of simple bodies, a mathematical foundation for the theory of continuous distributions of dislocations. Archive for Rational Mechanics and Analysis. 1967;27(1):33–94.

[110] Gurtin ME, Murdoch AI. A continuum theory of elastic material surfaces. Archive for Rational Mechanics and Analysis. 1975;57(4):291–323.

[111] Adachi M. Embeddings and immersions. Providence, Rhode Island: American Mathematical Soc.; 2012.

[112] Hirsch MW. Differential topology. New York: Springer Science & Business Media; 2012.

[113] Franzoni G. The Klein bottle: Variations on a theme. Notices of the AMS. 2012;59(8):1094–1099.

[114] Cauchy AL. Mémoire sur les dilatations, les condensations et les rotations produits par un changement de forme dans un systéme de points matériels. Oeuvres (2) Gauthier–Villars. 1841;12:343–377.

[115] Schield RT. Inverse deformation results in finite elasticity. Zeitschrift f́ur angewandte Mathematik und Physik ZAMP. 1967 jul;18(4):490–500.

[116] Poisson SD. Memoire sur les equations generates de l'equilibre et du mouvement des corps elastiques et des fluids. J Ecole Poly. 1831;.

[117] Lagrange JL. Analytical Mechanics. Netherlands: Springer; 1997.

[118] Cosserat EeF. Sur la théorie de l'élasticite. Ann Toulouse. 1896;10:1–116.

[119] Rudolph G, Schmidt M. Differential Geometry and Mathematical Physics. Part I. Manifolds, Lie Groups and Hamiltonian Systems. Springer Science + Business Media Dordrecht; 2013.

[120] Michal AD. Matrix and Tensor Calculus With Applications to Mechanics, Elasticity and Aeronautics. New York: John Wiley & Sons; London: Chapman-Hall, Ltd.; 1947.

[121] Kollmann FG, Hackenberg HP. On the Algebra of Two-Point Tensors on Manifolds with Applications in Nonlinear Solid Mechanics. ZAMM-Journal of Applied Mathematics and Mechanics/Zeitschrift für Angewandte Mathematik und Mechanik. 1993;73(11):307–314.

[122] Frankel T. The geometry of physics: an introduction. Cambridge: Cambridge University Press; 2003.

[123] Abraham R, Marsden JE, Ratiu T. Manifolds, tensor analysis, and applications. New York: Springer Science & Business Media; 2012.

[124] Segev R, Rodnay G. Cauchy's Theorem on Manifolds. Journal of Elasticity. 1999;56(2):129–144.

[125] Milnor J, Stasheff JD. Characteristic Classes. Princeton: Princeton University Press; 1974.

[126] Kanso E, Arroyo M, Tong Y, Yavari A, Marsden JG, Desbrun M. On the geometric character of stress in continuum mechanics. Zeitschrift für angewandte Mathematik und Physik. 2007;58(5):843–856.

[127] Lychev SA, Manzhirov AV. The mathematical theory of growing bodies. Finite deformations. Journal of Applied Mathematics and Mechanics. 2013;77(4):421–432.

[128] Lychev SA. Geometric Aspects of the Theory of Incompatible Deformations in Growing Solids. In: Advanced Structured Materials. Springer International Publishing; 2017. p. 327–347.

[129] Lychev SA, Kostin GV, Koifman KG, Lycheva TN. Modeling and Optimization of Layer-by-Layer Structures. Journal of Physics: Conference Series. 2018;1009:012014.

[130] Ciarletta P, Destrade M, Gower A, Taffetani M. Morphology of residually-stressed tubular tissues: Beyond the elastic multiplicative decomposition. Journal of the Mechanics and Physics of Solids. 2016;90.

[131] Chern SS, Chen WH, Lam KS. Lectures on Differential Geometry. Singapore: World Scientific Publishing; 1999.

[132] Choquet-Bruhat Y, Dewitt-Morette C, Dillard-Bleick M. Analysis, manifolds and physics. Part 1. Basics. North-Holland; 1982.

[133] Taubes CH. Differential Geometry: Bundles, Connections, Metrics and Curvature. Oxford Graduate Texts in Mathematics. OUP Oxford; 2011.

[134] Lee JM. Riemannian Manifolds. An Introduction to Curvature. New York: Springer-Verlag; 1997.

[135] Goldberg V, Chern SS. Riemannian Geometry In An Orthogonal Frame. Singapore: World Scientific Publishing Company; 2001.

[136] Weitzenböck R. Invariantentheorie. P Noordhoff, Groningen. 1923;.

[137] Fernandez OE, Bloch AM. The Weitzenböck Connection and Time Reparameterization in Nonholonomic Mechanics. J Math Phys. 2011;52(1):012901.

[138] Saa A. Volume-forms and minimal action principles in affine manifolds. Journal of Geometry and Physics. 1995;15:102–108.

[139] Norton JD. General Covariance and the Foundations of General Relativity: Eight Decades of Dispute. Reports of Progress in Physics. 1993;56:791–861.

[140] Yavari A, Marsden JE, Ortiz M. On spatial and material covariant balance laws in elasticity. Journal of Mathematical Physics. 2006;47(4):042903.

[141] Lychev SA, Koifman KG. Geometric Aspects of the Theory of Incompatible Deformations. Part II. Strain and Stress Measures. Nanoscience and Technology: An International Journal. 2018;In Print. Available from: http://dx.doi.org/10.1615/NanoSciTechnolIntJ.2018024573.

[142] Ciesielski K. Set Theory for the Working Mathematician. Cambridge: Cambridge University Press; 1997.

[143] Bourbaki N. Theory of Sets. Berlin Heidelberg: Springer-Verlag; 2004.

[144] Bourbaki N. Algébre. Chapitres 1 á 3. Berlin Heidelberg: Springer-Verlag; 2007.

[145] Kobayashi S, Nomizu K. Foundations of differential geometry. vol. 1, 2. New York: John Wiley & Sons; 1963.

[146] Noll W. A mathematical theory of the mechanical behavior of continuous media. Archive for rational Mechanics and Analysis. 1958;2(1):197–226.

[147] Berger M. Geometry I. Berlin Heidelberg: Springer-Verlag; 1987.

[148] Schwartz L. Analyse: I, Théorie des ensembles et topologie. Collection Enseignement des Sciences. Paris: Hermann; 1991.

[149] Bourbaki N. General Topology. Chapters 1-4. Berlin Heidelberg: Springer-Verlag; 1995.

[150] Cartan EJ. Sur certaines expressions différentielles et le probléme de Pfaff. Annales Scientifiques de l'Ecole Normale Supérieure. 1899;16:239–332.

[151] Mangiarotti L, Sardanashvili GA. Connections in Classical and Quantum Field Theory. Singapore: World Scientific; 2000.

[152] Steenrod NE. The topology of fibre bundles. vol. 14. Princeton: Princeton University Press; 1951.

[153] Husemoller D. Fibre Bundles. New York: Springer; 1994.

[154] Rudolph G, Schmidt M. Differential Geometry and Mathematical Physics. Part II. Fibre Bundles, Topology and Gauge Fields. Springer Science + Business Media Dordrecht; 2017.

# Index



https://doi.org/10.1515/9783110563214-016

## Also of Interest

**Volume 49**
*Vladimir K. Dobrev*
Invariant Differential Operators: Volume 3: Supersymmetry, 2018
ISBN 978-3-11-052663-9, e-ISBN (PDF) 978-3-11-052749-0, e-ISBN (EPUB) 978-3-11-052669-1

**Volume 48**
*Jared Maruskin*
Dynamical Systems and Geometric Mechanics: An Introduction, 2018
ISBN 978-3-11-059729-5, e-ISBN (PDF) 978-3-11-059780-6,
e-ISBN (EPUB) 978-3-11-059803-2

**Volume 47**
*Eugene Stefanovich*
Elementary Particle Theory: Volume 3: Relativistic Quantum Dynamics, 2018
ISBN 978-3-11-049090-9, e-ISBN (PDF) 978-3-11-049322-1,
e-ISBN (EPUB) 978-3-11-049139-5

**Volume 46**
*Eugene Stefanovich*
Elementary Particle Theory: Volume 2: Quantum Electrodynamics, 2018
ISBN 978-3-11-049089-3, e-ISBN (PDF) 978-3-11-049320-7,
e-ISBN (EPUB) 978-3-11-049143-2

**Volume 45**
*Eugene Stefanovich*
Elementary Particle Theory: Volume 1: Quantum Mechanics, 2018
ISBN 978-3-11-049088-6, e-ISBN (PDF) 978-3-11-049213-2,
e-ISBN (EPUB) 978-3-11-049103-6

**Volume 44**
*Vladimir V. Kiselev*
Collective Effects in Condensed Matter Physics, 2018
ISBN 978-3-11-058509-4, e-ISBN (PDF) 978-3-11-058618-3,
e-ISBN (EPUB) 978-3-11-058513-1